Phytoremediation Potential of Perennial Grasses

Phytoremediation Potential of Perennial Grasses

Vimal Chandra Pandey
**Department of Environmental Science,
Babasaheb Bhimrao Ambedkar University,
Lucknow, India**

D.P. Singh
**Department of Environmental Science,
Babasaheb Bhimrao Ambedkar University,
Lucknow, India**

ELSEVIER

Elsevier
Radarweg 29, PO Box 211, 1000 AE Amsterdam, Netherlands
The Boulevard, Langford Lane, Kidlington, Oxford OX5 1GB, United Kingdom
50 Hampshire Street, 5th Floor, Cambridge, MA 02139, United States

Notices
Knowledge and best practice in this field are constantly changing. As new research and experience broaden our understanding, changes in research methods, professional practices, or medical treatment may become necessary.

Practitioners and researchers must always rely on their own experience and knowledge in evaluating and using any information, methods, compounds, or experiments described herein. In using such information or methods they should be mindful of their own safety and the safety of others, including parties for whom they have a professional responsibility.

To the fullest extent of the law, neither the Publisher nor the authors, contributors, or editors, assume any liability for any injury and/or damage to persons or property as a matter of products liability, negligence or otherwise, or from any use or operation of any methods, products, instructions, or ideas contained in the material herein.

Library of Congress Cataloging-in-Publication Data
A catalog record for this book is available from the Library of Congress

British Library Cataloguing-in-Publication Data
A catalogue record for this book is available from the British Library

ISBN: 978-0-12-817732-7

For information on all Elsevier publications visit
our website at https://www.elsevier.com/books-and-journals

Publisher: Joe Hayton
Acquisitions Editor: Marisa LaFleur
Editorial Project Manager: Lena Sparks
Production Project Manager: Omer Mukthar
Cover Designer: Matthew Limbert

Typeset by Thomson Digital

Working together
to grow libraries in
developing countries

www.elsevier.com • www.bookaid.org

Contents

Contributors   xi
About the authors   xiii
Foreword   xv
Preface   xvii
Acknowledgments   xix

1. Perennial grasses in phytoremediation—challenges and opportunities   1
Vimal Chandra Pandey, Deblina Maiti
  1 Introduction to phytoremediation   1
  2 Perennial grass genetic resources: what can they contribute toward phytoremediation?   4
  3 Importance of perennial grasses   10
  4 Why perennial grasses in phytoremediation?   18
  5 Coupling phytoremediation with perennial native grasses   18
  6 Perennial growth—an essential aspect for sustainable biomass source   19
  7 Improvement of perennial grasses for enhanced phytoremediation   20
  8 Perennial grass-based phytoremediation practices   21
  9 Policy framework   21
  10 Conclusions and future prospects   21
  References   22

2. *Vetiveria zizanioides* (L.) Nash – more than a promising crop in phytoremediation   31
Vimal Chandra Pandey, Ashish Praveen
  1 Introduction   31
  2 Morphology, reproduction, and propagation   33
  3 Ecology and physiology   35
  4 Geographical distribution and expansion   35
  5 Multipurpose usage of vetiver grass   36
  6 Limitations   51
  7 Potential features of vetiver grass: the reason of vetiver's success   52
  8 Conclusions   53
  References   53

3. The potential of Sewan grass (*Lasiurus sindicus* Henrard) in phytoremediation—an endangered grass species of desert   63
Vimal Chandra Pandey, D.P. Singh
  1 Introduction to Sewan grass   63
  2 Origin and geographical distribution   64

3	Ecology	64
4	Morphological description	65
5	Propagation	66
6	Important features of Sewan grass	66
7	Multiple uses	67
8	Phytoremediation	67
9	Biomass productivity of Sewan grass	70
10	Genetic diversity and conservation	73
11	Rhizospheric microbiology of Sewan grass	74
12	Conclusion and future prospects	75
	References	76

4. *Miscanthus*–a perennial energy grass in phytoremediation   79
Ashish Praveen, Vimal Chandra Pandey

1	Introduction	79
2	*Miscanthus* biology and taxonomy	81
3	Propagation	81
4	Easy harvesting	83
5	*Miscanthus* grass as a biofuel crop	83
6	Phytoremediation	84
7	Environmental consideration	88
8	Multiple uses	88
9	Merits and demerits of *Miscanthus* with SWOT analysis	89
10	Conclusion	91
	References	91

**5. *Phragmites* species—promising perennial grasses for phytoremediation
and biofuel production**   97
Vimal Chandra Pandey, Deblina Maiti

1	Introduction	97
2	General aspects of *Phragmites* species	98
3	Important features of *Phragmites* species	103
4	Multiple uses and management consideration	103
5	Conclusion	109
6	Future perspectives	110
	References	110

6. Feasibility of *Festuca rubra* L. native grass in phytoremediation   115
Gordana Gajić, Miroslava Mitrović, Pavle Pavlović

1	Introduction	115
2	General aspects of *F. rubra* L.	123
3	Ecorestoration techniques	128
4	The role of *F. rubra* L. in phytoremediation of contaminated sites	132

5   Physiological and morphological response of *F. rubra* L.   139
6   Conclusion and future outlook   148
References   150

**7.   Reed canary grass (*Phalaris arundinacea* L.): coupling
       phytoremediation with biofuel production**   165
Vimal Chandra Pandey, Ambuj Mishra, Sudhish Kumar Shukla, D.P. Singh
1   Introduction   165
2   Origin and geographical distribution   166
3   Ecology   167
4   Botanical description   167
5   Propagation   168
6   Main features of reed canary grass in relation to phytoremediation   168
7   Multiple uses of reed canary grass   171
8   Conclusions and future prospects   172
References   173

8.   Switchgrass—an asset for phytoremediation and bioenergy production   179
Divya Patel, Vimal Chandra Pandey
1   Introduction   179
2   General aspect of switchgrass   180
3   Multiple uses   183
4   Limiting factors   184
5   Phytoremediation   184
6   Bioenergy production   187
7   Carbon sequestration   188
8   Physiological adaptation   188
9   Conclusion and future perspectives   189
References   189

**9.   *Cymbopogon flexuosus*—an essential oil-bearing aromatic grass
       for phytoremediation**   195
Vimal Chandra Pandey, Apurva Rai, Anuradha Kumari, D.P. Singh
1   Introduction   195
2   Ecology   196
3   Origin and distribution   197
4   Botanical description   198
5   Propagation   198
6   Important aspects in relation to phytoremediation   198
7   Multiple uses of lemongrass   199
8   Medicinal use   204
9   Other commercial uses   204
10   Socio-economic development   204
11   Implementation strategies   205
12   Conclusion and future prospects   205
References   205

10. *Saccharum* spp.—potential role in ecorestoration and biomass
    production   211
    Vimal Chandra Pandey, Ashutosh Kumar Singh
    1   Introduction   211
    2   Ecology   212
    3   Morphological description   213
    4   Geographic distribution   215
    5   Propagation   215
    6   Multiple uses   216
    7   Role of *Saccharum* spp. in ecological restoration
        of waste land   218
    8   Role of *Saccharum* spp. in ecological restoration of fly ash dumps   220
    9   Biomass and bioenergy production   220
    10 Conclusion   221
    References   222

11. Bermuda grass –its role in ecological restoration and biomass
    production   227
    Vimal Chandra Pandey, Jitendra Ahirwal
    1   Introduction   227
    2   Origin, geographical distribution, and occurrence   228
    3   Ecology   228
    4   Morphology and propagation   229
    5   Abiotic stress tolerance of Bermuda grass   230
    6   Multiple uses   232
    7   Conclusion   240
    References   240

12. Moso bamboo (*Phyllostachys edulis* (Carrière) J.Houz.)–one of the
    most valuable bamboo species for phytoremediation   245
    Purabi Saikia, Vimal Chandra Pandey
    1   Introduction   245
    2   Bamboo-provisioned ecosystem services   247
    3   Major role of bamboo toward nature sustainability   248
    4   Future research prospects   252
    5   Conclusions   253
    References   253

13. The application of *Calamagrostis epigejos* (L.) Roth. in
    phytoremediation technologies   259
    Dragana Ranđelović, Ksenija Jakovljević, Slobodan Jovanović
    1   Introduction   259
    2   Morphology, propagation, and reproduction   261
    3   Ecology   261
    4   Distribution and expansion   265

5 Suppression and control 267
6 Phytoremediation 268
7 Other uses 273
References 275

14. Potential of Napier grass (*Pennisetum purpureum* Schumach.) for phytoremediation and biofuel production 283
Vimal Chandra Pandey, Divya Patel, Shivakshi Jasrotia, D.P. Singh
1 Introduction 283
2 Origin and geographical distribution 284
3 Ecology 286
4 Taxonomy and morphological description 286
5 Propagation 287
6 Important features of Napier grass 288
7 Multiple uses 288
8 Phytoremediation 289
9 Bioenergy production 294
10 Conclusion and future prospects 296
References 297

15. Role of microbes in grass-based phytoremediation 303
Madhumita Roy, Vimal Chandra Pandey
1 Introduction 303
2 Perennial grasses: suitable agents for phytomanagement 304
3 Phytoremediation strategies 308
4 Importance of microbial role in grass–based phytoremediation 312
5 Phytoremediation of different types of pollutants through perennial grass species 313
6 Pros and cons of phytoremediation with perennial grasses 321
7 Conclusions 326
References 327

16. Case studies of perennial grasses—phytoremediation (holistic approach) 337
Vimal Chandra Pandey, Divya Patel, Deblina Maiti, D.P. Singh
1 Introduction 337
2 Potential case studies of perennial grasses in phytoremediation 338
3 Conclusion and future prospects 345
References 345

Index 349

Contributors

Jitendra Ahirwal   Department of Environmental Science and Engineering, Indian Institute of Technology (Indian School of Mines), Dhanbad, Jharkhand, India

Gordana Gajić   Department of Ecology, National Institute of Republic of Serbia, University of Belgrade, Belgrade, Serbia

Ksenija Jakovljević   University of Belgrade, Faculty of Biology, Institute of Botany and Botanical Garden, Belgrade, Serbia

Shivakshi Jasrotia   Department of Clinical Research, Delhi Institute of Pharmaceutical Sciences and Research, Government of N.C.T. of Delhi, India

Slobodan Jovanović   University of Belgrade, Faculty of Biology, Institute of Botany and Botanical Garden, Belgrade, Serbia

Anuradha Kumari   School of Environmental Sciences, Jawaharlal Nehru University, New Delhi, India

Deblina Maiti   Central Institute of Mining and Fuel Research, Dhanbad, Jharkhand, India

Ambuj Mishra   School of Environmental Sciences, Jawaharlal Nehru University, New Delhi, India

Miroslava Mitrović   Department of Ecology, National Institute of Republic of Serbia, University of Belgrade, Belgrade, Serbia

Vimal Chandra Pandey   Department of Environmental Science, Babasaheb Bhimrao Ambedkar University, Lucknow, Uttar Pradesh, India

Divya Patel   Department of Biotechnology, Sant Gadge Baba Amravati University, Amravati, Maharashtra, India

Pavle Pavlović   Department of Ecology, National Institute of Republic of Serbia, University of Belgrade, Belgrade, Serbia

Ashish Praveen   Plant Ecology and Environmental Science Division, National Botanical Research Institute, Lucknow , Uttar Pradesh; Department of Botany, Markham college of Commerce, VBU, Hazaribag, India

Apurva Rai   Plant Ecology and Environmental Science Division, National Botanical Research Institute, Lucknow, Uttar Pradesh, India

Dragana Ranđelović   University of Belgrade, Faculty of Mining and Geology, Department for Mineralogy, Crystallography, Petrology and Geochemistry; Institute for Technology of Nuclear and other Mineral Raw Materials, Belgrade, Serbia Belgrade, Serbia

Madhumita Roy   Department of Microbiology, Bose Institute, Kolkata, India

Purabi Saikia   Department of Environmental Sciences, School of Natural Resource Management, Central University of Jharkhand, Ranchi, India

Sudhish Kumar Shukla   Department of Chemistry, Manav Rachna University, Faridabad, Haryana, India

Ashutosh Kumar Singh   CAS Key Laboratory of Tropical Forest Ecology, Xishuangbanna Tropical Botanical Garden, Chinese Academy of Sciences, Menglun, Yunan, China

D.P. Singh   Department of Environmental Science, Babasaheb Bhimrao Ambedkar University, Lucknow, Uttar Pradesh, India

About the authors

Vimal Chandra Pandey Dr. Vimal Chandra Pandey is currently a CSIR-Senior Research Associate (CSIR-Pool Scientist) in the Department of Environmental Science at Babasaheb Bhimrao Ambedkar University, Lucknow, India. Dr. Pandey is well recognized internationally in the field of phytomanagement of fly ash/polluted sites. His research includes phytoremediation and revegetation of fly ash dumpsites, heavy metal polluted sites, and restoration of degraded lands with special reference to raising rural livelihoods and maintaining ecosystem services. He is a recipient of a number of awards/honors/fellowships such as (CSTUP-Young Scientist award, DST SERB-Young scientist award, UGC-Dr. DS Kothari Postdoctoral Fellowship and CSIR-SRA (Pool Scientist) Award, and a member (MNASc) of National Academy of Sciences, India (NASI), Commission Member of IUCN-CEM Ecosystem Restoration, and a member of the BECT's Editorial Board. He has published more than 50 peer-reviewed articles in reputed international journals and 11 book chapters. He is the author and editor of two books published by Elsevier with several more forthcoming. He is also serving as a potential reviewer for several journals from Elsevier, Springer, Wiley, Taylor & Francis, etc. and received Elsevier Reviewer Recognition Awards from the editors of many journals. ORCID iD: https://orcid.org/0000-0003-2250-6726, Google Scholar: https://scholar.google.co.in/citations?user=B-5sDCoAAAAJ&hl.

D.P. Singh Dr. D.P. Singh is a professor of Environmental Science at Babasaheb Bhimrao Ambedkar University, Lucknow, and obtained his PhD from Department of Botany, Banaras Hindu University, Varanasi, India. He has been the recipient of Commonwealth post-doctoral fellowship and worked in the University College of Swansea (UK). His research has majorly focused in the area of wastewater treatment, microbiology, stress physiology, bioremediation, and bioenergy options. He has received several honors and awards to his credit. Dr. Singh has published more than 135 research publications in high impact factor journals of national and international repute. He has supervised more than 24 PhD students and several MSc and MTech students for their research work. He has delivered invited lectures in different seminars and symposia and served as a principal investigator for several governments funded projects. Dr. Singh has published five books in the field of Environmental Microbiology and Biotechnology, Stress Physiology, and Sustainable Management of Soil and Water.

Foreword

"Phytoremediation Potential of Perennial Grasses" is an important collection of the precise literature in the form of chapters that reveal applied and feasible ways in remediating polluted lands through multipurpose perennial grasses from the ecologically to socio-economically. The tactics are also such that offer many co-benefits, including the stabilization of pollutants. Aid to the green economy is mainly notable. Phytoproducts, for instance, biofuels (biogas and bioethanol), essential oils of aromatic grasses, ornamental grasses, and pulp-paper grasses as well as different derived grass products, are significant additional benefits. Presently, there is an urgent need to develop phytomanagement of polluted lands through perennial grasses that benefit society directly and indirectly. A multipurpose perennial grass-based phytoremediation is a fantastic example of environmental sustainability that has a constructive future across the nations. Native perennial grasses are suitable candidates for use in sustainable phytoremediation of polluted sites due to being nurse species and having good adaptive strategies.

This book explores the phytoremediation potential of perennial grasses that will tackle to link the gap between phytoremediation and grass bioeconomy through phytomanagement, to address the challenge of remediation of the increasing number and area of polluted sites through valuable grasses with least inputs, low risk, and minimum maintenance toward maintaining ecosystem services and raising rural livelihoods. This book is therefore focused on recent findings and strategies of the most valuable perennial grasses toward phytoremediation of polluted sites with multiple benefits. Moreover, it offers the description of morphological; ecology, physiology, origin, geographical distribution, expansion, and propagation as well as important features of the grass regarding phytoremediation.

I appreciate the efforts of authors Dr. Vimal Chandra Pandey and Prof. D.P. Singh, in bringing out this valuable edition through the leading global publisher, Elsevier Publishing, with 16 chapters covering various aspects of the Phytoremediation Potential of Perennial Grasses. I hope the book will be a noteworthy asset for PhD scholars, environmental scientists, researchers, practitioners, policy makers, entrepreneurs, and other stakeholders alike.

Prof. D.P. Singh
Chairman
University Grants Commission
New Delhi
10th January, 2020

Preface

Ever-increasing numbers and areas of industrially polluted sites is a major concern across the nations and have resulted in environmental pollution. The major challenge is to develop new and cost-effective solutions to decontaminate polluted sites. In this regard, the plants-based remediation especially grasses is a promising and cost-effective approach for the environmental clean up on a large scale.

Over two decades of scientific development in phytoremediation, its reliability among stakeholders has not yet been achieved. They remain doubtful about its current applicability or future prospects. To overcome this challenge, the book will address grass-based phytoremediation with green economy. It is thus desirable to explore commercial and perennial grass-based phytoremediation. Grasses are well known for their work as nurse plants. Nurse plants are considered not only to play a key role in recovering the properties and functions of the primary ecosystem, but also to drive succession in poor environments on the early stage of restoration. Thus, perennial grass-based phytoremediation can play a key role in remediation of polluted sites as well as maintaining ecosystem services.

This book will be useful for practitioners to select specific perennial grass species according to site-specificity of the contaminated site. As this book will show, there are clearly some potential opportunities in grass-based phytoremediation that also yield phytoproducts especially bioenergy (biomass) and aromatic essential oils as green economy while remediating contaminated sites. This book will provide new knowledge and insights of the phytoremediation potential of perennial grasses across the nations with economic returns. In this book, an attempt has been made to popularize grass-based phytoremediation among ecological engineers, environmental scientists, practitioners, policy makers, and stakeholders. It is first book on *"Phytoremediation Potential of Perennial Grasses"* and will be a valuable asset to the students, researchers, practitioners, policy makers, stakeholders alike.

Vimal Chandra Pandey, Author;
D.P. Singh, Author

Acknowledgments

We sincerely wish to thank Candice Janco and Marisa LaFleur (Acquisitions Editor), Lena Sparks (Editorial Project Manager), and Swapna Praveen (Copyrights Coordinator) from Elsevier for their excellent support, guidance, and coordination of this fascinating project. We would like to thank all the reviewers for their time and expertise in reviewing the chapters of this book. Vimal Chandra Pandey is grateful to the Council of Scientific and Industrial Research, Government of India, New Delhi for Senior Research Associateship under Scientist's Pool Scheme (Pool No. 13 (8931-A)/2017). Finally, we must thank our respective families for their unending support, interest, and encouragement, and apologize for the many missed dinners!

Perennial grasses in phytoremediation—challenges and opportunities

1

Vimal Chandra Pandey[a],*, Deblina Maiti[b]
[a]Department of Environmental Science, Babasaheb Bhimrao Ambedkar University, Lucknow, Uttar Pradesh, India; [b]Central Institute of Mining and Fuel Research, Dhanbad, Jharkhand, India
*Corresponding author

1 Introduction to phytoremediation

Phytoremediation is a cost-effective green technology which exploits the ability of some plant species to uptake, metabolize, accumulate, and detoxify heavy metals or other harmful organic or inorganic pollutants from contaminated soil and waste materials (Besalatpour et al., 2008; Langella et al., 2014; Gołda and Korzeniowska, 2016; Pandey and Bajpai, 2019). This process is often a part of the integrated approach used during ecological restoration programs of degraded lands. The plants which are preferred for this process are chosen on the basis of their growth potential, massive biomass development, invasive nature, high pollutant accumulation potential in their roots or shoots, tolerance from the pollutants, and pollutant detoxification potential (De Koe, 1994; Pandey, 2012a, b; Pandey et al., 2014; Pandey et al., 2015a; Niknahad Gharmakher et al., 2018). Advantageously, the plants of the grass (Poaceae) family have demonstrated a huge potential in this aspect as various studies in literature using grasses have shown successful results in all of the earlier-mentioned criteria (Pandey et al., 2012b, 2015c; Pandey and Singh, 2015; Pastor et al., 2015; Verma et al., 2014). These grasses are often classified as accumulators and excluders on the basis of their tolerance potential to metals or pollutants. Accumulator plants have the potential to remain metabolically stable in conditions in which the concentrations of metals in their photosynthetic parts are relatively higher than the maximum allowable limits (Elekes and Busuioc, 2011; Houben and Sonnet, 2015; Gołda and Korzeniowska, 2016). This is because these plants efficiently counter the oxidative stress generated due to high metal accumulation (Sharma et al., 2018). On the other hand, excluder grass plants accumulate pollutants either in their roots or change them into complexes in the root zone, but restrict the pollutant movement up to the aerial parts. Roots of these plants are more tolerant to metals than shoots; the up taken metals are often accumulated inside the root cell vacuoles after detoxification (Fatima et al., 2018; Pastor et al., 2015). The processes concerning the rhizospheric metal bioavailability and their uptake by the adventitious roots of the grasses have been shown in Fig. 1.1. The bio-available metals are partitioned between the solid and solution phase around the soil particles

Phytoremediation Potential of Perennial Grasses. http://dx.doi.org/10.1016/B978-0-12-817732-7.00001-8

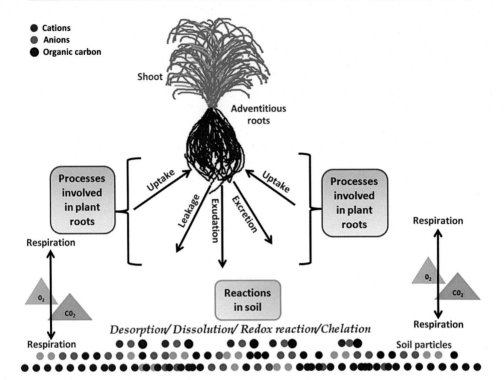

Figure 1.1 Mechanisms of metal uptake by the grass roots from soil.
Modified from Adriano et al., 2004.

by chemical reactions such as adsorption, complexation, precipitation, and redox reactions. The metals in solution phase are up taken by the roots through the root membranes. Metal uptake is also guided by active (symplastic movement of metals across cell membranes) or passive processes (movement of metal along cell walls) and various membrane protein metal transporters which transport the non-essential metals in addition to essential metals which are useful for biosynthetic processes. The transport process is then followed by metabolization, detoxification, and compartmentalization of the metals and varies from species to species (Adriano et al., 2004; Besalatpour et al., 2008; De Koe, 1994; Sharma et al., 2018).

Additionally, grasses possess an elaborate root structure, highly propagating underground rhizomatous stems which bind the soil to prevent erosion and leaching of contaminants. These structures also assist in rapidly covering vast acres of contaminated lands to form a thick green cover (Aprill and Sims, 1990; Langella et al., 2014; Subhashini and Swamy, 2013; Verma et al., 2014). In addition to having elaborate fibrous roots and rhizomes, some other general characteristics are simple, exstipulate, sessile leaves having parallel venation, tubular sheath at leaf base surrounding the internodes, sessile inflorescences having compound spikes with bracteates containing incomplete, zygomorphic, hypogynous flowers, 3–6 stamens having long filaments, and dithecous anthers in androecium, single carpel with a short style, bifid stigma, and a superior ovary having a single ovule with basal placentation in gynoecium, fruits having pericarp fused with seed coat and monocot seeds embedded in endosperm (Jeguirim et al., 2010). Thus, utilization of grasses as initial colonizers ultimately aids in developing an aesthetic landscape (Subhashini and Swamy, 2013). Specifically, aromatic grasses have proven to be much more efficient subjects for phytoremediation of a range of pollutants from waste lands and in turn benefits from ecological to socio-economic importance (Verma et al., 2014; Pandey and Singh, 2015; Pandey et al., 2019). Some widely cultivated aromatic grasses belong to *Vetiveria* sp. (Vetiver grass) and *Cymbopogon* sp. (Lemon grass) genera (Desai et al., 2014). One of the products yielded by aromatic grasses is essential oils which advantageously do not get contaminated by the pollutants even if the metal has been up taken in the plant body. The reason for such results is that oil extraction from the plant parts is done through steam distillation. Many studies have also proven that oil production by such plants is not affected significantly under various abiotic stresses (Gupta et al., 2015; Aftab et al., 2011; Verma et al., 2014; Pandey and Singh, 2015; Pandey et al., 2019).

Globally, millions of hectares of land is currently lying barren and has been extensively contaminated due to various industrial activities (Zurek et al., 2013). However, very meagre studies have been done in the aspect of possibilities of grass growth on such areas as phytoremediating agents. Till now, the hyperaccumulation property or phytoextraction of metals has mostly been identified in dicotyledonous herbs and shrubs, which are mainly annual in their life cycle (Adriano et al., 2004; Antoniadis et al., 2017). Biomass harvest of these species will thus be a single event with confined potential for further growth as it mainly occurs from the apical meristems which gets harvested during biomass cropping (Antoniadis et al., 2017). However, the regeneration potential of the grasses is confined to the base of the shoot origination above the root system, which doesn't get diminished even after biomass harvest. The perennial life cycle of these plants are an advantage to the phytoremediation process as the shoot biomass can be harvested a number of times which adds to continuous phytoextraction of pollutants from the contaminated substrate (Awasthi et al., 2017; Gołda and Korzeniowska, 2016). Natural colonization in an ecosystem is often witnessed by the presence of grasses as initial successors; which are also termed as native to the particular area. They have the potential to regrow and adaptable to a variety of environmental stresses; such plants should constitute the pioneer species for plantation on contaminated lands which outweighs the concept of planting edible plants, as the inclusion of latter plants doesn't declines the transmission of pollutants to the food chain through grazing livestock. Subsequently over years of grass plantation, the remediated land could be utilized to end uses like a productive agricultural land, esthetic tourist place, or left as such to continue to become a part of the landscape (Awasthi et al., 2016; De Koe, 1994; Subhashini and Swamy, 2013).

2 Perennial grass genetic resources: what can they contribute toward phytoremediation?

Continuous increase in the percentage of industrially degraded lands across the world requires an effective strategy of phytoremediation (Zurek et al., 2013). Sometimes, monoculture of grass species is less efficient in remediating a contaminated site compared to a mixed culture. Mixed crops of the perennial grasses can restore the soil quality, biodiversity, and energy flow of degraded ecosystems (Creutzig et al., 2015; Saha and Kukal, 2015). In this context, an appropriate genetic resource material is necessary for developing a perpetuating vegetation cover on such sites (Zorica et al., 2005). Though indigenous and site-specific grasses have often been used, yet specific exotic grasses can also perform better on many degraded sites (Sharma et al., 2018). Thus before a phytoremediation program supply of planting material should be organized for specific sites which may include a mixture of grass species. Efforts should be carried out to document the genetic resources of the grasses for their utilization in remediation programs (Sharma et al., 2018; Zorica et al., 2005). Some studies have denoted that perennial grasses are most important plants for phytoremediation due to their huge biomass production potential, low requirement of fertilizers, and ability to grow even on less fertilized soils (Zurek et al., 2013; Gołda and Korzeniowska, 2016). In addition they provide multiple benefits like improvement in soil structure and quality, reduction in erosion, and increase in biodiversity. Growing perennial grasses on degraded lands are much more sustainable than regular agricultural practices because the later may promote soil erosion and would not yield the adequate economic return. Additionally, on an ecosystem point of view, grasses are a group of ecologically dominant plant species, because they are found covering extensively on any barren or degraded land (Awasthi et al., 2017; Singh et al., 2018; Subhashini and Swamy, 2013).

The overall genetic resources of the perennial grass family are categorized into (1) wild varieties of the domesticated grass species of same genus, (2) seminatural form of the domesticated grass species, and (3) domesticated populations of the plant adapted in a specific region, also known as ecotype and the cultivars undergoing constant breeding techniques (Sokolovic et al., 2017). It is the largest, monocotyledonous angiosperm family of Poales order, with a variable number of species as per the previous literature. The number of species varies from 7500 to 11000, grouped into 650–785 genera, 25–50 tribes, and 3–12 subfamilies (Boller and Greene, 2010; Gibson, 2009). As also stated earlier, the grass family is also categorized into aromatic and non-aromatic grasses. The later ones are also known as pasture grasses (*Cynodon dactylon*); some are also included in food crops (*Triticum aestivum, Zea mays, Oryza sativa, Hordeum vulgare*) and in groups of trees (*Dendrocalamus* sp., *Bambusa* sp.) (Boller and Greene, 2010; Delgado-Caballero et al., 2017; Sokolovic et al., 2017). Diverse list of perennial grasses which can be used as genetic resources for phytoremediation of waste lands are given in Table 1.1. Here, we are describing only 13 most important grasses for phytoremediation along with their multiple uses (Table 1.2). They are widely known by their common names which are: Vetiver, Sewan grass, Miscanthus, Common reed, Red fescue, Reed canary grass, Switch grass, Lemon grass, Wild cane,

Table 1.1 Perennial grasses which are potential genetic resources for phytoremediation.

Scientific name of the grass	Common name	Contaminants	References
Agropyron smithii	Western wheat grass	Hydrocarbons	Besalatpour et al. (2008); Niknahad Gharmakher et al., 2018
Agrostis castellana	Colonial bent grass	As, Pb, Zn, Mn, Al	De Koe (1994); Pastor et al. (2015)
Agrostis alba	European grass	Ti, Ni, Mo, Cr, Cu, Sn	Elekes and Busuioc (2011)
Agrostis tenuis	Colonial bent grass	Cu, Cr, Cu, Sn	Houben and Sonnet (2015); Sharma et al. (2018)
Brachiaria mutica	Paragrass	Cr, Hydrocarbons	Fatima et al. (2018)
Bouteloua gracilis	Blue gamma grass	Hydrocarbons	Aprill and Sims (1990); Subhashini and Swamy (2013)
Buchloe dactyloides	Buffalo grass	Hydrocarbons	Delgado-Caballero et al. (2017)
Cynodon dactylon	Bermuda grass	Cr, Cu, Mo, Ni, Pb, Sn	Maiti and Prasad (2017)
Elymus canadensis	Canadian wild rye	Hydrocarbons	Aprill and Sims (1990)
Festuca arundinacea	Tall fescue	Ni, Pb, Cd, As, Atrazine	Sun et al. (2011)
Festuca rubra	Red fescue	Hydrocarbons, Cd, Cu, Zn, Ni	Langella et al. (2014); Gołda and Korzeniowska (2016)
Festuca pratensis	Meadow fescue	Cd, hydrocarbons	Soleimani et al. (2010a, b)
Poa pratensis	Kentucky bluegrass	Cd, Ni, Pb, creosote	Huang et al. (2004); Mahmoudzadeh et al. (2016)
Lolium perenne	English ryegrass	Cd, Pb, Cr, Cu, Sn, Atrazine, anthracene	Gołda and Korzeniowska (2016); Cui et al. (2018)
Leersia oryzoides	Rice cutgrass	As	Ampiah-Bonney et al. (2007)
Panicum maximum	Guinea grass	Trinitrotoulene	Jiamjitrpanich et al. (2013)
Panicum virgatum	Switchgrass	Atrazine, Pb, radionuclieds	Shen et al. (2013); Trocsanyi et al. (2009)
Sorghum halepense	Johnson grass	Pb, Cd, Co, Cu, Cr, Ni, Zn, Cu	Ziarati et al. (2015)
Stenotaphrum secundatum	St. Augustine grass	Hydrocarbons	Nedunuri et al. (2000)
Stipa capillata	Needle grass	Sn	Elekes and Busuioc (2011)
Triticum aestivum	Wheat	Se, Cd, Cu, Ni, Pb, Zn, Cr, As, Hg	Suchkova et al. (2010); Yasin et al. (2015)
Vetiveria zizanioides	Vetiver grass	As, Cd, Cr, Pb, Cu, Zn, Fe, Mn, endosulfan, Fly ash	Ghosh et al. (2015); Verma et al. (2014)
Saccharum spontaneum	Kans grass	Fly ash	Maiti and Prasad (2016, 2017); Pandey et al. (2015c)

Table 1.2 Catalog of the fifteen potential grass species dealt in the book in the context of phytoremediation.

Grass species	English name	Characteristics	References
Arundo donax L.	Giant reed	Perennial, warm season grass, C3 plant, Rhizomatous shrubby plant, grows in damp saline or fresh soils, sand dunes, wetlands, used as chemical feedstock for energy production, pulp and paper production, alternative to woody plant cultivation	Jeguirim et al. (2010); Shatalov and Pereira (2002)
Calamagrostis epigejos L.	Eurasian grass	Cool season grass, Grows across the globe on sand dunes, flood-plains, steppes, sub-alpine grasslands, Invasive, grows on low nutrient soil, regenerate from rhizome fragments, photophilous species, indicator of low groundwater levels, Used in re-vegetation of fly ash deposits, exhibits metal phytostabilization and tolerates salinity	Mitrovic et al. (2008); Pruchniewicz et al. (2017); Somodi et al. (2008); Talik et al. (2018)
Cymbopogon flexuosus	Lemon grass	Perennial grass, has vigorous root system, binds soil particles, stops erosion, leaves yield citronellal, myrcene, citronellic acid, borneol, which have medicinal properties, aromatic oil is used in foods, flavors, cosmetics, perfumes, soaps, detergents, and as insect repellents	Aftab et al. (2011); Desai et al. (2014); Gupta et al. (2015)
Cynodon dactylon	Bermuda grass	Perennial, creeping runner, has green flowers, tiny, grayish fruit grains, found throughout the globe, initial colonizer in barren lands and tolerates salt and heavy metal stress propagates by seeds, runners, rhizomes, cures wounds, cough, cramps, epilepsy, hemorrhage	Hameed et al. (2010); Nagori and Solanki (2011)
Festuca arundinacea Schreb	Tall fescue	Perennial, cool season, allohexaploid grows in semi-arid to very wet conditions, responds to fertile as well as impoverished soils, used as forage, useful for phytoremediation of metals and hydrocarbons, soil conservation	Sun et al. (2011); Majidi et al., 2009
Panicum virgatum L.	Switch grass	Perennial, warm season grass, C4 crop, can produce huge biomass by elongation of tillers in summer months, used for pulp and paper production, as a lignocellulosic feedstock for bioenergy production, forage, removes harmful residues from soil, suitable for wildlife shelter, prevents ground water nutrient contamination	Madakadze et al. (2010); Shen et al. (2013); Trocsanyi et al. (2009)

Grass species	English name	Characteristics	References
Pennisetum purpureum Schum.	Elephant grass, Mott grass, Napier grass	Perennial, warm season grass, tall, indigenous to tropical Africa, grows up to an altitude of 2000 m, also grows in sub-tropical areas, withstands repeated cutting and re-grows rapidly, has high productive potential, carrying capacity, nutrient quality, mostly used forage, used for pulp and paper production	Madakadze et al. (2010); Techio et al. (2002)
Phalaris arundinacea L.	Reed canary grass	Perennial, cool season grass, promising energy plant, roots have nutritional value, used as medicine and forage, competitive nature under wide range of ecological conditions and abiotic stresses, invader of wetlands, reproduces by seeds, tillers, and rhizomes, but requires light for establishment	Caunii et al. (2015); Lavergne and Molofsky (2004)
Phragmites australis	Common reed	Grows in temperate, tropic or coastal regions, C3–C4 photosynthetic system, vegetatively propagated through rhizomes and stolons, has rhizospheric, bacterial or mycorrhizal association, deep root system, high biomass production, useful for ecorestoration, improving soil fertility, heavy metal uptake, bioenergy feed stock, fencing purposes, pulp and paper production	Brix et al. (2014); Gilson (2017); Hafliger et al. (2006)
Phyllostachys pubescens	Moso bamboo	Perennial, occurs in tropical, subtropical and temperate regions, originally found in China, largest bamboo species, leptomorph root system, fast-growing plants, culms grow to a height of 3–30 m within few months due to expansion of internodes, high economic value young sprouts and culm woods, total above-ground biomass is highest among all bamboo communities, used for traditional handiwork; alternative to timber in all wooden works and furtinures.	Isagi et al. (1997); Lin et al. (2002); Obataya et al. (2007); Yu et al. (2008)
Lolium perenne L.	English rye grass	Perennial, found in temperate areas, rapid establishment, high yields, long growing season, tolerance to grazing, high palatability for ruminants, used as hay, silage and pasture, phystabilizes toxic metals, like Cd	Golda and Korzeniowska (2016); Grinberg et al. (2016)

(Continued)

Table 1.2 Catalog of the fifteen potential grass species dealt in the book in the context of phytoremediation. (*Cont.*)

Grass species	English name	Characteristics	References
Lasiurus sindicus Henrard	Sewan grass	Perennial, highly nutritive, drought tolerant, harbors bacteria with potential for nitrogen fixation, endemic to sub-tropical arid and semi-arid parts of Thar Desert of India, known as the "king of desert grasses," can live up to 20 years, Propagation is done by sowing root slips, grows best on alluvial soils or light brown sandy soils with a pH of 8.5, consumed by cattles, develops rangeland, stabilizes blowing sand dunes	Chowdhury et al. (2009); Sanadya et al. (2018)
Miscanthus sp.	Giant miscanthus	Perennial, tall, rhizomatous grass, C4 photosynthesis, originated in East Asia, propagates vegetatively by rhizomes, promising "energy plant," used in short rotation forestry, shelter for wild life, increases soil fertility, induces carbon sequestration	Lewandowski et al. (2000); Jeguirim et al. (2010)
Saccharum spontaneum	Wild cane	Perennial, C4 plant, robust and tall grass, deep rhizomatous root system, grows in marginal ecosystems, young shoot is edible; mature shoots used for making papers, fibers, thatching material, ropes, mats, baskets, brooms, hats, fodder, mulching material, ethanol production, soil barrier; phytostabilize metals from soil	Graham et al. (2014); Pandey et al. (2012b)
Vetiveria zizanioides L. Nash	Vetiver	Perennial, C4 plant used in perfumery industry, vegetatively propagated, resistant to various abiotic and biotic stresses, especially intensive grazing, heavy metals, abundantly occurs on marginal used lands, reduces fly ash genotoxicity	Ghosh et al. (2015)

Bermuda grass, Eurasian grass, Moso bamboo, and Elephant grass. All these grass species have been reported as a potential candidate for phytoremediation in various studies. The marketable products from these grasses and multiple utilizations have been explored in the present chapter, while their sustainability toward phytoremediation of waste lands will be analyzed in details in the chapters.

2.1 As a phytoremediator

Grasses specifically perennial grasses have been reported to reduce metal toxicity and induce phytoremediation by various authors in literature (Cui et al., 2018; Ghosh et al., 2015; Gołda and Korzeniowska, 2016; Suchkova et al., 2010; Yasin et al., 2015; Ziarati et al., 2015). Perennial grasses are metal-tolerant plants and are generally terrestrial, herbaceous or but may be woody (Bamboo) or aquatic in nature. Reportedly they accumulate or phytostabilize metals in their roots and eventually reduces the concentration of metals from the soil. For example, Madejon et al. (2002) reported that Bermuda grass (*Cynodon dactylon*) can stabilize spill affected soils while Tall fescue showed Pb phytoextraction and grew effectively in Pb contaminated soil without reduction in biomass (Begonia et al., 2005). Thomas et al. (2014) reported that Bermuda grass roots release exudates which can form complexes with Pb. Apart from accumulating heavy metals, various grasses have been reported to degrade hydrocarbons. For example, Efe Sunday and Ephram (2014) showed that *Axonopus* grass species can phytoremediate hydrocarbons from soil and can be effectively grown throughout the globe. Such characteristics of tolerance to wide range of climatic conditions are an advantage of the grass community over herbs, shrubs, and trees for utilization in phytoremediation. Some more characteristics which make grasses suitable for phytoremediation are (1) presence of phytoliths in some grasses which are siliceous compounds present in epidermal cells; they impart resistance against many abiotic stresses like metal toxicity, (2) drought-resistant nature, and (3) feedstock for bio-energy and bioproducts (Singh et al., 2018; Talik et al., 2018).

2.2 Ornamental grasses in park

Ornamental grasses have become increasingly popular for the past few years because they are adaptable to a wide range of climatic conditions including exceptionally wet or dry environment, and have wider utilization area. In addition, these grasses require minimum maintenance, have less pest problems compared to other herbs, and able to grow in problematic soils on which other ornamental plants are unable to grow (Davidson and Gobin, 1998). Their myriads of useful properties have attracted the community of Landscape architects, designers, and nurserymen for developing an aesthetic landscape (Gao et al., 2008). They can be used for developing ground covers, edging borders, container planting, and erosion control. Some of them have special foliage colors which can draw the attention of wildlife specially butterflies and birds (Liu et al., 2012). Studies have also shown that ornamental grasses can also be adopted for phytoremediation of contaminated soils. For example, *Festuca arundinacea* can efficiently reduce total petroleum hydrocarbons (TPH) from 10,000 mg kg^{-1} TPH contaminated soil. Italian ryegrass (*Lolium multiflorum*) could grow effectively in soil having increasing Cd concentration (Liu et al., 2013). Some other grasses which have been screened as per their adaptability, utilization in terms of growth, stress resistance, and ornamental value are *Nassella tenuissima, Cortaderia selloana, Saccharum arundinaceum, Panicum virgatum, Carex albula, C. chungii, Dianthus gratianopolitanus, S. ravennae, C. muskingumensis, Miscanthus sinensis, Arundo donax, M. sinensis, Chasmanthium latifolium* (Gao et al., 2008). Davidson and Gobin (1998) reported *Andropogon gerardii, Calamagrostis epigejos, Carex* sp., *Erianthus ravennae, Festuca hervierri, Festuca lemanii, Molinia caerulea, Panicum virgatum, Saccharum ravennae,* and *Spartina pectinata* to have very high horticultural value and visual appeal. They were evaluated under field conditions for a period of 4 years and assessed on the basis of their survival, growth, and development in colder regions.

3 Importance of perennial grasses

The perennial grasses which will be discussed in the present book have a myriad of utilization aspects which have been classified into ecological, societal, and economic aspects. In short, ecological aspects comprise the use of the grasses for restoration, phytoremediation, climate change mitigation, biodiversity conservation, wild life shelter, soil erosion control carbon sequestration, providing ecological corridors. Societal aspects include their utilization as raw material for crafts, huts and animal shades, fodder, cultural programs, rope manufacturing; while economic aspects include their role in bioenergy, medicine, essential oil, pulp and paper manufacturing industry compared to their cultivation costs (Fig. 1.2). These aspects have been discussed in detail as follows:

3.1 Ecological aspects

3.1.1 Restoration

Restoration of degraded environments often gets limited due to harsh environmental conditions. These effects can be minimized by using "nurse plants" which can improve the performance of nearby target species. The later procedure is known as facilitation, which is the benefit caused to some plants toward their establishment from the closely associated neighbors. Grasses have enormous potential to be included in nurse plants in restoration management procedures as per promising experimental results (Padilla and Pugnaire, 2006). To cite a few examples, in one study the grass *Calamagrostis epigejos* showed greater dominance and survival potential on bare ash deposits due to its multiple tolerances to the conditions and competitive ability (Mitrovic et al., 2008). Another grass *Miscanthus* can be used to cover the soil for a longer period as because the inputs of organic matter from the shedded leaves are expected to increase the soil organic matter and improve soil structure, compared to other arable crops. Studies have reported that the humus content, cation exchange capacity, and water retention capacity of the soil increased under a 4–8 year old *Miscanthus* plantation.

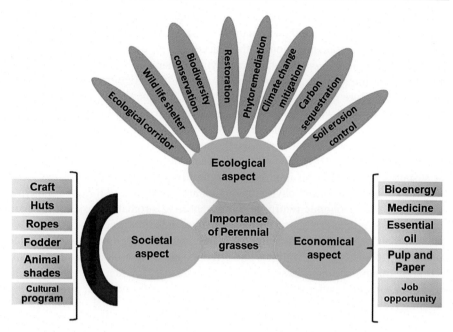

Figure 1.2 Ecosystem services provided by perennial grasses.

A full-grown stand of *Miscanthus* accumulates after 2–3 years yields 8–20 tons of below-ground (0–40 cm) and above-ground dry matter per hectare respectively. The dense root mat can penetrate up to a depth of 2.5 m and prevent leaching of ions to the groundwater table (Jeguirim et al., 2010). In another study, Maiti and Maiti (2015) could successfully grow *Cymbopogon citratus* in a grass-legume mixture for ecorestoration of a steel industry waste dump. The grass species having potential for fly ash dumps' restoration is identified on the basis of their ecological and socio-economic significance as well as dominance at ash dumpsites to support rural livelihoods. These are *Saccharum spontaneum, Cynodon dactylon, Saccharum bengalense* (syn. *Saccharum munja*), *Dactyloctenium aegyptium, Cyperus esculentus, Fimbristylis bisumbellata, and Eragrostis nutans. S. spontaneum* and *S. munja* have great ability to colonize on bare fly ash deposits. Therefore, they can be exploited as a valuable genetic grass resource for ecological restoration (Pandey, 2015; Pandey et al., 2015b, 2015c).

3.1.2 Phytoremediation

Various studies have also suggested the phytoremediation potential of the grass species. Randelovic et al. (2018) reported that *Calamagrostis epigejos* can uptake a significant amount of the available metals from soil in its roots, which shows phytostabilization. *Phragmites* grass species are one of the best plants which can remediate pollutants from contaminated soil or wastewater. Nayyef Alanbary et al. (2018) reported *P. karka*'s ability to phytoremediate sand contaminated with hydrocarbons of crude oil sludge. Aromatic grasses on the other hand are recently being promoted for phytoremediation due to their economic value, adaptability, remediation potential, and negligible contamination of the marketable products. China et al. (2014) reported that *Cymbopogon citratus* can phytostabilize Cu from its mine tailings in presence of amendments. Gautam and Agarwal (2017) reported that *C. citratus* grown on a mixture of soil-sludge-red mud could give better oil yields while the metals in the oil were within Food Safety and Standards Authority of India limits in spite of the higher metal levels in leaves. Similarly, Tall Fescue intercropped with alfalfa showed better removal of soil polyaromatic hydrocarbons than monoculture of alfalfa (Sun et al., 2011). The remediation and utilization of contaminated sites for sustainable development is possible through phytomanagement, desired globally. Market opportunities in phytoremediation are pressing need for environmental remediation with economic return. In this perspective, aromatic grasses have been considered more than promising crop for phytoremediation programs. As they are safe and their main product essential oil is free from the metal toxicity risk and is unpalatable to herbivores. Profitable and effective phytoremediation requires a collection of suitable aromatic grass species that are stress tolerant and produce high yield of essential oil under such stressed environments (Verma et al., 2014; Pandey and Singh, 2015; Pandey et al., 2019).

3.1.3 Climate change mitigation

The most prominent factor, which is driving climate change and global warming are the increased atmospheric concentrations of greenhouse gases (GHGs). The main GHGs after water vapor is carbon dioxide (CO_2). The present atmospheric concentration of CO_2 is 38% higher than that existed during the preindustrial era, that is, 387 ppm. The average global surface temperature has been projected to increase by 1.5–5.8°C by the end of the 21st century (Stavi and Lal, 2013). In this context, the immediate benefit of biomass production by growing perennial grasses is the sequestration of atmospheric CO_2 in the shoot and root. Additionally, the grasses, which are an important source of bio-energy feedstock, are an added advantage which can replace fossil fuels. Further, these perennial grass biomasses are also renewable or appropriately renewable source of energy (Creutzig et al., 2015). It has also been reported that perennial grass bio-fuel crops have more carbon storage potential and lower N_2O emission potential than agricultural crops. Thus, the climate changes mitigation potential of perennial grasses showing bio-energy production is the most suitable option and can accomplish many solutions simultaneously, namely, abatement of CO_2 emissions, mitigation of global warming, tropical deforestation, and fossil fuel depletion (Anderson-Teixeira et al., 2012).

3.1.4 Biodiversity conservation

Cultivation of perennial grasses and biomass production can also increase biodiversity of the area by developing the habitat heterogeneity (Berendse et al., 2015). Perennial bio-energy grass cultivation can gradually develop diversity of flora and fauna in the area, apart from other ecosystem services to the local people. Rendering of ecosystem services by these cultivation systems will indirectly reduce biodiversity loss from the nearby natural ecosystems by decreasing the anthropogenic pressure. Dominance of rhizomatous perennial grasses can mitigate nutrient imbalance, which is one of the major cause which affects biodiversity in degraded ecosystems (Bai et al., 2010).

3.1.5 Wild life shelter

According to reports, dense field of perennial grasses such as *Saccharum spontaneum, Cynodon dactylon, Cyperus esculentus, and Fimbristylis umbellata* also increase countryside faunal biodiversity and provides home to wild animals like Indian rhinoceros (Singh et al., 2018). Similar studies have been reported for *Panicum virgatum* (Trocsanyi et al., 2009). Tall stands of *Miscanthus*, which have attained a height of more than 4 m have a considerable visual impact and serves as a habitat for more number of larger birds and mammals than other herbaceous crops as the grass vegetation harbor more ecological niches (Lewandowski et al., 2000).

3.1.6 Soil erosion control

Perennial grasses provide ground cover throughout the year and have enormous potential in soil conservation and erosion control. This is in virtue of their massive root system, which binds soil particles which act as porous vegetative barrier on sloppy lands and are an alternative to cost-intensive soil physical structures. Functionally, the vegetative barriers perform similar to soil physical structures by reducing soil erosion (Chowdhury et al., 2009). *Festuca rubra* has been extensively used to prevent erosion and stabilize slopes, hillsides, irrigation channels, and banks (Strausbaugh and Core, 1977). *Phalaris arundinacea* is also a cool-season, perennial grass, which is also used for soil stabilization as well as forage production (Lavergne and Molofsky, 2004). Alternatively, *Lasiurus sindicus* is a warm season grass of the Indian deserts, which develops a good rangeland and stabilizes blowing sand dunes (Chowdhury et al., 2009).

3.1.7 Carbon sequestration

Higher biomass developing perennial grasses reduces carbon in the atmosphere by rendering a dual role of substituting fossil fuels as well as carbon sequestration in the soil through their deep root system. Higher biomass production of such grasses is desirable not only for bio-energy but also for carbon sequestration because more biomass leads to more carbon storage (Creutzig et al., 2015; Saha and Kukal, 2015). For example, *Miscanthus x giganteus* plantation which is a bio-energy yielding grass can give 13 ton shoot dry matter ha^{-1} $year^{-1}$ and 21 ton root dry matter ha^{-1} $year^{-1}$; while the total carbon mitigated by this crop is 7 ton ha^{-1} $year^{-1}$ depending on the time of harvest (Clifton-Brown et al., 2007). Similarly, *Phalaris arundinacea* gives a biomass of 8 ton ha^{-1} (Zhou et al., 2011). *Phyllostachys edulis* yields an aboveground biomass of 138 ton ha^{-1} which is the highest range of above-ground biomass yielded among all the bamboo communities in the world. The net production of this bamboo species is also similar to the forest productivity under same climate conditions, which again proves the fact that perennial grasses can not only phytoremediate polluted sites but are also important for inducing carbon sequestration (Isagi et al., 1997). *Vetiveria zizanoides* is also an indispensable plant with respect to soil carbon sequestration as well as erosion control due to its deep root system. Comparatively, the amount of carbon sequestered by the grass in its biomass is 15 ton ha^{-1} year^{-1}. Additionally, it also gives an economically important essential oil from its roots, which is used in perfumery industry. The deep root system of the grass contributes in nutrient sequestration from wetlands and also prevents their leaching. Lignocellulosic biomass of high yielding perennial grasses can also be pyrolyzed to obtain high carbon bio-char which can be used as a soil fertilizer and incur permanent carbon sequestration in the soil. The bio-char has adsorption properties, which reduces the possibilities of nitrous oxide emissions into atmosphere or nitrate leaching into the ground water which are also a problem in some ecosystems (Georgescua et al., 2010; Singh et al., 2018).

3.1.8 Providing ecological corridors

Dense vegetation and canopy structures of perennial grasses serve as habitat corridors in fragmented ecosystems and connect the habitat patches (Lewandowski et al., 2000).

3.2 Societal aspects

3.2.1 Grasses as raw material for crafts

Apart from the earlier-mentioned significant characteristics of perennial grasses, they have multitude of other uses. For example, the biomass of *Phragmites australis* is used for thatching and building rafts. Stems of the plant are used in preparing arrow shafts mats and nets (Batterson and Hall, 1984). Flexible bamboo sticks have also found their wide use in traditional handiwork and is even being used till the present day (Obataya et al., 2007). Fibers from sabai grass have been used by tribal people for making goods for their own use and other craft items which add to their source of income (Khandual and Sahu, 2016).

3.2.2 Huts and animal shades

With the diminishing timber resources across the world, woody monocotyledons perennial grasses, such as bamboo have gained considerable importance (Lin et al., 2002). Various properties of the grass are rapid growth rate, high flexibility, and tensile strength, short rotation age which make it a perfect raw material for a wide variety of products used in our day to day lives. Some examples of the products made from bamboo are domestic household products, floors, concrete molding boards, furniture, cooking items pulp, and handicraft (Yu et al., 2008).

3.2.3 Grasses as a fodder

Fodder for livestock is another important area where perennial grasses have a wide usage. For example, grasses such as *Phragmites australis*, Red fescue, Switch grass produce huge biomass and can be used for forage in cool and warm seasons (Batterson and Hall, 1984; Madakadze et al., 2010; Majidi et al., 2009; Shen et al., 2013; Sun et al., 2011; Trocsanyi et al., 2009). Sewan grass produces a dry biomass of 3 ton ha^{-1} year^{-1} in arid environments which not only maintains the soil cover but also the wild mammals and livestock (Sanadya et al., 2018). Nutrient quality of Elephant grass has also been highlighted which has made it as one of the most important forages (Techio et al., 2002). Properties of Ryegrass (*Lolium perenne* L.) like faster establishment, higher yields, tolerance to grazing, higher palatability to ruminants also make it a widely grown forage grass in temperate zone (Grinberg et al., 2016).

3.2.4 Cultural programs

Some esthetic uses of the grasses are their use in preparing prayer sticks, weaving sticks, and musical instruments. *Phragmites australis* shoot parts have been reported to be used for the earlier-mentioned purposes (Shaltout et al., 2006). Flexible bamboo sticks have many useful acoustic properties which make it useful for making sound-boards of musical instruments which is an alternative to some rare wood species used for the same purpose (Obataya et al., 2007).

3.2.5 Rope manufacturing

Textiles prepared from natural fibers have attracted the interest of the scientific community since a long period. Present research in this area is targeted to develop techniques for the utilization of unconventional plant fibers unlike cotton, silk, and wool. Utilization of fibers obtained from perennial grasses will not only enrich the textile industry but also open new avenues for job opportunities for the local people. Grasses like sabai (*Eulaliapsis binata*) have thin long leaves containing high quality fiber which has been in paper industry till now. However, they also have huge potential for preparing ropes and rope-like items due to the flexible, strong nature, and high cellulose content of the leaves. In addition to conserving soil, the fibers from the grass are made into yarn for knitting (Khandual and Sahu, 2016).

3.3 Economic aspects

3.3.1 Low input and minimum maintenance

Perennial grasses are widely adapted to degraded soils, which have low moisture and available nutrients. Some of the grasses can grow more than 5 m and produce dense biomass, which is produced using minimum cost. These biomasses can be used for thermo-chemical conversions to produce low molecular weight compounds for bio-fuel generation. The harvested biomass of the grasses which plays the role for restoring the waste lands and cleaning them can also be used for producing sugar by enzymolysis and other marketable chemicals provided the heavy metals accumulated in the biomass is not harmful to the processes. In rare cases, the accumulated metals can also be recovered for utilization (Jeguirim et al., 2010; Shen et al., 2013; Trocsanyi et al., 2009; Zhou et al., 2011). The goal of low input and minimum maintenance in phytoremediation programs can be easily obtained through phytomanagement of polluted sites (Pandey and Bajpai, 2019). It is described in details in the chapter of Pandey and Bajpai (2019).

3.3.2 Bioenergy feedstock

In view of a sustainable energy sources, bioenergy feedstocks have received great attention as an alternative to fossil fuels. They are renewable in nature and if produced sustainably, they can diminish oil demand, environmental problems, and boost economic growth of a country. The application of energy plants for the remediation and utilization of contaminated sites is a vital approach toward ecological and socio-economic sustainability (Pandey, 2013, 2017; Pandey et al., 2012a, 2016). In addition, a number of perennial grasses with high biomass production throughout the year can be used for the earlier-mentioned purpose, for example, switch grass, Bermuda grass, elephant grass, and timothy grass. These perennial grasses often grow in wild conditions and termed as low maintenance high biomass yielding plants. Vast acres of degraded lands can be used sustainably for the generation of biomass feed stocks. According to an estimate around 1400 Mha of degraded land is available throughout the globe, which if used for production of bio-energy feedstock can mitigate up to 50% of the world's energy requirement. Energy crop biomass yield on degraded lands can attain a maximum value of 10 ton ha^{-1} $year^{-1}$ which can be an efficient way to conserve natural resources. Though woody biomasses have many useful properties, which influence bio-energy generation yet their annual biomass production is less due to the lesser growth rate after establishment. Some other grasses like *Arundo donax, Miscanthus, Phalaris arundinacea, Cenchrus ciliaris,* and *Pennisetum pedicellatum* have also been tested for energy biomass production on degraded lands which will also serve the purpose for reclamation (Jeguirim et al., 2010; Shen et al., 2013; Trocsanyi et al., 2009; Zhou et al., 2011). *C. flexuosus* was also revealed as a promising biofuel for diesel engine in which 20% raw oil obtained from the grass was mixed with 80% diesel (Dhinesh et al., 2016). In another approach, the methane content generated in biogas, which is obtained from animal manures, can also be improved by codigestion with *C. citratus* biomass for 30 days at 33°C which will yield approximately 2 L kg^{-1} day^{-1} of better quality of biogas (Owamah et al., 2014).

3.3.3 Medicinal use

Grasses like *Phragmites australis* has been traditionally used for the treatment of cancer, arthritis, bronchitis, cholera, diabetes, hemorrhage, jaundice, and typhoid (Srivastava et al., 2014). Essential oil of lemon grass has citral as a constituent, which inhibits growth of bacteria and fungus. Commercially, 2.5% of Lemon grass oil is used for preparing antifungal creams (Wannissorn et al., 1996). *Phalaris arundinacea* extracts also has possible medicinal uses. It contains inulin, which is beneficial for health as it acts as a probiotic agent and stimulates the development useful intestinal bacteria. Solubility of the compound in hot water allows its incorporation in food (Caunii et al., 2015).

3.3.4 Essential oil

Aromatic perennial grass species of *Cymbopogon* (Lemon grass) and *Vetiveria* genus not only stabilizes soil but are also economic due to the oil obtained from their shoots and roots respectively. In India, Lemon grass cultivation and oil extraction from its leaves through leaf distillation had started in 1958 (Lal, 2012). Presently, the grass is grown in 4000 ha area in our country with a yield of 250 ton; most importantly India leads in Lemon grass oil production (Gupta et al., 2015) and exports oil to countries such as Europe, USA, Japan. Possibilities of utilization of the degraded areas for Lemon grass cultivation and oil extraction can open up new avenues of income generation for local people besides ameliorating the waste and boosting economic growth of the country (Verma et al., 2014; Pandey and Singh, 2015; Pandey et al., 2019).

3.3.5 Pulp and paper manufacturing

Pulp and paper production industry is another important area where grasses like *Arundo donax*, Sabai grass, Elephant grass, and switchgrass are used extensively (Madakadze et al., 2010; Shatalov and Pereira, 2002; Trocsanyi et al., 2009). Dry biomass productivity of the grass is 37 ton ha^{-1} year^{-1} which can be increased through more intensive cultivation on the otherwise barren degraded lands. Bagasse of *Cymbopogon nardus* is used for paper and pulp industries (Dhinesh et al., 2016).

3.3.6 Industrialization

In view of the earlier ecosystem services, massive cultivation of the grasses on degraded sites can initiate industrialization in the area for obtaining the following most important products, which in turn will boost economic growth of the country (Dhinesh et al., 2016; Gupta et al., 2015; Khandual and Sahu, 2016; Yu et al., 2008; Zhou et al., 2011).

- Paper and pulp
- Essential oil
- Bio-fuel
- Textile
- Usage of bamboo as wood replacement

3.3.7 Job creation and poverty alleviation

Presently, extraction of bioenergy from biomass is the most popular technology being used by many countries including India. This can be used as an alternate livelihood generation for poor people and increase the village economy, apart from its beneficial role toward ecorestoration of degraded lands (Jeguirim et al., 2010; Shen et al., 2013; Trocsanyi et al., 2009; Zhou et al., 2011). Switchgrass management in Europe as a bio-energy crop is relatively a new intervention. Results to date from various European studies have suggested that switchgrass is broadly adapted to many European countries. It is currently introduced into temperate regions of Europe, that is, regions that lie between latitudes 40 degree 09 N (Greece) and 52 degree 38 N (the Netherlands) (Parrish et al., 2012).

4 Why perennial grasses in phytoremediation?

Perennial grasses have been found more suitable for phytoremediation because they have increased capacity for surface stabilization, building soil organic matter in a small period of time, and holding nutrients as well as ions efficiently along with their recycling. Additionally, they show greater availability of grazing season, forage, and their dry mass is less flammable than other herbaceous plants. The established vegetation has a visual appeal and is esthetically more pleasing which increases the biodiversity of the area. Alternatively, they can also be used for horticultural purposes on degraded areas. Incorporation of perennial grass in a developing ecosystem substantially reduces the management due to their extended growing season and requirement of fertilizers inputs for initial establishment are less; which is an advantage over annually growing plants (Aftab et al., 2011; Fernando et al., 2012; Gupta et al., 2015; Jeguirim et al., 2010; Pruchniewicz et al., 2017; Shatalov and Pereira, 2002). Major factors, which further support the use of perennial grasses for phytoremediation of polluted sites, are summarized as follows:

- Ability to sustain harsh conditions
- Reproduction of fresh shoots by tillers recovered during cutting
- Give rapid ground coverage through spreading of rhizomes forming adventitious roots
- High biomass development
- Deep root systems
- Low greenhouse gas emissions
- Carbon sequestration
- Phytostabilization capacity
- Permanent green cover and sustained productivity
- Reduction of soil erosion and ability to increase site fertility

5 Coupling phytoremediation with perennial native grasses

Phytoremediation of degraded lands involve remediation of pollutants from the contaminated sites. However, addition to this aim restoration of original flora of the landscape should also be incorporated in the goals of such programs. This is efficiently done by using native plants, especially pioneer species, which are already adapted on such sites. Incorporation of such plants helps in developing a sustainable ecosystem and also facilitates succession of other species in the long term (Aftab et al., 2011; Pidlisnyuk et al., 2014). This is unlike to the introduction of other commercial species, which often hinder invasion by native species due to the former's competitive nature and often lead to creation of "weed." Native species which are already adapted to the existing nutrient conditions can not only remediate the pollutants but also flourish with low maintenance as well as show genetic variability to adapt as per the environment (Fernando et al., 2012; Sonowal et al., 2018). Though many native plants are not metal hyperaccumulators, but are metal tolerant; they work by up taking metals in slow rates from the substrate, detoxify them and also produce a huge biomass. The plant characteristics such as fibrous root systems, tolerance of low available nitrogen and phosphorus in the soil make them more suitable for the purpose (Gupta et al., 2015; Jeguirim et al., 2010). Studies have reported many pioneer or native species which have invaded degraded sites and become dominant plants; for example, *Arctagrostis latifolia, Arctophila fulva, Calamagrostis canadensis, Carex aquatalis, Epilobium angustifolium, Equisetum arvense, Eriophorum augustifolium, Hierochloe alpine, H. ordorata, Hordeum jubatum, Poa arctica, P. lanata,* and *Luzula confusa.* The grass *C. epigejos* showed greater dominance and survival potential on bare ash deposits due to its multiple tolerances to the metals in fly ash and its competitive ability (Mitrovic et al., 2008; Pruchniewicz et al., 2017; Shatalov and Pereira, 2002). These types of species can also incur stabilization of such loose wastes through penetration of the plant roots. Integration of phytoremediation with perennial native grasses also gives the key to a sustainable future by solving the problem of bio-energy demand as well as gradual biological restoration at a much smaller time period (Tripathi et al., 2016).

6 Perennial growth—an essential aspect for sustainable biomass source

Generally grasses show both type (i.e., C3 and C4) of photosynthesis. It is well observed that C4 grasses have more capacity for carbon fixation, have ability to grow in poor soil-moisture circumstances with a consequent higher biomass yield (Mitrovic et al., 2008; Somodi et al., 2008). This is the reason why C4 grasses are better known as energy crops than C3 grasses. Moreover, comparing with C3 grasses, they do not want recurrent mowing for higher biomass production (Jeguirim et al., 2010). There are some C3 and C4 grasses on which constantly researches have been done, are giant reed (*Arundo donax*), energy cane (*Saccharum* spp.), common reed (*Phragmites communis*), reed canary grass (*Phalaris arundinacea*), prairie cord grass (*Spartina pectinata*), big bluestem (*Andropogon gerardi* Vitman), cocksfoot grass (*Dactylis glomerata*), and weeping lovegrass (*Eragrostis curvula*) (Shatalov and Pereira, 2002; Jeguirim et al., 2010; Aftab et al., 2011; Pandey et al., 2012b, 2015; Gupta et al., 2015; Pruchniewicz et al., 2017). *Miscanthus* grass is naturally present in temperate climate region, having C4 photosynthetic pathway. C4 grasses have extra potential to produce more yield than C3 grasses due to higher radiation, water and nitrogen-use efficiencies, but they need warmer region than C3 grasses to start their growth in spring time. A number of benefits of C4 grasses have been emphasized than other energy plants such as higher tolerance to abiotic stresses (Mitchell et al., 2014), higher soil carbon storage potential (Liebig et al., 2008), and higher use efficiencies of water, nitrogen, and light energy (Taylor et al., 2010).

Due to earlier-mentioned features, C4 grasses have potential to deliver consistent and sustainable yields over a wide range of agricultural sites, can offer to produce large-scale biomass in uncertain environmental situations without causing competition with food crop production (Kryževičienė et al., 2011; Robbins et al., 2012). Consequently, in Europe, the progress of C4-based perennial grasses toward biomass production is currently on the rise (Searle and Malins, 2014). *A. donax* (Giant Reed) has a longer growing season than other perennial grasses of temperate regions like *M. giganteus* and *P. arundinacea*. In England, *A. donax* prolongs its growth by maintaining higher moisture in their tissues and green leaves during dry periods as well as at the end of growing period (Smith and Slater, 2010). Though, *A. donax* is a C3 grass but having higher photosynthetic rates and productivity similar to C4 grass (Angelini et al., 2009). *A. donax* attained high biomass production due to its continuous growth. In addition, the root system of *A. donax* forms a huge network in the soil system, that is, its root system can spread up to a depth of >1 m, which offers suitable surface and root exudates for microbial affection and growth (Sharma et al., 2005; Nsanganwimana et al., 2014).

7 Improvement of perennial grasses for enhanced phytoremediation

The aims of upgrading of perennial grasses toward enhanced phytoremediation depend on ecological conditions of growing and way of exploitation. Researchers involved in the grass improvement for enhanced phytoremediation worldwide. Improved grasses via genetically or new cultivars must fulfill some of the features toward grass phytoremediation as:

- High growth rate
- Developed root system via increase of root length
- Higher biomass yield
- Resistance against major diseases
- Resistance against pests
- Resistance against environmental stress factors
- Adaptability of grass species toward growing in association (i.e., grass-to-grass, grass-to-herbs, and grass-to-tree)
- Adaptability to certain agro-ecological conditions
- Designer grass ecosystem for enhanced phytoremediation

Various techniques have been developed that can be used to improve physiological, genetic, reproductive traits of the grass species and which will be resulted in the creation of potentially genetically modified grass species with high biomass production, high remediation potential as well as high tolerance against pathogens and environmental stresses. Researchers involved in the improvement of grasses worldwide, via modern technologies of improvement have created a number of cultivars and hybrids with high biomass production potential. In addition, it is required to direct attention to creation of perennial grass cultivars of specific traits, such as high phytoremediation potential, high regeneration potential, and ability for fast regeneration after cutting and resistance to stress factors (toxicants, drought, and low temperatures), major diseases, and pests. Genetic and genome engineering biotechnologies will be useful in the development of improved grass species for enhanced phytoremediation and biomass production.

8 Perennial grass-based phytoremediation practices

The practices of perennial grass-based phytoremediation are very limited. It is pressing need to develop grass-based phytoremediation practices across the nations particularly both hot and cold countries. Linking knowledge of grass-based phytoremediation into action is in promising stage and has been assessed in some potential grasses such as Vetiver, *Miscanthus*, Giant reed, Switch grass, Elephant grass, Reed canary grass, Common reed, Moso bamboo, Sewan grass, etc. Commercialization of perennial grass-based phytoremediation could be achieved if we include market opportunities in phytoremediation programs (Pandey and Souza-Alonso, 2019).

9 Policy framework

Policy framework should be implemented for perennial grasses to be utilized in phytoremediation programs. There are several important points given here to be implemented in policy of national and international forum for promoting the utilization of perennial grasses in phytoremediation programs. These are:

- Three main key factors of bioeconomy is feedstock (adopting a sustainable biomass supply), bio-refineries (optimizing efficient processing through R & D), and supporting market (developing market for bio-based products) which will help toward implementation of perennial grass system in phytoremediation programs.
- Bringing togather diverse yet complementary stakeholders across the nations.
- It will create new jobs, especially in rural and tribal regions.
- Utilizing waste dumpsites or polluted sites for biomass production of perennial grasses.
- Optimization of perennial grasses for biomass production as bio-energy feedstock from polluted sites during phytoremediation.
- Optimization of aromatic perennial grasses for essential oil production in polluted sites during phytoremediation.
- Assessing perennial grasses as fodder utility, those are used in phytoremediation of polluted sites to check any risk in animal.
- Assessing the potential of perennial grasses to provide other ecosystem services in phytoremediation of polluted sites.
- Assessing the ecological suitability and adaptability of grasses in phytoremediation programs with the aim of biofuel production.

10 Conclusions and future prospects

Most of the perennial grasses have fast-growing nature with aggressive root system that helps in its regeneration. Grasses are extremely tolerant to different stress conditions (i.e., drought resistant, salt tolerant, highly versatile, heavy metal tolerant and adapted to hostile environmental conditions) and valuable for vegetation cover on industrial waste dumpsites and soil conservation. However, site-specific factors should be perfectly assessed to assess the suitability between grass species and polluted site.

Additionally, the main research gaps and limitations are lack of knowledge regarding the effect of grass-microbe-pollutant interaction on phytoremediation and biomass production. A wide range of potential perennial grasses have been utilized for multiple uses such as lawn cover, beautification, fodder, golf courses, remediation, vegetation cover on industrial waste dump sites, and restoration of degraded and polluted sites. In addition, ornamental grasses can be used as potential genetic resources in park development on waste dump sites. Thus, perennial grasses—a potential asset for livelihood security and climate change mitigation has been proved. However, native perennial grass species have great level of diversity, potential tolerance against multiple stress conditions, and potential ability for regeneration, high biomass potential, and remediation potential in polluted sites.

Acknowledgments

Financial assistance given to Dr. V.C. Pandey under the Scientist's Pool Scheme (Pool No. 13 (8931-A)/2017) by the Council of Scientific and Industrial Research, Government of India is gratefully acknowledged. The authors declare no conflicts of interest.

References

Adriano, D.C., Wenzel, W.W., Vangronsveld, J., Bolan, N.S., 2004. Role of assisted natural remediation in environmental cleanup. Geoderma 122, 121–142.

Aftab, K., Ali, M.D., Aijaz, P., Beena, N., Gulzar, H.J., Sheikh, K., Sofia, Q., Tahir Abbas, S., 2011. Determination of different trace and essential element in lemon grass samples by x-ray fluorescence spectroscopy technique. Int. Food Res. J. 18 (1), 265–270.

Ampiah-Bonney, R.J., Tyson, J.F., Lanza, G.R., 2007. Phytoextraction of arsenic from soil by *Leersia oryzoides*. Int. J. Phytoremediation 9 (1), 31–40.

Anderson-Teixeira, K.J., Snyder, P.K., Twine, T.E., Cuadra, S.V., Costa, M.H., DeLucia, E.H., 2012. Climate-regulation services of natural and agricultural ecoregions of the Americas. Nat. Clim. Change 2 (3), 177–181.

Angelini, L.G., Ceccarini, L., o Di Nasso, N.N., Bonari, E., 2009. Comparison of Arundo donax L. and Miscanthus x giganteus in a long-term field experiment in Central Italy: Analysis of productive characteristics and energy balance. Biomass Bioenergy 33, 635–643.

Antoniadis, V., Levizou, E., Shaheen, S.M., Ok, Y.S., Sebastian, A., Baum, C., Prasad, M.N., Wenzel, W.W., Rinklebe, J., 2017. Trace elements in the soil-plant interface: phytoavailability, translocation, and phytoremediation–a review. Earth-Sci. Rev. 171, 621–645.

Aprill, W., Sims, R.C., 1990. Evaluation of the use of prairie grasses for stimulating polycyclic aromatic hydrocarbon treatment in soils. Chemosphere 20, 253–265.

Awasthi, A., Singh, K., Singh, R.P., 2017. A concept of diverse perennial cropping systems for integrated bioenergy production and ecological restoration of marginal lands in India. Ecol. Eng. 105, 58–65.

Awasthi, A., Singh, K., O Grady, A., Courtney, R., Kalra, A., Singh, R.P, Cerda, A., Steinberger, Y., Patra, D.D., 2016. Designer ecosystems: a solution for the conservation-exploitation dilemma. Ecol. Eng. 93, 73–75.

Bai, Y., Wu, J., Clark, C.M., Naeem, S., Pan, Q., Huang, J., Zhang, L., Han, X., 2010. Tradeoffs and thresholds in the effects of nitrogen addition on biodiversity and ecosystem functioning: evidence from inner Mongolia Grasslands. Glob. Change Biol. 16, 358–372.

Batterson, T.R., Hall, D.W., 1984. Common reed Phragmites australis (Cav.) Trin. ex Steudel. Aquatics 6 (2), 16–20.

Begonia, M.F.T., Begonia, G.B., Ighoavodha, M., Gilliard, D., 2005. Lead accumulation by Tall Fescue (Festuca arundinacea Schreb.) grown on a lead-contaminated soil. Int. J. Environ. Res. Public Health 2 (2), 228–233.

Berendse, F., van Ruijven, J., Jongejans, E., Keesstra, S.D., 2015. Loss of plant species diversity reduces soil erosion resistance of embankments that are crucial for the safety of human societies in low-lying areas. Ecosystems 18, 881–888.

Besalatpour, A.A., Hajabbasi, M.A., Khoshgoftarmanesh, A.H., Afyuni, M., 2008. Remediation of petroleum contaminated soils around the Tehran oil refinery using phytostimulation method. J. Agric. Sci. Nat. Res. 15 (4), 22–37.

Boller, B., Greene, S.L., 2010. Genetic resources. In: Boller, B., Posselt, U.K., Veronesi, F. (Eds.), Fodder Crops and Amenity Grasses. Handbook of Plant Breeding 5. Springer Science + Business Media, NY, USA, pp. 13–37.

Brix, H., Ye, S., Laws, E.A., Sun, D., Li, G., Ding, X., Yuan, H., Zhao, G., Wang, J., Pei, S., 2014. Large-scale management of common reed, Phragmites australis, for paper production: a case study from the Liaohe Delta, China. Ecol. Eng. 73, 760–769.

Caunii, A., Butu, M., Rodino, S., Motoc, M., Negrea, A., Samfira, I., Butnariu, M., 2015. Isolation and separation of Inulin from Phalaris arundinacea roots. Revista de chimie 66 (4), 472–476.

China, S.P., Das, M., Maiti, S.K., 2014. Phytostabilization of Mosaboni copper mine tailings: a green step towards waste management. Appl. Ecol. Environ. Res. 12 (1), 25–32.

Chowdhury, S.P., Schmid, M., Hartmann, A., Tripathi, A.K., 2009. Diversity of 16S-rRNA and nifH genes derived from rhizosphere soil and roots of an endemic drought tolerant grass, Lasiurus sindicus. Eur. J. Soil Biol. 45 (1), 114–122.

Clifton-Brown, J.C., Breuer, J., Jones, M.B., 2007. Carbon mitigation by the energy crop, Miscanthus. Glob. Change Biol. 13 (11), 2296–2307.

Creutzig, F., Ravindranath, N.H., Berndes, G., Bolwig, S., Bright, R., Cherubini, F., et al., 2015. Bioenergy and climate change mitigation: an assessment. GCB Bioenergy 7, 916–944.

Cui, T., Fang, L., Wang, M., Jiang, M., Shen, G., 2018. Intercropping of gramineous pasture ryegrass (Lolium perenne L.) and leguminous forage alfalfa (Medicago sativa L.) increases the resistance of plants to heavy metals. J. Chem., 2018. Article ID 7803408, 11. Available from: https://doi.org/10.1155/2018/7803408.

Davidson, C.G., Gobin, S.M., 1998. Evaluation of ornamental grasses for the Northern Great Plains. J. Environ. Horticult. 16 (4), 18–229.

De Koe, T., 1994. Agrostis castellana and Agrostis delicatula on heavy metal and arsenic enriched sites in NE Portugal. Sci. Total Environ. 145 (1–2), 103–109.

Delgado-Caballero, M., Alarcon-Herrera, M., Valles-Aragon, M., Melgoza-Castillo, A., Ojeda-Barrios, D., Leyva-Chávez, A., 2017. Germination of Bouteloua dactyloides and Cynodon dactylon in a multi-polluted soil. Sustainability 9 (81).

Desai, M.A., Parikh, J., De, A.K., 2014. Modelling and optimization studies on extraction of lemongrass oil from Cymbopogon flexuosus (Steud.) Wats. Chem. Eng. Res. Design 92 (5), 793–803.

Dhinesh, B., Lalvani, J.I.J., Parthasarathy, M., Annamalai, K., 2016. An assessment on performance, emission and combustion characteristics of single cylinder diesel engine powered by Cymbopogon flexuosus biofuel. Energ. Convers. Manage. 117, 466–474.

Efe Sunday, I., Ephram, I.E., 2014. Phytoremediation of crude oil contaminated soil with Axonopus compressus in the Niger Delta Region of Nigeria. Nat. Resour. 5, 59–67.

Elekes, C.C., Busuioc, G., 2011. The modelling of phytoremediation process for soils polluted with heavy metals. Lucraristiintifice 54, 133–136.

Fatima, K., Imran, A., Amin, I., Khan, Q.M., Afzal, M., 2018. Successful phytoremediation of crude-oil contaminated soil at an oil exploration and production company by plants-bacterial synergism. Int. J. Phytoremediation 20 (7), 675–681.

Fernando, A.L., Boleo, S., Barbosa, B., Costa, J., Sidella, S., Nocentini, A., Duarte, M.P., Mendes, B., Monti, A., Cosentino, S.L., 2012. Perennial grasses: environmental benefits and constraints of its cultivation in Europe. In 20th European Biomass Conference and Exhibition, pp. 18–22.

Gao, H., Liu, J.X., Guo, A.G., 2008. Preliminary evaluation of adaptability and utilization value of ornamental grasses in Nanjing [J]. Pratacultural Sci. 25, 8.

Gautam, M., Agrawal, M., 2017. Influence of metals on essential oil content and composition of lemongrass (Cymbopogon citratus (DC) Stapf.) grown under different levels of red mud in sewage sludge amended soil. Chemosphere 175, 315–322.

Georgescua, M., Lobellb, D.B., Fieldc, C.B., 2010. Direct climate effects of perennial bioenergy crops in the United States. PNAS USA 108, 4307–4312.

Ghosh, M., Paul, J., Jana, A., De, A., Mukherjee, A., 2015. Use of the grass, *Vetiveria zizanioides* (L.) Nash for detoxification and phytoremediation of soils contaminated with fly ash from thermal power plants. Ecol. Eng. 74, 258–265.

Gibson, D.J., 2009. Grasses and Grassland Ecology. OUP Oxford, New York.

Gilson, E., 2017. Biogas production potential and cost-benefit analysis of harvesting wetland plants (*Phragmites australis* and *Glyceria maxima*). Master's Thesis, Halmstad University.

Gołda, S., Korzeniowska, J., 2016. Comparison of phytoremediation potential of three grass species in soil contaminated with cadmium. Ochrona Srodowiska i Zasobów Naturalnych 27 (1), 8–14.

Graham, D.B., Josef, N.S.K., Kristin, S., 2014. The reproductive biology of *Saccharum spontaneum* L.: implications for management of this invasive weed in Panama. NeoBiota 20, 61–79.

Grinberg, N.F., Lovatt, A., Hegarty, M., Lovatt, A., Skot, K.P., Kelly, R., Blackmore, T., Thorogood, D., King, R.D., Armstead, I., Powell, W., 2016. Implementation of genomic prediction in *Lolium perenne* (L.) breeding populations. Front. Plant Sci. 7, 133.

Gupta, P., Dhawan, S.S., Lal, R.K., 2015. Adaptability and stability based differentiation and selection in aromatic grasses (*Cymbopogon species*) germplasm. Ind. Crops Products 78, 1–8.

Gupta, P., Dhawan, S.S., Lal, R.K., 2015. Adaptability and stability based differentiation and selection in aromatic grasses (*Cymbopogon species*) germplasm. Ind. Crop. Prod. 78, 1–8.

Gajić, G., Djurdjević, L., Kostić, O., Jarić, S., Mitrović, M., Stevanović, B., Pavlović, P., 2016. Assessment of the phytoremediation potential and an adaptive response of *Festuca rubra* L. Sown on fly ash deposits: native grass has a pivotal role in ecorestoration management. Ecol. Eng. 93, 250–261.

Hafliger, P., Schwarzlander, M., Blossey, B., 2006. Impact of *Archanara geminipuncta* (Lepidoptera: Noctuidae) on aboveground biomass production of *Phragmites australis*. Biol. Control 38 (3), 413–421.

Hameed, M., Ashraf, M., Naz, N., Al-Qurainy, F., 2010. Anatomical adaptations of *Cynodon dactylon* (L.) Pers. from the salt range (Pakistan) to salinity stress. I. Root and stem anatomy. Pakistan J. Bot. 42 (1), 279–289.

Houben, D., Sonnet, P., 2015. Impact of biochar and root-induced changes on metal dynamics in the rhizosphere of *Agrostis capillaris* and *Lupinus albus*. Chemosphere 139, 644–651.

Huang, X.D., El-Alawi, Y., Penrose, D.M., Glick, B.R., Greenberg, B.M., 2004. Responses of three grass species to creosote during phytoremediation. Environ. Pollut. 130 (3), 453–463.

Isagi, Y., Kawahara, T., Kamo, K., Ito, H., 1997. Net production and carbon cycling in a bamboo *Phyllostachys pubescens* stand. Plant Ecol. 130 (1), 41–52.

Jeguirim, M., Dorge, S., Trouve, G., 2010. Thermogravimetric analysis and emission characteristics of two energy crops in air atmosphere: *Arundo donax* and *Miscanthus giganthus*. Bioresour. Technol. 101 (2), 788–793.

Jiamjitrpanich, W., Parkpian, P., Polprasert, C., Kosanlavit, R., 2013. Trinitrotoluene and its metabolites in shoots and roots of *Panicum maximum* in nano-phytoremediation. Int. J. Environ. Sci. Dev. 4 (1), 7.

Khandual, A., Sahu, S., 2016. Sabai grass: possibility of becoming a potential textile. In: Sustainable Fibres for Fashion Industry, Springer, Singapore, pp. 45–60.

Kojić, M., Popović, R., Karadžić, B., 1997. Vascular plants of Serbia as indicators of habitats. Agricultural Research Institute "Serbia", Institute for Biological Research "Siniša Stanković", Belgrade, p. 160.

Kryževičienė, A., Kadžiulienė, Z., Sarunaite, L., Dabkevičius, Z., Tilvikiene, V., Šlepetys, J., 2011. Cultivation of Miscanthus × giganteus for biofuel and its tolerance of Lithuania's climate. Zemdirbyste-Agriculture 98(3), 267-274.

Lal, R.K., 2012. Stability for oil yield and variety recommendations' using AMMI (additive main effects and multiplicative interactions) model in Lemongrass (*Cymbopogon* species). Ind. Crop. Prod. 40, 296–301.

Langella, F., Grawunder, A., Stark, R., Weist, A., Merten, D., Haferburg, G., Büchel, G., Kothe, E., 2014. Microbially assisted phytoremediation approaches for two multi-element contaminated sites. Environ. Sci. Poll. Res. 21 (11), 6845–6858.

Lavergne, S., Molofsky, J., 2004. Reed canary grass (*Phalaris arundinacea*) as a biological model in the study of plant invasions. Crit. Rev. Plant Sci. 23 (5), 415–429.

Lewandowski, I., Clifton-Brown, J.C., Scurlock, J.M.O., Huisman, W., 2000. Miscanthus: European experience with a novel energy crop. Biomass Bioenergy 19 (4), 209–227.

Liebig, M.A., Schmer, M.R., Vogel, K.P., Mitchell, R.B., 2008. Soil carbon storage by switchgrass grown for bioenergy. Bioenergy Res. 1 (3–4), 215–222.

Lin, J., He, X., Hu, Y., Kuang, T., Ceulemans, R., 2002. Lignification and lignin heterogeneity for various age classes of bamboo (*Phyllostachys pubescens*) stems. Physiol. Plant. 114 (2), 296–302.

Liu, Z., He, X., Chen, W., Zha, M., 2013. Eco toxicological responses of three ornamental herb species to cadmium. Environ. Toxicol. Chem. 32, 8.

Liu, R., Jadeja, R.N., Zhou, Q., Liu, Z., 2012. Treatment and remediation of petroleum-contaminated soils using selective ornamental plants. Environ. Eng. Sci. 29 (6), 494–501.

Madakadze, I.C., Masamvu, T.M., Radiotis, T., Li, J., Smith, D.L., 2010. Evaluation of pulp and paper making characteristics of elephant grass (*Pennisetum purpureum* Schum) and switchgrass (*Panicum virgatum* L.). Afr. J. Environ. Sci. Technol. 4 (7), 465–470.

Madejon, P., Murillo, J.M., Maranon, T., Cabera, F., Lopez, R., 2002. Bioaccumulation of As, Cd, Cu, Fe and Pb in wild grasses affected by Aznal collar mine spill (SW Spain). Sci. Total Environ. 290, 105–120.

Majidi, M.M., Mirlohi, A., Amini, F., 2009. Genetic variation, heritability and correlations of agro-morphological traits in tall fescue (*Festuca arundinacea* Schreb.). Euphytica 167 (3), 323–331.

Mahmoudzadeh, M., Tehranifar, A., Fotvat, A., Shour, M., Kazemi, F., 2016. Phytoremediation using Poa pratensis balin lawn and sludge in Mashad industrial town treatment plant. IIOAB J. 7 (1), 224–231.

Maiti, D., Prasad, B., 2017. Studies on colonisation of fly ash disposal sites using invasive species and aromatic grasses. J. Environ. Eng. Landsc. Manag. 25 (3), 251–263.

Maiti, D., Prasad, B., 2016. Revegetation of fly ash–a review with emphasis on grass-legume plantation and bioaccumulation of metals. Appl. Ecol. Environ. Res. 14 (2), 185–212.

Maiti, S.K., Maiti, D., 2015. Ecological restoration of waste dumps by topsoil blanketing, coir-matting and seeding with grass–legume mixture. Ecol. Eng. 77, 74–84.

Mitchell, R., Lee, D. K., Casler, M., 2014. Switchgrass. In: D.L. Karlen (Ed.), Cellulosic Energy Cropping Systems. West Sussex, Wiley, pp. 75-89.

Mitrovic, M., Pavlovic, P., Lakusic, D., Djurdjevic, L., Stevanovic, B., Kostic, O., Gajic, G., 2008. The potential of *Festuca rubra* and *Calamagrostis epigejos* for the revegetation of fly ash deposits. Sci. Total Environ. 407 (1), 338–347.

Nagori, B.P., Solanki, R., 2011. *Cynodon dactylon* (L.) Pers.: a valuable medicinal plant. Res. J. Med. Plant 5, 508–514.

Nayyef Alanbary, S., Rozaimah Sheikh Abdullah, S., Hassan, H., Razi Othman Malaysia, A., 2018. Screening for resistant and tolerable plants (*Ludwigia octovalvis* and *Phragmites karka*) in crude oil dludge for phytoremediation of hydrocarbons. Iran. J. Energy Environ. 9 (1).

Nedunuri, K.V., Govindaraju, R.S., Banks, M.K., Schwab, A.P., Chen, Z., 2000. Evaluation of phytoremediation for field-scale degradation of total petroleum hydrocarbons. J. Environ. Eng. 126 (6), 483–490.

Niknahad Gharmakher, H., Esfandyari, A., Rezaei, H., 2018. Phytoremediation of cadmium and nickel using *Vetiveria zizanioides*. Environ. Resourc. Res. 6 (1), 57–66.

Nsanganwimana, F., Marchand, L., Douay, F., Mench, M., 2014. *Arundo donax* L., a candidate for phytomanaging water and soils contaminated by trace elements and producing plant-based feedstock. A review. Int. J. Phytoremediation 16 (10), 982–1017.

Obataya, E., Kitin, P., Yamauchi, H., 2007. Bending characteristics of bamboo (*Phyllostachys pubescens*) with respect to its fiber–foam composite structure. Wood Sci. Technol. 41 (5), 385–400.

Owamah, H.I., Alfa, M.I., Dahunsi, S.O., 2014. Optimization of biogas from chicken droppings with *Cymbopogon citratus*. Renewable Energy 68, 366–371.

Padilla, F.M., Pugnaire, F.I., 2006. The role of nurse plants in the restoration of degraded environments. Front. Ecol. Environ. 4 (4), 196–202.

Pandey, V.C., 2012a. Phytoremediation of heavy metals from fly ash pond by *Azolla caroliniana*. Ecotoxicol. Environ. Saf. 82, 8–12.

Pandey, V.C., 2013. Suitability of *Ricinus communis* L. cultivation for phytoremediation of fly ash disposal sites. Ecol. Eng. 57, 336–341.

Pandey, V.C., 2015. Assisted phytoremediation of fly ash dumps through naturally colonized plants. Ecol. Eng. 82, 1–5.

Pandey, V.C., 2017. Managing waste dumpsites through energy plantations. In: Bauddh, K., Singh, B., Korstad, J. (Eds.), Phytoremediation Potential of Bioenergy Plants. Springer, Singapore, pp. 371–386.

Pandey, V.C., Bajpal, O., Singh, N., 2016. Energy crops in sustainable phytoremediation. Renew. Sust. Energ. Rev. 54, 58–73.

Pandey, V.C., Pandey, D.N., Singh, N., 2015a. Sustainable phytoremediation based on naturally colonizing and economically valuable plants. J. Clean Prod. 86, 37–39.

Pandey, V.C., Prakash, P., Bajpai, O., Kumar, A., Singh, N., 2015b. Phytodiversity on fly ash deposits: evaluation of naturally colonized species for sustainable phytorestoration. Environ. Sci. Pollut. Res. 22, 2776–2787.

Pandey, V.C., Singh, N., Singh, R.P., Singh, D.P., 2014. Rhizoremediation potential of spontaneously grown *Typha latifolia* on fly ash basins: study from the field. Ecol. Eng. 71, 722–727.

Pandey, V.C., 2012b. Invasive species based efficient green technology for phytoremediation of fly ash deposits. J. Geochem. Explor. 123, 13–18.

Pandey, V.C., Abhilash, V.C., Singh, N., 2009. The Indian perspective of utilizing fly ash in phytoremediation, photomanagement and biomass production. J. Environ. Manage. 90, 2943–2958.

Pandey, V.C., Bajpai, O., 2019. Phytoremediation: from theory toward practice. In: Pandey, V.C., Bauddh, K. (Eds.), Phytomanagement of Polluted Sites. Elsevier, Netherlands, pp. 1–49.

Pandey, V.C., Bajpai, O., Pandey, D.N., Singh, N., 2015c. *Saccharum spontaneum*: an underutilized tall grass for revegetation and restoration programs. Genet. Resour. Crop Evol. 62 (3), 443–450.

Pandey, V.C., Rai, A., Korstad, J., 2019. Aromatic crops in phytoremediation: from contaminated to waste dumpsites. In: Pandey, V.C., Bauddh, K. (Eds.), Phytomanagement of Polluted Sites. Elsevier, Netherlands, pp. 255–275.

Pandey, V.C., Singh, K., Singh, J.S., Kumar, A., Singh, B., Singh, R.P., 2012a. Jatropha curcas: a potential biofuel plant for sustainable environmental development. Renew. Sust. Energ. Rev. 16, 2870–2883.

Pandey, V.C., Singh, K., Singh, R.P., Singh, B., 2012b. Naturally growing *Saccharum munja* on the fly ash lagoons: a potential ecological engineer for the revegetation and stabilization. Ecol. Eng. 40, 95–99.

Pandey, V.C., Singh, N., 2015. Aromatic plants versus arsenic hazards in soils. J. Geochem. Explor. 157, 77–80.

Pandey, V.C., Souza-Alonso, P., 2019. Market opportunities in sustainable phytoremediation. In: Pandey, V.C., Bauddh, K. (Eds.), Phytomanagement of Polluted Sites. Elsevier, Netherlands, pp. 51–82.

Parrish, D.J., Casler, M.D., Monti, A., 2012. The evolution of switchgrass as an energy crop. In: Switchgrass, Springer, London, pp. 1-28.

Pastor, J., Gutierrez-gines, M.J., Hernandez, A.J., 2015. Heavy-metal phytostabilizing potential of *Agrostis castellana* Boiss. & Reuter. Int. J. Phytoremediation 17 (10), 988–998.

Pidlisnyuk, V., Stefanovska, T., Lewi, E.E., Erickson, L.E., Davis, L.C., 2014. Miscanthus as a productive biofuel crop for phytoremediation. Crit. Rev. Plant Sci. 33 (1), 1–19.

Pruchniewicz, D., Zolnierz, L., Andonovski, V., 2017. Habitat factors influencing the competitive ability of *Calamagrostis epigejos* (L.) Roth in mountain plant communities. Turk. J. Bot. 41 (6), 579–587.

Randelovic, D., Jakovljevic, K., Mihailovic, N., Jovanovic, S., 2018. Metal accumulation in populations of *Calamagrostis epigejos* (L.) Roth from diverse anthropogenically degraded sites (SE Europe, Serbia). Environ. Monitor. Assess. 190 (4), 183.

Robbins, M.P., Evans, G., Valentine, J., Donnison, I.S., Allison, G.G., 2012. New opportunities for the exploitation of energy crops by thermochemical conversion in Northern Europe and the UK. Progr. Energy Combust. Sci. 38 (2), 138–155.

Saha, D., Kukal, S.S., 2015. Soil structural stability and water retention characteristics under different land uses of degraded lower Himalayas of North-West India. Land Degrad. Dev. 26, 263–271.

Sanadya, S.K., Shekhawat, S.S., Sahoo, S., Kumar, A., 2018. Variability and inter-relationships of quantitative traits in Sewan grass (*Lasiurus sindicus* Henr.) accessions. IJCS 6 (6), 1843–1846.

Searle, S.Y., Malins, C.J., 2014. Will energy crop yields meet expectations? Biomass Bioenergy 65, 3–12.

Shaltout, K.H., Al-Sodany, Y.M., Eid, E.M., 2006. Biology of common reed *Phragmites australis* (cav.) trin. ex steud: review and inquiry. Assiut University Center for Environmental Studies (AUCES).

Sharma, G.P., Singh, J.S., Raghubanshi, A.S., 2005. Plant invasions: emerging trends and future implications. Curr. Sci. 88 (5), 726–734.

Sharma, R., Bhardwaj, R., Gautam, V., Bali, S., Kaur, R., Kaur, P., Sharma, M., Kumar, V., Sharma, A., Thukral, A.K., Vig, A.P., 2018. Phytoremediation in waste management: hyperaccumulation diversity and techniques. In: Plants Under Metal and Metalloid Stress, Springer, Singapore, pp. 277–302.

Shatalov, A.A., Pereira, H., 2002. Influence of stem morphology on pulp and paper properties of *Arundo donax* L. reed. Ind. Crop. Prod. 15 (1), 77–83.

Shen, H., Poovaiah, C.R., Ziebell, A., Tschaplinski, T.J., Pattathil, S., Gjersing, E., Engle, N.L., Katahira, R., Pu, Y., Sykes, R., Chen, F., 2013. Enhanced characteristics of genetically modified switchgrass (*Panicum virgatum* L.) for high biofuel production. Biotechnol. Biofuels 6 (1), 71.

Singh, K., Awasthi, A., Sharma, S.K., Singh, S., Tewari, S.K., 2018. Biomass production from neglected and underutilized tall perennial grasses on marginal lands in India: a brief review. Energ. Ecol. Environ. 3 (4), 207–215.

Smith, R., Slater, F.M., 2010. The effects of organic and inorganic fertilizer applications to Miscanthus × giganteus, Arundo donax and Phalaris arundinacea, when grown as energy crops in Wales, UK. Gcb Bioenergy 2 (4), 169–179.

Sokolovic, D., Babic, S., Radovic, J., Lugic, Z., Simic, A., Zornic, V., Petrovic, M., 2017. Genetic resources of perennial forage grasses in Serbia: current state, broadening and evaluation. Selekcija i semenarstvo 23 (1), 69–82.

Soleimani, M., Afyuni, M., Hajabbasi, M.A., Nourbakhsh, F., Sabzalian, M.R., Christensen, J.H., 2010a. Phytoremediation of an aged petroleum contaminated soil using endophyte infected and non-infected grasses. Chemosphere 81 (9), 1084–1090.

Soleimani, M., Hajabbasi, M.A., Afyuni, M., Mirlohi, A., Borggaard, O.K., Holm, P.E., 2010b. Effect of endophytic fungi on cadmium tolerance and bioaccumulation by *Festuca arundinacea* and *Festuca pratensis*. Int. J. Phytoremediation 12 (6), 535–549.

Somodi, I., Viragh, K., Podani, J., 2008. The effect of the expansion of the clonal grass Calamagrostis epigejos on the species turnover of a semi arid grassland. Appl. Veg. Sci. 11 (2), 187–192.

Sonowal, S., Prasad, M.N.V., Sarma, H., 2018. C3 and C4 plants as potential phytoremediation and bioenergy crops for stabilization of crude oil and heavy metal co-contaminated soils-response of antioxidative enzymes. Trop. Plant Res. 5 (3), 306–314.

Srivastava, J., Kalra, S.J., Naraian, R., 2014. Environmental perspectives of *Phragmites australis* (Cav.) Trin. Ex. Steudel. Appl. Water Sci. 4 (3), 193–202.

Stavi, I., Lal, R., 2013. Agroforestry and biochar to offset climate change: a review. Agron. Sustain. Dev. 33 (1), 81–96.

Subhashini, V., Swamy, A.V.V.S., 2013. Phytoremediation of Pb and Ni contaminated soils using *Catharanthus roseus* (L.). Univer. J. Environ. Res. Technol. 3, 465–472.

Suchkova, N., Darakas, E., Ganoulis, J., 2010. Phytoremediation as a prospective method for rehabilitation of areas contaminated by long-term sewage sludge storage: a Ukrainian–Greek case study. Ecol. Eng. 36 (4), 373–378.

Sun, M., Fu, D., Teng, Y., Shen, Y., Luo, Y., Li, Z., Christie, P., 2011. In situ phytoremediation of PAH-contaminated soil by intercropping alfalfa (*Medicago sativa* L.) with tall fescue (*Festuca arundinacea* Schreb.) and associated soil microbial activity. J. Soil. Sediment. 11 (6), 980–989.

Strausbaugh, P.D., Core, E.L., 1977. Flora of West Virginia. In: Morgantown, W.V. (Ed.), second ed., Seneka Books, Inc., Grantsville, West Virginia, p. 1079.

Sampoux, J.P., Huyghe, C., 2009. Contribution of ploidy level variation and adaptive trait diversity to the environmental distribution of taxa in the 'fine-leaved' lineage (genus *Festuca* subg. Festuca). J. Biogeogr. 36, 1978–1993.

Talik, E., Guzik, A., Małkowski, E., Wozniak, G., Sierka, E., 2018. Biominerals and waxes of *Calamagrostis epigejos* and *Phragmites australis* leaves from post-industrial habitats. Protoplasma 255 (3), 773–784.

Taylor, S.H., Hulme, S.P., Rees, M., Ripley, B.S., Ian Woodward, F., Osborne, C.P., 2010. Ecophysiological traits in C3 and C4 grasses: a phylogenetically controlled screening experiment. New Phytol. 185 (3), 780–791.

Techio, V.H., Davide, L.C., Pereira, A.V., Bearzoti, E., 2002. Cytotaxonomy of some species and of interspecific hybrids of *Pennisetum* (Poaceae, Poales). Genetics Mol. Biol. 25 (2), 203–209.

Thomas, C., Butler, A., Larson, S., Medina, V., Begonia, M., 2014. Complexation of lead by bermuda grass root exudates in aqueous media. Int. J. Phytoremediation 16, 634–640.

Tripathi, V., Edrisi, S.A., Abhilash, P.C., 2016. Towards the coupling of phytoremediation with bioenergy production. Renew. Sustain. Energ. Rev. 57, 1386–1389.

Trocsanyi, Z.K., Fieldsend, A.F., Wolf, D.D., 2009. Yield and canopy characteristics of switchgrass (*Panicum virgatum* L.) as influenced by cutting management. Biomass Bioenergy 33 (3), 442–448.

Verma, S.K., Singh, K., Gupta, A.K., Pandey, V.C., Trivedi, P., Verma, R.K., Patra, D.D., 2014. Aromatic grasses for phytomanagement of coal fly ash hazards. Ecol. Eng. 73, 425–428.

Wannissorn, B., Jarikasem, S., Soontorntanasart, T., 1996. Antifungal activity of lemon grass oil and lemon grass oil cream. Phytother. Res. 10 (7), 551–554.

Yasin, M., El-Mehdawi, A.F., Anwar, A., Pilon-Smits, E.A., Faisal, M., 2015. Microbial-enhanced selenium and iron biofortification of wheat (*Triticum aestivum* L.)-applications in phytoremediation and biofortification. Int. J. Phytoremediation 17 (4), 341–347.

Yu, H.Q., Jiang, Z.H., Hse, C.Y., Shupe, T.F., 2008. Selected physical and mechanical properties of Moso bamboo (*Phyllostachys pubescens*). J. Trop. Forest Sci. 20 (4), 258–263.

Zhou, X., GE, Z.M, KellomÄki, S., WANG, K.Y., Peltola, H., Martikainen, P., 2011. Effects of elevated CO_2 and temperature on leaf characteristics, photosynthesis and carbon storage in aboveground biomass of a boreal bioenergy crop (*Phalaris arundinacea* L.) under varying water regimes. GCB Bioenergy 3 (3), 223–234.

Ziarati, P., Ziarati, N.N., Nazeri, S., Saber-Germi, M., 2015. Phytoextraction of heavy metals by two Sorghum spices in treated soil "using black tea residue for cleaning-Uo the contaminated soil". Orient. J. Chem. 31 (1), 317–326.

Zorica, T., Dukic, D., Katic, S., Sanja, V., Mikic, A., Milic, D., Lugic, Z., Jasmina, R., Sokolovic, D., Stanisavljevic, R., 2005. Genetic resources and improvement of forage plants in Serbia and Montenegro. Acta Agriculturae Serbica 10 (19), 3–16.

Zurek, G., Pogrzeba, M., Rybka, K., Prokopiuk, K., 2013. Suitability of grass species for phytoremediation of soils polluted with heavy-metals. In: Breeding Strategies for Sustainable Forage and Turf Grass Improvement. Springer, Dordrecht, pp. 245–248.

Vetiveria zizanioides (L.) Nash – more than a promising crop in phytoremediation

2

Vimal Chandra Pandey[a, *], *Ashish Praveen*[b,c]
[a]Department of Environmental Science, Babasaheb Bhimrao Ambedkar University, Lucknow, Uttar Pradesh, India; [b]Plant Ecology and Environmental Science Division, National Botanical Research Institute, Lucknow, Uttar Pradesh, India; [c]Department of Botany, Markham college of Commerce, VBU, Hazaribag, India
[*]Corresponding author

1 Introduction

Industrial revolution has laid to the improvement of socio-economic condition of people, but it also bought a by-product with it which is the environmental contamination and pollution. Undoubtedly, at that time most of the land was covered with forest so the impact of industrialization on ecosystem was not realized. However, with increased population and industrialization, deforestation increased and also greed for having more had caused a great loss to the environment. The level of contaminants in soil, water, and air has increased to limits that are unsafe for present and future generations (Ayangbenro and Babalola, 2017). The impact of industrialization was realized in later years of the 19th century when the effects were visible with the several ailments in human as well as the cattle's and plant diseases and toxicity in them. Now realizing the adverse effects of pollution, several steps have been taken to reduce contamination of environment. The steps can be plant based, mechanical, chemical, or physical ways for the removal of contaminants. The plant-based remediation, that is, phytoremediation is the cheap and environment friendly method for reducing the contamination (Salt et al., 1998; Valderrama et al., 2013). Bioremediation includes the remediation through plants, microbes, fungus, and their interaction with the adjoining environment. The method of remediation through different available methods varies, like the edaphic condition (soil, pH, EC, etc.) and also the climatic conditions. The variation in climatic conditions means the suitable conditions for plants and microbes to grow, as different plants and microbes grow in different seasons and climate. The combined interaction of microbes and plants have shown to enhance the process of remediation of contaminants as the microbes make available some of the contaminants which in normal conditions are not available to plants. There is also an important saying that if a small work can be done through a needle then why to use the sword for the same, here vetiver is like a needle being small herb but having various ecosystem services. Several plants (*Pteris, Vetiver,*

Phytoremediation Potential of Perennial Grasses. http://dx.doi.org/10.1016/B978-0-12-817732-7.00002-X

Thlapsi, etc.) have been identified that can remove the contaminants at a faster rate. However, among the known plants, vetiver stood tall with its role in phytomanagement of ecosystem such as controlling erosion of soil, water conservation, phytoremediation, and also some economic return like essential oil, the use of roots, and shoots for various purposes (energy, domestic use, etc.).

Vetiveria zizanioides L. Nash synonymously known as *Chrysopogon zizanioides* L. Roberty (Family: Poaceae/Graminae), is widely cultivated in the tropical regions of the world. Popularly known as "KHUS," it is the major source of the well-known vetiver oil with world-wide demand of 250 metric tons annually. It is a miraculous grass native to India first developed for soil and water conservation by the World Bank during mid-1980s. The grass holds application in medicine, cosmetics, and in perfumery industries. Our focus for the Vetiver plant is in concern due to its pharmacological, phytoremedial, and phytoeconomic activities. There is also a need to discover plant-based biologically active molecules for the scientific validation of its traditional medicinal and phytoremedial value. The importance of vetiver increases with its capability to grow both in terrestrial as well as wetland and water systems. This plant has shown its efficiency in all the conditions with considerable removal of contaminants. The plant not only does the role of contaminants removal but also has proven its role in agriculture. Vetiver has been reported to reduce the concentration of nutrients like N, P, K in soil and water so in cases of excess nutrients where there is dominance of species and reduced diversity can be controlled. The root system of vetiver has a natural ability to establish symbiotic association with the microorganisms in rhizosphere (Singh et al., 2014). Vetiver is also known to reduce the termite population through its oil in roots. Thus, vetiver is "Plant for all trades" in plants for environment clean-up and it can be correctly called as ideal grass/magic grass for its services to the ecosystem and mankind.

Vetiver although seems to be a small herb but its considerable role in phytoremediation is unchallenged. The importance of vetiver is not only with its essential oil and phytoremediation potential but also for its role in soil and water conservation. This plant has been proved to reduce soil erosion with its extensive root system. This also improves the water retention in soil and increased ground water recharge. Vetiver does not require extra attention to grow so it can be grown in households also for waste water treatment. This will reduce the contaminants in household wastewater thus reducing the further contamination in nearby areas. Vetiver has good root and shoot biomass. Essential oil is extracted from the roots of vetiver and has been used in perfumery, pharmaceuticals, and cosmetics. The leftover parts of root are not waste rather it can further be used in production of biogas and energy. The shoots of vetiver have also good biomass and contain cellulose, hemicelluloses, and lignin. The processing of the shoots will yield energy like bio-ethanol production (Rao et al., 2015). The biogas production from the shoots of vetiver has been reported (Li et al., 2014). The vetiver grass is a plant of all purpose and an ideal plant; this will be more evident to the readers in this chapter as they will go through it. Readers will also be amazed to know the various facts about this plant. At the time of writing and collecting materials about the role of vetiver in ecosystem services, even we (authors) were amazed of the various functions this plant does to the humans on this planet Earth. We have tried to gather as much information about vetiver plant and hope it to be interesting and knowledgeable to the readers. This plant is in reality a gift of nature to the mankind. In this chapter, the important part that vetiver contributes in phytoremediation and phytocommerce has been discussed in detail.

2 Morphology, reproduction, and propagation

Leaves—Vetiver is a grass with long leaves, many tillers, and deep roots. Fig. 2.1 depicts vetiver grass growing in patches on the degraded soil. The leaves of vetiver have a strong mid rib and also the edges of leaves are quite sharp such that it can cut skin when touched sharply to its edges (Chomchalow, 2001). The leaves fold like a flap along its midrib when there is too much of drought and during long humid conditions it opens for transpiration (Liao et al., 2003). There is presence of stomata on both the surfaces of leaf, thus, this adaptation has a great importance during drought when the leaves fold to reduce water loss as discussed earlier. Vetiver appears V-shaped and this has an important implication with exposure to sunlight and wind velocity (Truong and Danh, 2015). *Stems*—The stems of vetiver or the culms are usually sheathed with leaves. The stem is a small region above the rhizome and the origin of leaf. The stem that bears the inflorescence are the strongest. The stem has nodes due to which even when buried gives rise to the root and the leaves. The complete structure (stem and leaf) help the plant to cope up with the unfavorable conditions. *Flowers*—The flower of vetiver is actually an inflorescence and the complete length of inflorescence may be up to 1.5 m long. The lower florals are hermaphrodite, that is, both male and female. The color of flower is generally brown and both the male and female parts of flower are separate. *Roots*— The roots of vetiver grass are the most important part because of the presence of essential oil, which is the source of economy. The roots are very extensive which anchors soil and reduces soil erosion. The roots also retain moisture and thus conserve water. The root length has been the success for growth of vetiver even in the drought areas. The roots are angled steeply downward which does not

Figure 2.1 Vetiver grass growing in patches on the degraded soil.

Table 2.1 Feasibility of vetiver root system for remediating both surface + deeper contaminants in soil compared to other plants.

Plant	Maximum root depth
Indian mustard	To 12 inches
Grasses	To 48 inches
Poplar tree	To 15 feet
Vetiver grass	To 19.68 feet (6 m)

Source: USEPA, 2002; Movahed and Maeiyat, 2009; Lavania and Lavania, 2009.

interfere with the roots of other plants. Thus, crops can be grown along with vetiver in cases of phytoremediation and reducing the uptake of contaminants by crops (Praveen et al., 2017). The roots can grow extremely fast and due to its spread and anchorage ability, it prevents soil erosion in heavy downpour from slopes or terrains. The roots can reach a length up to 3–4 m and more underground (up to 6 m) (Danh et al., 2009; Lavania, 2003; Truong, 2000; Nilaweera and Hengchaovanich, 1996). The feasibility of vetiver root system for remediating both surface + deeper contaminants in soil compared to other plants is presented in Table 2.1.

Reproduction and propagation—It has been observed that vetiver flowers but do not form seeds, being sterile (Dahn et al., 2012). Thus being sterile it can't propagate through seeds and reproduces vegetatively, however, there are different methods for propagation of vetiver. Vetiver propagates through tillers, that is, by root divisions or slips. Vetiver propagates vegetatively through tillers. The vetiver flowers but being sterile does not form seeds so, it also does not have rhizome and stolon. Thus the only way it can propagate is through the tillers. The tillers grow at fast rate and reach 2 m in height within a few months. It is easy to build up large numbers of vetiver slips. The tillers are grown generally on light soil so that the plants can be pulled easily. Besides tillers, there are also other ways to propagate vetiver. Some of the methods are as (1) tissue culture is one of the methods through which various plants have been propagated. Likewise micro-propagation of vetiver has been reported in the late 1980s. Thus tissue culture can pave the way of large scale propagation of vetiver and also through this, the quality can also be improved, (2) Ratooning is a method that is applied to sugarcane and other plants. In this method, the plant is cut to the ground and left to sprout. This method also works with vetiver; the shoots are cut to the ground and left to sprout. New saplings come of the underground part of vetiver, (3) budding is an important method of propagation in plants and common in lower plants (algae, bryophytes). In vetiver, intercalary buds on surface of crown also known as "eyes" have been grown by researchers in South Africa, (4) cuttings and culms have been an important method of propagation in plants such as rose, chrysanthemum, sugarcane, banana, etc. This method has been also applied to vetiver where culms were kept on moist soil and finally rapid formation of shoots on each node. In China, a farmer grew vetiver from stem cuttings with addition of root hormone indole acetic acid (IAA). It was achieved with 70% survival of the vetiver.

3 Ecology and physiology

Vetiver belongs to *Poaceae* family. It grows well in sandy loam soil and is perennial. The name derives from the Tamil "vetti" (khus-khus) and "ver" (root), referring to aromatic roots. It is mentioned in the ancient Sanskrit writings and it is a part of Hindu mythology (Lal, 2013; Sujatha et al., 2011; Singh et al., 2014; Lavania, 2000). Vetiver has been found to grow in the tropical and subtropical regions of the world and is native to India (Darajeh et al., 2014). The range of distribution of vetiver is also vast from riverbanks, marshy lands, wastelands, degrades land to fly-ash dumped lands. It can be said that vetiver is omnipresent with its wide ecological distribution. Vetiver is an "ecological-climax" species. It outlasts its neighbors and seems to survive for decades while (at least under normal conditions) showing little or no aggressiveness or colonizing ability, thus it is not invasive. There are some variations with the type of vetiver found in the world like in India in the northern and southern regions variations have been observed. It grows very rapidly and becomes effective for environmental restoration works in only 4–5 months as compared to 2–3 years taken by trees and shrubs for the same job. It can tolerate very high acidity and alkalinity conditions (pH from 3.0 to 10.5); high soil salinity (EC = 8 dScm), sodicity (ESP = 33%), and magnesium; very high concentrations of heavy metals Al, Mn, Mg, As, Cd, Cr, Ni, Cu, Pb, Hg, Se, Zn, and the herbicides and pesticides in soils. Ecologically and physiologically, vetiver grass suits to all environments but there are some conditions and limitations also. As reported, the plant's environmental limits are unknown; however, it is surprisingly broad. If we talk about the precipitation and moisture, the vetiver can grow with a range of 200–3000 mm. It has been reported that in some places of Sri Lanka, it can even grow in 5000 mm rainfall. If temperature is taken in consideration then it can grow in tropical as well as in temperate regions. There are reports of this plant surviving at temperatures of −10°C in the United States. Thus from this, the range of this plant with respect to precipitation and temperature can be judged.

Like its relatives maize, sorghum, and sugarcane, vetiver is among the group of plants that uses a specialized photosynthesis with Kranz anatomy present in C4 plants. These plants are C4 because there first product is a four-carbon compound called oxaloacetic acid (OAA). Plants employing this pathway use carbon dioxide more efficiently than those with the normal (C3 or Calvin cycle) photosynthesis. These plants are also efficient in water use and thus can thrive in dry conditions also. The vetiver plant is insensitive to photoperiod and grows and flowers year-round where temperatures permit. It is best suited to open sunlight and will not establish easily under shady conditions. However, once established, plants can survive in deep shade for decades.

4 Geographical distribution and expansion

Vetiver has been found to grow in the tropical and subtropical regions of the world. Vetiver is widely distributed in the continents such as Africa, Asia, North America, South America, and Europe (National Research Council, 1993). African countries are

Algeria, Angola, Burundi, Tonga, Western Samoa, Comoro, Central African Republic, Ethiopia, Gabon, Ghana, Kenya, Madagascar, Malawi, Mauritius, Nigeria, Rwanda, Seychelles, Somalia, South Africa, Tanzania, Tunisia, Uganda, Zaire, Zambia, and Zimbabwe. Asian countries are India, Malaysia, Japan, Nepal, Pakistan, Philippines, Singapore, Sri Lanka, Thailand, Indonesia, China, Burma, Bangladesh, and Russia. North American countries are Costa Rica, Haiti, Jamaica, Cuba, Dominican Republic, Barbados, Antigua, Guatemala, Trinidad, Martinique, St. Lucia, St. Vincent, Puerto Rico, and United States of America. South American countries are Paraguay, Guyana, Brazil, Argentina, and Columbia. European countries are France and Russia. The range of expansion of vetiver is also vast from riverbanks, marshy lands, wastelands, degraded land to fly ash dumped lands over the world. Fig. 2.2 shows global distribution of vetiver grass, which is presented with green color.

5 Multipurpose usage of vetiver grass

5.1 Phytoremediation of different types of pollutants

Phytoremediation is a process of removal of contaminants from soil, water, and air through plants. The process is cheap and environment friendly. Vetiver is an ideal plant that has the capability of various functions. Vetiver is a plant that performs all sorts of remediation such as phytoextraction, phytovolatilization, phytostabilization, phytodegradation, and rhizoremediation. Vetiver has an important role in remediation of different types of pollutants such as heavy metals, organic pollutants, pharmaceutical wastes, nutrient load, radioactive compounds, etc., which is discussed in the following subsections.

Heavy metals—Vetiver is a choice plant for various functions and not a plant that has by chance so many functions. Thus it is a plant not chosen by chance but it is the plant selected by nature. Vetiver can grow in various soil types ranging from sodic, alkaline to acidic, coal and gold mines (Truong, 2000). The ability to accumulate contaminants has added an importance to these plants (Danh et al., 2009). Recently vetiver grass, due to its eco-friendly nature, found a new use for phytoremediation of

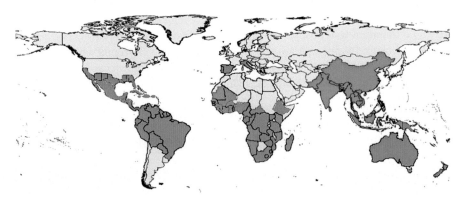

Figure 2.2 Global distribution of vetiver grass (shown with *green*).

contaminants from soil as well as water. Vetiver is an excellent gift of nature, it provides essential oil, its root and shoot are used in various ways and besides these, it accumulates heavy metals and contaminants, thus, performing an important role in phytoremediation. A complete list of various works on phytoremediation, conservation, and other applications of vetiver have been given in Table. 2.2. The choice of vetiver over other available plants for remediation has been due to its easy growing ability in harsher conditions compared to other plants. It also has higher threshold levels to accumulate metals compared to other plants (Danh et al., 2011). Vetiver can grow on soil contaminated with heavy metal such as Pb, Zn, Cu, Fe, Mn, etc. The accumulation of these has been found to be both in roots and shoots with higher percent in root compared to shoot (Danh et al., 2011; Truong, 1999). Increased use of radioactive substances for electricity and also in bombs, also increased use of fertilizers had led to the high levels of radioactive substances in soil (Yamamoto et al., 2012). Thus this creates a serious situation where soil is already burdened with heavy metal. So, there is an urgent need to remediate soil. In relation to remediation of radioactive substances, Hung et al. (2010) found that vetiver grows well in soils spiked with uranium and did not show any symptoms of toxicity even at highest levels of contamination. The presence of uranium had not any effect on the biomass of vetiver and accumulation was more in root compared to shoot.

Organic pollutants—Singh et al. (2008a) observed that vetiver can remediate phenol with 200 ppm within a short time of 4 days. This removal of phenol was the sole act of vetiver without involving any microbial action as the experiment was performed in aseptic condition. Phenrat et al. (2015) in an experiment with vetiver on floating platform and wetland with horizontal flow found that the degradation of phenol was slow with floating platform compared to the horizontal flow. However, the removal of phenol was quite effective in both the setups. Similarly, Makris et al. (2007) set up a hydroponic experiment to observe the removal of TNT with vetiver. The concentration of TNT was 40 mg/L in the solution. It was observed that vetiver completely removed TNT from the solution and it was also observed that the removal rate was enhanced with addition of urea in the medium (Makris et al., 2007). Das et al. (2010) also observed same result in soil with vetiver grass on TNT remediation. One of the important lab waste, ethidium bromide (EtBr) used for DNA staining is mutagenic and toxic. This compound is used in labs all over the world and it should be disposed properly as if released as such can cause problems to the organisms in the locality. Similarly, acrylamide used to prepare polyacrylamide gels for RNA and DNA analysis is also very toxic. Acrylamide is neurotoxin and carcinogenic and thus harmful to the flora and fauna if reaches soil and water (Paz-Alberto et al., 2011). The safe removal of these compounds is thus of utmost importance and it has been shown that vetiver grass can remediate these compounds (Paz-Alberto et al., 2011). Vetiver has been shown to accumulate 0.7 µg of EtBr from 1 kg (Danh et al., 2009).

Agrochemicals and antibiotics—In a hydroponic experiment with vetiver, Marcacci et al. (2006) observed that it easily grew in 20 ppm of atrazine and had no visible symptoms which were because of the detoxification mechanism present in it. Atrazine was detoxified through conjugation and dealkylation. In leaves of vetiver, the conjugated atrazine was observed while alkylated atrazine was found in both root and

Table 2.2 A complete list of the phytoremediation and conservation properties of vetiver system.

S. No.	Plant parts where contaminants sequestered	Experiment site	Contaminants remediated	Outcome	Experiment medium	Amendments	References
1	Root and shoot but maximum in root	Pot experiment in green house	Iron ore mine soil	Phytostabilizer	Soil		Banerjee et al. (2016)
2	Root	Field	Iron ore mine	Phytostabilizer	Soil	Primary nutrients and Fe, Zn, Mn, Cu	Roongtanakiat et al. (2008)
3	Root	Dump sites	Coal mine	Phytoremediation with increased P uptake	Soil	Mycorrhiza	Meyer et al. (2017)
4	Root	Coal dump sites	Coal heaps Cu, Zn, Cd, Pb, As	Phytoremediation for ecological restoration	Soil		Chen et al. (2016)
5	Root	Acid mine drainage	Zn, Mn, Ni, Cu	Phytoremediation	Soil		Gwenzi et al. (2017)
6	Root	Hydroponic, mine and saline groundwater	Na, K, Ca, Cl, Mg	Phytoremediation	Water		Keshtkar et al. (2016)
7	Root	Gold ores	Arsenic and salinity	Phytoremediation and revegetation	Soil		Guimarães et al. (2016)
8	Root	Mine soil	Cu, Zn	Phytostabilization	Soil	Humic acid	Vargas et al. (2016)
9	Root	Gold mine trailing	Zn, Cu, Ni and other metals	Phytoremediation	Soil		Melato et al. (2016)
10	As in shoot and Cr, Mn in root	Field	As, Cr, Cu, Mn, Pb, Zn	Phytoremediation	Soil	Rice husk ash Fe coated RHA	Tariq et al. (2016)

S. No.	Plant parts where contaminants sequestered	Experiment site	Contaminants remediated	Outcome	Experiment medium	Amendments	References
11	Root	Chromite-asbestos mines	Heavy metals	Phytostabilization	Soil	Chicken manure, farmyard manure, garden soil	Kumar and Maiti (2015)
12	As-root, Pb-leaf	Soil contaminants Soil	Arsenic and Lead	Phytoremediation	Soil	Fertilizer (Osmocote), cow manure, organic fertilizer	Hao et al. (2015)
13	Root	Pot and field	Lead	Phytostabilization	Soil		Meeinkuirt et al. (2013)
14	Root and shoot	Cu waste land	Cu	Phytoremediation and ecological rehabilitation	Soil	Increased microbial biomass, soil enzyme activity	Xu et al. (2012)
15	Root	Heavy metal	Pb, Zn	Phytostabilization	Soil	AM fungi	Wu et al. (2010)
16	Root and Shoot	Fly ash	Contaminants in fly ash	Phytoremediation	Soil		Chakrabarty and Mukherjee (2011)
17	Root	Zn mine soil	Zn, Mn, Cu, Fe	Phytoremediation	Soil	Compost, chelating agent (EDTA, DPTA)	Roongtanakiat et al. (2009)
18	Root	Iron ore tailings	Fe, Zn, Mn, Cu	Phytostabilizer	Soil	Compost, chelating agent (EDTA, DPTA)	Roongtanakiat et al. (2008)
19	Root	Pb/Zn and Cu mine tailings	Pb, Zn, Cu	Phytoremediation	Soil	Manure compost, sewage sludge	Chiu et al. (2006)
20	Root	Pb/Zn mine	Pb, Zn	Phytoremediation	Soil	Domestic refuse, NPK fertilizer	Yang et al. (2005)
21	Root	Coal dumps	Heavy metals	Phytoremediation	Soil		Nikhil (2004)

(Continued)

Table 2.2 A complete list of the phytoremediation and conservation properties of vetiver system. (*Cont.*)

S. No.	Plant parts where contaminants sequestered	Experiment site	Contaminants remediated	Outcome	Experiment medium	Amendments	References
22	Root	Metalliferous mine wastelands	Pb, Zn	Phytoremediation	Soil		Pang et al. (2003)
23	Root	Pb, Zn mine tailings	Pb, Zn, Cu, N, P, K, and organic matter	Phytoremediation	Soil	Domestic refuse and fertilizer	Yang et al. (2003)
24	Root	Pb, Zn tailings	Pb, Zn, Cu, Cd, organic matter	Phytoremediation and revegetation	Soil	Domestic refuse and N, P, K	Shu et al. (2002)
25	Root	Pb, Zn tailings	Pb, Zn, Cu, and N, P, K	Phytoremediation	Soil		Shu et al. (2000)
26	Root	Mining wastes and tailings	Heavy metals, Al, salinity	Phytoremediation and erosion control	Soil		Truong (1999)
27	Root	Cd, Pb contamination	Cd, Pb	Phytoremediation	Soil	EDTA, elemental S and N fertilizers	Ng et al. (2016)
28	Root	Heavy metals	Pb, Cd	Phytoremediation	Soil	Microbe	Xueyan et al. (2016)
29	Root	Radioactive waste	239 Pu	Phytoremediation	Soil	Chelating agents, citric acid and DPTA	Singh et al. (2016)
30	Root	Cd contamination	Cd	Phytoremediation and Phytostabilization	Soil	Cow manure, Pig manure, bat manure, organic fertilizer	Phusantisampam et al. (2016)
31	Root	Heavy metal	Cu, Cd, Endosulfan	Phytoremediation	Soil		Dousset et al. (2016)
32	Root	Pb	Pb	Phytoremediation	Soil	Arbuscular mycorrhizal fungi	Bahraminia et al. (2016)

S. No.	Plant parts where contaminants sequestered	Experiment site	Contaminants remediated	Outcome	Experiment medium	Amendments	References
33	Root and shoot	Pb	Pb	Stress response in plant	Soil		Pidatala et al. (2016)
34	Root	Heavy metals	Zn, Cd, Pb	Phytoremediation	Soil		Roongtanakiat and Sanoh (2015)
35	Root and shoot	Fly ash contaminants	Heavy metals	Phytoremediation and restoration	Soil		Ghosh et al. (2015)
36	Root	Heavy metals	N, P, Zn, Mn, Ni, and sewage effluents	Phytoremediation and Phytoextraction	Soil and water	Microorganisms	Mudhiriza et al. (2015)
37	Cr-stem Pb-Root	Heavy metals	Pb, Cr	Phytoremediation	Soil		Singh et al. (2016)
38	Root	Heavy metal	Cd	Phytoremediation	Soil		Chen et al. (2015)
39	Root	Soil and water	Arsenic	Phytoremediation	Soil and water	AMF	Caporale et al. (2014)
40	Root	Heavy metal	Cu, Cd	Phytoremediation	Soil		Abaga et al. (2014)
41	Root	Cr	Cr	Phytoremediation	Soil	Earthworm (*Eisenia foelide*)	Zhang et al. (2014a,b)
42	Root	Contaminated soil	Cr	Phytoremediation	Soil	Organic manure	Pillai et al. (2013)
43	Root	Contaminated soil	Pb	Phytoremediation	Soil	Ca amended river sand soil	Danh et al. (2011)
44	Root	Contaminated soil	Zn, Cd, Pb	Phytoremediation	Soil	Vermicompost	Roongtanakiat and Sanoh (2011)
45	Root	Heavy metal from textile waste water soil	Pb, Cd, Cu	Phytoremediation	Soil		Jayashree et al. (2011)

(Continued)

Table 2.2 A complete list of the phytoremediation and conservation properties of vetiver system. (*Cont.*)

S. No.	Plant parts where contaminants sequestered	Experiment site	Contaminants remediated	Outcome	Experiment medium	Amendments	References
46	Root	Heavy metal	Hg	Phytoremediation	Soil	Simulated rainfall, sawdust, humus soil	Wang et al. (2011)
47	Root	Heavy metal	Hg	Phytoremediation	Soil	Compost addition	Mangkoedihardjo and Triastuti (2011)
48	Root	Pesticide contaminated soil	As	Phytoremediation	Soil		Datta et al. (2011)
49	Root	Heavy metal	Pb, Cu, Zn	Phytoremediation	Soil		Danh et al. (2011)
50	Root	Heavy metal	Pb	Phytoremediation	Water		Rotkittikhun et al. (2010)
51	Increased oil production Root	Uranium	Uranium from yellow cake solution	Phytoremediation	Soil		Roongtanakiat et al. (2010)
52	Root	Heavy metal	Pb	Phytoremediation	Water	Phosphorus, role of Phytochelatins	Andra et al. (2010)
53	Root	Heavy metal	Cd and Pb	Phytoextraction and Phytostabilisation	Soil		Minh and Khoa (2009)
54	Root	Heavy metal	Lead	Phytoremediation	Soil	EDTA	Gupta et al. (2008)
55	Root	Pb	Pb	Phytoremediation	Soil		Paz-Alberto et al. (2007)
56	Root	Heavy metal	As, Cu, Pb, Zn	Phytoremediation	Soil	Chelating agents (EDTA, HEDTA, NTA, OA AND PA)	Lou et al. (2007)

S. No.	Plant parts where contaminants sequestered	Experiment site	Contaminants remediated	Outcome	Experiment medium	Amendments	References
57	Root	Heavy metal	As	Phytoremediation	Soil	Organic amendments, dairy sludge, Azotobactor, Mycorrhizae	Singh et al. (2007)
58	Root	Pb	Pb	Phytostabilisation	Soil	Organic matters, nutrients (N, P, K), pig manure, inorganic fertilizer	Rotkittikhun et al. (2007)
59	Root	Heavy metal	N, P, Zn, Pb	Phytoremediation	Soil	Mycorrhizae	Wong et al. (2007)
60	Root	Heavy metal	Pb, Zn, Cu, Fe	Phytoremediation	Soil	Fertilizer (Osmocote and N,P,K), EDTA	Wilde et al. (2005)
61	Root	Heavy metal	As, Zn, Cu	Phytoremediation	Soil	Chelating agents(NTA)	Chiu et al. (2005)
62	Root	Heavy metal	Pb	Phytoextraction	Soil		Chantachon et al. (2004)
63	Root and shoot	Conservation	Soil loss and erosion	Conservation	Soil		Donjadee and Tingsanchali (2016)
64	Root and shoot	Bonding reinforcing soil	Slope stabilization	Conservation	Soil		Noorasyikin and Zainab (2016)
65	Root and shoot	Field	Slope stabilization	Conservation	Soil		Noorasyikin and Mohamed (2015)
66	Root	Field	Runoff and soil loss	Conservation	Soil		Donjadee and Tingsanchali (2013)

(Continued)

Table 2.2 A complete list of the phytoremediation and conservation properties of vetiver system. (*Cont.*)

S. No.	Plant parts where contaminants sequestered	Experiment site	Contaminants remediated	Outcome	Experiment medium	Amendments	References
67	Root OA-As EDTA-Cu, Pb HEDTA-Zn	Vetiver hedge row	Runoff and soil loss, steep slopes	Conservation	Soil		Donjadee and Tingsanchali (2013)
68	Root	Agrochemical	Atrazine	Phytoremediation	Water		Marcacci et al. (2006)
69	Root	Antibiotics	Tetracyclene	Phytoremediation	Water		Datta et al. (2013)
70	Root	Organic wastes	Phenol	Phytoremediation	Water		Singh et al. (2008)
71	Root	Organic wastes	Phenol	Phytoremediation	Water		Phenrat et al. (2015)
72	Root	Organic wastes	TNT	Phytoremediation	Water	Urea	Makris et al. (2007)

shoot. It has been observed that atrazine accumulated is sequestered in the lipid part of the plant. Thus atrazine concentration can be found in oil of vetiver. The action of microbes has an important role in association with the roots of vetiver in degrading atrazine (Winter, 1999).

Tetracycline is one of the important antibiotics with increased use in medicine and also in laboratories. The unsafe release of this causes contamination of soil and water and also sometimes causes tolerance to the microbial community or affects the microbial diversity (Depledge, 2011; Luo et al., 2014; Sengupta et al., 2016). Thus its remediation is necessary and timely removal should be the first step. Datta et al. (2013) performed hydroponic experiment with vetiver to observe the effects on the accumulation or degradation of tetracycline in the solution. He observed that vetiver removes tetracycline from the hydroponic solution. Naproxen is one of the major micro-pollutant. It is a drug that is administered for the treatment of inflammation and pain. Vetiver grass has been used to remediate this drug from wastewater and soil to about 60% at different concentrations (Marsidi et al., 2016).

Radioactive substances and oil contaminated soil—Nuclear power plants are now a solution for the power problems in the developing and developed countries because of its effective power generation. However, this also produces radioactive wastes that are harmful to the flora and fauna. The aftermath effects of the nuclear bombing in Hiroshima and Nagasaki, Chernobyl disaster are the examples of the ailments and hazardous impacts of nuclear wastes. The nuclear wastes have its impacts for a very long time in vast areas, thus there is an urgent need to remediate these wastes from the environment. Shaw and Bell (1991) and Singh et al. (2008b) have reported that the remediation of nuclear waste through vetiver plantation. It was observed that vetiver was effective in remediating strontium (^{90}Sr) and cesium (^{137}Cs) from a solution spiked with these radioactive compounds. Mahmood et al. (2014) reported that vetiver remediates Cs contaminated soil to 95%. Similarly, Hung et al., 2010 reported that vetiver can accumulate about 50% of uranium from a solution of 250 mg/kg uranium. Thus, vetiver grass can be an effective plant in remediation of nuclear wastes.

The soil nearby to the oil drilling site is highly contaminated with oil in contaminated sites. Vetiver is found to grow on the heavily polluted oil sites and has the ability to degrade the hydrocarbon in oil (Rajaei and Seyedi, 2018). Although there are some reports where by-products of oil, have been found to be toxic to vetiver growth. In Australia, a greenhouse trial with vetiver on industrial oil was conducted, to observe the tolerance limit of Vetiver on various types of oil. This experiment gave two results (1) the growth of vetiver was highly reduced with diesel fuel and it was concluded that it is toxic to the plant, (2) hydraulic oils were also deterrent for growth of vetiver but to a moderate level. Thus from these observations, it was concluded that oil contaminated sites can be remediated with vetiver but it will depend on the type of fuel and the level of contamination.

5.2 Phytomanagement of fly ash deposits

Fly ash as we all know is a by-product of coal combustion. Electricity has become the necessity of human life and in order to fulfill the demands various power plants have been built and out of which most of them are coal-based power plants. During the combustion of coal, various by-products are generated and the major one is fly ash (15%–30%). Although some of the fly ash is used in making cements, bricks, road, land reclamation, and also amendment in agriculture (Pandey et al., 2011; Pandey and Singh, 2010). The rest leftover fly ash is disposed in open landfills and this becomes the major source of contamination of soil and water (Pandey et al., 2009). This is the reason that sometimes these landfills are called as fly ash ponds. The fly ash contains various contaminates and heavy metals and these leach to the nearby water and underground sources causing a great risk to the livelihood at those areas. Thus, it becomes an urgent need to remediate these fly ash ponds to protect the livelihood from the ill effects of contamination (Pandey and Singh, 2012). There are mechanical as well as plant-based system for the removal of contaminants, while the former is costly and the later one is cheap and environment friendly. The plant-based methods will require the screening and identification of the plants that can grow well in fly ash deposits and heavy metal contaminated sites, and also accumulate these with efficient and fast (Pandey et al. 2015a–c). Vetiver grass has been found to grow in the fly ash ponds (Pandey and Singh, 2015; Verma et al., 2014). So, vetiver was selected as plant of importance in remediation of fly ash ponds. Ghosh et al. (2015) found that the vetiver grass remediates heavy metal from fly ash with in a period of 1 year and 6 months in a pot experiment. Chakraborty and Mukherjee (2011) reported the phytoremediation of fly ash dumps near the sites. They grew vetiver for 3 months and observed that vetiver has a phytostabilization effect in fly ash dumps. It was also observed that maximum concentration of heavy metals remains in the root and a very little amount is translocated to the shoot. Thus, if grazed by animals it will not have damaging effect in animals. Thus, here again, vetiver the ideal plant became a plant of choice for phytoremediation.

5.3   Removing nutrient loads

The level of pollutants in soil and water are alarming and the effect can be seen through the loss of diversity and dominance of species. The addition of contaminant to the aquatic system causes eutrophication also known as dying of the aquatic ecosystem. Although this can be averted with stopping the flow of effluent but those that have already been added should be removed. Here, again plant-based method is effective and cheap and the vetiver grass comes to the rescue. Vetiver grass has been found to remove nitrogen and phosphorous faster from the water compared to other plants (Hart et al., 2003; Truong, 2002; Zheng et al., 1997). Similarly boron, fluoride, aluminum was also shown to be removed by vetiver (Ruiz et al., 2013; Angin et al., 2008; Aldana Arcila, 2014). Thus, with the presence of such an ideal plant or it can be rightly said as natures gift to humankind to remove the leftovers of humans is the boon to whole civilization.

5.4   Carbon sequestration

Today with increasing environmental pollution and global warming, the effect on the ecosystem has been devastating with loss of biodiversity. Processes through which the atmospheric carbon can be sequestered will be a boon in the present scenario. There are various plants and microorganisms that sequester the atmospheric carbon and a combination of both will have a great effect in combating the global warming. There are various factors that favor and also negate the process of carbon sequestration such as climate, soil properties, and regional land pattern. Lemon grass, Palmasora, and some trees are known to sequester carbon but the rate of sequestration is high through Vetiver (Lavania, 2011). Singh et al. (2014) reported that if all the estimated degraded land in India if put to vetiver plantation then this will sequester nearly 150 Tg C per year which is approximately 46% of the total carbon emissions in India. Thus to identify a particular plant that best suits to the process and is sustainable, is the key to resolve the global warming. Vetiver is one of the best suited plants with great economic importance and also its suitability to grow without any extra effort. It has been shown through studies that vetiver has high carbon sequestration capacity (Lavania and Lavania, 2009). Thus, vetiver with its multipurpose use and multidirectional approach has proved itself to be the best in countering climate change and the global warming issue. Comparing the complete 1-year-cycle of carbon sequestered through vetiver and other trees show that vetiver has a very high capability to sequester carbon compared to others. The significance of vetiver also increased with its essential oil, shoot, and root that are used in thatching. Thus, besides sequestering carbon, it will also provide revenue to the growers. There is a very high demand of vetiver oil. The leftover shoot and root is used for thatching houses in rural areas. This plant is also rarely grazed so the danger of herbivory is also mitigated. Vetiver grows well in poor fertile soil like wastelands and degraded lands.

5.5 Adaptive agricultural practices

Vetiver grass has been used for soil and water conservation. Intercropping with other plants helped in improving the soil structure, reduced erosion, and also increased the water holding in soil. In addition, intercropping of vetiver with crop plants in arsenic contaminated soil helped in reduced uptake of arsenic in crop plants. The accumulation of arsenic was in the root and shoots of vetiver (Praveen et al., 2017). Yaseen et al. (2014) has observed that the intercropping of vetiver increases the land use efficiency by upto 130% in vetiver-sweet basil-radish intercropping. Similarly, Sujatha et al. (2011) observed increased productivity of Arecanut (*Areca catechu* L.) in combination of vetiver arecanut system. Vetiver has multidiverse roles from remediation to insect control. Vetiver has been successfully used to control termites with its essential oil which is effective in termite control (Verma et al., 2009). In the roots of vetiver, a compound called nootkatone is present which has the capacity of disturbing the behavior in termites (Van Du and Truc, 2008; Maistrello et al., 2001). This is because the nootkatone causes the death of protozoa in the gut of termite (Maistrello et al., 2001). Maistrello et al. (2001) has found reduction in the length and number of tunnels constructed by termites when the vetiver oil is used. Van Du and Truc (2008) observed increased plant growth of cacao and also controlled termite population with vetiver grass compost. Mulching of vetiver on plants reduces the attack of fungus (Greenfield, 2002).

5.6   Soil and water conservation

Soil and water are the most important components in ecosystem. Disturbance to them affects both the flora and fauna of ecosystem. Soil erosion is common during heavy rains on slopes or on bare land (Mekonnen et al., 2015). Soil which has plantation, erosion is less but with the increasing population and demands from the ecosystem, the deforestation and contamination has increased and now bare lands are common thus increasing erosion (Mekonnen et al., 2015). The equilibrium in ecosystem is disturbed and the effects are visible with reduced diversity, erratic rainfall, global warming, loss of fertile soil, and reduced ground water recharge (Morgan, 2009; Nearing et al., 2005). The problem can be solved through afforestation but the areas where earthquake and hurricane are frequent and also slope areas, the erosion can only be avoided through the plantation of those plants that can withstand these effects. Vetiver, a magic plant, is a sturdy grassy herb with extensive root system that anchors soil and a good shoot biomass thus, plays an important role in ecosystem services (Fig. 2.3). There have been reports of increased soil nutrients and also increased water conservation in areas of vetiver plantation (Babalola et al., 2007). Rao et al. (1992) reported that vetiver population reduces soil runoff to about 69% compared to the one without any vetiver plantation. Babalola et al. (2003) carried out an experiment to estimate the reduction in soil loss and run off in plots without any vetiver plantation. The soil loss and runoff in the non-planted plots had 70% and 130% higher rate as compared to the planted plots and also the nitrogen use efficiency was enhanced. In some of the countries (China, Malaysia, Thailand, Venezuela, and Argentina), Vetiver is planted for road side stabilization (Smyle, 2011). In India, Thailand, Vietnam, Madagascar, South Africa, Uganda etc., Vetiver is planted in fields, coastal areas, slopes, for road side stabilization, near pond to reduce soil erosion and runoff and for water conservation (Prakasa Rao et al., 2008; Panklang, 2011; Van and Truong, 2011; Quang et al., 2014).

5.7   Economic return (phytocommerce/phytoeconomics)

Vetiver is a plant for all purpose, so it has returns also with an effect on the socioeconomic condition of the people (Fig. 2.4). The services are economic and useful to the human beings as by selling the products one gets revenue. The plants that pay off in form of revenue through various ways for instance bioremediation, conservation, agroeconomics, and valuable products to the human beings can be called phytocommerce or phytoeconomics. The vetiver has wide range of use to make products that generate revenue are discussed here as:

Essential oil—The oil is extracted from the roots of vetiver. Vetiver has an extensive root system and thus a good yield of oil can be obtained from it. The expected oil market of vetiver has been estimated to reach around $170 million by 2022 (Grand View Research, 2016). The average oil that can be obtained from root on dry weight basis is around 1%–3%. In case of a hectare of land about 15–30 kg of oil can be obtained (Lal, 2013). The major producers of vetiver oil in the world are India, Indonesia, and Haiti (Lal, 2013). The oil has multifaceted use, that is, it is used in perfumery, cosmetics, and medicine. The odor of oil is somewhat earthy and woody type fragrance and

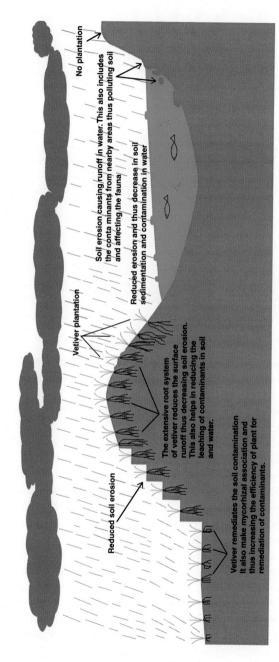

Figure 2.3 Schematic representation to show the control of soil erosion and phytoremediation with vetiver system.

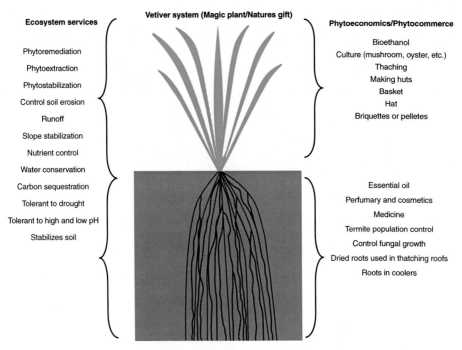

Figure 2.4 Schematic representation of Vetiver system with its various properties.

this adds to its value with high demand in global market. According to an estimate about 250 kg of root can give 1 kg of oil (Sarrazin, 2017). The constituents of oil are hydrocarbons, alcohols, carboxylic acid, and sesquiterpenes (bi and tricyclic). About 150 sesquiterpenoides has been reported of which main are vetiverols, carbonyl compound, and esters. Demole et al., 1995 reported that the carbonyl compounds, namely, A-Vetivone, β-Vetivone, and khusinol are the primary compounds that influences the odor of oil.

Perfumery and cosmetics—Oil of vetiver have a soothing fragrance and thus is used in perfumery. The use of vetiver oil as perfumes and in cosmetics has been since ancient times (Moussou et al., 2016). Thwaites (2010) reported that about 36% of western perfumes and 20% of men colognes and fragrances have vetiver oil as major ingredient. In the cosmetic industries also, vetiver oil is used. In southern India, vetiver oil mixed with coconut oil is used as cosmetics. The use of oil as perfumes and cosmetics has been increasing due to its soothing fragrance; it is natural and also safe to skin.

Medicines—Vetiver is a multipurpose ideal grass and has its use in all sorts of work. Vetiver as a medicine sounds a little different but it has been in use since long time as stomachic, in antimicrobial and digestive purposes. In balms also, it is used for its soothing properties and fragrance during headache (Jain, 1991). In India, the tribal population has been using it to treat mouth ulcers, burns, snake bite, fever, headache, etc. and in addition, the root vapor and root ash is given during malarial fever and acidity, respectively (Rao and Suseela, 2000; Jain, 1991). Vetiver use as medicine doesn't end here rather it is also used for skin ailments like remove acne. The skin acne is removed with the use of vetiver oil as it is reported that it balances the activity of sebaceous glands and thus normalizing the skin. The pregnancy marks are also prevented with regular use of vetiver oil (Lavania, 2003a). The cooling and soothing properties of oil are helpful in overcoming tension, stress, depression, insomnia, etc. (Wilson, 1995). In sports, it is massaged to relax muscles. Hair loss is one of the biggest problem in today's life due to pollution and irregular diet but it can be controlled with the use of vetiver oil.

Energy—Vetiver has a huge biomass with its above ground shoot and underground extensive root system. In 1 hectare of field approximately 120 tons of biomass can be obtained annually (Tomar and Minhas, 2004; Troung and Smeal, 2003), which is much more than other plants with high biomass such as, *Miscanthus* grass (Cliftonbrown et al., 2004), *Eucalyptus spp.*, and *Salix* spp. (Wright, 1994). The leaves of vetiver have cellulose, hemicelluloses, lignin which is considered to be a good source for energy (Rao et al., 2015; Raman and Gnansounou, 2015). Rao et al., 2015 reported the use of vetiver in bioethanol production. Wongwatanapaiboon et al. (2012) reported that 1.14 g/L of ethanol can be obtained from Sri Lankan ecotype of vetiver. Li et al. (2006) in China investigated the biogas production from the shoots of vetiver. The leftover digest (of leaves) can be used as soil amendment. Briquettes or pellets are also produced from the dried vetivers which are used for generation of heat (Pinners, 2014). The pellets have a good energy value of 16.3 GJ/C compared to the sugarcane bagasse and drywood (9.3 and 19.8 GJ/C). Unikode enterprise in Haiti is producing briquettes from leaf and root of vetiver (http://www.unikodesa.com/biomass, UNEP, 2015).

Other uses—Vetiver leaves with its cellulose, hemicelluloses, lignin, and proteins are a good medium for culture and cultivation of mushroom, oyster, shiitake (Chomchalow and Chapman, 2003).The shoot and root of vetiver after drying is used for domestic purposes such as thatching roofs, making huts, baskets, and hats. Vetiver is also used to make beverages as herbal drink (Akther et al., 2010) and it is sold by Usher industries (Balasankar et al., 2013).The ash of vetiver grass can be used as cementing material or in making bricks (Nimityongskul et al., 2003).

6   Limitations

Vetiver is an ideal plant with various applications in different fields so also called as plant for all trades. However, having so many importances, it also has some limitations. The root of vetiver is extensive and goes deep in the soil. The roots when dug out leaves soil that are vulnerable to soil erosion and runoff. So, there must be a proper vigilance to avoid uncontrolled cultivation by the organizations involved in vetiver trade. Rules have been setup by Natural Resources Stewardship Circle (NRSC, 2014) and the organizations must abide to the rules. For phytoremediation purposes, also strict vigilance should be in place for the disposal of the biomass (Van Aken, 2008). There must also be check for the oil extracted from the vetiver of contaminated sites as there have been reports of lower concentrations of contaminants in the oil.

7 Potential features of vetiver grass: the reason of vetiver's success

Minimum competition for nutrient and moisture—Vetiver has extensive root system so sometimes there has been misconception that it can cause competition and hinder the growth of nearby plants. The roots of Vetiver grow vertically underground up to 12–15 inches without any lateral or horizontal growth. This reduces the competition for moisture and nutrients with the other accompanying plants. The growth of lateral or horizontal roots has been found to occur at higher depths.

Symbiotic association with microbes—The roots of vetiver are extensive and it also forms microbial association with the microbes present in the rhizosphere (Monteiro et al., 2009). This helps vetiver in surviving even at very less concentration of nutrient in soil. The range of microbial association is also very wide, that is, it forms association with wide range of microorganisms (Leaungvutiviroj et al., 2010).

High biomass production—Vetiver has a high biomass with bushy shoot growth and extensive root system. Vetiver is a C4 plant thus at high light intensity, it has high rate of CO_2 utilization with efficient Hatch Slack pathway (Hatch, 1987). It also has high growth rate, thus high biomass in less time (Truong, 2002). In a hectare of land about 120 tons of biomass cab be obtained annually (Tomar and Minhas, 2004).

Resistance to disease and pest—There is a major problem with most of the plants of infection to diseases and pests. It has been observed that vetiver is resistant to various diseases and pests (Danh et al., 2011). There has been occurrence of leaf blight (USDA) and also some fungus.

Sterile and non-invasive—It has been observed that vetiver flowers but do not form seeds, being sterile. Thus it can't invade the native species as the spread becomes limited due to no seed formation. This will also have limited impact on the competition for nutrients, space, and light. The roots also penetrate straight downward and do not spread sideways.

Long life span—Vetiver has a long life and can survive up to 150 years (Don Miller, Pers.com). This is a very important requirement with respect to phytoremediation as it will help in long-term removal of contaminants at highly contaminated sites. In addition, the plant requires much less attention to grow.

Extensive root system—Vetiver has long roots. This also has an important role in reducing soil erosion. The roots can extend up to the length of 3–7 m (Lavania, 2003). This extensive root system also helps in anchoring the soil and thus reducing soil erosion. This also helps in water conservation. The roots sometimes form mycorrhizal associations thus it also help the plant in better extraction of nutrients and water.

Xerophytic and hydrophytic properties—Plants that have both the xerophytic and hydrophytic characters can survive at different conditions and thus increase the range of its growing regions. The vetiver leafs are folded and looks like needle thus reducing transpiration. It has a high biomass both aboveground and underground. The roots are also extensive and deep penetrating. The most important character of vetiver is being a C4 plant it will lose less water and better utilize the carbon.

8 Conclusions

Vetiver, a common grass with uncommon properties, is a boon to the mankind due to its remedial and economic properties. This is the reason we have called it a magic plant/magic grass/natures gift. It is a plant for all purposes, be it remediation, agriculture, domestic, cosmetic, medicine, etc. It is a plant of phytocommerce because in addition to the environmental clean-up it also adds up some useful product (oil, medicine, cosmetics, etc.) which gives revenue to the grower. Phytoremediation properties of vetiver are also of phytocommerce as compared to the other processes (mechanical, chemical) of remediation, it is cheap and environment friendly. So, indirectly it is saving money in comparison to other methods and also contributing in environmental clean-up. There have been various works on the remediation properties and phytoeconomics of vetiver and now it is of urgent need to apply this magic grass in action. The land and water that are contaminated with pollutants must be cleaned up in a manner to have socioeconomic benefits because there have been questions on the feasibility of the phytoremediation process as it requires a long time. So, this is plant to answer the every question of the critics as with phytoremediation it is also giving revenue which will improve the socioeconomic condition of the growers in question.

Acknowledgment

Financial assistance given to Dr. V.C. Pandey under the Scientist's Pool Scheme (Pool No. 13 (8931-A)/2017) by the Council of Scientific and Industrial Research, Government of India is gratefully acknowledged. The authors declare no conflicts of interest.

References

Abaga, N.O.Z., Dousset, S., Mbengue, S., Munier-Lamy, C., 2014. Is vetiver grass of interest for the remediation of Cu and Cd to protect marketing gardens in Burkina Faso? Chemosphere 113, 42–47.

Akther, S., Hussain, A., Iman, S., 2010. Preparation and evaluation of physicochemical characteristics of herbal drink concentrate. Pak. J. Biochem. Mol. Biol. 43, 149–152.

Aldana Arcila, E.J., 2014. Remoción de aluminio en aguas residuales industriales usando especies macrófitas.

Andra, S.S., Datta, R., Sarkar, D., Makris, K.C., Mullens, C.P., Sahi, S.V., Bach, S.B., 2010. Synthesis of phytochelatins in vetiver grass upon lead exposure in the presence of phosphorus. Plant Soil 326 (1–2), 171–185.

Angin, I., Turan, M., Ketterings, Q.M., Cakici, A., 2008. Humic acid addition enhances B and Pb phytoextraction by vetiver grass (*Vetiveria zizanioides* (L.) Nash). Water Air Soil Poll. 188 (1–4), 335–343.

Ayangbenro, A., Babalola, O., 2017. A new strategy for heavy metal polluted environments: a review of microbial biosorbents. Int. J. Environ. Res. Public Health 14 (1), 94.

Babalola, O., Jimba, J.C., Maduakolam, O., Dada, O.A., 2003. Use of vetiver grass for soil and water conservation in Nigeria. In: Proceedings of the Third International Conference on Vetiver and Exhibition. Guangzhou, China, pp. 293–309.

Babalola, O., Oshunsanya, S.O., Are, K., 2007. Effects of vetiver grass (*Vetiveria nigritana*) strips, vetiver grass mulch and an organomineral fertilizer on soil, water and nutrient losses and maize (*Zea mays*, L.) yields. SoilTillage Res. 96 (1–2), 6–18.

Bahraminia, M., Zarei, M., Ronaghi, A., Ghasemi-Fasaei, R., 2016. Effectiveness of arbuscular mycorrhizal fungi in phytoremediation of lead-contaminated soil by vetiver grass. Int. J. Phytoremediation 18 (7), 730–737.

Balasankar, D., Vanilarasu, K., Preetha, P.S., Rajeswari, S., Umadevi, M., Bhowmik, D., 2013. Traditional and medicinal uses of vetiver. J. Med. Plant Stud. 1 (3), 191–200.

Banerjee, R., Goswami, P., Pathak, K., Mukherjee, A., 2016. Vetiver grass: an environment clean-up tool for heavy metal contaminated iron ore mine-soil. Ecol. Eng. 90, 25–34.

Caporale, A.G., Sarkar, D., Datta, R., Punamiya, P., Violante, A., 2014. Effect of arbuscular mycorrhizal fungi (*Glomus spp.*) on growth and arsenic uptake of vetiver grass (*Chrysopogon zizanioides* L.) from contaminated soil and water systems. J. Soil Sci. Plant Nutr. 14 (4), 955–972.

Chakraborty, R., Mukherjee, A., 2011. Vetiver can grow on coal fly ash without DNA damage. Int. J. Phytoremediation 13 (2), 206–214.

Chantachon, S., Kruatrachue, M., Pokethitiyook, P., Upatham, S., Tantanasarit, S., Soonthornsarathool, V., 2004. Phytoextraction and accumulation of lead from contaminated soil by vetiver grass: laboratory and simulated field study. Water Air Soil Poll. 154 (1–4), 37–55.

Chen, X.C., Liu, Y.G., Zeng, G.M., Duan, G.F., Hu, X.J., Hu, X., Xu, W.H., Zou, M., 2015. The optimal root length for *Vetiveria zizanioides* when transplanted to Cd polluted soil. Int. J. Phytoremediation 17 (6), 563–567.

Chen, C., Liu, H., Hao, J., Zhao, L.-L., Cheng, W., 2016. Effect of different planting years of *Vetiveria zizanioides L.* on the distribution of heavy metals in coal spoil-heap soil. J. China Coal Soc. 41, 3101–3107.

Chiu, K.K., Ye, Z.H., Wong, M.H., 2005. Enhanced uptake of As, Zn, and Cu by *Vetiveria zizanioides* and *Zea mays* using chelating agents. Chemosphere 60 (10), 1365–1375.

Chiu, K.K., Ye, Z.H., Wong, M.H., 2006. Growth of *Vetiveria zizanioides* and *Phragmities australis* on Pb/Zn and Cu mine tailings amended with manure compost and sewage sludge: a greenhouse study. Bioresourc. Technol. 97 (1), 158–170.

Chomchalow, N., 2001. The Utilization of Vetiver as Medicinal and Aromatic Plants with Special Reference to Thailand. Tech. Bull. No. 2001/1, PRVN / ORDPB, Bangkok, Thailand.

Chomchalow, N., Chapman, K., 2003. Other uses and utilization of vetiver. Proceedings of the Third Conferences of Vetiver and Exhibition, pp. 6–9.

Clifton-brown, J.C., Stampfl, P.F., Jones, M.B., 2004. Miscanthus biomass production for energy in Europe and its potential contribution to decreasing fossil fuel carbon emissions. Global Change Biol. 10 (4), 509–518.

Danh, L.T., Truong, P., Mammucari, R., Tran, T., Foster, N., 2009. Vetiver grass, *Vetiveria zizanioides*: a choice plant for phytoremediation of heavy metals and organic wastes. Int. J. Phytoremediation 11 (8), 664–691.

Danh, L.T., Truong, P., Mammucari, R., Foster, N., 2011. Effect of calcium on growth performance and essential oil of vetiver grass (*Chrysopogon zizanioides*) grown on lead contaminated soils. Int. J. Phytoremediation 13 (Suppl. 1), 154–165.

Darajeh, N., Idris, A., Truong, P., Abdul Aziz, A., Abu Bakar, R., Che Man, H., 2014. Phytoremediation potential of vetiver system technology for improving the quality of palm oil mill effluent. Adv. Mat. Sci. Eng. 4, 1–10.

Das, P., Datta, R., Makris, K.C., Sarkar, D., 2010. Vetiver grass is capable of removing TNT from soil in the presence of urea. Environ. Pollut. 158, 1980–1983.

Datta, R., Quispe, M.A., Sarkar, D., 2011. Greenhouse study on the phytoremediation potential of vetiver grass, *Chrysopogon zizanioides* L., in arsenic-contaminated soils. Bull. Environ. Contamin. Toxicol. 86 (1), 124–128.

Datta, R., Das, P., Smith, S., Punamiya, P., Ramanathan, D.M., Reddy, R., Sarkar, D., 2013. Phytoremediation potential of vetiver grass [*Chrysopogon zizanioides* (L.)] for tetracycline. Int. J. Phytoremediation 15 (4), 343–351.

Demole, E.P., Holzner, G.W., Youssefi, M.J., 1995. Malodor formation in alcoholic perfumes containing vetiveryl acetate and vetiver oil. Perfum. Flavorist 20 (1), 35–40.

Depledge, M., 2011. Reduce drug waste in the environment. Nature 478 (7367), 36–136.

Donjadee, S., Tingsanchali, T., 2013. Reduction of runoff and soil loss over steep slopes by using vetiver hedgerow systems. Paddy Water Environ. 11 (1–4), 573–581.

Donjadee, S., Tingsanchali, T., 2016. Soil and water conservation on steep slopes by mulching using rice straw and vetiver grass clippings. Agric. Nat. Resourc. 50 (1), 75–79.

Dousset, S., Abaga, N.O.Z., Billet, D., 2016. Vetiver grass and micropollutant leaching through structured soil columns under outdoor conditions. Pedosphere 26 (4), 522–532.

Ghosh, M., Paul, J., Jana, A., De, A., Mukherjee, A., 2015. Use of the grass, *Vetiveria zizanioides* (L.) Nash for detoxification and phytoremediation of soils contaminated with fly ash from thermal power plants. Eco. Eng. 74, 258–265.

Grand View Research, 2016. Vetiver oil market expected to reach $169.5 million by 2022, Grand View Research, Inc. Available from: http://www.grandviewresearch.com/industry-analysis/vetiver-oil-market.

Greenfield, J.C., 2002. Vetiver Grass: An Essential Grass for the Conservation of the Planet Earth. Infinity Publishing, Haverford, PA, USA.

Guimarães, L.A., Dias, L.E., Assis, I.R.D., Cordeiro, A.L., 2016. Cultivation of vetiver in saline tailings contaminated with arsenic under phosphorus doses. Rev. Bras. Eng. Agr. Amb. 20 (10), 891–896.

Gupta, D.K., Srivastava, A., Singh, V.P., 2008. EDTA enhances lead uptake and facilitates phytoremediation by vetiver grass. J. Environ. Biol. 29 (6), 903–906.

Gwenzi, W., Mushaike, C.C., Chaukura, N., Bunhu, T., 2017. Removal of trace metals from acid mine drainage using a sequential combination of coal ash-based adsorbents and phytoremediation by bunchgrass (Vetiver [*Vetiveria zizanioides* L]). Mine Water Environ. 36 (4), 520–531.

Hao, G., Yin, Y., Fangfang, L., Sen, Y., Jiangyong, A., Yun, F., 2015. Removal of arsenic and lead from soils contaminated with coal gangue using *Vetiveria zizanioides*. Int. J. Mining Reclamation Environ. 29 (1), 47–61.

Hart, B., Cody, R., Truong, P., 2003, October. Hydroponic vetiver treatment of post septic tank effluent. Int. Conf. Vetiver Exhibition 3, 121–131.

Hatch, M.D., 1987. C4 photosynthesis: a unique blend of modified biochemistry, anatomy and ultrastructure. Biochim. Biophys. Acta 895 (2), 81–106.

Hung, L.V., Maslov, O.D., Nhan, D.D., My, T.T., Ho, P.K., 2010. Uranium Uptake of *Vetiveria zizaniodes* Nash. JINR Commun. E18-2010-71. Dubna.

Jain, S.K., 1991. Dictionary of Indian folk medicine and ethnobotany. Deep Publications.

Jayashree, S., Rathinamala, J., Lakshmanaperumalsamy, P., 2011. Determination of heavy metal removal efficiency of *Chrysopogon zizanioides*(Vetiver) using textile wastewater contaminated soil. J. Environ. Sci. Technol. 4 (5), 543–551.

Keshtkar, A.R., Ahmadi, M.R., Naseri, H.R., Atashi, H., Hamidifar, H., Razavi, S.M., Yazdanpanah, A., Karimpour Reihan, M., Moazami, N., 2016. Application of a vetiver system for unconventional water treatment. Desalin. Water Treat. 57 (53), 25474–25483.

Kumar, A., Maiti, S.K., 2015. Effect of organic manures on the growth of *Cymbopogon citratus* and *Chrysopogon zizanioides* for the phytoremediation of Chromite-Asbestos mine waste: a pot scale experiment. Int. J. Phytoremediation 17 (5), 437–447.

Lal, R.K., 2013. On genetic diversity in germplasm of vetiver '*Veteveria zizanioides* (L.) Nash'. Ind. Crops Prod. 43, 93–98.

Lavania, U.C., 2000. Primary and secondary centres of origin of vetiver and its dispersion. In: Second International Proceedings on Vetiver and the Environment Conference, Thailand, TVN(424-426).

Lavania, S., 2003. Vetiver root system: search for the ideotype. In: Proceedings of the Third International Conference on Vetiver and Exhibition, Guangzhou, China, pp. 526–530.

Lavania, U.C., 2003. Vetiver root-oil and its utilization. Office of the Royal Development Projects Board.

Lavania, U.C., 2011. Vetiver grass model for long term carbon sequestration and development of designer genotypes for implementation. In: ICV5 Fifth International Conference on Vetiver, Lucknow, India.

Lavania, U.C., Lavania, S., 2009. Sequestration of atmospheric carbon into subsoil horizons through deep-rooted grasses-vetiver grass model. Cur. Sci. 97, 618–619.

Leaungvutiviroj, C., Piriyaprin, S., Limtong, P., Sasaki, K., 2010. Relationships between soil microorganisms and nutrient contents of *Vetiveria zizanioides* (L.) Nash and *Vetiveria nemoralis* (A.) Camus in some problem soils from Thailand. Appl. Soil Ecol. 46 (1), 95–102.

Li, H., Luo, Y.M., Song, J., Wu, L.H., Christie, P., 2006. Degradation of benzo [a] pyrene in an experimentally contaminated paddy soil by vetiver grass (*Vetiveria zizanioides*). Environ. Geochem. Health 28 (1–2), 183–188.

Li, Y., Ren, X., Dahlquist, E., Fan, P., Chao, T., 2014. Biogas potential from *Vetiveria zizaniodes* (L.) planted for ecological restoration in China. Energy Procedia 61, 2733–2736.

Liao, X., Luo, S., Wu, Y., Wang, Z., 2003. Studies on the abilities of *Vetiveria zizanioides* and *Cyperus alternifolius* for pig farm wastewater treatment. In: International Conference on Vetiver and Exhibition, 3, 174–181.

Lou, L.Q., Ye, Z.H., Wong, M.H., 2007. Solubility and accumulation of metals in Chinese brake fern, vetiver and *Rostrate sesbania* using chelating agents. Int. J. Phytoremediation 9 (4), 325–343.

Luo, Y., Guo, W., Ngo, H.H., Nghiem, L.D., Hai, F.I., Zhang, J., Liang, S., Wang, X.C., 2014. A review on the occurrence of micropollutants in the aquatic environment and their fate and removal during wastewater treatment. Sci. Total Environ. 473, 619–641.

Mahmood, Z.U., Abdullah, N.H., Rahim, K.A., Mohamed, N., Muhamad, N.A., Wahid, A.N., Desa, N.D., Ibrahim, M.Z., Othman, N.A., Anuar, A.A., Ishak, K., 2014. Uptake evaluation of caesium by glasshouse grown grasses for radiophytoremediation of contaminated soil. Research and Development Seminar, Bangi (Malaysia).

Maistrello, L., Henderson, G., Laine, R.A., 2001. Efficacy of vetiver oil and nootkatone as soil barriers against Formosan subterranean termite (Isoptera: Rhinotermitidae). J. Econ. Entomol. 94 (6), 1532–1537.

Makris, K.C., Shakya, K.M., Datta, R., Sarkar, D., Pachanoor, D., 2007. High uptake of 2, 4, 6-trinitrotoluene by vetiver grass–potential for phytoremediation? Environ. Poll. 146 (1), 1–4.

Mangkoedihardjo, S., Triastuti, Y., 2011. Vetiver in phytoremediation of mercury polluted soil with the addition of compost. J. Appl. Sci. Res. V 7 (4), 465–469.

Marcacci, S., Raveton, M., Ravanel, P., Schwitzguébel, J.P., 2006. Conjugation of atrazine in vetiver (*Chrysopogon zizanioides* Nash) grown in hydroponics. Environ. Exp. Bot. 56 (2), 205–215.

Marsidi, N., Nye, C.K., Abdullah, S.R., Abu Hassan, H., Halmi, M.I., 2016. Phytoremediation of naproxen in waste water using *Vetiver zizaniodes*. J. Eng. Sci. Technol. V 11 (8), 1086–1097.

Meeinkuirt, W., Kruatrachue, M., Tanhan, P., Chaiyarat, R., Pokethitiyook, P., 2013. Phytostabilization potential of Pb mine tailings by two grass species, *Thysanolaena maxima* and *Vetiveria zizanioides*. Water Air Soil Poll. 224 (10), 1750.

Mekonnen, M., Saskia, D., Keesstra, L., Stroosnijder, J., Baartman, E.M., Jerry, M., 2015. Soil conservation through sediment, trapping: a review. Land Degr. Dev. 26 (6), 544–556.

Melato, F.A., Mokgalaka, N.S., McCrindle, R.I., 2016. Adaptation and detoxification mechanisms of Vetiver grass (*Chrysopogon zizanioides*) growing on gold mine tailings. Int. J. Phytoremediation 18 (5), 509–520.

Meyer, E., Londoño, D.M.M., de Armas, R.D., Giachini, A.J., Rossi, M.J., Stoffel, S.C.G., Soares, C.R.F.S., 2017. Arbuscular mycorrhizal fungi in the growth and extraction of trace elements by *Chrysopogon zizanioides* (vetiver) in a substrate containing coal mine wastes. Int. J. Phytoremediation 19 (2), 113–120.

Minh, V.V., Khoa, L.V., 2009. Phytoremediation of cadmium and lead contaminated soil types by vetiver grass. VNU J. Sci. Earth Sci. 25, 98–103.

Monteiro, J.M., Vollu, R.E., Coelho, M.R.R., Alviano, C.S., Blank, A.F., Seldin, L., 2009. Comparison of the bacterial community and characterization of plant growth-promoting rhizobacteria from different genotypes of *Chrysopogon zizanioides* (L.) Roberty (vetiver) rhizospheres. J. Microbiol. 47 (4), 363–370.

Morgan, R.P.C., 2009. Soil erosion and conservation. third ed. John Wiley & Sons.

Moussou, P., Danoux, L., Bardey, V., Jeanmaire, C., Pauly, G., BASF Beauty Care Solutions France SAS, 2016. Cosmetic preparations containing PTH fragments. U.S. Patent 9,387,235.

Movahed, N., Maeiyat, M.M., 2009. Phytoremediation and sustainable urban design methods (Low carbon cities through phytoremediation). In: 45th ISOCARP Congress. http://www.isocarp.net.

Mudhiriza, T., Mapanda, F., Mvumi, B.M., Wuta, M., 2015. Removal of nutrient and heavy metal loads from sewage effluent using vetiver grass, *Chrysopogon zizanioides* (L.) Roberty. Water SA 41(4), 457–493.

National Research Council, 1993. Vetiver Grass: A Thin Green Line Against Erosion. National Academy Press, Washington DC.

Nearing, M.A., Jetten, V., Baffaut, C., Cerdan, O., Couturier, A., Hernandez, M., Le Bissonnais, Y., Nichols, M.H., Nunes, J.P., Renschler, C.S., Souchère, V., 2005. Modeling response of soil erosion and runoff to changes in precipitation and cover. Catena 61 (2–3), 131–154.

Ng, C.C., Boyce, A.N., Rahman, M.M., Abas, M.R., 2016. Effects of different soil amendments on mixed heavy metals contamination in Vetiver grass. Bull. Environ. Contamin. Toxicol. 97 (5), 695–701.

Nikhil, K., 2004. Vetiver grass for the bio-reclamation of coal overburden dumps. Int. J. Ecol. Environ. Conserv. 10 (4).

Nilaweera, N.S., Hengchaovanich, D., 1996. Assessment of strength properties of vetiver grass roots in relation to slope stabilization. In: Vetiver: A Miracle Grass, Chiang Rai (Thailand).

Nimityongskul, P., Panichnava, S., Hengsadeekul, T., 2003. Use of vetiver grass ash as cement replacement materials. Proc. ICV-3,148–159.

Noorasyikin, M.N., Mohamed, Z., 2015. Assessment of soil-root matrix of vetiver and bermuda grass for mitigation of shallow slope failure. Electron. J. Geotech. Eng. 20 (15), 6569–6580.

Noorasyikin, M.N., Zainab, M., 2016. A tensile strength of Bermuda grass and Vetiver grass in terms of root reinforcement ability toward soil slope stabilization. In: IOP Conference Series: Materials Science and Engineering, IOP Publishing, 136(1), 012029.

NRSC, 2014. Specifications as applied to the Haitian vetiver supply chain. The Natural Resources Stewardship Circle, France.

Pandey, V.C., Singh, N., 2010. Impact of fly ash incorporation in soil systems. Agric. Ecosyst. Environ. 136, 16–27.

Pandey, V.C., Singh, B., 2012. Rehabilitation of coal fly ash basins: current need to use ecological engineering. Ecol. Eng. 49, 190–192.

Pandey, V.C., Singh, N., 2015. Aromatic plants versus arsenic hazards in soils. J. Geochem. Explor. 157, 77–80.

Pandey, V.C., Abhilash, V.C., Singh, N., 2009. The Indian perspective of utilizing fly ash in phytoremediation, photomanagement and biomass production. J. Environ. Manage. 90, 2943–2958.

Pandey, V.C., Singh, J.S., Singh, R.P., Singh, N., Yunus, M., 2011. Arsenic hazards in coal fly ash and its fate in Indian scenario. Resourc. Conserv. Recy. 55, 819–835.

Pandey, V.C., Pandey, D.N., Singh, N., 2015a. Sustainable phytoremediation based on naturally colonizing and economically valuable plants. J. Clean. Prod. 86, 37–39.

Pandey, V.C., Prakash, P., Bajpai, O., Kumar, A., Singh, N., 2015b. Phytodiversity on fly ash deposits: evaluation of naturally colonized species for sustainable phytorestoration. Environ. Sci. Pollut. Res. 22, 2776–2787.

Pandey, V.C., Bajpai, O., Pandey, D.N., Singh, N., 2015c. *Saccharum spontaneum*: an underutilized tall grass for revegetation and restoration programs. Genet. Resour. Crop Evol. 62 (3), 443–450.

Pang, J., Chan, G.S.Y., Zhang, J., Liang, J., Wong, M.H., 2003. Physiological aspects of vetiver grass for rehabilitation in abandoned metalliferous mine wastes. Chemosphere 52 (9), 1559–1570.

Panklang, P., 2011. The promotion of vetiver grass cultivation for soil and water conservation by farmer's participation. In: The ICV5 Fifth International Conference on Vetiver, Lucknow, India.

Paz-Alberto, A.M., Sigua, G.C., Baui, B.G., Prudente, J.A., 2007. Phytoextraction of lead-contaminated soil using vetivergrass (*Vetiveria zizanioides* L.), cogongrass (*Imperata cylindrica* L.) and carabaograss (*Paspalum conjugatum* L.). Environ. Sci. Pollut. Res. Int. 14(7), 498–504.

Paz-Alberto, A.M., De Dios, M.J.J., Alberto, R.T., Sigua, G.C., 2011. Assessing phytoremediation potentials of selected tropical plants for acrylamide. J. Soil. Sediment. 11 (7), 1190–1198.

Phenrat, T., Teeratitayangkul, P., Imthiang, T., Sawasdee, Y., Wichai, S., Piangpia, T., Naowaopas, J., Supanpaiboon, W., 2015. Laboratory-scaled Developments and Field-scaled Implementations of Using Vetiver Grass to Remediate Water and Soil Contaminated with Phenol and Other Hazardous Substances from Illegal Dumping at Nong-Nea Subdistrict, Phanom Sarakham District, Chachoengsao Province, Thailand. In: The Sixth International Conference on Vetiver, Vietnam, Danang, pp. 3–5.

Phusantisampan, T., Meeinkuirt, W., Saengwilai, P., Pichtel, J., Chaiyarat, R., 2016. Phytostabilization potential of two ecotypes of *Vetiveria zizanioides* in cadmium-contaminated soils: greenhouse and field experiments. Environ. Sci. Pollut. Res. 23 (19), 20027–20038.

Pidatala, V.R., Li, K., Sarkar, D., Ramakrishna, W., Datta, R., 2016. Identification of biochemical pathways associated with lead tolerance and detoxification in *Chrysopogon zizanioides* L. Nash (Vetiver) by metabolic profiling. Environ. Sci. Technol. 50 (5), 2530–2537.

Pillai, S.S., Girija, N., Williams, G.P., Koshy, M., 2013. Impact of organic manure on the phytoremediation potential of *Vetiveria zizanioides* in chromium-contaminated soil. Chem. Ecol. 29 (3), 270–279.

Pinners, E., 2014. Vetiver system: reversing degradation on and off farm to keep soil carbon in place, build up root biomass, and turn degraded areas into biofuel sources. In: Geotherapy: Innovative Methods of Soil Fertility Restoration, Carbon Sequestration, and Reversing CO_2 Increase, CRC Press, pp. 301–324.

Porras, S., Francisco, J., 2002. Hair restorer containing vetiver grass extract. U.S. Patent Application 10/121, 795.

Prakasa Rao, E.V.S., Gopinath, C.T., Khanuja, S.P.S., 2008. Environmental, economic and equity aspects of vetiver in south India. In: First National Indian Vetiver Workshop. The Vetiver International. Kochi, India, pp. 21–23.

Praveen, A., Mehrotra, S., Singh, N., 2017. Rice planted along with accumulators in arsenic amended plots reduced arsenic uptake in grains and shoots. Chemosphere 184, 1327–1333.

Quang, D.V., Schreinemachers, P., Berger, T., 2014. Ex-ante assessment of soil conservation methods in the uplands of Vietnam: an agent-based modeling approach. Agric. Syst. 123, 108–119.

Rajaei, S., Seyedi, S.M., 2018. Phytoremediation of petroleum contaminated soils by *Vetiveria zizanioides* (L.) Nash. Clean–Soil, Air, Water, 1800244.

Raman, J.K., Gnansounou, E., 2015. LCA of bioethanol and furfural production from vetiver. Bioresour. Technol. 185, 202–210.

Rao, R.R., Suseela, M.R., 2000. *Vetiveria zizanioides* (Linn.) Nash–a multipurpose eco-friendly grass of India. ICV-2 held in Cha-am, Phetchaburi, Thailand, pp. 18–22.

Rao, K.P.C., Cogle, A.L., Srivastava, K.L., 1992. ICRISAT Annual Report 1992. Andhra Pradesh, India.

Rao, E.P., Akshata, S., Gopinath, C.T., Ravindra, N.S., Hebbar, A., Prasad, N., 2015. Vetiver Production for Small Farmers in India. In: Sustainable Agriculture Reviews, Springer, Cham, pp. 337–355.

Roongtanakiat, N., Sanoh, S., 2011. Phytoextraction of zinc, cadmium and lead from contaminated soil by vetiver grass. Kasetsart J. (Nat. Sci.) 45, 603–612.

Roongtanakiat, N., Sanoh, S., 2015. Comparative growth and distribution of Zn, Cd and Pb in rice, vetiver and sunflower grown in contaminated soils. Kasetsart J. Nat. Sci. 49, 663–675.

Roongtanakiat, N., Osotsapar, Y., Yindiram, C., 2008. Effects of soil amendment on growth and heavy metals content in vetiver grown on iron ore tailings. Kasetsart J. (Nat. Sci.) 42, 397–406.

Roongtanakiat, N., Osotsapar, Y., Yindiram, C., 2009. Influence of heavy metals and soil amendments on vetiver (*Chrysopogon zizanioides*) grown in zinc mine soil. Kasetsart J. (Nat. Sci.) 43, 37–49.

Roongtanakiat, N., Sudsawad, P., Ngernvijit, N., 2010. Uranium absorption ability of sunflower, vetiver and purple guinea grass. Kasetsart J. (Nat. Sci.) 44, 182–190.

Rotkittikhun, P., Chaiyarat, R., Kruatrachue, M., Pokethitiyook, P., Baker, A.J.M., 2007. Growth and lead accumulation by the grasses *Vetiveria zizanioides* and *Thysanolaena maxima* in lead-contaminated soil amended with pig manure and fertilizer: a glasshouse study. Chemosphere 66 (1), 45–53.

Rotkittikhun, P., Kruatrachue, M., Pokethitiyook, P., Baker, A.J.M., 2010. Tolerance and accumulation of lead in *Vetiveria zizanioides* and its effect on oil production. J. Environ. Biol. V 31 (3), 329.

Ruiz C., Rodríguez O., Luque O., Alarcón M., 2013. Effecto del vetiver (*Chrysopogon zizanioides* L.) en la reducción del flúor y otros compuestos contaminantes en aguas de consumo humano. Caso: Caserío Guarataro, estado Yaracuy, Venezuela. The second Latin America International Conference on the Vetiver System, October 3–5, 2013, Medillin, Colombia.

Salt, D.E., Smith, R.D., Raskin, I.,1998. Phytoremediation. Annu. Rev. Plant. Physiol. Plant Mol. Biol., 49, 643–668.

Sarrazin, E., 2017. The Scent Creation Process. In: Springer Handbook of Odor. Springer, Cham, pp. 137–138.

Sengupta, A., Sarkar, D., Das, P., Panja, S., Parikh, C., Ramanathan, D., Bagley, S., Datta, R., 2016. Tetracycline uptake and metabolism by vetiver grass (*Chrysopogon zizanioides* L. Nash). Environ. Sci. Pollut. Res. 23 (24), 24880–24889.

Shaw, G., Bell, J.N.B., 1991. Competitive effects of potassium and ammonium on caesium uptake kinetics in wheat. J. Environ. Radioactiv. 13 (4), 283–296.

Shu, W.S., Lan, C.Y., Zhang, Z.Q., Wong, M.H., 2000. Use of vetiver and other three grasses for revegetation of Pb/Zn mine tailings at Lechang, Guangdong province: field experiment. Int. J. Phytoremediation 4 (1), 47–57.

Shu, W.S., Xia, H.P., Zhang, Z.Q., Lan, C.Y., Wong, M.H., 2002. Use of vetiver and three other grasses for revegetation of Pb/Zn mine tailings: field experiment. Int. J. Phytoremediation 4 (1), 47–57.

Singh, S.K., Juwarkar, A.A., Kumar, S., Meshram, J., Fan, M., 2007. Effect of amendment on phytoextraction of arsenic by *Vetiveria zizanioides* from soil. Int. J. Environ. Sci. Technol. 4 (3), 339–344.

Singh, S., Eapen, S., Thorat, V., Kaushik, C.P., Raj, K., D'souza, S.F., 2008a. Phytoremediation of 137cesium and 90strontium from solutions and low-level nuclear waste by *Vetiveria zizanoides*. Ecotoxicol. Environ. Safety 69 (2), 306–311.

Singh, S., Melo, J.S., Eapen, S., D'souza, S.F., 2008b. Potential of vetiver (*Vetiveria zizanoides* L. Nash) for phytoremediation of phenol. Ecotoxicol. Environ. Safety 71 (3), 671–676.

Singh, M., Guleria, N., Rao, E.V.P., Goswami, P., 2014. Efficient C sequestration and benefits of medicinal vetiver cropping in tropical regions. Agron. Sustain. Dev. 34 (3), 603–607.

Singh, S., Fulzele, D.P., Kaushik, C.P., 2016. Potential of *Vetiveria zizanoides* L. Nash for phytoremediation of plutonium (239Pu): chelate assisted uptake and translocation. Ecotoxicol. Environ. Safety 132, 140–144.

Smyle, J., 2011. Confronting climate change: vetiver system applications. In Libro de resúmenes. The Fifth International Conference on Vetiver (ICV-5) Vetiver and Climate Change CSIR-Central Institute of Medicinal and Aromatic Plants, Lucknow, India, pp. 28–30.

Sujatha, S., Bhat, R., Kannan, C., Balasimha, D., 2011. Impact of intercropping of medicinal and aromatic plants with organic farming approach on resource use efficiency in arecanut (*Areca catechu* L.) plantation in India. Ind. Crop. Prod. 33(1), 78–83.

Tariq, F.S., Samsuri, A.W., Karam, D.S., Aris, A.Z., 2016. Phytoremediation of gold mine tailings amended with iron-coated and uncoated rice husk ash by vetiver grass (*Vetiveria zizanioides* (Linn.) Nash). Appl. Environ. Soil Sci. 2016, 12.

Thwaites, C., 2010. Haitian vetiver: uprooted? Perfumer Flavorist 35 (5), 22–23.

Tomar, O.S., Minhas, P.S., 2004. Ind. J. Agron. 49, 2017–2208.

Troung, P.N., Smeal, C., 2003. Pacific Rim Vetiver Network Tech. Bull, (2003/3), 19.

Truong, P., 1999. Vetiver grass technology for mine rehabilitation. Office of the Royal Development Projects Board.

Truong, P., 2000. Application of the vetiver system for phytoremediation of mercury pollution in the Lake and Yolo Counties, Northern California. Proceedings of pollution solutions, pp. 1–13.

Truong, P., 2002. Internet communication on the Vetiver Network Discussion Board, Vetiver Grass Nursery, Planting Techniques, Propagation and General Management Issues: Fertilizer for experiments with vetiver in greenhouse (09.03.2002).

Truong, P., 2011. Global review of contribution of VST in alleviating climate change disasters. In: The Fifth International Conference on Vetiver, Lucknow, India.

Truong, P., Danh, L.T., 2015. The Vetiver System for Improving Water Quality. second ed. The Vetiver Network International, San Antonio, TX, USA.

UNEP, 2015. Vetiver Briquette: Feasibility Report, Carbon Roots International, Inc., Haiti, pp. 1–38.

Unikode, in press. Haiti Alternative Fuels: Using vetiver biomass and other agricultural waste to reduce deforestation. Available from: http://www.unikodesa.com/biomass.

US EPA. (2002). Passive Treatment Technologies. Coal Remining BMP Guidance Manual, 4, 1–40. Retrieved from http://water.epa.gov/scitech/wastetech/guide/coal/upload/2002_01_24.

Valderrama, A., Tapia, J., Peñailillo, P., Carvajal, D.E., 2013. Water phytoremediation of cadmium and copper using *Azolla filiculoides* Lam. in a hydroponic system. Water Environ. J. 27 (3), 293–300.

Van Aken, B., 2008. Transgenic plants for phytoremediation: helping nature to clean up environmental pollution. Trends Biotechnol. 26 (5), 225–227.

Van Du, L., Truc, N.H., 2008. Termite Biocontrol on Cacao Seedling: Vetiver Grass Application. Nong Lam University, Ho Chi Minh City, Vietnam.

Vargas, C., Pérez-Esteban, J., Escolástico, C., Masaguer, A., Moliner, A., 2016. Phytoremediation of Cu and Zn by vetiver grass in mine soils amended with humic acids. Environ. Sci. Pollut. Res. 23 (13), 13521–13530.

Verma, M., Sharma, S., Prasad, R., 2009. Biological alternatives for termite control: a review. Int. Biodeterior. Biodegradation 63 (8), p. 959–972.

Verma, S.K., Singh, K., Gupta, A.K., Pandey, V.C., Trivedi, P., Verma, R.K., Patra, D.D., 2014. Aromatic grasses for phytomanagement of coal fly ash hazards. Ecol. Eng. 73, 425–428.

Wang, H., Feng, X.B., Wang, J.X., et al., 2011. Mitigation effects of *Vetiveria zizanioides* and additives on surface runoff mercury concentration from mercury-contaminated soil and slag under simulated rainfall. Chin. J. Ecol. 5, 922–927.

Wilde, E.W., Brigmon, R.L., Dunn, D.L., Heitkamp, M.A., Dagnan, D.C., 2005. Phytoextraction of lead from firing range soil by Vetiver grass. Chemosphere 61 (10), 1451–1457.

Wilson, R., 1995. Aromatherapy for vibrant health & beauty. Avery.

Winter, S., 1999. Plants reduce atrazine levels in wetlands. Final year report. School of Land and Food, University of Queensland, Brisbane, Queensland, Australia.

Wong, C.C., Wu, S.C., Kuek, C., Khan, A.G., Wong, M.H., 2007. The role of mycorrhizae associated with vetiver grown in Pb/Zn contaminated soils: greenhouse study. Restor. Ecol. 15 (1), 60–67.

Wongwatanapaiboon, J., Kangvansaichol, K., Burapatana, V., Inochanon, R., Winayanuwattikun, P., Yongvanich, T., Chulalaksananukul, W., 2012. The potential of cellulosic ethanol production from grasses in Thailand. J. BioMed. Res. 2012, 10.

Wright, L.L., 1994. Production technology status of woody and herbaceous crops. Biomass Bioenergy 6 (3), 191–209.

Wu, S.C., Wong, C.C., Shu, W.S., Khan, A.G., Wong, M.H., 2010. Mycorrhizo-remediation of lead/zinc mine tailings using vetiver: a field study. Int. J. Phytoremediation 13 (1), 61–74.

Xu, D.C., Zhan, J., Chen, Z., Gao, Y., Xie, X.Z., Sun, Q.Y., Dou, C.M., 2012. Effects of *Vetiveria zizanioides* L. growth on chemical and biological properties of copper mine tailing wastelands. Acta Ecol. Sinica 32 (18), 5683–5691.

Xueyan, Y., Daihua, J., Jinna, S., et al., 2016. Remediation mechanism of "double resistant" bacteria – *Vetiveria zizanioides* on Cd and Pb contaminated soil. Chin. J. Appl. Environ. Biol. 22, 884–890.

Yamamoto, M., Takada, T., Nagao, S., Koike, T., Shimada, K., Hoshi, M., Zhumadilov, K., Shima, T., Fukuoka, M., Imanaka, T., Endo, S., 2012. An early survey of the radioactive contamination of soil due to the Fukushima Dai-ichi Nuclear Power Plant accident, with emphasis on plutonium analysis. Geochem. J. 46 (4), 341–353.

Yang, B., Shu, W.S., Ye, Z.H., Lan, C.Y., Wong, M.H., 2003. Growth and metal accumulation in vetiver and two Sesbania species on lead/zinc mine tailings. Chemosphere 52 (9), 1593–1600.

Yang, B., Lan, C.Y., Shu, W.S., 2005. Growth and heavy metal accumulation of *Vetiveria zizanioides* grown on lead/zinc mine tailings. Acta Ecol. Sinica 25 (1), 45–50.

Yaseen, M., Singh, M., Ram, D., 2014. Growth, yield and economics of vetiver (*Vetiveria zizanioides* L. Nash) under intercropping system. Ind. Crop. Prod. 61, 417–421.

Zhang, K., Qiang, C.H.E.N., Hongbing, L.U.O., Li, X., 2014. Application of *Vetiveria Zizanioides* assisted by different species of earthworm in chromium-contaminated soil remediation. Adv. Mater. Res. 1010–1012, 564–569.

Zhang, K., Qiang, C.H.E.N., Hongbing, L.U.O., Xiaoting, L.I., 2014. Accumulation of chromium in *Vetiveria zizanioides* assisted by earthworm (*Eisenia foelide*) in contaminated soil. Adv. Mater. Res., 989–994, 1313–1318.

Zheng, C.R., Tu, C., Chen, H.M., 1997. Preliminary study on purification of eutrophic water with vetiver. In: Proceedings of International Vetiver Workshop, Fuzhou, China.

The potential of Sewan grass (*Lasiurus sindicus* Henrard) in phytoremediation—an endangered grass species of desert

3

Vimal Chandra Pandey[*], *D.P. Singh*
Department of Environmental Science, Babasaheb Bhimrao Ambedkar University, Lucknow, Uttar Pradesh, India
[*]Corresponding author

1 Introduction to Sewan grass

Grasses aid as an important source of fodder in arid environments because of their palatable and nutritive value (Haase et al., 1995; Mansoor et al., 2002). Grasses especially perennial grasses are valuable components of rangelands. Some important desert perennial grasses such as *Chloris virgate* Sw., *Cenchrus ciliaris* Linn., *Coelachyrum brevifolium* Hochst. & Nees., *Pennisetum divisum* (Gmel.) Henrard, *Stipagrostis drarii* (Täckh.) De Winter and *Lasiurus sindicus* Henrard have a C4 photosynthetic pathway, which make them drought tolerant to grow in arid condition (Waramit, 2010). All desert perennial grasses are well used as a fodder in India, Pakistan, and Arab countries (Khan and Ansari, 2008; Patel et al., 2012; El-Keblawy et al., 2013). But in Rajasthan state of India, *L. sindicus* attracts more attention to provide forage for livestock and cover to the soil. Moreover, native forage species has more potential as an opportunity for fodder production in a sustainable way with minimum cost. Besides fodder use, *L. sindicus* can be used to increase the rangeland productivity through rehabilitation or restoration of degraded desert grasslands and control the soil erosion (Peacock et al., 2003; Osman et al., 2008). Therefore, *L. sindicus* is one of the famous and most promising grass species in hot dry desert areas for forage production. It flourishes well in desert area and produces high forage yield. It is famous as the "King grass" due to its decent ground cover and has capacity to survive extreme arid condition and remain productive for a period of 6–10 years (Rao et al., 1989). It is also known as daah in Arabic. It is highly xeric in nature, drought tolerant, and nutritious grass of the desert area at the arrival of the monsoon and spring, it is one of the earliest sprouting grass species offered for livestock (Rao et al., 1989). After rainfall, its green patches occur on sand dunes and interdunal sandy areas indicating its ecophysiological potential, excelling other parent grasses (Arshad et al., 2007). The main problems of the Indian desert are the shifting of sand dunes and dust storms during summer periods. It is an urgent need to control these problems. The selection of native grass species for stabilizing hot dry desert area is a sustainable approach to reclaim

Phytoremediation Potential of Perennial Grasses. http://dx.doi.org/10.1016/B978-0-12-817732-7.00003-1

and management of this zone toward a higher level of grass productivity. *L. sindicus* perennial grass is a characteristics of the Indian desert and has been recognized as one of the most common and natural inhabitant of hot arid ecosystem of Thar Desert. It has a high nutritive value and is favorably consumed by livestock in the desert. To explore this hidden genetic treasure, the present chapter describes its origin and geographical distribution, ecology, morphological description, propagation, important features of this grass, multiple uses, phytoremediation, sewan grass productivity, and future prospects.

2 Origin and geographical distribution

It is native to India and found mostly in arid zones of Rajasthan, covering to the parts of Haryana and Punjab. It is the main grass of extremely arid parts of Jaisalmer, Barmer, and Bikaner districts of western Rajasthan in the Indian Thar Desert. It is also found in dry areas of Arabia, Africa, Mall, Niger, Ethiopia, Egypt, and Pakistan. It is distributed between 25 and 27°N latitude in dry open plains, rocky ground, and gravelly soils (Quattrocchi, 2006). Optimal growth needs annual rainfall below 250 mm and alluvial or light sandy soils with pH 8.5. It is highly tolerant against drought conditions but should be protected from wind in the early stages of establishment (FAO, 2010).

3 Ecology

L. sindicus Henrard (locally known as Sewan grass "King of desert grasses") is a dominating grass species of *Dichanthium-Cenchrus–Lasiurus* type grass lands of hot arid ecosystem of Great Indian Desert, covering Rajasthan, the parts of Haryana, and Punjab. It is noted as one of the most common and endemic perennial grasses of the Thar Desert, India. It grows naturally in wide range of dry areas covering Africa and Asia. Sewan grass (daah in Arabic) belongs to poaceae family. It flourishes well in dry climate getting annual rainfall below 250 mm existing between 25 and 27°N latitude on well aerated light brown sandy alluvial soils with a pH of 8.5. The phenology of *L. sindicus* can be divided in three phase such as growing period (July–August), growth period (September–October), and senescence period (November–January). It height ranges from 30 to 60 cm but it can attain a height of 90–100 cm in good conditions. This agronomically important grass tolerates prolonged droughts (Khan and Frost, 2001) and faces a severe threat of becoming an endangered due to changes in the land use pattern, increase in soil moisture regime, and overgrazing. It is specially consumed by livestock in the desert zone due to its high nutritive value. It helps in the expansion of good grasslands and in stabilizing the sand dunes (Chauhan, 2003; Khan and Frost, 2001). Severely overgrazed pasture, land-use change and enhanced aridity pose a severe threat to this grass for their survival in the hot arid ecosystem of Thar Desert (Chowdhury et al., 2009). There is less information on rhizospheric microbial studies of desert plants, while the rhizospheric effect is qualitatively and quantitatively more noticeable in desert soils as compared to soils of humid regions (Buyanovsky et al., 1982; Yechieli et al., 1995). As desert plants have capability to survive in extreme conditions of temperature and moisture as well as degraded soils with poor organic content and limited amounts of bioavailable inorganic nutrients. Rhizosphere microorganisms of desert plants should keep adaptive mechanisms to handle some harsh conditions such as frequent droughts, starvation, desiccation, high temperature, and high osmolarity (Chowdhury et al., 2009). The rhizospheric study of Sewan grass is an important job to explore in situ conservation of biodiversity associated with such niches to sustain ecological processes in hot arid desert ecosystem. Being highly tolerable in nature but it needs protection in the early stages of establishment. The sewan grass can be found in heavy metal contaminated sites. It flourishes well under moisture and temperature stress on sandy plains, low dunes, and hills of this region. The Sewan grasslands supported continuous increasing livestock population in the Indian desert for decades, where animal husbandry stayed as the main profession of the residents (Mertia, 1992). Severe overgrazing of grassland disturbs root reserves, which assist regeneration. Therefore, for avoiding the deterioration of root reserves of arid rangelands, Mertia (1992) proposed controlled grazing in drought-affected rangelands. So, the study of regenerative behavior of Sewan grass with regard to soil moisture levels in the rhizosphere is urgent needed.

4 Morphological description

It is an erect, tufted, and branched perennial C_4 grass and attains a height of about 1.2 m. The stem is stout and smooth. A mature grass contains 15–20 numbers of tillers with 23–30 leaves, up to 10 nodes, and 4–6 reproductive branches on the main tiller. It has flat, acuminate, nonauriculate leaves, with hairy ligules and open leaf sheath margins that fuse toward base. These are the key characters, including presence of distinct mid vein, in identification of *L. sindicus* in vegetative stage. The narrow, long, thick, amphistomatic leaves (Pearson et al., 1995) are more common in xeric habitats (Parkhurst, 1978) and exhibit higher photosynthetic rates than hypostomatic leaves (Peat and Fitter, 1994). Leaves are linear, silver-green-yellow, and 20–45 cm long with setaceous tip. Leaves are alternate with a thin leaf-blade. The leaves show characteristic C4 NADP-ME type of anatomy and have developed sclerenchyma to impart mechanical strength during drought and high wind. The inflorescence is a silky, white, densely villous, 10 cm long raceme bearing hairy spikelets, and 3 spikelets at each node. Two spikelets are sessile and one is pedicelled. Its flowers appear in early summer with 10–14 cm long and 1 cm broad silver-silky spikes. Fruit is a caryopsis (FAO, 2010). Seeds are dimorphic. All these morphological characters add to biomass of this grass. Being drought resistant and xeric in nature, *L. scindicus* can live up to 20 years. It is tufted, bushy, and multi-branched desert grass with ascending to erect wiry stems, forming a more or less oblique and woody rhizomatous rootstock with many shoots arising from the base, height up to 1–1.6 m. During dominant phase of drought and winter season, Sewan grass shows high root to shoot ratio and confirms less sensitivity of roots to these harsh conditions.

5 Propagation

L. sindicus (Sewan) grass has deep, extensive, and fibrous root system that can penetrate in middle part of sand dunes that contain moisture. Its roots can lie and wait for years during recurring periods of drought. It is propagated by vegetatively through the tillers from the older generation. Its propagation is also done by sowing and pricking out method. Vegetative growth has several ecological advantages originating from tillers (Pitelka and Ashmun, 1985). A study has been done in arid grasslands of Jaisalmer district of Rajasthan state, India, evaluated the effect of rhizospheric soil moisture and grazing pressure on regenerative potential and fodder yields of sewan grass. It is observed that uncontrolled grazing considerably diminished stand density of grass tussocks and their regeneration compared to controlled grazing or no grazing (Mertia et al., 2006). In cultural practice, it needs a well prepared soil for sowing through broadcast or in line at 1.5 cm depth and 50 cm distance with a seed rate of 5–7 kg per hectare. In addition, it is also established through transplanting of rooted slips/seedlings at the distance of 2×2 or 2×3 m. Sowing or transplanting is done always during the rains or with 24 hours of rains. The plantlets are protected from winds till to their establishment. Rhizospheric moisture and grazing are two main factors that controlled the sprouting of new shoots and regeneration. For this, Mertia et al. (2006) executed a study to evaluate the regenerative potential of Sewan grass under different levels of grazing closely after the worst drought in 100 years. The rhizospheric moisture at different depths (i.e., 0–30 cm and 30–60 cm) controlled sprouting of new shoots while moisture at 60–90 cm depth had the maximum impact on new shoot growth and biomass production (Mertia et al., 2006). The grazing of grasslands must be controlled to let a rest phase for regeneration, if the Sewan grasslands are to be retained on a sustainable basis (Mertia et al., 2006). The extensive root system of Sewan tussocks, exploring 4–5 m^3 of soil, enables it to withstand severe drought (Singh and Singh, 1997). However, resting of pasture, particularly in drought years, is essential to maintain its vigor and the capacity to regenerate and yield forage on a sustainable basis (Mertia et al., 2006).

6 Important features of Sewan grass

L. hirsutus is highly drought-tolerant, and mostly, propagated by sowing and pricking out. This bushy desert perennial grass has deep root systems that prefers a non-alkaline sandy soil and attains up to 1 m height in good conditions. This foraging grass is of utmost importance in regions where yearly precipitation is below 250 mm (Ecocrop, 2010). It is liked by ruminants but vanishes when overgrazed (El-Keblawy et al., 2013). Fertilization is not essential. This grass has developed a number of morphological, anatomical, and biochemical strategies to withstand the extreme climatic conditions. Being native, Sewan grass should be planted more in landscape design of hot dry regions, because they have good adaptation potential against desert conditions. All of these features make it a very valuable and suitable grass for landscape design in hot arid region.

7 Multiple uses

L. scindicus (Sewan grass) is very rich in protein and has been popularized as a fodder. This grass is quite palatable and nutritious for the livestock. Crude protein found in young leaves which ranges from 7%–14% and make it suitable for potential utilization in agri-horti-pastoral production system for animal. It is very important and dominant grass in hyper arid regions because they provide forage, which maintains both wild mammals and livestock, and soil cover (Assaeed, 1997). Sewan grass can be used to stabilize desert sandy dunes (Ecocrop, 2010; FAO, 2010). Reseeding arid range-lands with sewan grass would improve the forage resource, which are more palatable than native species (Khan et al., 1999). In Bikaner, Barmer, and Jaisalmer districts of Rajasthan, the sustainability and productivity of livestock mostly depends on the Sewan grass based pasture system. It is an extremely palatable grass and is used for hay, silage, and grazing. It also establishes sand dunes in desert areas. *L. hirsutus* can be planted as a colonizer, slope stabilizer for hillside plantation and phytoremediation programs. In urban planning, it may be used in planting programs for public open spaces, parks, car parks, street planting, pedestrian precincts, and private gardens.

8 Phytoremediation

It is well reported that more than 400 plant species have been identified to have the potential for remediation of soil and water (Robinson et al., 2009). The use of green plants in phytoremediation is a good option and must be aimed with three main goals (low input, economic yield, and least risk) toward removing/stabilizing metals (Pandey and Souza-Alonso, 2018) and is a 50%–80% low cost method compared to classi-cal method. Grasses belong to poaceae family and recently it attracts more attention toward phytoremediation due to its stabilization potential. *L. scindicus* is a drought tolerant grass and has potential to grow in harsh condition. *L. scindicus* have some important features in order to phytoremediation which make it a potential candidate in remediation of metals from contaminated desert area. There is a pressing need for an update of the current knowledge on phytoremediation of *L. scindicus*. Currently, Sharma and Pandey (2017) studied a phytoremediation experiments on *L. scindicus* (where its seeds are grown in industrial polluted soil) and results showed that *L. scin-dicus* grass can be used for remediation of lead from industrial contaminated site. It accumulates more lead in roots compared to leaves that make it potential candidate for phytostabilization program because lead is low translocated through the vascular system from roots to shoots. Fig. 3.1 depicts the accumulation of Pb in the roots and the leaf of *L. scindicus* at 45, 65, 85, and 105 days under the different treatments viz. 80%WS + 20%NS, 60%WS + 40%NS, 40%WS + 60%NS, 20%WS + 80%NS, and control. Fig. 3.1 shows that *L. scindicus* can be used in Pb phytostabilization because high amount of lead accumulated in root compare to leaf. At 105 days, it was observed that Pb causes adverse effects on growth and metabolism of *L. scindicus*. Tarafdar and Rao (1997) reported that mycorrhizal colonization, nutrient and heavy metals concentration of naturally grown grass *L. scindicus* on gypsum mine spoil. Percent

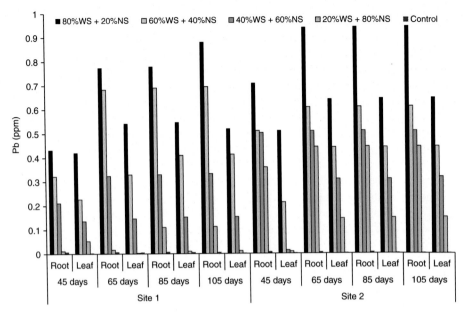

Figure 3.1 Accumulation of lead in root and leaf of *L. scindicus* under different treatments.
Source: Data from Sharma and Pandey, 2017.

root infection by arbuscular mycorrhizal fungi (AMF) was significantly higher on mine spoil than on normal soil in *L. scindicus* grass (Fig. 3.2). *L. scindicus* grass on mined spoil had significantly higher K, Ca, Mg, and Na nutrient concentrations and lower concentrations of N and P nutrient (Fig. 3.3) than normal soil. While mine spoil grown *L. scindicus* had significantly higher Mn and Fe concentrations and lower concentrations of Cu and Zn (Fig. 3.4) than normal soil. The presence of N and P in lower concentration has been attributed to their poor availability in mine spoil (Rao et al., 1996). The results proved the possibility of using AMF for rehabilitating mine spoils. As AMF are well recognized to help the mobilization and uptake of plants nutrients (Tarafdar and Marschner, 1994).

Grass communities associated with *L. sindicus* rangelands on different habitats of western Rajasthan are depicted in Fig. 3.5 (Gupta and Saxena, 1970). These grass communities are *Lasiurus sindicus-Eleusine compressa, Lasiurus sindicus-Cymbopogon jawarancusa-Eleusine compressa, Lasiurus sindicus-Dactyloctenium sindicum-Aristida adscensionis, Lasiurus sindicus-Cymbopogon schoenanthus, Lasiurus sindicus-Panicum turgidum, Lasiurus sindicus-Cenchrus biflorus* and found on different locations, that is, Chandan (Jaisalmer), Beechwal (Bikaner), Sobhala (Barmer), Binjasar (Barmer), Udsisar (Barmer), Hingola (Jodhpur), respectively. *Lasiurus sindicus-Panicum turgidum* grass community showed maximum basal cover with 43.5%, while *Lasiurus sindicus-Cenchrus biflorus* exhibited minimum basal cover with 4.2% (Gupta and Saxena, 1970). 95 species, 55 genera, 26 families, 3.03 H' and 0.34 Simpson Index are reported in *Lasiurus-Panicum* grassland of Rahasthan. *L. scindicus*

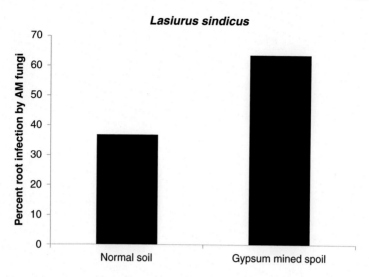

Figure 3.2 Percent root infection by AM fungi in *L. scindicus* grass grown on normal soil and gypsum mined spoil.
Source: Data from Tarafdar and Rao, 1997.

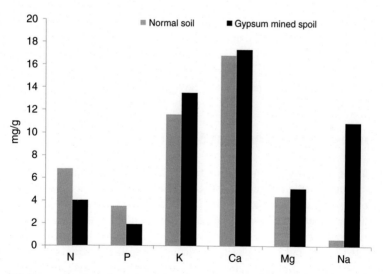

Figure 3.3 Concentration of nutrients (mg/g) in *L. scindicus* grass grown on normal soil and gypsum mined spoil.
Source: Data from Tarafdar and Rao, 1997.

showed highest IVI (63.6) in *Lasiurus-Panicum* grassland followed by *Panicum turgidum* (32.9), *Ochthochloa compressa* (17.3), *Indigofera cordifolia* (12.6), and *Aerva persica* (6.9). Species density was 115 individuals/m^2 in *Lasiurus-Panicum* grassland (Krishna et al., 2014). Therefore, it seems that *L. sindicus* is a well-adapted grass

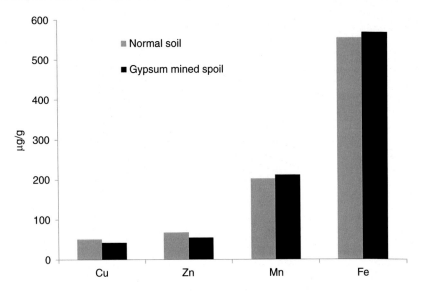

Figure 3.4 Concentration of heavy metals (μg/g) in *L. scindicus* grass grown on normal soil and gypsum mined spoil.
Source: Data from Tarafdar and Rao, 1997.

species of hot Indian arid desert areas that has the potential to stabilize sand dune, and to reduce the loss in soil fertility due to being aridity resistant. The high quality of *L. sindicus* rangelands in western Rajasthan is restored by Rajasthan Forest Department (Personal communication by Dr. Deep Narayan Pandey, Indian Forest Service, Secretary to Government, Department of Environment, Government of Rajasthan).

9 Biomass productivity of Sewan grass

L. sindicus, Sewan grass, is one of the grasses that provide high nutritional fodder to livestock in the extreme dry regions of Rajasthan desert. Now, it is well remarked as a wonder grass for desert area due to its hardy nature. Maximum fodder yield of *L. sindicus* at dry stage/tussock was reported 176, 106, and 62 g at no grazing, controlled grazing, and open grazing sites, respectively. According to stand density, this grass yield was 1617, 877, and 353 kg/ha at no grazing, controlled grazing, and open grazing sites, respectively (Mertia et al., 2006). Its biomass studies indicate that average shoot and root weight is 296 and 414 g/m^2, respectively. Based on biomass and related studies, it acted as a carbon sink to atmospheric CO_2 and fixed carbon to approximately 3.2 tons/hectare. Dry forage production of *L. sindicus* ranges from 0.25 to 0.38 tons/hectare (Raheja, 1962). In a man-made grassland, *L. sindicus* produces 3.99 g/unit/day of above ground biomass while the average dry matter production is 1.32 tons/hectare, in 1968 (Gupta and Saxena, 1970). A 30-day cutting break at 15 cm height offers the best dry matter yields. *L. sindicus* grass yields 2.7–10.5 tons fresh forage/ha/year and up to 3.4 tons dry matter/ha in well-established swards (FAO, 2010).

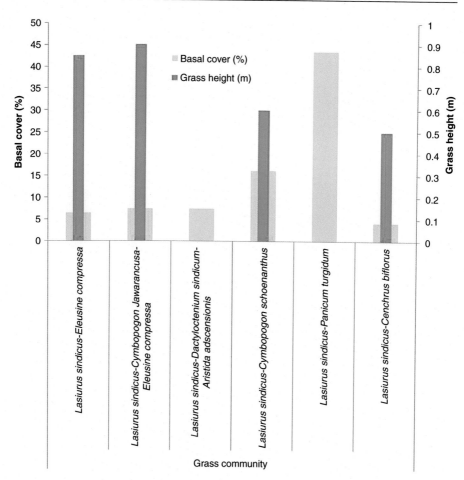

Figure 3.5 Grass communities associated with *L. scindicus* grasslands in Rajasthan state of India.
Source: Data from Gupta and Saxena, 1970.

A field research was carried out by Sharma (2013) to assess the effect of row spacing and nitrogen levels on the productivity and quality of sewan (*L. sindicus*) grass in hot dry region of Bikaner, Rajasthan. Results revealed that maximum green fodder yield (15.4 tons/ha) and dry matter yield (5.1 tons/ha) were recorded at 25 cm row spacing, while these values were similar with spacing 50 cm but significantly more over 75 cm (Fig. 3.6). In nitrogen case, the highest dose 60 kg N/ha recorded maximum green fodder yield (17.8 tons/ha) and dry matter yield (5.9 tons/ha), while both values were in statistical par with N dose of 40 kg/ha but significantly greater over other lower doses (Fig. 3.7). Generally, it may be concluded that for getting more economical fodder yields, *L. sindicus* should be propagated at 50 cm row spacing and nourished with 40 kg N/ha in hot dry area of Rajasthan (Sharma, 2013).

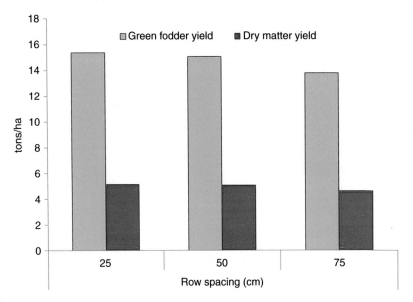

Figure 3.6 Forage yields (tons/hectare) of *L. sindicus* grass as influenced by row spacing (cm).
Source: Data from Sharma, 2013.

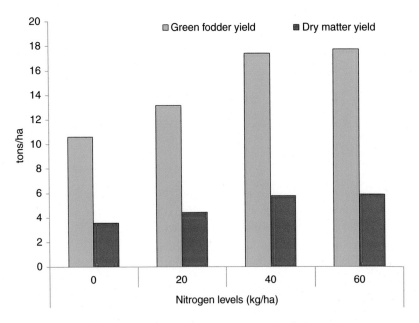

Figure 3.7 Forage yields (tons/hectare) of *L. sindicus* grass as influenced by nitrogen doses (kg/ha).
Source: Data from Sharma, 2013.

10   Genetic diversity and conservation

Being most valuable grass in the Thar Desert area of India, Sewan grass has been recognized as an endangered grass species of desert. The rangelands under Sewan grass is constantly shrinking due to severe overgrazing, agricultural modernization, alternative land-use patterns, frequent droughts, and increased human activities which pose a serious threat to its survival in the Thar ecosystem of India. Hence, it is utmost important to conserve its biodiversity to achieve the forthcoming desires under the changing environments. "Effective conservation of vulnerable species depends largely on the knowledge of extend and patterns of genetic variation" (Sharma et al., 2017). There are several methods viz. morphological, biochemical, and molecular which are used for the identification of genetic variability in vulnerable species. Morphological and biochemical traits are less effective in variability studies of endangered species in comparison to molecular marker technique due to lack of polymorphism. Genetic diversity of endangered species has been of great attention to biologists and conservationist (Young and Clarke, 2000; Hedrick, 2001). The genetic diversity analysis is essential not only to completely assess the impact of the endangered status on genetic variation of the population, but also knowledge on the genetic diversity of the species can be applied toward the conservation goals (Godt and Hamrick, 1998). Therefore, molecular markers tools can be valuable for the assessment of genetic diversity pattern in endangered species, and clarifying demographic and ecological issues primary in species management in order to plan long-term conservation or restoration projects.

Recently, molecular markers technique is applied to explore the genetic diversity of Sewan grass for the conservation of diverse germplasm for future needs. In this context, a study was done from the districts viz. Bikaner, Barmer, and Jaisalmer of Rajasthan (the diversity rich districts of hyper-arid Rajasthan) to evaluate the genetic variability among Sewan grass germplasm for its importance in determining survival under changing climate through ISSR and RAPD markers (Sharma et al., 2017). The 27 genotypes of Sewan grass were collected from its grasslands of western arid Rajasthan [from the districts Jaisalmer (10 accessions), Barmer (9 accessions), and Bikaner (7 accessions) and Jodhpur (1 old collection maintained at CAZRI)]. There are number of molecular biology tools have been used for biodiversity analysis. In this study, Shannon index for diversity was higher for Bikaner region (0.3570) followed by Jaisalmer (0.3304) and Barmer (0.3047). This is in agreement with the average similarities that were calculated using Jaccard's similarity matrix, 0.65, 0.70, and 0.71 respectively. RAPD analysis showed the diversity among Barmer-Jaisalmer region was lower (40%) than the Bikaner-Jaisalmer (46%) and Bikaner-Barmer (46%). Similarly, ISSR markers have been found with lower levels of diversity such as 25%, 27%, and 30%, respectively for Barmer-Jaisalmer, Bikaner-Barmer, and Bikaner-Jaisalmer. The higher diversity based on allele content and its frequency in Bikaner collections was found because of less harsh conditions and availability of proper irrigation, whereas the adaptive pressures might be limiting factor for diversity in harsher climate changes of Jaywalker and Barmer regions. Closed relationship between the collections from Barmer and Jaisalmer may be found because of same climatic conditions. While the collection sites from

Bikaner were geographically separated and more localized. Both AMOVA and Nei's gene diversity index detected considerable genetic diversity among the population, however, diversity within the population was higher. Both RAPD and ISSR markers independently detect population diversity in the range of 14% and 17%, respectively, while collectively show 15% population diversity. In this study, the Nei's genetic diversity based on combined RAPD-ISSR analysis, among population was also obtained similar range (18%). Genetic diversity plays an important role for survival and considerable importance for sustainability of plants (Wang et al., 2007). Moreover, the geographically isolated individuals tend to accumulate the genetic variations during the course of environmental adaptations (Sarwat et al., 2008). Higher level of within population diversity is important for the survival of *L. sindicus* under unpredictable fragile ecosystem and diversity among populations that might have been important in niche specific adaptations.

The present molecular study has disclosed a great deal of genetic variability in *L. sindicus* that was largely unrevealed through the morphological observations. Being a perennial rhizomatous crop limited variations are depicted for most traits. Leaf, stem, and spike morphology tend to vary with phenology, season, and age of clone. A pressing need has arisen to use and establish more molecular tools for discrimination and identification of stable traits of plant species while changing the environmental and stress conditions.

11   Rhizospheric microbiology of Sewan grass

There is very scarce information on rhizospheric microbiology of Sewan grass, while the rhizosphere effect is qualitatively and quantitatively more noticeable in hot arid desert soils (Buyanovsky et al., 1982). As desert grasses survive extreme conditions of temperature and moisture, and develop in usually poor soils with low organic content, limited amounts of bioavailable inorganic nutrients, and rhizospheric microbes associated with desert grasses should possess adaptive tools to cope with some stresses, namely, frequent droughts, high temperature, starvation, high osmolarity, and desiccation. Many investigations have addressed the significance and role of nitrogen fixation in ecologically unique terrestrial habitats by aiming on the diversity of nifH sequences (Zehr et al., 2003). Such studies have provided a rapidly expanding database of nifH sequences and revealed a wide diversity of uncultured diazotrophs (Tan et al., 2003). Sharma et al. (2015) explored association of *Lasiurus* roots with *Azotobacter* and *Azospirillum*. They found that the population of *Azotobacter* in the rhizosphere *of* Sewan grass was highest in Jaisalmer soil, followed in decreasing order by Bikaner, Pali, Bhilwara, Jodhpur, Jaipur, Hanumangarh, Sikar, Ajmer, and Alwar soils (Fig. 3.8). While the population *of Azospirillum* in the rhizosphere of Sewan grass was maximum in the Bikaner soil, followed by Pali, Jaisalmer, Bhilwara, Jaipur, Ajmer, Hanumangarh, Sikar, Alwar, and Jodhpur soils (Fig. 3.8). It is proved that the rhizosphere of Sewan grass have good populations of free living nitrogen fixing bacteria, such as *Azotobacter* and *Azospirillum* which are responsible for its high productivity in the arid and semi-arid regions of Rajasthan. Thus, the rhizospheric

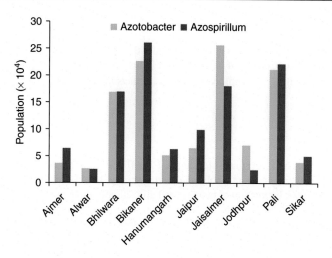

Figure 3.8 The population of *Azotobacter* and *Azospirillum* in the rhizosphere *of* Sewan grass.
Source: Data from Sharma et al., 2015.

microbiology of Sewan grass is significant in order to conservation and biodiversity associated with such niches to tolerate gentle ecological processes in the oligotrophic desert ecosystem.

12 Conclusion and future prospects

Perennial grasses are appropriate candidates for ecorerstoration and phytoremediation of degraded sites or waste dumps sites toward the development of self-sustaining vegetation cover. In this regards, Sewan grass (*L. sindicus*) is a most significant and suitable perennial pasture grass of hot dry arid region because of its high ecological plasticity nature. It is extensively used as forage grass in pastures especially extreme dry regions of desert. Great effort is necessary to understand its high potential for survivability and ecophysiological adaptation to hostile conditions.

However, there is a need to explore mechanisms of accumulation, distribution, translocation, and sequestration of inorganic and organic pollutants in roots and shoots of *L. sindicus* growing on the polluted sites. In addition, it is also necessary to determine how this grass in polluted site can stimulate microbial activity by root exudates. Genetic studies can be significant for a better understanding of the link between novel genes and adaptive response of *L. sindicus* to pollutants. Therefore, genetic engineering and breeding of *L. sindicus* with the help of physiology, biochemistry, molecular biology, metabolomes, and proteomes can improve the information about the selection of phytoremediation technologies and biomass production. The exploitation of *L. sindicus* in phytomanagement practice offers great prospects in developing policy outlines to promote remediation of polluted sites along with biomass production in hot dry arid region worldwide.

Acknowledgment

Financial assistance given to Dr. V.C. Pandey under the Scientist's Pool Scheme (Pool No. 13 (8931-A)/2017) by the Council of Scientific and Industrial Research, Government of India is gratefully acknowledged. The authors declare no conflicts of interest.

References

Arshad, M., Ashraf, M.Y., Ahmad, M., Zaman, F., 2007. Morpho-genetic variability potential of *Cenchrus ciliaris L.* from Cholistan desert, Pakistan. Pak. J. Bot. 39 (5), 1481–1488.

Assaeed, A.M., 1997. Estimation of biomass and utilization of three perennial range grasses in Saudi Arabia. J. Arid Environ. 36, 103–111.

Buyanovsky, G., Dicke, M., Berwick, P., 1982. Soil environment and activity of soil microflora in the Negev desert. J. Arid Environ. 5, 13–28.

Chauhan, S.S., 2003. Desertification control and management of land degradation in the Thar desert of India. Environmentalist 23, 219–227.

Chowdhury, S.P., Schmid, M., Hartmann, A., Tripathi, A.K., 2009. Diversity of 16S-rRNA and nifH genes derived from rhizosphere soil and roots of an endemic drought tolerant grass, *Lasiurus sindicus*. Eur. J. Soil Biol. 45 (2009), 114–122.

Ecocrop, 2010. Ecocrop database. FAO.

El-Keblawy, A., Bhatt, A., Gairola, S., 2013. Perienth colour affect germination behavior in the wind pollinated *Salsola rubescens* in the Arabian Deserts. Can. J. Bot. 92, 69–75.

FAO, 2010. Grassland Index. A searchable catalogue of grass and forage legumes.

Godt, M.J.W., Hamrick, J.L., 1998. Allozyme diversity in the endangeredpitcher plant *Sarracenia rubra* ssp. *alabamensis* (Sarraceniaceae) and its close relative *S. rubra* ssp. *rubra*. Am. J. Bot. 85, 802–810.

Gupta, R.K., Saxena, S.K., 1970. Some ecological aspects of improvement and management of Sewan *(Lasiurus sindicus)* rangelands. Ann. Arid Zone 9, 193–208.

Haase, P., Pugnaire, F.I., Incoll, L.D., 1995. Seed production and dispersal in the semi-arid tussock grass *Stipa tenacissima* L. during masting. J. Arid Environ. 31, 55–65.

Hedrick, P.W., 2001. Conservation genetics: where are we now? Trends Ecol. Evol. 16, 629–636.

Khan, M.F., Anderson, D.M., Nutkani, M.I., Butt, N.M., 1999. Preliminary results from reseeding degraded Dera Ghazi Khan rangeland to improve small ruminant production in Pakistan. Small Ruminant Res. 32, 43–49.

Khan, M.A., Ansari, R., 2008. Potential use of halophytes with emphasis on fodder production in coastal areas of Pakistan. Biosaline Agriculture and High Salinity Tolerance. Birkhä user Basel, pp. 157–162.

Khan, T.I., Frost, S., 2001. Floral biodiversity: a question of survival in the Indian Thar desert. Environmentalist 21, 231–236.

Krishna, P.H., Reddy, C.S., Meena, S.L., Katewa, S.S., 2014. Pattern of plant species diversity in grasslands of Rajasthan, India. Taiwania 59 (2), 111–118., DOI: 10.6165/tai.2014.59.111.

Mansoor, U., Hameed, M., Wahid, A., Rao, A.R., 2002. Ecotypic variability for drought resistance in *Cenchrus ciliaris* L. germplasm from Cholistan desert in Pakistan. Int. J. Agr. Biol. 4, 392–397.

Mertia, R.S., 1992. Studies on improvement and utilization of rangelands of Jaisalmer region. CAZRI Monograph No. 38. pp. 1–42.

Mertia, R.S., Prasad, R., Khandpal, B.K., Narain, P., 2006. Regeneration of *Lasiurus sindicus* in relation to grazing pressure and root-zone soil moisture in arid rangelands of western Rajasthan (India). Trop. Grasslands 40, 40–44.

Osman, A.E., Makawi, M., Ahmed, R., 2008. Potential of the indigenous desert grasses of the Arabian Peninsula for forage production in a water-scarce region. Grass Forage Sci. 63, 495–503.

Pandey,V.C., Souza-Alonso, P., 2018. Market opportunities in sustainable phytoremediation. In: Pandey, V.C., Bauddh, K., (Eds.), Phytomanagement of Polluted Sites. Elsevier. Doi. org/10.1016/B978-0-012-813912-7.00002-8.

Parkhurst, D.F., 1978. The adaptive significance of stomatal occurrence on one or both surfaces of leaves. J. Ecol. 66, 367–383.

Patel, Y., Dabgar, Y.B., Joshi, P.N., 2012. Distribution and diversity of grass species in Banni Grassland, Kachchh District, Gujarat, India. Int. J. Sci. Res. Rev. 1, 43–56.

Peacock, J.M.F., Erguson, M.E., Alhadrami, G.A., Mccann, I.R., Al-Hajoj, A.S., Aleh, A., Karnik, R., 2003. Conservation through utilization: a case study of the indigenous forage grasses of the Arabian Peninsula. J. Arid Environ. 54, 15–28.

Pearson, M., Davies, W.J., Mansfield, T.A., 1995. Asymmetric responses of adaxial and abaxial stomata to elevated CO_2: impacts on the control of gas exchange by leaves. Plant Cell Environ. 18, 837–843.

Peat, H.J., Fitter, A.H., 1994. A comparative study of the distribution and density of stomata in the British flora. Biol. J. Linnean Soc. 52, 377–393.

Pitelka, L.F., Ashmun, J.W., 1985. Physiology and integration of ramets in clonal plants. In: Jackson, J.B.C., Buss, L.W. Cook, R.E. (Eds.), Population Biology and Evolution of Clonal Organisms, Yale University Press, New Haven, CT, pp. 399–435.

Quattrocchi, U., 2006. CRC World Dictionary of Grasses: Common Names, Scientific Names, Eponyms, Synonyms and Etymology. CRC Press, Taylor & Francis Group, Boca Raton, FL, USA.

Raheja, P.C. 1962. Research and development to the Indian arid zone. Arid Zone, UNESCO, 15, pp. 7–12.

Rao, A.R., Arshad, M. Shafiq, M., 1989. Perennial grasses germplasm of Cholistan desert and its phytosociology. Cholistan Institute of Desert Studies, Islamia University, Bahawalpur, pp. 162.

Rao, A.V., Tarafdar, J.C., Sharma, B.K., 1996. Characteristics of gypsum mined spoils. J. Indian Soc. Soil Sci. 44, 544–546.

Robinson, B.H., Banuelos, G., Conesa, H.M., Evangelou, M.W.H., Schulin, R., 2009. The phytomanagement of trace elements in soil. Crit. Rev. Plant Sci. 28, 240–266, DOI: 10.1080/07352680903035424.

Sarwat, M., Das, S., Srivastava, P.S., 2008. Analysis of genetic diversity through AFLP, SAMPL, ISSR and RAPD markers in *Tribulus terrestris*, a medicinal herb. Plant Cell Rep. 27, 519–528.

Sharma, K.C., 2013. Effect of row spacings and nitrogen levels on the productivity of sewan (*Lasiurus sindicus*) grass in hot arid region of Western Rajasthan. Forage Res. 39 (3), 140–143.

Sharma, P., Pandey, S., 2017. Phytomanagement of heavy metal lead by fodder grass *Lasiurus scindicus* in polluted soil and water of Dravyawati river. Am. J. Environ. Sci. 13 (2), 167–171, DOI: 10.3844/ajessp.2017.167.171.

Sharma, R., Rajora, M.P., Dadheech, R., Bhatt, R.K., Kalia, R.K., 2017. Genetic diversity in Sewan grass (*Lasiurus sindicus* Henr.) in the hot arid ecosystem of Thar Desert of Rajasthan, India. J. Environ. Biol. 38, 419–426, Doi.org/10.22438/jeb/38/3/MS-265.

Sharma, R., Tilawat, A.K., Sharma, S.K., 2015. Free living nitrogen fixing diazotrophs in rhizosphere of *Lasiurus sindicus*. Indian J. Environ. Sci. 19 (1&2), 15–17.

Singh, K.C., Singh, S.D., 1997. Water use and production potential of Sewan (*Lasiurus sindicus* Henr.) pasture in the Thar desert. In: Yadav, M.S., Singh, Manjit, Sharma, S.K., Tewari, J.C., Burman, U. (Eds.), Silvipastoral Systems in Arid and Semi-Arid Ecosystem, Central Arid Zone Research Institute, Jodhpur, pp. 157–162.

Tan, Z., Hurek, T., Reinhold-Hurek, B., 2003. Effect of N fertilization, plant genotype and environmental conditions on nifH gene pools in roots of rice. Environ. Microbiol. 5, 1009–1015.

Tarafdar, J.C., Marschner, H., 1994. Efficiency of VAM hyphae in utilization of organic phosphorus by wheat plants. Soil Sci. Plant Nutr. 40, 593–600.

Tarafdar, J.C., Rao, A.V., 1997. Mycorrhizal colonization and nutrient concentration of naturally grown plants on gypsum mine spoils in India. Agr. Ecosyst. Environ. 61, 13–18.

Wang, W., Chen, L., Yang, P., Hou, L., He, C., Gu, Z., Liu, Z., 2007. Assessing genetic diversity of populations of topmouth culter (*Culter albumus*) in China using AFLP markers. Biochem. Syst. Ecol. 35, 662–669.

Waramit, N., 2010. Native warm-season grasses: species, nitrogen fertilization, and harvest date effects on biomass yield and composition. Graduate Theses and Dissertations. Iowa State University.

Yechieli, A., Oren, A., Yair, A., 1995. The effect of water distribution on bacterial numbers and microbial activity along a hill slope, northern Negev, Israel. Adv. Geoecol. 28, 193–207.

Young, A.G., Clarke, G.M., 2000. Genetics, Demography and Viability of Fragmented Populations. Cambridge University Press, Cambridge.

Zehr, J.P., Jenkins, B.D., Short, M.S., Steward, G.F., 2003. Nitrogenase gene diversity and microbial community structure: a cross-system comparison. Environ. Microbiol. 5, 539–554.

Miscanthus–a perennial energy grass in phytoremediation

Ashish Praveen[a,b], Vimal Chandra Pandey[c,*]
[a]Plant Ecology and Environmental Science Division, National Botanical Research Institute, Lucknow, Uttar Pradesh, India; [b]Department of Botany, Markham college of Commerce, VBU, Hazaribag, India; [c]Department of Environmental Science, Babasaheb Bhimrao Ambedkar University, Lucknow, Uttar Pradesh, India
*Corresponding author

1 Introduction

Pollution and population are increasing proportionately with its effect on ecosystem and the carrying capacity of planet earth. Pollution is the result of rapid industrialization with greed of every country to be the most prosperous. In this race of being on top, every one forgot the value of environment. Now the outcomes of greed are in the form of contamination and pollution. It is true that we can't compromise with our comfort and stop industrialization but we can do this in a way that will leave some time for nature to reconstruct. The contaminants are being added in soil and water (Pandey et al., 2015a) and are affecting the natural biogeochemical cycles. The problem of heavy metal contamination is serious as these do not decompose and also changeable to other forms making it more toxic. Thus, it is necessary to reduce and remediate these heavy metals. There are ways to reduce these contaminants like mechanical, chemical, and plant based, that is, phytoremediation. The methods discussed earlier have its benefits and limitation. The mechanical and chemical methods of removal are costly, and are not a permanent solution of the problem. Although, contaminants from these two methods can be removed in short time but those are costly. The removal of contaminants through phytoremediation is cost-effective and also environmentally-friendly but it require long time to remediate. Thus what is the option? Having the pros and cons of the methods known for removal of contaminants, the method of phytoremediation can be considered to be a low-cost and environmentally-friendly (Pandey and Singh, 2019; Pandey and Bajpai, 2019). The limitation of long time can be negated with use of plants that in addition to phytoremediation also yields products like energy (bioethanol or biogas or biofuel) or commercial products (essential oil, etc.) (Pandey and Souza-Alonso, 2019). Furthermore, *Miscanthus* being grass will take less time in phytoremediation than trees. In this regards, some grasses such as *Vetiver, Saccharum, Miscanthus,* etc. that fulfills the requirement of present problem (Pandey et al., 2015b, 2019). The present chapter describes about *Miscanthus* plant as a bioenergy grass and also an accumulator of contaminants. The bioenergy grass

Phytoremediation Potential of Perennial Grasses. http://dx.doi.org/10.1016/B978-0-12-817732-7.00004-3

Miscanthus has phytoremediation potential for the remediation of contaminated sites with economic returns (Pandey et al., 2016; Pandey and Souza-Alonso, 2019).

In order to combat the present scenario of climate change and the various effects of pollution to the flora and fauna, it is the time to rethink for other sources of energy. In addition, the demand for energy is also increasing in world. The use of renewable source of energy will solve the present issues of energy and climate change. The renewable source of energy like solar, wind have been in use throughout the world but not sufficient to all the countries as these require setups that are costly and poor countries are not in a position to afford it. Also, if some cheap source of energy production is available that it will be affordable by poor countries and the developed one as every country wants to have more with less spent. The generation of renewable source of energy in question here are plants that have good biomass as the biomass can be converted to fuel for uses. If the plant used is also a good accumulator of contaminants than it would be a boon as both the problems are solved. In this regard, *Miscanthus* species are excellent plant that has both the properties of phytoremediation and enough biomass for bioenergy production (Pandey, 2017). *Miscanthus* species is widely used in European countries for bioenergy production. *Miscanthus* has a good biomass thus it also fulfills one of the basic requirement for plants required in phytoremediation. One will point out that there are plants that are hyperaccumulators of heavy metals than why not to use them when *Miscanthus* is not a hyperaccumulator. So it is true that *Miscanthus* is not a hyperaccumulator but the comparable amount of metal accumulated in its biomass far exceeds to that of hyperaccumulators and in addition, it is also a bioenergy crop. *Miscanthus* is perennial, herbaceous, fast growing, and has a good root and shoots biomass, thus best suits to the phytoremediation (Nsanganwimana et al., 2014; Pandey et al., 2016).

Can the biomass of perennial plants fulfill the energy needs? The solution to this question lies in how much biomass is needed and how much is in use now. Again what is the gap in other sources of energy that these biomass generated method have to fill. So, this will be clearer from the available data from some of the countries that have started the use of plant biomass for energy and fuel production. In the year 2008 in the United States, the total 105 exajoules (EJ) of energy is consumed and only 4.1 EJ, that is, 4% came from biomass sources (DOE, 2009). After 24–30 years, this energy need will be approximately 121 EJ (DOE, 2010). To fulfill the demand, there is an urgent need to increase energy production, which is possible through other costly sources of energy production but may have impact on climate. Now in order to have minimum impact on environment, the use of plant biomass for energy production is a good approach. The additional benefit of using these plants is that they are also helping in phytoremediation so, this is acting like a pleiotropic gene, that is, a gene controlling various characters, here it is the multifunctional role that *Miscanthus* does. *Miscanthus* species are being used for both phytoremediation and bioenergy from biomass. There are various species in *Miscanthus* of which two of the species, namely, *M. sinensis* and *M. sacchariflorus* are known for phytoremediation and energy production. *Miscanthus* × *giganteus* is a sterile hybrid produced by crossing the earlier-mentioned two species and is a high biomass crop than both so used as energy crop in Europe.

2 *Miscanthus* biology and taxonomy

Miscanthus is a perennial grass with C4 strategy and belonging to family poaceae. *Miscanthus* was first described by Anderson (1855). The inflorescence of *Miscanthus* holds importance with its stalked pedicilate spikelet, which is the basis of its botanical name (mischis: pedicilate, anthos: flower). *Miscanthus* is native to India, Japan, East Asia, Malayasia, Philipines, and Southern Africa. It can grow on a variety of soil types, such as acidic, nutrient deficient, well drained, alkaline, etc. It can be found growing along roadside or at places which are generally not suitable for agriculture. It is widely distributed in tropical and subtropical areas and also in disturbed land, hence also known as weed of disturbed regions. Thus, the *Miscanthus* does not need any extra effort to grow which is a benefit for phytoremediation. There are about 20 species in the genus *Miscanthus*. Some of the *Miscanthus* genus show similarity with *Saccharum* and can form intergeneric hybrids with it (Jones and Walsh, 2001).

Miscanthus is a tall cane like reed which may attain a height of 2 m. The leaves are 0.8–1.2 cm in width and can be 40–60 cm long. Inflorescence is open and it is spikelet, the pedicels of spikelet are unequal in length with silky hairs surrounding them (USDA Forest Service, 2006). The flowering starts in the beginning of September and remains throughout the winter (Watson and Dallwitz, 1992; Gilman, 1999). *Miscanthus* has rhizomes which are short and inconspicuous (Meyer et al., 2017). Sun et al. (2010) has made a detailed taxonomic revision of the *Miscanthus* from China, where the distribution, phenology, habitat, and morphological description of each species of *Miscanthus* have been given in detail. Deuter (2000) and Anderson et al. (2011) described that *Miscanthus* forms a monophyletic group with about 20 species and the highest chromosome number being 19. The *Miscanthus* that is widely cultivated for economic purposes include *M. sinensis, M. sacchariflorus, M. floridulos,* and *M. × giganteus*. The *M. giganteus* have been developed from *M. sacchariflorus* (tetraploid and *M. sinensis* (diploid) and it is an interspecific hybrid (Deuter, 2000).

3 Propagation

Propagation is one of the important characters in all the living beings to increase in number and survive. Plants can propagate through various ways, such as stem cuttings, nodes, leaves, roots, rhizomes, and seeds. Of all the methods, propagation through seeds is the most common. *Miscanthus* species also have these methods of propagation but some hybrid forms like *M. giganteus* cannot propagate through seeds because it is a triploid and do not form viable seeds. The importance of *Miscanthus* has increased in recent years with the knowledge of its use as energy crops and thus the demand increased but with limited knowledge of its propagation, the demand remained unfilled. However, with enhanced knowledge of the various conditions required for a plant to grow and survive successfully, the number has increased. *Miscanthus* species and its propagation are discussed as follows: rhizomes, nodal stem cuttings, micropropagation, seeds, CEED, and collar propagation.

Rhizome—Rhizome is an underground stem with node and buds (Stevens, 1966). It has roots and also forms shoots. Propagation through rhizomes depends upon various factors, such as size of rhizome, temperature, weather type, soil conditions, and also on the genotype (Xue et al., 2015). Success of this method of propagation also depends upon the selection of rhizome (healthy and of good quality) and protection from pests (wireworm larvae, microtus, etc.).

Nodal stem cutting—Nodes are the parts in plant from where leaves, branches, and roots emerge. The nodal stem cutting here means a part of stem with nodes (Fig. 4.1A). These are capable to grow when planted in soil. This method of propagation is already in practice for sugarcane (James, 2008), bamboo (Hirimburegama and Gamage, 1995), and various other plants (Guse and Larsen, 1975). Although a suitable way of propagation but care is needed as harsh environmental conditions will reduce the growing percentage. Thus, it is better to first grow the nodal stem cuttings in optimal environmental conditions and then planted to fields.

Micropropagation—It is a method in tissue culture to rapidly increase the quantity of plants. The plants are developed for germplasm conservation and also these are less susceptible to disease. It includes: (1) Direct shoot regeneration: nodal stem or rhizomes are selected for this purpose; (2) Callus culture: it includes somatic embryos from inflorescence, leaf, shoot apices, buds, and leaves (Lewandowski and Kahnt, 1993; Nielsen et al., 1993; Lewandowski et al., 2000; Kim et al., 2010; Gubišová et al., 2013).

Seed propagation—Propagation through seeds is natural and most common method of multiplication in plants. The problem arises when the seeds are not viable (sterile). *M. sinensis* and *M. sacchariflorus* "Robustus" are the two species which easily propagate through seeds. *M. giganteus* a hybrid is triploid and thus do not form viable seeds (Wilson and Knox, 2006; Anderson et al., 2015). The propagation through seeds has its own benefit and limits, benefit in terms of easy growing and limit in terms of invasiveness.

CEED—Crop, Expansion, Encapsulation, and Delivery system called CEED is developed. *M. giganteus* is grown through this method. It is a type of somatic

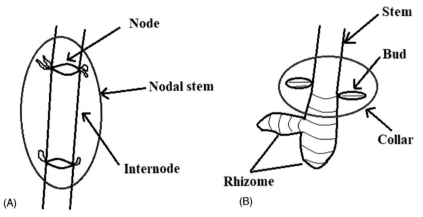

Figure 4.1 The propagation of *Miscanthus* by two ways, that is, (A) Nodal stem and (B) Collar.

embryogenesis in which plant material is encapsulated and then put in soil. This encapsulated plant material than give rise to new plantlets. The New Energy Farms has been working in commercial production of *M. giganteus* through CEED (NFF, Ceeds TM, 2014).

Collar propagation—It is a type of vegetative propagation generally used for sterile genotypes of the *Miscanthus*. This method of propagation was suggested by Mangold et al. (2018) as an alternative to the rhizome propagation. Collars are the junction point of stem and rhizome with presence of buds (Fig. 4.1B). The harvesting of collars is less destructive than that of rhizome. Harvesting is simply pulling the stems of senescence plants, while leaving the rhizomes behind, that is, in soil. The buds present at collars are generating a new shoot. However, of the various methods available for propagation of the *Miscanthus*, the propagation through rhizomes is the most suitable, successful, and commercially used propagation tool for this plant.

4 Easy harvesting

Harvesting of the *Miscanthus* can be done annually from the second year of planting. The harvesting is carried out during spring as the moisture content of the plant at this season is low and thus easy to store and also high calorific value is obtained. *Miscanthus* grows vigorously during summer and growth retards during autumn. If harvesting is done during January or February then there is problem of moisture as the plant has high moisture content and is not suitable for various applications. Harvesting between last of April or beginning of May is also not suggested as it renders the growth of emerging plant. The best suitable time for harvesting is in March or beginning of April. To ensure best quality, it is suggested to harvest plant at low moisture and store in dry place. Litters can also be harvested but the litters during winters are avoided due to high moisture content. The left over litter in winter is beneficial for plants as it adds organic matter to soil and thus help the plant in growing.

5 *Miscanthus* grass as a biofuel crop

The problem of increased fuel demand has led to pressure on the natural reserves and as is certain, it is not forever. The natural reserves are limited and soon it will vanish. So, there has to be search for other sources of fuel requirement. There are other sources through which we can reduce the pressure on natural resource and the best source is the plants. Thus, nature itself has the solutions to our problems, only we have to search for these. Plants with good biomass can be used for production of biofuel. Production of biofuel has been in progress through *Jatropha curcas* and various other plants where its seeds are being used for the purpose. *Miscanthus,* a C4 grass with good biomass, has been used in production of biofuel across the nations. In the European countries, *Miscanthus* is widely used to produce heat, electricity, and fuel. In England and Ireland, most of the *Miscanthus* is grown for power generation. Trials have been done by Edenderry power station for power generation. *Miscanthus* can be directly used in stoves, boiler, or power station and its processed products like pellets, chips, or briquettes are use in heat generation. Pasat and farm 2000 is a boiler that can be used for combustion of *Miscanthus* biomass. Danish manufacturers (REKA and LINKA) are also using bales for combustion. Various other manufacturers are also using *Miscanthus* bales, pellets, and chips for combustion (Caslin et al., 2010). *Miscanthus* has good biomass and it can be of great value in biogas production, however there is need for further exploration. The production depends on the harvest time which is favorable for anaerobic digestion of *Miscanthus* biomass. In Germany, sufficient biogas (methane) production has been obtained (247 mL (g oDM)$^{-1}$) (Kiesel and Lewandowski, 2017). October is the best harvesting time as CH_4 production is around 6000 m^3 (Purdy et al., 2015; Kiesel and Lewandowski, 2017). *M. giganteus* is the best genotype in regards of biomass and yield.

The advantage of *Miscanthus* is that being a C4 plant, it can abide harsh environmental conditions which are the situation now with climate change (Byrt et al., 2011). Among C4, many crops are food crops, such as maize, sorghum, sugarcane, etc. with good biomass and used for bioethanol production (Bennetzen, 2009). But the demands are high and to fulfill these demands, we need to search other plants that can be used for this purpose. As discussed, *Miscanthus* has good biomass and it is a C4 plant, in addition it is widely used in phytoremediation also. So why not grow *Miscanthus* at contaminated sites where it will serve both purposes of remediation and production of biofuel. In the era of increasing population, there has been massive increase in contamination, and works are in progress to combat the problems. *Miscanthus* is a good choice as it will serve this world with its phytoremediation property and biofuel generation from its biomass. *Miscanthus* species is widely used as energy crop (Lewandowski et al., 2000; Heaton et al., 2008). We have discussed that food crops are in use for biofuel production but it can directly affect food security (Eisentraut, 2010). There will be monotony of particular crop and it will affect the growing of other crops, this will affect the production of other crops as the choice of farmers will be such crops that have good return. The use of such food crops for biofuel production was called the first generation biofuels. But now with the knowledge of other crops that are not food crops but have biomass and potential of bioethanol production are used. These are second generation biofuel plants. So, the selection in biofuel must be that plant which is not a food crop but has good biomass and also serving some benefits (phytoremediation). Market value of *Miscanthus*: Kim et al. (2015) has described the content of *Miscanthus* (22% lignin, 36% alpha cellulose, and 24% hemicelluloses). The annual energy production from per kilogram of *Miscanthus* is estimated at 17 MJ (Collura et al., 2006). Biomass is about 20,000 kg/ha. *Miscanthus* produces more biomass thus more energy per hectare (Fischer et al., 2007; Heaton et al., 2010).

6 Phytoremediation

Phytoremediation is an important feature in some of the plants. This is a boon to the mankind as with increasing pollution, there is need for sources that can clean the environment. There are plants that can accumulate contaminants but if a plant in

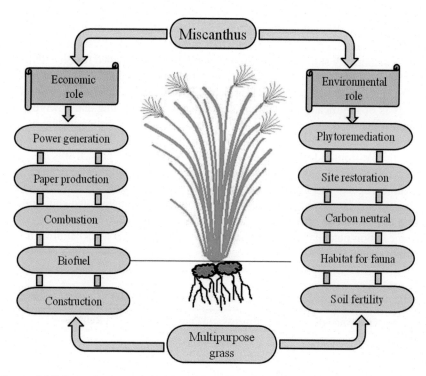

Figure 4.2 *Miscanthus* **grass with its multipurpose services.**

addition of remediating contaminants is also serving other services (from ecologically to socio-economically importance) than it is the most suitable plant for phytoremediation (Fig. 4.2). *Miscanthus* is one such plant that in addition to remediation is also an energy crop. The phytoremediation property of *Miscanthus* has been discussed as follows:

Metal(loid)s—The problem of metal(loid)s contamination in soil and water has increased to levels that are affecting the population surviving nearby. The conditions are grim but with use of plants that can remediate the contaminants but also has economic value will be effective. *Miscanthus* species are a good solution to the problem as they remediate contaminants and are energy crops with good economic return. Krzyzak et al. reported that *Miscanthus* species accumulate Pb, Zn, and Cd. *Miscanthus* has a benefit of being C4 system and so can survive at higher and stress condition. It can also grow at low temperature of 5°C with its effective absorption of solar radiation (Swaminathan et al., 2010). For successful phytoremediation, the plant must be fast growing with good biomass but low growth and low ability to produce biomass is a hinderance (Macková and Macek, 2005). Metal(loid)s are the most abundant pollutant. The remediation depends on the plant ability to overcome the toxicity caused by metal(loid)s. Jones et al. (2006) reported treatment of wastewater and dump leachants through *miscanthus* (Formanek and Ambus, 2004). In fly ash dumps, the *Miscanthus* has a good growth rate (Techer et al., 2012). *Miscanthus* also grows well in industrial contaminated sites

(Wanat et al., 2013). At heavy metal contaminated sites where the concentration are high, *Miscanthus* grows well with good biomass (Arduini et al., 2006).The phytoremediation ability of *miscanthus* has been shown in Table 4.1. Effect of chromium in *M. sinensis* was elaborately studied by Sharmin et al. (2012). They studied the physiological and biochemical response of this grass due to Cr toxicity. They identified some protein during the proteomic analysis that might be playing some role in metal homeostasis. This study is of significance as with the knowledge of upregulating and downregulating proteins, the needed modification can be made in plants for better adaptation and phytoremediation of contaminants. In a study by Gordana et al. (2017) for remediation of heavy metals from industrial wastes through *M. giganteus* in pots. It was observed that the contaminants (mainly Cd, Cr, and Pb) were accumulated by *M. giganteus* and the maximum uptake was in root while in shoots the translocation was less. Therefore, *M. giganteus* can be suitable for renediation of industrialy contaminated sites and the overhead bimass can be utilized for enrgy production (Rascio and Navari-Izzo, 2011). Phytoremediation potential of *Miscanthus* can be enhanced with association with fungus. Firmin et al. (2015) has shown that how association of fungus with roots of *Miscan-*

Table 4.1 Remediation of contaminated sites with *Miscanthus* species by various authors.

S. no.	Process involved	Amendments	Pollutants	References
1.	Phytostabilization	AMF inoculums	Cd, Pb, Zn	Nsanganwimana et al. (2014)
2.	Phytostabilization Phytodegradation	Soil microorganism	Organic and Inorganic pollutants	Nsanganwimana et al., 2014
3.	Rhizoremediation		PAH pollutants	Techer et al., 2012
4.	Phytoremediation Phytostabilization	*Pseudomonas koreensis* AGB-1	Contaminated mine soil	Babu et al. (2015)
5.	Phytoremediation	Bacteria	PAH soil	Techer et al., 2012
6.	Phytostabilization		Lead smelter site	Al Souki et al. (2017)
7.	Phytoremediation		Military contaminated soil	Pidlisnyuk et al., 2019
8.	Phytoextraction		Zn	Barbosa et al., 2015
9.	Phytoremediation	AMF	Trace elements	Firmin et al., 2015
10.	Phytoremediation		Zn	Krzyzak et al., 2017
11.	Phytoremediation		Zn	Korzeniowska and Glubiak, 2015
12.	Phytoremediation	Sewage sludge, municipal compost	Post mining soil	Placek et al., 2018
13.	Phytoremediation		Heavy metal from industrial disposal	Dražić et al. (2017)
14.	Imparts heavy metal tolerance		In presence of Cr	Sharmin et al., 2012
15.	Phytoremediation		Coal mine soil	Jezowski et al., 2017

thus enhances the uptake of heavy metal contaminants and also helps in growth of plant. They planted *M. giganteus* in association with arbuscular mychorrhizal fungi in trace metal contaminated soil. They reported that with the AMF assistance, the metal uptake is augmented in *Miscanthus*. *M. giganteus* is also a good phytoremediator of zinc (Zn) and phytostabilizes of heavy metal thus preventing its spread underground or elsewhere (Barbosa et al., 2015). *M. giganteus* is a hybrid of two species of *Miscanthus* and it is most popular among researchers for phytoremediation and energy production. Pidlisnyuk et al. (2019) designed a pot experiment to check the phytoremediation potential of *M. giganteus* in polluted military soil. The aim of the experiment was to examine the phytoremediation potential of contaminats (As, Cu, Mn, Pb, Zn, and Cr) from military soil. The contamination in soil is due to the various equipments and substances (explosives, etc.) used by military. The remediation of contaminats is necessary from soil as the contaminantion will affect the flora and faunna in and nearby areas. They observed that these heavy metal contaminants were taken by *Miscanthus* and the accumulation was more in root compared to shoot. *M. gigantus* has been also utilized for the restoration of metal contaminated sites. In Northern France, polluted soil near a lead smelter was checked with *M. giganteus* and it was noticed that it is acting as a phytostabilizer of the contaminants. Thus *Miscanthus* by phytostabilizing the contaminants and is helping in restoring the soil flora and properties of soil at the contaminated sites (Al Souki et al., 2017).

Organic pollutants—*Miscanthus* remediate organic pollutants which was shown by Techer et al. (2012) in both lab and field conditions. Techer et al. (2012) reported that metabolites from root of *Miscanthus* leads to association of bacteria present in soil and helps in biostimulation of PAHs. Soil bacteria is playing an important role in plant growth as they are making some nutrients available to plants. In addition to this, the bacteria also changes the state of heavy metals making it available to plants. *Pseudomonas koreansis* AGB-1, a bacteria, isolated from the roots of *M. sinensis* growing at mine contaminated site by Babu et al. (2015). They found that AGB-1 has a high tolerance to heavy metals. Thus AGB-1 in association with *M. sinensis* could be used for phytoremediation of heavy metals from mine contaminated sites. This bacteria is also promoting the growth of *M. sinensis*. The remediation of organic contaminants is, however, less complex as compared to trace elements or metalloids. Living organism releases various enzymes which degrade organic contaminants (Gerhardt et al., 2009). It has also been established that roots of plants release chemicals that attract microorganisms, which in turn increases and make available various nutrients to plant that further help in growth of plant. This also help in increasing the plants ability to withstand toxicity of contaminants (Megharaj et al., 2011). The consortia from rhizospheric zone of *Miscanthus* showed presence of microorganism that were identified as degraders of PAH (Techer et al., 2012). Also, the enzymes released by plants, such as laccase, peroxidase, etc. are found to degrade organic compounds (Gerhardt et al., 2009). However, more works are needed for a better idea of phytoremediation potential of *Miscanthus*.

Thus, various works on the phytoremediation of contaminants with *Miscanthus* species have been documented and it is a promising plant for remediation of contaminated sites (Table 4.1). *Miscanthus* species has also good biomass and the accumulation of contaminants is in roots as compared to the shoots so the overhead part of the plant is safe for use. Thus, the plant biomass after phytoremediation is being successfully used in enrgy production in European countries and the United States. Thus, *Miscanthus* species is giving a dual role of phytoremediation and as a energy crop and in addition it is non-edible, easily growing, and require less maintenance.

7 Environmental consideration

Miscanthus is sometimes also called a carbon neutral plant because these grasses have C4 cycle which enables them to utilize atmospheric CO_2. So, how it is a carbon neutral plant? Since this plant has good biomass and is being used for energy production while CO_2 released during this energy production processes is utilized by the young growing plant, thus the amount of CO_2 produced is also utilized by the plant. Thus, the *Miscanthus* is storing carbon and preventing its release in environment thus reducing global warming. In Ireland, *Miscanthus* plantation was done in a hectare of land, this was done to calculate the CO_2 sequestration of *Miscanthus* (Caslin et al., 2010). It was observed that in 1 ha of land when *Miscanthus* was left to grow for 12 years; about 8.8 tons of carbon was stored. Thus, *Miscanthus* has a good CO_2 sequestration. This is one of the important environmental benefits of *Miscanthus*. There are also other environmental benefits like reduction in use of herbicides. How *Miscanthus* is reducing the use of herbicides? It was seen that the canopy of *Miscanthus* led to competition for light, nutrient, and moisture to the herbs and thus herbs do not grow. This uses the use of herbicides and as we know those herbicides are pollutant and causes harm to human health. Therefore, reducing the herbicide pollution and contamination from spreading nearby sites is also a benefit of using *Miscanthus* in phytoremediation programs. The good biomass of *Miscanthus* is also preparing a ground for habitat of various fauna. The plantation of *Miscanthus* on contaminated sites will have a dual benefit of remediation of contaminants and on the other hand habitat for wildlife, thus site restoration. This newly formed *Miscanthus* forest will provide a good habitat for various small and large animals, thus stopping habitat destruction and fragmentation. In a study in Germany and Rothamsted plantation of *Miscanthus* and cereal crop (wheat) was done to observe site restoration and reducing habitat loss (Caslin et al., 2010). It was observed that the number of earthworms and spiders (in terms of diversity) was increased as compared to wheat plantation. In addition to these organisms, some species of mammals and birds were also increased. Thus, the plantation of *Miscanthus* is restoring the lost habitat of organisms. There are also other benefits of *Miscanthus* plantation as this is providing habitat for small animals (mice, vole, shrew, rabbit, etc.) due to its dense canopy and in turn it is also providing a feeding ground for predators. *Miscanthus* is also providing foraging and cover for birds, nesting ground for organisms. Thus, the plantation of the *Miscanthus* species on contaminated sites is a good choice for site restoration and environmental protection.

8 Multiple uses

Miscanthus is a crop of need to fulfill the demands and let the fossil reserves secure. The overall area under *Miscanthus* cultivation throughout the globe is approximately 1.23 lakhs ha, and largest is in China. *Miscanthus* is an important and capable energy crop,

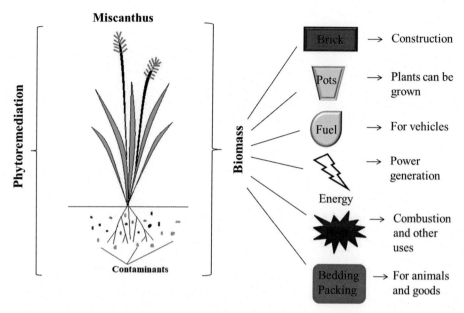

Figure 4.3 Schematic representation of *Miscanthus* for phytoremediation and its biomass utilization.

and has been used successfully in phytoremediation programs (Fig. 4.3). Besides, the *Miscanthus* species is serving various purposes, such as in paper making, building materials, pickles, power generation, biofuel, bio-compost, combustion, animal bedding, etc. Its uses are various which cannot be undermined. Some of its important uses are: *building and packing*–GiebereitehnikUwe Kuehn, Germany, manufactures building materials (Pots, bricks) from the *Miscanthus* biomass (Waldmann et al., 2016), *animal bedding*–*Miscanthus* grass is used for making animal beds. Bedding is very important for overall growth of any animal and this has been proved by DeBruyn (2015) with healthy feet of turkey. The work on cow and horses has also been done with *Miscanthus* bedding and there also the hoofs of these animals were better as compared to normal beddings (Van Weyenberg et al., 2015; Rauscher and Lewandowski, 2016), *feed*–*Miscanthus* is used as feed in dog food (Heaton, 2017, personal communication) and in cows in Netherlands (http://www.bkcbv.nl/miscanthus-rantsoem).

9 Merits and demerits of *Miscanthus* with SWOT analysis

There are many properties of *Miscanthus* which are being utilized for various purposes and thus making it an important grass for welfare of society. However, there are two sides of plant like which has its own benefit and disadvantages but merits outnumbers the demerits (Fig. 4.4). *Miscanthus* plant has its own strength, weakness, opportunity, and threat which have been discussed in Table 4.2, although there are many but some of them have been dealt in this chapter and the table.

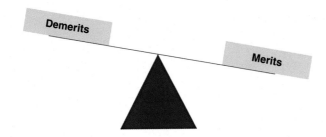

Figure 4.4 A comparison of pros and cons of *Miscanthus* on a balance.

Table 4.2 The various inputs and outputs of *Miscanthus* with SWOT analysis.

S. no.	Strength	Weakness	Opportunity	Threat
1.	*Miscanthus* does not require much attention to grow and can be easily established with rhizome (Xue et al., 2015)	In the beginning, the selection of healthy rhizome is required	The accumulated mass is used for energy production	Whether act as invasive species should be ascertained (Quinn et al., 2010)
2.	Phytoremediation potential (Nsanganwimana et al., 2014)	Good biomass, require a wait for 2–3 years (Zub et al., 2012)	The bioethanol can be used as fuel and give money	
3.	Good lignocellulose and hemicellulose content that are base for bioethanol (Byrt et al., 2011)	In Asian regions, there is difficulty in growing *Miscanthus*	Bedding material for animals and also as transportation material	
4.	Less chance of disease and pest infection		Site restoration	
5.	It is a carbon neutral grass		Less inputs in production	
6.	Nesting ground for fauna		Heat and energy (Brosse et al., 2012) Conserving biodiversity Jørgensen (2011)	

10 Conclusion

Miscanthus is an important C4 grass that has good biomass and grows well without any maintenance. *Miscanthus* is both an energy crop and phytoremediator. *Miscanthus* also provides habitat to various organisms. Thus, it is an important plant which can be solution to the emerging problem of pollution and habitat destruction. The increasing contamination and ruthless cutting of tress has resulted in huge loss of both flora and fauna. The problem of global warming is another issue with which the whole world is struggling and in this regard *Miscanthus* is a boon to solve the problem. Therefore, *Miscanthus* can be suggested for remediation and management of polluted sites with multiple benefits. Habitat destruction is causing a great loss to the species diversity and the *Miscanthus* with its good canopy also provides a habitat for various organisms and in a way restoring the habitat and thus giving the lost home of various organisms.

Acknowledgment

Financial assistance given to Dr. V.C. Pandey under the Scientist's Pool Scheme (Pool No. 13 (8931-A)/2017) by the Council of Scientific and Industrial Research, Government of India is gratefully acknowledged. The authors declare no conflicts of interest.

References

Al Souki, K.S., Louvel, B., Douay, F., Pourrut, B., 2017. Assessment of *Miscanthus* × *giganteus* capacity to restore the functionality of metal-contaminated soils: ex situ experiment. Appl. Soil Ecol. 115, 44–52.

Anderson, E., Arundale, R., Maughan, M., Oladeinde, A., Wycislo, A., Voigt, T., 2011. Growth and agronomy of *Miscanthus* × *giganteus* for biomass production. Biofuels 2 (1), 71–87.

Anderson, E.K., Lee, D., Allen, D.J., Voigt, T.B., 2015. Agronomic factors in the establishment of tetraploid seeded *Miscanthus* × giganteus. GCB Bioenergy 7 (5), 1075–1083.

Andersson, N.J., 1855. Om de med Saccharumbeslägtade genera. Öfversigtaf Förhandlingar: Kongliga. Svenska Vetenskaps-Akademien 12, 151–168.

Arduini, I., Ercoli, L., Mariotti, M., Masoni, A., 2006. Response of *Miscanthus* to toxic cadmium applications during the period of maximum growth. Environ. Exp. Bot. 55 (1–2), 29–40.

Babu, A.G., Shea, P.J., Sudhakar, D., Jung, I.B., Oh, B.T., 2015. Potential use of *Pseudomonas koreensis* AGB-1 in association with *Miscanthus sinensis* to remediate heavy metal(loid)-contaminated mining site soil. J. Environ. Manage. 151, 160–166.

Barbosa, B., Boléo, S., Sidella, S., Costa, J., Duarte, M.P., Mendes, B., Cosentino, S.L., Fernando, A.L., 2015. Phytoremediation of heavy metal-contaminated soils using the perennial energy crops *Miscanthus* spp. and *Arundo donax* L. Bioenergy Res. 8 (4), 1500–1511.

Bennetzen, J.L., 2009. The future of maize. In: Handbook of Maize. Springer, New York, NY, pp. 771–779.

Brosse, N., Dufour, A., Meng, X., Sun, Q., Ragauskas, A., 2012. *Miscanthus*: a fast-growing crop for biofuels and chemicals production. Biofuels Bioprod. Bioref. 6 (5), 580–598.

Byrt, C.S., Grof, C.P., Furbank, R.T., 2011. C4 Plants as biofuel feedstocks: optimising biomass production and feedstock quality from a lignocellulosic perspective free access. J. Integr. Plant Biol. 53 (2), 120–135.

Caslin, B., Finnan, J., Easson, L., 2010. Miscanthus Best Practice Guidelines. Agriculture and Food Development Authority, Teagasc, and Agri-Food and Bioscience Institute, Hillsborough, Northern Ireland.

Collura, S., Azambre, B., Finqueneisel, G., Zimny, T., Weber, J.V., 2006. Miscanthus× giganteus straw and pellets as sustainable fuels. Environ. Chem. Lett. 4 (2), 75–78.

DeBruyn, J., 2015. Livestock farm uses of switchgrass and Miscanthus. In: Presentation: Environmental Management Branch, OMAFRA, at the OBPC Ag Biomass Day (27th March).

Deuter, M., 2000. Breeding approaches to improvement of yield and quality in Miscanthus grown in Europe. EMI Project, Final report, pp. 28–52.

DOE, 2009. U.S. Energy Consumption by Source US Energy Information Administration. Available from: http://www.eia.doe.gov/cneaf/alternate/page/renew_energy_consump/table1.html.

DOE, 2010. 2010 Annual Energy Outlook (DOE/EIA 0383(2010)) U.S. Department of Energy, Energy Information Administration, Office of Integrated Analysis and Forecasting, Washington, DC. Available from: http://www.eia.doe.gov/oiaf/aeo/.

Dražić, G., Milovanović, J., Stefanović, S., Petrić, I., 2017. Potential of Miscanthus × giganteus for heavy metals removing from Industrial deposol. Acta Regionalia et Environmentalica 14 (2), 56–58.

Eisentraut, A., 2010. Sustainable Production of Second-generation Biofuels: Potential and Perspectives in Major Economies and Developing Countries: Information Paper. OECD/IEA.

Firmin, S., Labidi, S., Fontaine, J., Laruelle, F., Tisserant, B., Nsanganwimana, F., Pourrut, B., Dalpé, Y., Grandmougin, A., Douay, F., Shirali, P., 2015. Arbuscular mycorrhizal fungal inoculation protects Miscanthus × giganteus against trace element toxicity in a highly metal-contaminated site. Sci. Total Environ. 527, 91–99.

Fischer, G., Hizsnyik, E., Prieler, S., van Velthuizen, H., 2007. Assessment of biomass potentials for biofuel feedstock production in Europe: Methodology and results. REFUEL project report, July, Available from: https://refuel.eu/.

Formánek, P., Ambus, P., 2004. Assessing the use 13C natural abundance in separation of root and microbial respiration in a Danish beech (Fagus sylvatica L.) forest. Rapid Commun. Mass Spec. 18, 897–902.

Gerhardt, K.E., Huang, X.D., Glick, B.R., Greenberg, B.M., 2009. Phytoremediation and rhizoremediation of organic soil contaminants: potential and challenges. Plant Sci. 176 (1), 20–30.

Gilman, E., 1999. Miscanthus sinensis. Fact Sheet FPS-405. University of Florida. Cooperative Extension Service, Institute of Food and Agricultural Services.

Gordana, D., Jelena, M., Jela, I., Ivana, P., 2017. Influence of fertilization on Miscanthus × giganteus (Greef et Deu) yield and biomass traits in three experiments in Serbia. Plant Soil Environ. 63 (4), 189–193.

Gubišová, M., Gubiš, J., Žofajová, A., Mihálik, D., Kraic, J., 2013. Enhanced in vitro propagation of Miscanthus × giganteus. Ind. Crops Prod. 41, 279–282.

Guse, W.E., Larsen, F.E., 1975. Propagating herbaceous plants from cuttings. PNW Bull Pac Northwest Coop Ext.

Heaton, E.A., Dohleman, F.G., Miguez, A.F., Juvik, J.A., Lozovaya, V., Widholm, J., Zabotina, O.A., McIsaac, G.F., David, M.B., Voigt, T.B., Boersma, N.N., 2010. Miscanthus: a promising biomass crop. In: Advances in Botanical Research (56), Academic Press, pp. 75–137.

Heaton, E.A., Flavell, R.B., Mascia, P.N., Thomas, S.R., Dohleman, F.G., Long, S.P., 2008. Herbaceous energy crop development: recent progress and future prospects. Curr. Opin. Biotechnol. 19 (3), 202–209.

Hirimburegama, K., Gamage, N., 1995. Propagation of *Bambusa vulgaris* (yellow bamboo) through nodal bud culture. J. Hortic. Sci. 70 (3), 469–475.

James, G. (Ed.), 2008. Sugarcane. John Wiley & Sons.

Jeżowski, S., Mos, M., Buckby, S., Cerazy-Waliszewska, J., Owczarzak, W., Mocek, A., Kaczmarek, Z., McCalmont, J.P., 2017. Establishment, growth, and yield potential of the perennial grass *Miscanthus* × *giganteus* on degraded coal mine soils. Front. Plant Sci. 8, 726.

Jones, D.L., Williamson, K.L., Owen, A.G., 2006. Phytoremediation of landfill leachate. Waste Manag. 26 (8), 825–837.

Jones, M.B., Walsh, M. (Eds.), 2001. *Miscanthus* for energy and fibre. Technology and engineering, Earthscan, p. 192.

Jørgensen, U., 2011. Benefits versus risks of growing biofuel crops: the case of *Miscanthus*. Curr. Opin. Environ. Sustain. 3 (1–2), 24–30.

Kiesel, A., Lewandowski, I., 2017. *Miscanthus* as biogas substrate–cutting tolerance and potential for anaerobic digestion. GCB Bioenergy 9 (1), 153–167.

Kim, H.S., Zhang, G., Juvik, J.A., Widholm, J.M., 2010. *Miscanthus* × *giganteus* plant regeneration: effect of callus types, ages and culture methods on regeneration competence. GCB Bioenergy 2 (4), 192–200.

Kim, R.Y., Yoon, J.K., Kim, T.S., Yang, J.E., Owens, G., Kim, K.R., 2015. Bioavailability of heavy metals in soils: definitions and practical implementation—a critical review. Environ. Geochem. Health 37 (6), 1041–1061.

Krzyżak, J., Pogrzeba, M., Rusinowski, S., Clifton-Brown, J., McCalmont, J.P., Kiesel, A., Mangold, A., Mos, M., 2017. Heavy metal uptake by novel miscanthus seed-based hybrids cultivated in heavy metal contaminated soil. Civil Environ. Eng. Rep. 26 (3), 121–132.

Korzeniowska, J., Stanislawska-Gulbiak, E., 2015. Phytoremediation potential of *Miscanthus* × *giganteus* and *Spartina pectinata* in soil contaminated with heavy metals. Environ. Sci. Pollut. Res. 22 (15), 11648–11667.

Lewandowski, I., Clifton-Brown, J.C., Scurlock, J.M.O., Huisman, W., 2000. *Miscanthus*: European experience with a novel energy crop. Biomass Bioenergy 19 (4), 209–227.

Lewandowski, I., Kahnt, G., 1993. Development of a tissue culture system with unemerged inflorescences of *Miscanthus* 'Giganteus' for the induction and regeneration of somatic embryoids. Beitr. Biol. Pflanzen. 67, 439–451.

Macková, M., Macek, T., 2005. Využití rostlin k eliminaci xenobiotik z životního prostředí. VŠCHT, ÚOCHB.

Mangold, A., Lewandowski, I., Xue, S., Kiesel, A., 2018. 'Collar propagation'as an alternative propagation method for rhizomatous *Miscanthus*. GCB Bioenergy 10 (3), 186–198.

Megharaj, M., Ramakrishnan, B., Venkateswarlu, K., Sethunathan, N., Naidu, R., 2011. Bioremediation approaches for organic pollutants: a critical perspective. Environ. Int. 37 (8), 1362–1375.

Meyer, F., Wagner, M., Lewandowski, I., 2017. Optimizing GHG emission and energy-saving performance of *Miscanthus*-based value chains. Biomass Conv. Bioref. 7 (2), 139–152.

NEF, 2014. CeedsTM-the easy way to establish energy crops. Available from: http://newenergyfarms.com/site/ceeds.html.

Nielsen, J.M., Brandt, K., Hansen, J., 1993. Long-term effects of thidiazuron are intermediate between benzyladenine, kinetin or isopentenyladenine in *Miscanthus sinensis*. Plant Cell Tissue Organ Cult. 35 (2), 173–179.

Nsanganwimana, F., Pourrut, B., Mench, M., Douay, F., 2014. Suitability of *Miscanthus* species for managing inorganic and organic contaminated land and restoring ecosystem services. A review. J. Environ. Manag. 143, 123–134.

Pandey, V.C., 2017. Managing waste dumpsites through energy plantations. In: Bauddh, K., Singh, B., Korstad, J. (Eds.), Phytoremediation Potential of Bioenergy Plants. Springer, Singapore, pp. 371–386.

Pandey, V.C., Bajpai, O., 2019. Phytoremediation: From Theory Toward Practice. In: Pandey, V.C., Bauddh, K. (Eds.), Phytomanagement of Polluted Sites. Elsevier, Amsterdam Netherlands, pp. 1–49.

Pandey, V.C., Bajpal, O., Singh, N., 2016. Energy crops in sustainable phytoremediation. Renew. Sust. Energ. Rev. 54, 58–73.

Pandey, V.C., Pandey, D.N., Singh, N., 2015a. Sustainable phytoremediation based on naturally colonizing and economically valuable plants. J. Clean. Prod. 86, 37–39.

Pandey, V.C., Bajpai, O., Pandey, D.N., Singh, N., 2015b. *Saccharum spontaneum*: an underutilized tall grass for revegetation and restoration programs. Genet. Resour. Crop Evol. 62 (3), 443–450.

Pandey, V.C., Rai, A., Korstad, J., 2019. Aromatic crops in phytoremediation: from contaminated to waste dumpsites. In: Pandey, V.C., Bauddh, K. (Eds.), Phytomanagement of Polluted Sites. Elsevier, Amsterdam, Netherlands, pp. 255–275.

Pandey, V.C., Singh, V., 2019. Exploring the potential and opportunities of current tools for removal of hazardous materials from environments. In: Pandey, V.C., Bauddh, K. (Eds.), Phytomanagement of Polluted Sites. Elsevier, Amsterdam, Netherlands, pp. 501–516.

Pandey, V.C., Souza-Alonso, P., 2019. Market opportunities in sustainable phytoremediation. In: Pandey, V.C., Bauddh, K. (Eds.), Phytomanagement of Polluted Sites. Elsevier, Amsterdam, Netherlands, pp. 51–82.

Pidlisnyuk, V., Erickson, L., Stefanovska, T., Popelka, J., Hettiarachchi, G., Davis, L., Trögl, J., 2019. Potential phytomanagement of military polluted sites and biomass production using biofuel crop *Miscanthus* × *giganteus*. Environ. Poll. 249, 330–337.

Placek, A., Grobelak, A., Włóka, D., Kowalska, A., Singh, B.L., Almas, A.R., Kacprzak, M., 2018. Methods for calculating carbon sequestration in degraded soil of zinc smelter and post-mining areas. Desalin. Water Treat. 134, 233–243.

Purdy, S.J., Cunniff, J., Maddison, A.L., Jones, L.E., Barraclough, T., Castle, M., Davey, C.L., Jones, C.M., Shield, I., Gallagher, J., Donnison, I., 2015. Seasonal carbohydrate dynamics and climatic regulation of senescence in the perennial grass, *Miscanthus*. BioEnergy Res. 8 (1), 28–41.

Quinn, L.D., Allen, D.J., Stewart, J.R., 2010. Invasiveness potential of *Miscanthus* sinensis: implications for bioenergy production in the United States. GCB Bioenergy 2 (6), 310–320.

Rascio, N., Navari-Izzo, F., 2011. Heavy metal hyperaccumulating plants: how and why do they do it? And what makes them so interesting? Plant Sci. 180 (2), 169–181.

Rauscher, B., Lewandowski, I., 2016. *Miscanthus* horse bedding compares well to alternatives. In: Perennial Biomass Crops for a Resource-Constrained World, Springer, Cham, pp. 297–305.

Sharmin, S.A., Alam, I., Kim, K.H., Kim, Y.G., Kim, P.J., Bahk, J.D., Lee, B.H., 2012. Chromium-induced physiological and proteomic alterations in roots of *Miscanthus* sinensis. Plant Sci. 187, 113–126.

Stevens, O.A., 1966. Rhizomes, stolons and roots. Castanea 31 (2), 140–145, https://www.jstor.org/stable/i375990.

Sun, Q., Lin, Q., Yi, Z.L., Yang, Z.R., Zhou, F.S., 2010. A taxonomic revision of *Miscanthus* s.l. (Poaceae) from China. Bot. J. Linn. Soc. 164 (2), 178–220.

Swaminathan, K., Alabady, M.S., Varala, K., De Paoli, E., Ho, I., Rokhsar, D.S., Arumugana-than, A.K., Ming, R., Green, P.J., Meyers, B.C., Moose, S.P., 2010. Genomic and small RNA sequencing of *Miscanthus* × *giganteus* shows the utility of sorghum as a reference genome sequence for *Andropogoneae* grasses. Genome Biol. 11 (2), R12.

Techer, D., D'Innocenzo, M., Laval-Gilly, P., Henry, S., Bennasroune, A., Martinez-Chois, C., Falla, J., 2012. Assessment of *Miscanthus* × *giganteus* secondary root metabolites for the biostimulation of PAH-utilizing soil bacteria. Appl. Soil Ecol. 62, 142–146.

USDA Forest Service, Forest Health Staff. 2006. Chinese Silvergrass-*Miscanthus* sinensis An-derss. Weed of the Week.

Van Weyenberg, S., Ulens, T., De Reu, K., Zwertvaegher, I., Demeyer, P., Pluym, L., 2015. Fea-sibility of *Miscanthus* as alternative bedding for dairy cows. Vet. Med. 60 (3).

Waldmann, D., Thapa, V., Dahm, F., Faltz, C., 2016. Masonry blocks from lightweight concrete on the basis of *Miscanthus* as aggregates. In: Perennial Biomass Crops for a Resource-Constrained World, Springer, Cham, pp. 273–295.

Wanat, N., Austruy, A., Joussein, E., Soubrand, M., Hitmi, A., Gauthier-Moussard, C., Lenain, J.F., Vernay, P., Munch, J.C., Pichon, M., 2013. Potentials of *Miscanthus*× *giganteus* grown on highly contaminated Technosols. J. Geochem. Explor. 126, 78–84.

Watson, L., Dallwitz, M.J., 1992. The Grass Genera of the World. CAB international. Cabdi-rectorg, pp. 1038.

Wilson, S.B., Knox, G.W., 2006. Landscape performance, flowering, and seed viability of 15 Japanese silver grass cultivars grown in northern and southern Florida. Hort. Technol. 16 (4), 686–693.

Xue, S., Kalinina, O., Lewandowski, I., 2015. Present and future options for *Miscanthus* propa-gation and establishment. Renew. Sustain. Energ. Rev. 49, 1233–1246.

Zub, H.W., Arnoult, S., Younous, J., Lejeune-Hénaut, I., Brancourt-Hulmel, M., 2012. The frost tolerance of *Miscanthus* at the juvenile stage: differences between clones are influenced by leaf-stage and acclimation. Eur. J. Agron. 36 (1), 32–40.

Phragmites species—promising perennial grasses for phytoremediation and biofuel production

Vimal Chandra Pandey[a],[*], Deblina Maiti[b]
[a]Department of Environmental Science, Babasaheb Bhimrao Ambedkar University, Lucknow, Uttar Pradesh, India; [b]Central Institute of Mining and Fuel research, Dhanbad, Jharkhand, India
[*]Corresponding author

1 Introduction

Phragmites is a genus of perennial grasses from *Poaceae* family in which the widely known species is *Phragmites australis*, commonly called common reed. In total, the genus includes five species namely *P. frutescens* H. Scholz, *P. japonicus* Steud., *P. karka* (Retz.) Trin. ex Steud., *P. australis* (Cav.) Trin. ex Steud., and *P. mauritianus* Kunth. All the species are morphologically very similar but show minute cytological variation (Clevering and Lissner, 1999). The lack of difference in the genetic structure among different *Phragmites* species indicates that there persists substantial reproductive contact among them through cross-pollination; added to this, the activities such as seed dispersal by wind, migratory birds, and human, rafting of eroded plant fragments such as rhizomes may also aid to the consequence of minute genetic differences among the plant species. However, the plants have restricted geographical distributions, from the tropics till the cold temperate areas across the globe; while *P. australis* is widely distributed in temperate and subtropical areas (Lambertini et al., 2012). From decades, immense research is being carried out globally to develop efficient techniques to mitigate the impacts of pollution on terrestrial land and improve the already degraded areas. In this context, *Phragmites* has proven its effectivity to reduce the environmental degradation in its surroundings (Srivastava et al., 2014). A lot of literature shows that *P. australis* plant has the capability to grow massively at extreme environmental conditions like high CO_2 and temperature due to its ability to incur C3 or C4 cycle as and when required, root interaction with a huge diversity of microbes and multiple biochemical adaptations to the pollutants present in the soil. Thus it is also a preferred plant for ecological restoration of degraded lands. The following sections will review the present state of research conducted till now in connection to the suitability of *Phragmites* for environmental remediation with emphasis on *P. australis*. The review will also deal with topics which will help in knowing the grass in detail.

Phytoremediation Potential of Perennial Grasses. http://dx.doi.org/10.1016/B978-0-12-817732-7.00005-5

2   General aspects of *Phragmites* species

Taxonomy—"*Phragmites*" genus name is obtained from a Greek word "phragma" which denotes the zonation seen by the plant's growth along the water bodies. As stated earlier, the genus *Phragmites* consists of five species, out of which *P. australis* sub species like *P. australis* (Cav.) Trin. ex Steud is widely spread in temperate regions of both hemispheres and have a diploid chromosome number of 36, 48, 54, 84, and 96 in its cells; while another sub species *P. australis* altissimus (Benth.) Clayton having a diploid chromosome number of 36, 48, and 96 are found to be present on the shores of Mediterranean region and extending to Iran and southwards to Arabia, Ethiopia, Kenya as well as Sahara. Unlike the genus name "*Phragmites,*" the species name "*australis*" is a Latin word meaning southern as it was found growing in warmer regions of the globe (Batterson and Hall, 1984). *P. australis* is also referred to as *P. communis* Trin. *P. karka* having a diploid chromosome number 18, 36, 48 is usually found in tropical Asia, northern Australia, Ethiopia, Sudan, and West Africa. The diploid number of *P. mauritianus* and *P. japonicus* is 48 which are distributed in Tropical Africa, Mascarene Islands, Ethiopia, Sudan, Congo, Japan, and China, respectively (Shaltout et al., 2006).

Ecology—The plant *Phragmites* either grows in clumps or fully covers any fresh, brackish, or disturbed wetlands, littoral zones of close water bodies, springs, ditches, mined areas, and wastelands. Optimum plant growth is observed in mineral clays, soil rich in organic matter, and alkaline habitats. It can tolerate moderate salinity and soil conductivity up to 12 mS/cm, soil pH of 5–9, annual temperature of 7–27°C, annual precipitation of 3–24 dm, and water level up to a height and depth of 15 cm below or above the soil surface, respectively (Serag, 1996). Therefore, the plant also plays a role in bioindicating the water supply in a place. Seedling germination and establishment of the plant is rare because the plant seedlings show poor growth in unfavorable conditions, such as low light, temperature, phosphate, flooding, drought, and salinity (Karunaratne et al., 2003). However, vegetative reproduction of the plant is extensive and it rapidly grows by means of underground rhizomes and aboveground stolons which may reach a length of more than 10 m; the developed aerial shoots and culms are > 1 m tall; density of the culms vary from 13 to 125 per m^2; aboveground biomass ranges from 0.7–3.7 kg dry weight per m^2 which is far more than the biomass of other wetland plants. The root biomass is mostly 6 times greater than the shoot biomass and varies around 1.4–7.2 kg dry weight per m^2 (Windham, 2001). Shoot height and biomass increases from June to October, while the shoot density indicates decreases from June to October; which shows that shoot height and standing crop biomass are positively correlated while shoot density or productivity is negatively correlated with the earlier-mentioned parameters (Shaltout et al., 2004). Plant growth rate and aboveground height are correlated to the width of the emergent bud and finally determines the plant's basal diameter. The plant grows rapidly in warm, humid habitats, or in early bud stage and when buds with the greatest diameter are arising from. Generally, the buds burst in spring; however the growth characteristics depend on the climatic temperature present in different geographic regions. Plants growing in higher latitudes have larger growth rate and period as well as show an early flowering time. The aboveground biomass of the plant is enriched with silica which makes the leaves and culms stiff which indirectly helps the plant to protect from mechanical damage or from consumers (Clevering et al., 2001). The plant has an average lifetime of 4–6 years, and due to its growth abilities through rhizome strategy, the stands can survive for more than decades. Usually, the dried standing biomass stay firm for 2 years due to presence of silica in the aboveground biomass; the leaves or the *Phragmites* litter decomposes rapidly and helps in silica cycling and availability along with other nutrients which had been sequestered earlier in the biomass (Struyf et al., 2007). The plant is indirectly capable of removing huge amounts of nitrogen contaminated into surface waters or from the soil due to its greater biomass compared to other wetland plants (Gonzalez-Alcaraz et al., 2012). In return, the plant also gets the required nutrients for its growth; dissolved organic nitrogen, amino acids, and urea nitrogen are assimilated more rapidly as the plant has high affinity for this form of nitrogen. On the other way round, the plant growth increases with nutrient supply and eutrophication of a water body; while the mineral nutrition is mainly obtained from the sediment part mud of the water body (Mozdzer et al., 2010). The young leaves of is like C4 species but gradually the mature leaves fixes carbon by C3 mechanism. Most importantly, in terms of soil fertility development, the plant roots can oxygenate the rhizosphere efficiently. Its growth is often associated with common species, such as *Aeluropus lagopoides, Amaranthus ascendens, Arthrocnemum macrostachyum, Atriplex cane-scens, Aster squamatus, Azolla filiculoides, Bassia indica, Ceratophyllum demer-sum, Chenopodium* sp., *Convolvulus arvensis, Cynodon dactylon, Halocnemum strobilaceum, Imperata cylindrica, Inula crithmoides, Juncus acutus, Lemna gibba, Limbarda crithmoides, Limonium pruinosum,Malva parviflora, Mentha longifolia, Pluchea dioscoridis, Polypogon monspliensis, Sonchus oleraceus, Rumex dentatus, Silybum marianum,* and *Typha domingensis* (Shaltout et al., 2006; Kiviat, 2013; Srivastava et al., 2014).

Origin and geographical distribution—The preserved remains of *Phragmites* found in the United States and Colorado, dating a time 40,000 years ago suggests that the plant is native to this region. Preserved rhizome fragments dating back to 3000–4000 years in salt marsh sediments of Netherlands also indicated that the plant is native to this place. Further in late 1700s, European *Phragmites* was introduced in North America as shipping and ballast material. Gradually in 20th century, the plant occupied the Atlantic coast. In some places across the globe, the plant was intention-ally introduced for wastewater treatment in lagoons, erosion control, and stabilization of shorelines. It is the most widely distributed plant in the world being as it is found throughout America, Europe, Africa, Asia, and Australia (Fig. 5.1); it is also abun-dantly found growing in temperate regions, as well as sub-tropical areas (Batterson and Hall, 1984).

Botanical description—The dense *Phragmites* clonal stands are made up of both living stems and standing dead stems; round and hollow and are usually green with yellow nodes during the growing season. Fig. 5.2 shows *Phragmites* stands on the experimental field site. The stems turn yellow when dry in the winter. The hollow morphology of the stems help in transporting air to the plant parts when growing in

Figure 5.1 The distribution of *Phragmites* across the countries.

Figure 5.2 Close view of planted *Phragmites australis* on the experimental field site.
Source: Photo credits by V.C. Pandey.

water. The leaves are blue-green to yellow-green in color, have a flat shape, and have a length up to 20 inches and width up to 1.5 inches at the widest point which taper at the end. The leaves are attached to the stem through loose smooth sheaths. Flowers are observed in late July and August purple to gold in color and borne on branched inflorescences, with a length of 20–60 cm; in which 1–6 flowers are grouped into spikelets of 1–1.5 mm length. Fig. 5.3 represents the inflorescence of *Phragmites australis*. Root growth of the plant is profuse and grows up to a depth of more than 10 feet in one growing season. *Phragmites* spreads by horizontal above ground stolon and underground rhizome (Fig. 5.4). Thousands of seeds are produced by each plant and they appear grayish and fluffy due to presence of silky hairs (Shaltout et al., 2006; Kiviat, 2013; Srivastava et al., 2014).

Habitat and propagation—As stated earlier, *Phragmites* can grow efficiently in roadside ditches, wetlands, freshwater and brackish marshes, moderately saline water bodies, rivers, lakes, pond edges, and degraded areas. It prefers to grow in full sun and can show its invasive nature, through massive growth to alter habitats and form monocultures. The plant can compete with other plants to completely cover the area

Figure 5.3 Inflorescence of *Phragmites australis*.
Source: Photo credits by V.C. Pandey.

Figure 5.4 Rhizome and Stolon of *Phragmites* australis.
Source: Photo credits by V.C. Pandey.

by its aboveground as well as belowground biomass. Gradual growth of the plant in degraded lands and water bodies can actually improve the characteristics, that is, by decreasing the salinity in brackish wetlands and the changing local topography or marsh hydrology (Shaltout et al., 2006; Srivastava et al., 2014).

Colonization of new sites by extensive colonies of *Phragmites* occurs through seed dispersal. Seed dispersal of *P. australis* by humans has been reported in North America from Europe (Paul et al., 2010; Saltonstall, 2011). Even though the seed viability is very less, but the plant propagates massively through creeping rhizomes and stolons which can grow up to 13 m from the parent plant. Additionally, *Phragmites* colonies are perennial; however, life cycle of a single *Phragmites* plant is up to 8 years. It has also been reported that two lineages of a species of the plant can interbreed and produce intermediate genetic patterns (Paul et al., 2010). This particular observation has been noted in case of *P. australis* as it is a widespread species and distantly related genotypes of the plant can coexist (Saltonstall, 2011).

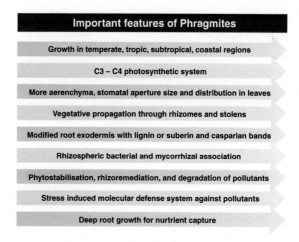

Figure 5.5 Adaptive features of *Phragmites*.

3 Important features of *Phragmites* species

The invasive nature of *Phragmites* is due its various adaptive features throughout its life cycle. Various important features of the plant have been summarized in Fig. 5.5. It colonizes disturbed lands by its seeds and vegetatively captures huge acres of land by its extensive rhizomes and stolons (Marks et al., 1994). It shows apical dominance and mostly found colonizing wetland habitats; the plant develops a lot of aerenchyma tissue whose volume in its stem and rhizomes varies from 33% to 50%. Added to the several features, it has a high stomatal density of 406 stomates per mm^2 on the adaxial surface and 633 stomates per mm^2 on the abaxial surface of the leaf. The plant produces huge stem density in 2–3 lifecycle as the reported number of total erect stems including living as well as dead is > 160 stems per m^2 (Ailstock and Center, 2000).

4 Multiple uses and management consideration

Various uses of *Phragmites* species in regard to the ecosystem services has been described later (Table 5.1), which show that this perennial grass can be effectively used to remediate various types of pollutants from degraded areas along with development of a sustainable ecosystem on them.

4.1 Phytoremediation

Phytoremediation is a technique which involves plants to remove contaminants like inorganic metals or organic pollutants like pesticides and other harmful chemicals from the environment (Pandey et al., 2015; Pandey and Bajpai, 2019). It is a suitable treatment procedure and experimentally involves the need to identify potential

Table 5.1 List of ecosystem services provided by *Phragmites* species.

Categories	Service provided	References
***Regulating* services**		
Ecological restoration	Recommended for restoration, found to develop habitat in mine degraded lands, produces a spontaneous vegetation cover on waste lands	Nawrot and Yaich (1982); Kiviat (2013); Kobbing et al. (2013)
Waste remediation	Phytoremediation/Rhizoremediation of metal(loid)s and other contaminants like organic from waste dumpsites and contaminated soil	Rungwa et al. (2013); Cui et al. (2015); McSorley et al. (2016); Fernandez et al. (2016, 2017); Su et al. (2018); Nayyef Alanbary et al. (2018)
Soil formation	Increases soil fertility, stops soil erosion	Soukup et al. (2002); Rooth et al. (2003); Dibble and Meyerson (2012)
Carbon sequestration	Has high carbon capture potential	Walters (2010); Ailstock and Center (2000)
Climate regulation	Evapotranspiration	Brix et al. (2001)
Water treatment	Highly efficient in removal of contaminants from water in a small period and improves water quality	Uka et al. (2013); Cui et al. (2015); Lyu et al. (2017); Bello et al. (2018); Soto-Rios et al. (2018)
Provisioning services		
Energy/biofuel	Compressed biomass bricks/pellets are burnt in furnace to generate energy and used for methane production	Gilson (2017)
Food	Sugar production	Denes et al. (2012)
Pharmaceuticals	Used for the treatment of carcinoma, arthritis, bronchitis, cholera, diabetes, hemorrhage, jaundice, and typhoid. Tested for production of hallucinogen	Schultes et al. (2001); Boulos (2005); Srivastava et al. (2014)
Fiber	Used for fencing, production of paper pulp, and packing material	Brix et al. (2014); Kraiem et al. (2013)
Cultural services		
Esthetic	Nest for bees, shelter for amphibians and reptiles, supports ecotourism	Kiviat (2013)
Ceremonial objects	Used for various ceremonies in Arizona, preparation of prayer sticks	Kiviat (2013)
Miscellaneous	Used as fishing poles, crafting material, thatching material, preparation of arrow shafts, weaving rods, mats, and nets	Kobbing et al. (2013)

plants which can tolerate the toxicity of the pollutants while also accumulating them or stabilizing them in the rhizosphere (Pandey et al., 2009). These plants can also be implemented on a full scale as an effective remedy for degraded lands. Various experimental results have proved *Phragmites* as a potential plant to remediate minesoil, metal contaminated soil, estuaries, wetlands, saltmarshes, and wastewater.

Heavy metals—Uka et al. (2013) reported that *P. karka* could uptake metals in the order Cu > Zn > Cd > Pb but efficiently stabilized the metals (mostly toxic metals, Cd and Pb) in the root system as reported by the bioconcentration and translocation factors from their study on polluted sites of a river in Nigeria. Native *P. karka* was also found to be a potential remediator and metal absorber from mine soil in Papua New Guinea (Rungwa et al., 2013). In another pilot plant experiment done by Cicero-Fernandez et al. (2016, 2017) on estuarine sediments, *P. australis* differentially showed phytoextraction and phytostabilization for metals, such as Co, Ni, Mo, Cd, Pb, Cr, Cu, Fe, Mn, Zn, Hg, As, Se, and Ba. The plant acts as a metal excluder for certain metals while for certain metals, the concentration increases along with the time. The studies reported that the plant can be used as a biomonitoring tool and efficient in remediating estuarine sediments. It has also been reported that bioaugmentation of metal contaminated saltmarshes with metal (like Cu and Cd), resistant microbes can increase the potential of metal uptake in the roots of *P. australis* and thus help in remediated moderately polluted estuaries (da Silva et al., 2014; Oliveira et al., 2014). It has been reported to be a hyperaccumulator for many metals like Cu from metal contaminated soil (Su et al., 2018). Native *P. australis* can also accumulate significant amounts of chloride (approximately 65 kg/km^2) from soil contaminated by cement kiln dust and remediate such landfills within 3–9 years by its phytoextraction potential (McSorley et al., 2016).

Organic pollutants—*Phragmites* is of the best plants which have the highest potential of remediation of pollutants from contaminated soil or wastewater. *P. karka* showed its ability for prolonged phytoremediation of sand contaminated with hydrocarbons present in crude oil sludge (Nayyef Alanbary et al., 2018). *P. australis* was reported to detoxify many harmful agrochemicals including xenobiotics and herbicides. Davies et al. (2005) also reported that the crude plant extract of *P. australis* could degrade azo dyes and aromatic amines. Reportedly, the plant induces conjugation of the compounds and shows induction of detoxification enzymes in its cells (Schroder et al., 2005). Sung et al. (2013) reported that addition of humic acid into wetland soil contaminated with total petroleum hydrocarbons and heavy metals can increase the remediation potential of *P. communis* and also prevent contaminant leaching. Sauvetre and Schroder (2015) reported that rhizobacteria, such as *Rhizobium daejeonense, Diaphorobacter nitroreducens, Achromobacter mucicolens, Pseudomonas veronii,* and *Pseudomonas lini* help in removal of recalcitrant drugs like carbamazepine by *P. australis* along with boosting plant growth, siderophore production, and phosphate solubilization. Low concentration of phenol in the range of 50 mg/L stimulated the growth of *P. australis* (Hubner et al., 2000). Toyama et al. (2009) reported that bacterial species like *Novosphingobium* and *Sphingobium yanoikuyae* interact with *P. australis* and help in removal of bisphenols from soil. Growth promoting substance released by the rhizobacteria increase the percentage of seed germination of

P. australis even in the presence of heavy metals like copper and organics like creosote (Reed et al., 2005). Fares et al. (2007) reported that *Phragmites* effectively removes phosphorus from water and also doesn't emit isoprene which is usually emitted by aquatic plants; while also isoprene biosynthesis requires large amounts of phosphorus compounds usually found in the leaves. Yanyu et al. (2010) reported 88% removal of nitrobenzene from a solution by *P. australis* within 2 days of growth.

Polluted water detoxification—Several studies have shown that *P. australis* can be used in treatment of waste water and can remove 80%–90% of metals like Pb, Cd, and Ni from hydroponic cultures (Bello et al., 2018). Soto-Rios et al. (2018) indicated that biomass of the plant can be used effectively to bio absorb Hg from water in high concentrations; thus cleaning the water comparatively efficiently to the conventional techniques. Cui et al. (2015) reported that *P. australis* can rhizoremediate metformin from polluted wastewater through organic cation transporters present in its roots. The plant has also shown the potential to degrade pesticides in its shoot and root parts; like as reported Tebuconazole and Imazalil from hydroponic cultures (Lyu et al., 2017). On view of such properties, the plant beds have been used as water filters. For example, Badejo et al. (2015) proved the efficiency of vegetated submerged bed constructed wetland for removal of toxic heavy metals like (98% of the Cr) from industrially polluted waste water from a steel manufacturing company. The creeping roots of the plant can also take up nitrates from water. Schroder et al. (2008) demonstrated uptake and detoxification of organic xenobiotics and pharmaceutical drugs, such as albendazole, flubendazole, and tetracycline by *P. australis* from polluted water. The detoxification process in the plant involves biotransformation of the chemicals into less harmless forms of glucosides (Podlipna et al., 2013; Topal, 2015). Vymazal and Brezinova (2016) reported that the amount of heavy metals uptaken in the shoots of the plant can be nearly 70%, calculated as a fraction of the total heavy metal removed. Al-Akeel et al. (2010) has also utilized the metal containing biomass for successful preparation of metallic nanoparticles, which adds value to the process of phytoremediation.

4.2 Ecological restoration

Studies have reported that apart from improving soil properties, stopping erosion, and carbon sequestration, *Phragmites* colonies support the colonization and habitat of various amphibians and reptiles. The later organisms seek a shelter for thermoregulation. The hollow internodes of dead stems are shelter for spiders (Kiviat, 2013). Moreover, ecorestoration programs involve incorporation of plant species which have an end use. In this context, *Phragmites* has shown its use for the production of a multitude of useful products depending on its time of harvest, nutrient as well as water availability, and climate. The various end use are feedstock for bio-fuel, paper production etc. Moreover with increasing industrial development, the demand for raw materials for these products is increasing; which can be mitigated by exploitation of these plant biomasses grown on vast tracts of degraded lands. This will offer a dual advantage of economic development and remediation of soil and water (Kobbing et al., 2013). However, more research is needed in the context of management of the grass growth on degraded lands which can't be used for food production, thus to evolve an economic benefit among the ever increasing demand for energy and biomass.

4.3   Soil formation

Phragmites can build and stabilizes soil and ultimately create a self-sustaining plant cover in areas where other plants are unable to survive. The roots of the plant colonies bind the soil to trap nutrients for its growth and in turn also add organic matter to the soil (Kiviat, 2013). Establishment of a sustainable ecosystem by the *Phragmites* helps in colonization of the areas with various small plants, insects, and smaller vertebrates. Reports have mentioned the plant's ability to build and stabilize tidal marsh soils, along with protection from erosion (Dibble and Meyerson, 2012). *Phragmites* has the ability to aerate the flooded soil and sediments through its rhizosphere and helps other small plants to survive on them which would have otherwise not survived due to anaerobic conditions (Callaway, 1995). Plants like *Phragmites* survive on flooded soil due to mechanisms like internal aeration from aerial parts and leakage of oxygen from its root to the soil (Armstrong and Armstrong, 1988). Hofmann (1990) observed large numbers of aerobic microbes in *Phragmites* rhizosphere compared to sewage sludge; moreover aeration capacity of the plant along with the oxygen released from the plant roots supports the oxidation of NH_4 to NO_3 by bacteria, which would have otherwise escaped as nitrogen gas leading to nitrogen deficit in flooded soils. This ability of *Phragmites*, builds an alternative oxidizing and reducing zones which successfully purifies such systems from contaminants present in them (Armstrong et al., 1992). Soukup et al. (2002) reported that under anaerobic conditions, the root development in *Phragmites* is characterized by modification of the exodermis with lignin or suberin and Casparian bands close to the root tip. This particular characteristic is commonly not observed in other plants.

4.4   Green cover development of derelict lands

The plant has the potential to develop thick large homogeneous beds in wetlands as well as degraded lands. The ratio of tall green stems to the dry stem is often 2:1 ratio. Continuous generation of new plants is also supported by the presence buried seeds, rhizomes, and stolons (Rudrappa et al., 2007; Baldwin et al., 2010). The plant has been reported to be more successful in covering a barren land than any other native species (Weis and Weis, 2004). Owing to the efficient propagation mechanisms of *Phragmites*, it is used for stabilization of sediments, banks as well as in wetland rehabilitation projects. Researchers have reported successful results from revegetation trials of disturbed riparian areas using *Phragmites* (Srivastava et al., 2014). The potential uses of *Phragmites* species for ecological and socioeconomic development are given in Table 5.1.

4.5   Carbon sequestration

Owing to high biomass production potential of the plant compared to other plants, it is capable of storing large amounts of carbon and nitrogen in its tissues which is ultimately helping in carbon sequestration. Brix et al. (2001) mentioned that *P. australis* populations act as carbon sink and storage house for greenhouse gases as it can assimilate CO_2 from the atmosphere through photosynthesis, and accumulate organic matter.

Though on yearly basis, 15% of the net carbon fixed by *Phragmites*, population is released as CH_4 which may be a cause for greenhouse effect yet on a longer time scale of > 100 years, the effect of CH_4 is less compared to CO_2 and thus *Phragmites* population are a sink for greenhouse gases. Adaptations like twice amounts of aerenchyma, size of stomatal aperture, and its distribution on *Phragmites* leaves makes it more productive in capturing CO_2 for photosynthesis compared to other plants (Ailstock and Center, 2000). Walters (2010) reported an average rate of net ecosystem exchange of carbon to be 920 mg $C/m^2/h$ for *P. australis* populations which is higher than any other wetland plants and can help in carbon sequestration as well as decrease global warming.

4.6 Biomass production

Phragmites colonies form dense stands of live as well as dead biomass (Kiviat, 2013). Kuusemets and Lohmus (2005) reported that biomass of *P. australis* can increase from 674 g to 3646 g/m^2 in a year's time and also accumulate nitrogen and phosphorus amount of 84 and 13 g/m^2 in its biomass. Baute et al. (2018) reported a biomass of 1.9 kg dry matter per m^2 which can be treated and used as a feedstock for increased methane yield and biogas production. The overall biogas and methane production is particularly higher from the biomass which is harvested in the month of June than September due to higher amounts of lignocelluloses in biomass harvested in September (Gilson, 2017). Commercialization of biogas/bioenergy production from the plant would be feasible because the biomass generation is suitably rapid even after harvesting it. Hansson and Fredriksson (2004) also reported that the biomass is often compressed into pellets to be burnt in furnaces for energy generation. Biomass production in the plant is not affected by the presence of pollutants, such as ammonia at a level of 21–82 mg/L in a wetland with a water level of 10–15 cm (Hill et al., 1997). However, larval feeding by the moth *Archanara geminipuncta* on the plants causes reduction in its height by 40%, aboveground biomass by 50%, flowering stems by 90%, clonal expansion, and competitive ability (Hafliger et al., 2006). During the growing period, the shoot biomass increases from April to May and rhizome biomass increases from May to August while the root biomass remains stable throughout the growing season. Moreover, the taller shoots show higher growth rate than the shorter shoots (Engloner, 2009). Owing to the high aboveground biomass of *P. australis*, some of the heavy metals like Zn, Cd, and Cr can accumulate at a percentage as high as 50%–70% with respect to total removed heavy metals in constructed wetlands (Vymazal and Brezinova, 2016). Brix et al. (2014) reported a biomass generation of 400,000 metric tons per year from an area of 786 km^2 through extensive management for production of pulp to be used in paper industry. Management strategies for yield increase included water table management, harvesting as well as burning to reduce pests control and irrigation by seawater to remove weeds. Management of such invasive species to generate a useful outcome not only renders ecosystem services but in a broader aspect, it also gives other associated benefits like carbon sequestration, global warming mitigation, and income generation for the local people (Fig. 5.6).

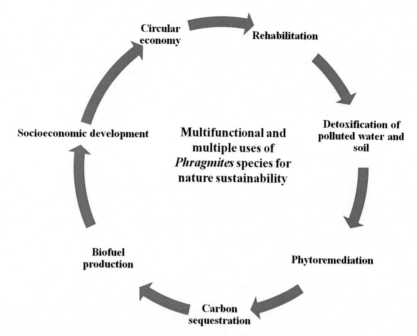

Figure 5.6 Linking *Phragmites* species with phytoremediation and biofuel production for environmental and socioeconomic development.

4.7   Other uses

*P. australis*has been traditionally used for the treatment of carcinoma and leukemia. It has also been used for the treatment of arthritis, bronchitis, cholera, diabetes, hemorrhage, jaundice, and typhoid (Boulos, 2005; Srivastava et al., 2014). Apart from the earlier-mentioned essential characteristics, the plants of this genus have a myriad of other uses like the biomass is used for thatching, as a building material for rafts, generating fodder for livestock, preparation of arrow shafts from stems, construction of mats and nets, and preparation of cellulose for textile industry (Table 5.1). Preparation of cemented reed blocks, insulation materials, and fertilizer has also been reported in the literature (Batterson and Hall, 1984). Edible products from the plant are preparation of alcohol and the young rootstocks as well as the seeds are used as food. Some aesthetic uses from the plants are preparation of prayer sticks for religious ceremonies, making of weaving rods, and musical instruments (Shaltout et al., 2006).

5   Conclusion

The plant's characteristics, such as high biomass, strong root structure, resistance to biotic/abiotic stresses make it a very efficient candidate for polluted soil remediation as well as restoration. From the literature, it was also observed that wastewater treatment by the plant was extensively proven as the plant could show excellent removal properties for nutrients, heavy metals, and other pollutants from wastewaters while also maintaining the oxygen level in it with the help of root-zone microbes. The up taken pollutants in the plant body are successfully degraded and complexed by its robust enzymatic systems. Advantageously, it can adapt to extreme environmental conditions while also offering alternative source of its biomass use for many commercial environmental benefits.

6 Future perspectives

The literature suggests that, there is still a gap in research in spite of the useful characters of the plant, for example, studies concerning management of wastelands, mine dumps, municipal solid waste dumps, fly ash dumps and ponds, and shorelines using the mass cultivation of this grass are needed. Studies are lacking which would ascertain the pollution removal capacity of the plant in presence of useful microbial consortia and at different climatic zones. Moreover, molecular level studies are a very interesting part in this case, which includes elucidating the mechanisms and role of the plant cellular machinery in degrading or complexing the pollutants inside the plant body; thus making the pollutants harmless and also makes the plant pollution tolerant. Further, harvesting of the plant biomass for various useful purposes like paper production, biofuel generation is the next field of research which should be focused on along with management of the plant biomass after contaminant uptake.

Acknowledgments

Financial assistance given to Dr. V.C. Pandey under the Scientist's Pool Scheme (Pool No. 13 (8931-A)/2017) by the Council of Scientific and Industrial Research, Government of India is gratefully acknowledged. The authors declare no conflicts of interest.

References

Ailstock, M.S., Center, E., 2000. Adaptive strategies of common reed Phragmites australis. Proceedings: The Role of Phragmites in the Mid-Atlantic Region. Environment Centre, Anne Arundel Community College, Arnold, MD, USA, pp. 1–7.

Al-Akeel,K., Reynolds, A.J.,Choudhary, A.J., 2010. Phytoremediation of waterways using reed plants. 15th International Conference on Heavy Metals in the Environment (15th ICH-MET), September 19–3, Gdansk University of Technology, Poland, pp. 430–433.

Armstrong, J., Armstrong, W., 1988. *Phragmites australis*–a preliminary study of soil oxidizing sites and internal gas transport pathways. New Phytol. 108 (4), 373–382.

Armstrong, J., Armstrong, W., Beckett, P.M., 1992. *Phragmites australis*: venturi and humidity induced pressure flows enhance rhizome aeration and rhizosphere oxidation. New Phytol. 120 (2), 197–207.

Badejo, A.A., Sridhar, M.K., Coker, A.O., Ndambuki, J.M., Kupolati, W.K., 2015. Phytoremediation of water using *Phragmites karka* and *Veteveria nigritana* in constructed wetland. Int. J. Phytoremediation 17 (9), 847–852.

Baldwin, A.H., Kettenring, K.M., Whigham, D.F., 2010. Seed banks of *Phragmites australis*-dominated brackish wetlands: relationships to seed viability, inundation, and land cover. Aquat. Bot. 93 (3), 163–169.

Batterson, T.R., Hall, D.W., 1984. Common reed, *Phragmites australis* (Cav.) Trin. ex Steudel. Aquatics 6 (2), 16–20.

Baute, K., Van Eerd, L., Robinson, D., Sikkema, P., Mushtaq, M., Gilroyed, B., 2018. Comparing the biomass yield and biogas potential of *Phragmites australis* with *Miscanthus* x *giganteus* and *Panicum virgatum* grown in Canada. Energies 11 (9), 2198.

Bello, A.O., Tawabini, B.S., Khalil, A.B., Boland, C.R., Saleh, T.A., 2018. Phytoremediation of cadmium-, lead-and nickel-contaminated water by *Phragmites australis* in hydroponic systems. Ecol. Eng. 120, 126–133.

Boulos, L., 2005. Flora of Egypt/ Vol. 4/ Monocotyledons (Altismataceae – Orchidaceae). Al-Hadara Publishing, Cairo, pp. 617.

Brix, H., Sorrell, B.K., Lorenzen, B., 2001. Are *Phragmites*-dominated wetlands a net source or net sink of greenhouse gases? Aquat. Bot. 69 (2–4), 313–324.

Brix, H., Ye, S., Laws, E.A., Sun, D., Li, G., Ding, X., Yuan, H., Zhao, G., Wang, J., Pei, S., 2014. Large-scale management of common reed, *Phragmites australis*, for paper production: a case study from the Liaohe Delta, China. Ecol. Eng. 73, 760–769.

Callaway, R., 1995. Positive interactions among plants. Bot. Rev. 61, 306–349.

Cicero-Fernandez, D., Pena-Fernandez, M., Exposito-Camargo, J.A., Antizar-Ladislao, B., 2016. Role of *Phragmites australis* (common reed) for heavy metals phytoremediation of estuarine sediments. Int. J. Phytoremediation 18 (6), 575–582.

Cicero-Fernandez, D., Pena-Fernandez, M., Exposito-Camargo, J.A., Antizar-Ladislao, B., 2017. Long-term (two annual cycles) phytoremediation of heavy metal-contaminated estuarine sediments by *Phragmites australis*. New Biotechnol. 38, 56–64.

Clevering, O.A., Brix, H., Lukavska, J., 2001. Geographic variations in growth responses in *Phragmites australis*. Aquat. Bot. 69, 89–108.

Clevering, O.A., Lissner, J., 1999. Taxonomy, chromosome numbers, clonal diversity and population dynamics of *Phragmites australis*. Aquat. Bot. 64, 175–208.

Cui, H., Hense, B.A., Müller, J., Schröder, P., 2015. Short term uptake and transport process for metformin in roots of *Phragmites australis* and *Typha latifolia*. Chemosphere 134, 307–312.

da Silva, M.N., Mucha, A.P., Rocha, A.C., Teixeira, C., Gomes, C.R., Almeida, C.M.R., 2014. A strategy to potentiate Cd phytoremediation by saltmarsh plants–autochthonous bioaugmentation. J. Environ. Manage. 134, 136–144.

Davies, L.C., Carias, C.C., Novais, J.M., Martins-Dias, S., 2005. Phytoremediation of textile effluents containing azo dye by using *Phragmites australis* in a vertical flow intermittent feeding constructed wetland. Ecol. Eng. 25 (5), 594–605.

Denes, A., Papp, N., Babai, D., Czucz, B., Molnar, Z., 2012. Wild plants used for food by Hungarian ethnic groups living in the Carpathian Basin. Acta Soc. Bot. Pol. 81 (4), 381–396.

Dibble, K.L., Meyerson, L.A., 2012. Tidal flushing restores the physiological condition of fish residing in degraded salt marshes. PLoS ONE 7, e46161.

Engloner, A.I., 2009. Structure, growth dynamics and biomass of reed (*Phragmites australis*)–a review. Flora Morphol. Distrib. Funct. Ecol. Plants 204 (5), 331–346.

Fares, S., Brilli, F., Nogues, I., Velikova, V., Tsonev, T., Dagli, S., Loreto, F., 2007. Isoprene emission and primary metabolism in *Phragmites australis* grown under different phosphorus levels. Plant Biol. 9 (1), 99–104.

Gilson, E., 2017. Biogas production potential and cost-benefit analysis of harvesting wetland plants (*Phragmites australis* and *Glyceria maxima*). Master's Thesis, Halmstad University.

Gonzalez-Alcaraz, M.N., Egea, C., Jimenez-Carceles, F.J., Parraga, I.A., Delgado, M.J., Alvarez-Rogel, J., 2012. Storage of organic carbon, nitrogen and phosphorus in the soil–plant system of *Phragmites australis* stands from a eutrophicated Mediterranean salt marsh. Geoderma 185-186, 61–72.

Hafliger, P., Schwarzlander, M., Blossey, B., 2006. Impact of *Archanara geminipuncta* (Lepidoptera: Noctuidae) on aboveground biomass production of *Phragmites australis*. Biol. Control 38 (3), 413–421.

Hansson, P.A., Fredriksson, H., 2004. Use of summer harvested common reed (*Phragmites australis*) as nutrient source for organic crop production in Sweden. Agric. Ecosyst. Environ. 102, 365–375.

Hill, D.T., Payne, V.W.E., Rogers, J.W., Kown, S.R., 1997. Ammonia effects on the biomass production of five constructed wetland plant species. Biores. Technol. 62 (3), 109–113.

Hofmann, K., 1990. Use of *Phragmites* in sewage sludge treatment. In: Cooper, P.F., Findlater, B.C. (Eds.), The Use of Constructed Wetlands in Water Pollution Control, Proceedings of the International Conference on the Use of Constructed Wetlands in Water Pollution Control, September 24–28, Cambridge, UK, pp. 269–278.

Hubner, T.M., Tischer, S., Tanneberg, H., Kuschk, P., 2000. Influence of phenol and phenanthrene on the growth of *Phalaris arundinacea* and *Phragmites australis*. Int. J. Phytoremediation 2 (4), 331–342.

Karunaratne, S., Asaeda, T., Yutani, K., 2003. Growth performance of *Phragmites australis* in Japan: influence of geographic gradient. Environ. Exp. Bot. 50, 51–66.

Kiviat, E., 2013. Ecosystem services of *Phragmites* in North America with emphasis on habitat functions. AoB Plants 5, plt008.

Kobbing, J.F., Thevs, N., Zerbe, S., 2013. The utilisation of reed (*Phragmites australis*): a review. Mires Peat 13, 1–14.

Kraiem, D., Pimbert, S., Ayadi, A., Bradai, C., 2013. Effect of low content reed (*Phragmite australis*) fibers on the mechanical properties of recycled HDPE composites. Compos. B Eng. 44 (1), 368–374.

Kuusemets, V., Lohmus, K., 2005. Nitrogen and phosphorus accumulation and biomass production by *Scirpus sylvaticus* and *Phragmites australis* in a horizontal subsurface flow constructed wetland. J. Environ. Sci. Health 40 (6–7), 1167–1175.

Lambertini, C., Mendelssohn, I.A., Gustafsson, M.H., Olesen, B., Riis, T., Sorrell, B.K., Brix, H., 2012. Tracing the origin of Gulf Coast *Phragmites* (Poaceae): a story of long distance dispersal and hybridization. Am. J. Bot. 99 (3), 538–551.

Lyu, T., Carvalho, P.N., Casas, M.E., Bollmann, U.E., Arias, C.A., Brix, H., Bester, K., 2017. Enantioselective uptake, translocation and degradation of the chiral pesticides tebuconazole and imazalil by *Phragmites australis*. Environ. Pollut. 229, 362–370.

Marks, M., Lapin, B., Randall, J., 1994. *Phragmites australis* (*P. communis*): threats, management and monitoring. Nat. Areas J. 14, 285–294.

McSorley, K., Rutter, A., Cumming, R., Zeeb, B.A., 2016. Phytoextraction of chloride from a cement kiln dust (CKD) contaminated landfill with *Phragmites australis*. Waste Manage. 51, 111–118.

Mozdzer, T.J., Zieman, J.C., McGlathery, K.J., 2010. Nitrogen uptake by native and invasive temperate coastal macrophytes: importance of dissolved organic nitrogen. Estuaries Coasts 33, 784–797.

Nawrot, J.R., Yaich, S.C., 1982. Wetland development potential of coal mine tailings basins. Wetlands 2, 179–190.

Nayyef Alanbary, S., Rozaimah Sheikh Abdullah, S., Hassan, H., Razi Othman Malaysia, A., 2018. Screening for resistant and tolerable plants (*Ludwigia octovalvis* and *Phragmites karka*) in crude oil sludge for phytoremediation of hydrocarbons. Iran. J. Energ. Environ. 9 (1), 48–51.

Oliveira, T., Mucha, A.P., Reis, I., Rodrigues, P., Gomes, C.R., Almeida, C.M.R., 2014. Copper phytoremediation by a salt marsh plant (*Phragmites australis*) enhanced by autochthonous bioaugmentation. Marine Pollut. Bull. 88 (1–2), 231–238.

Pandey, V.C., Bajpai, O., 2019. Phytoremediation: from theory toward practice. In: Pandey, V.C., Bauddh, K. (Eds.), Phytomanagement of Polluted Sites. Elsevier, Amsterdam, Netherlands, pp. 1–49.

Pandey, V.C., Pandey, D.N., Singh, N., 2015. Sustainable phytoremediation based on naturally colonizing and economically valuable plants. J. Clean. Prod. 86, 37–39.

Pandey, V.C., Abhilash, V.C., Singh, N., 2009. The Indian perspective of utilizing fly ash in phytoremediation, photomanagement and biomass production. J. Environ. Manage. 90, 2943–2958.

Paul, J., Vachon, N., Garroway, C.J., Freeland, J.R., 2010. Molecular data provide strong evidence of natural hybridization between native and introduced lineages of *Phragmites australis* in North America. Biol. Invasions 12, 2967–2973.

Podlipna, R., Skalova, L., Seidlova, H., Szotakova, B., Kubicek, V., Stuchlikova, L., Jirasko, R., Vanek, T., Vokral, I., 2013. Biotransformation of benzimidazole anthelmintics in reed (*Phragmites australis*) as a potential tool for their detoxification in environment. Bioresour. Technol. 144, 216–224.

Reed, M.L.E., Warner, B.G., Glick, B.R., 2005. Plant growth–promoting bacteria facilitate the growth of the common reed *Phragmites australis*in the presence of copper or polycyclic aromatic hydrocarbons. Curr. Microbiol. 51 (6), 425–429.

Rooth, J.E., Stevenson, J.C., Cornwell, J.C., 2003. Increased sediment accretion rates following invasion by *Phragmites australis*: the role of litter. Estuaries 26, 475–482.

Rudrappa, T., Bonsall, J., Gallagher, J.L., Seliskar, D.M., Bais HP, 2007. Root-secreted allelochemical in the noxious weed *Phragmites australis* deploys a reactive oxygen species response and microtubule assembly disruption to execute rhizotoxicity. J. Chem. Ecol. 33, 1898–1918.

Rungwa, S., Arpa, G., Sakulas, H.W., Harakuwe, A.H., Timi, D., 2013. Assessment of Phragmite karka (pitpit) as Possible Phytoremediation Plant Species for Heavy Metal Removal from Mining Environment in PNG. A Case Study on Closed Namie Mine Wau, Morobe Province. In The Proceedings, p. 9.

Saltonstall, K., 2011. Remnant native *Phragmites australis* maintains genetic diversity despite multiple threats. Conserv. Genetics 12, 1027–1033.

Sauvetre, A., Schroder, P., 2015. Uptake of carbamazepine by rhizomes and endophytic bacteria of *Phragmites australis*. Front. Plant Sci. 6, 83.

Schroder, P., Daubner, D., Maier, H., Neustifter, J., Debus, R., 2008. Phytoremediation of organic xenobiotics–Glutathione dependent detoxification in *Phragmites* plants from European treatment sites. Bioresour. Technol. 99 (15), 7183–7191.

Schroder, P., Maier, H., Debus, R., 2005. Detoxification of herbicides in *Phragmites australis*. Z. Naturforsch. C J Biosci. 60 (3-4), 317–324.

Schultes, R.E., Hofmann, A., Rätsch, C., 2001. Plants of the Gods: Their Sacred, Healing, and Hallucinogenic Powers, second ed. Healing Arts Press, Rochester, VT.

Serag, M.S., 1996. Ecology and biomass of *Phragmites australis* (Cav.) Trin. ex Steud. in the north-eastern region of the Nile Delta, Egypt. Ecoscience 3 (4), 473–482.

Shaltout, K.H., Al-Sodany, Y.M., Eid, E.M., 2006. Biology of common reed *Phragmites Australis* (cav.) trin. ex steud: review and inquiry. Assiut University Center for Environmental Studies (AUCES).

Shaltout, K.H., Al-Sodany, Y.M., El-Sheikh, M.A., 2004. *Phragmites australis* (Cav.) Trin. ex Steud.' in Lake Burullus, Egypt: is it an expanding or retreating population. Proceedings of Third International Conference on Biological Sciences (ICBS), Faculty of Science, Tanta University, 28–29 April 2004, Vol. 3, pp. 83–96.

Soto-Rios, P.C., Leon-Romero, M.A., Sukhbaatar, O., Nishimura, O., 2018. Biosorption of Mercury by Reed (*Phragmites australis*) as a potential clean water technology. Water Air Soil Pollut. 229 (10), 328.

Soukup, A., Votrubova, O., Cizkova, H., 2002. Development of anatomical structure of roots of *Phragmites australis*. New Phytol. 153 (2), 277–287.

Srivastava, J., Kalra, S.J., Naraian, R., 2014. Environmental perspectives of *Phragmites australis* (Cav.) Trin. Ex. Steudel. Appl. Water Sci. 4 (3), 193–202.

Struyf, E., Van Damme, S., Gribsholt, B., Bal, K., Beauchard, O., 2007. *Phragmites australis* and silica cycling in tidal wetlands. Aquat. Bot. 87, 134–140.

Su, F., Wang, T., Zhang, H., Song, Z., Feng, X., Zhang, K., 2018. The distribution and enrichment characteristics of copper in soil and *Phragmites australis* of Liao River estuary wetland. Environ. Monitor. Assess. 190 (6), 365.

Sung, K., Kim, K.S., Park, S., 2013. Enhancing degradation of total petroleum hydrocarbons and uptake of heavy metals in a wetland microcosm planted with *Phragmites communis* by humic acids addition. Int. J. Phytoremediation 15 (6), 536–549.

Topal, M., 2015. Uptake of tetracycline and degradation products by *Phragmites australis* grown in stream carrying secondary effluent. Ecol. Eng. 79, 80–85.

Toyama, T., Sato, Y., Inoue, D., Sei, K., Chang, Y.C., Kikuchi, S., Ike, M., 2009. Biodegradation of bisphenol A and bisphenol F in the rhizosphere sediment of *Phragmites australis*. J. Biosci. Bioeng. 108 (2), 147–150.

Uka, U.N., Mohammed, H.A., Aina, E., 2013. Preliminary studies on the phytoremediation potential of *Phragmites karka* (Retz.) in Asa river. J. Fish. Aquat. Sci. 8 (1), 87–93.

Vymazal, J., Brezinova, T., 2016. Accumulation of heavy metals in aboveground biomass of *Phragmites australis* in horizontal flow constructed wetlands for wastewater treatment: a review. Chem. Eng. J. 290, 232–242.

Walters, S., 2010. Carbon dynamics in a *Phragmites australis* invaded riparian wetland. Dissertations & Thesis in Natural Resources. University of Nebraska – Lincoln, p. 4.

Weis, J.S., Weis, P., 2004. Metal uptake, transport and release by wetland plants: implications for phytoremediation and restoration. Environ. Int. 30, 685–700.

Windham, L., 2001. Comparison of biomass production and decomposition between *Phragmites australis* (common reed) and *Spartina patens* (salt hay grass) in brackish tidal marshes of New Jersey, USA. Wetlands 21, 179–188.

Yanyu, S., Changchun, S., Songbai, J., Junhai, C., Jun, G., Quandong, Z., 2010. Hydroponic uptake and distribution of nitrobenzene in *Phragmites australis*: potential for phytoremediation. Int. J. Phytoremediation 12 (3), 217–225.

Feasibility of *Festuca rubra* L. native grass in phytoremediation

Gordana Gajić*, Miroslava Mitrović, Pavle Pavlović
Department of Ecology, National Institute of Republic of Serbia, University of Belgrade, Belgrade, Serbia
*Corresponding author

1 Introduction

1.1 Land contamination and effects

Rapid industrialization and urban development lead to the environmental pollution worldwide. Land contamination presents degradation of soils by xenobiotic chemicals that are produced as a result of anthropogenic activities and may pose a great threat to the environment, human health, flora, and fauna (Commission Proposal COM, 2006). The main sources of land contamination are: (1) industrial activities (mining, smelting, manufacturing); (2) urban construction activities; (3) agricultural practices (fertilizers, herbicides, pesticides, insecticides); (4) waste disposal (domestic, municipal, chemical, and nuclear wastes); (5) military activities and disposal of munitions; (6) accidental oil, gas, and chemical spills (European Commission, 2013) (Fig. 6.1A). Human activities contribute to the release of a significant amount of metal(loid)s and organic pollutants in the soil. The most common contaminants of public health concern are: arsenic (As), lead (Pb), cadmium (Cd), chromium (Cr), copper (Cu), mercury (Hg), nickel (Ni), and zinc (Zn) as well as persistent organic pollutants (POPs), such as polycyclic aromatic hydrocarbons (PAHs), polychlorinated biphenyls (PCBs), polybrominated biphenyls (PBBs), polychlorinated dibenzofurans (PCDFs), herbicides (trinitrotoluene, paraquat), insecticides (organochlorines, such as DDT, aldrin, dieldrin, and organophosphates, such as parathion, malathion, and methyl parathion), asbestos, benzene, fuels (gasoline, diesel), and pharmaceutical products (European Commission, 2013; WHO, 2013; Brevik and Burgess, 2013) (Fig. 6.1A).

Soil contamination adversely affects the physical, chemical, and biological characteristics of soil, decreases its quality, fertility, and productivity, reducing the growth and yields of many crops and plants species (Ranieri et al., 2016) (Fig. 6.1A). Furthermore, soil pollutants lead to the loss of arable and forest land, plant, and animal habitats causing the loss of biodiversity (Fig. 6.1A). Soil is the source of food for all people on the globe. Therefore, toxic chemicals in soils can affect the human health through foods (grain, vegetables, and fruit) (European Commission, 2013). Metal(loid)s and organic compounds have long-term consequences for human health leading to the cardiovascular and gastrointestinal diseases, skin, liver, and kidney damage, diabetes, neurological, renal, and bone damage, cancer, reproductive, and

Phytoremediation Potential of Perennial Grasses. http://dx.doi.org/10.1016/B978-0-12-817732-7.00006-7

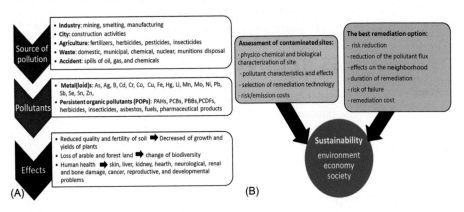

Figure 6.1 Source of pollution, pollutants, and their effects on environment and human health (A); assessment of contaminated sites and best remediation options together with three pillars of sustainability (environment, economy, and society) (B).

developmental problems (WHO, 2010; European Commission, 2013) (Fig. 6.1A). Resuspension of dust from contaminated soils can cause accumulation of chemicals in the crops and humans can intake them via inhalation and dermal absorption (WHO/UNECE, 2006).

1.2 Cleanup solutions, management, and assessment

Polluted sites should be carefully monitored and properly managed in order to reduce the risk for environment, human, and wildlife from exposure to hazardous contaminants. Therefore, decision-making and selection of relevant cleanup technology depends on the assessment of contaminated sites and characteristics of pollutants (Gruiz, 2009). Cost and environmentally efficient assessment of contaminated sites includes physico-chemical and biological characterization of sites (geochemical and hydro-geological properties, soil type, biota), risk assessment (pollutant characteristics, effects), and selection of remediation technology with risk/emission costs (Boivin and Ricour, 2005; Gruiz, 2009) (Fig. 6.1B). An integrated application of chemical, ecological, and biological models is the best monitoring tool for the estimation of the risk posed by contaminants providing environmental reliable results and a real field picture of the environment (Gruiz, 2009).

Key factors in remediation technology selection are: (1) risk management and assessment of land contamination that include the source of contamination (removal through destruction or extraction), the pathway control (stabilization, immobilization, and landfill) and protection of receptor (land use restriction); (2) project drivers which include protection of human health and environment and land regeneration where core and non-core stakeholders participate in the decision process; (3) technical suitability and feasibility (technical guidance about technology reliability and performance) (Nathanail et al., 2002; Colombano et al., 2009). Eco-efficiency and cost-benefit analyses are essential for the remediation technology selection because they connect the three pillars of sustainability, such as environment, economy, and society in order to assess the impacts and the benefits of the technologies and the most appropriate remediation option (Spira et al., 2007; Colombano et al., 2009) (Fig. 6.1B). The best remediation option should consider: risk reduction, reduction of the pollutant flux, environmental efficiency, effects on the neighborhood activities, remediation duration, risk of failure, and remediation cost. Sustainable management of contaminated land should integrate stakeholder participation and their acceptance of remediation options as well as the social consequences of the regeneration of the contaminated sites (RESCUE, 2005) (Fig. 6.1B).

1.3 Phytoremediation and ecorestoration

Phytoremediation is green, eco-friendly, low-cost technology that uses plants to clean contaminated sites from inorganic and organic pollutants (Salt et al., 1998; McCutcheon and Schnoor, 2003; Pilon-Smits, 2005). The main phytoremediation technologies are: phytostabilization, phytoextraction, rhizodegradation, and phyto-degradation (Raskin et al., 1997; Pilon-Smits, 2005; Gajić and Pavlović, 2018; Gajić et al., 2018). Metal(loid)s as inorganic pollutants can be reduced in the environment by phytostabilization and phytoextraction whereas organic contaminants can be re-moved by rhizodegradation and phytodegradation (McCutcheon and Schnoor, 2003; Pilon-Smits, 2005; Gajić and Pavlović, 2018).

Phytostabilization is a remediation technology which uses plants to reduce the mo-bility of pollutants, preventing their migration in the environment or their entry into the food web, especially in the areas with multi-element contamination, such as coalmine land, mine tailings, and fly ash deposits (Pandey et al., 2012; Gajić et al., 2018). This is a remediation technology that uses excluders, plant species which are capable to immobilize contaminants in the soil or plant roots, thus, they limit the pollutant's uptake in the high content inside the plants over a wide range of soil concentrations (Baker, 1981; Prasad and Freitas, 2006). Suitable plant species for phytostabilization should have extended root systems that prevent wind erosion and they improve the physico-chemical and biological characteristics of the contaminated soils by increas-ing the content of organic matter. Furthermore, the main mechanisms of phytostabi-lization are avoiding or excluding the pollutants at the root level by the presence of mycorrhizal fungi and root exudates in the rhizosphere or binding pollutants to the cell wall and/or preventing their transport through the plasma membrane (Pilon-Smits and LeDuc, 2009; Gajić et al., 2018). Phytoextraction is a remediation technology that uses plant accumulators or hyperaccumulators that accumulate metal(loid)s in the roots and aboveground parts of the plants (Reeves and Baker, 2000). These plant species should have a high root to shoot transfer and good tolerance to excess con-centrations of metal(loid)s in the plant tissue (Baker, 1981; Gajić et al., 2018). The main mechanisms of phytoextraction of pollutants is accumulation that is achieved through the formation of chelates (complex of metal(loid)s with glutathione, phytohe-latins, metallothioneins, amino acids, organic acids, and carbohydrates) and seques-tration of the metal(loid)s in the vacuole of roots and leaves (Tong et al., 2004; Kidd et al., 2009). Plants' phytoremediation potential and their tolerance to the pollutants can be determined by the bioconcentration factor (BCF) and translocation factor (TF) (Baker, 1981; Yoon et al., 2006). BCF provides information about metal(loid) accumulation in the roots and presents the relationship between the content of the metal(loid) in the root and its content in the soil whereas the TF gives information about the efficiency of plants to transport element from root to leaves and presents the relationship between the content of the metal(loid) in the root and leaves.

Rhizodegradation/phytostimulation is a remediation technology that uses plant's root to stimulate the microbial and fungal activity releasing exudes in the rhizosphere and breaking down organic pollutants (Pilon-Smits, 2005; Ma et al., 2011; Gajić et al., 2018). These compounds are hydrophobic, they are tightly bound to soil organic matter and do not easily dissolve in soil solutions. Stimulation of soil microbial communities by root exudates (organic acids, sugars, amino acids, phenolics, and enzymes dehalogenase, nitroreductase, peroxidase, laccase) increase the availability of nutrients around roots and that can lead to the changes in pH of soil solution (Kala, 2014). Plant growth promoting bacteria (PGPB) present a consortium of bacteria that colonize different niches of plant roots and may reduce the metal(lod) toxicity by biosorption and bioaccumulation (Khan et al., 2014; Tokala et al., 2002; Dimpka et al., 2009). Phytodegradation/phytotransformation is a remediation technology where degradation of organic compounds occurs through the plant enzymes inside the roots or leaves to the CO_2 and H_2O (Burken, 2003; Pilon-Smits, 2005). The degradation of organic compounds in plants can be achieved through the process of transformation, conjugation (complex of organic pollutants and carbohydrates, carboxylic acids, amino acids, and glutathione), and storage where conjugates become less toxic and they can be sequenced in cell vacuole, apoplast, and cell walls (Burken, 2003; Gajić et al., 2018).

Ecorestoration is the process of establishment of the self-sustaining vegetation cover in order to initiate the recovery of degraded ecosystem with respect to its health, integrity, and sustainability (SER, 2002). The plant species in grass-legume mixture on the contaminated sites begin the process of revegetation, thus improving the physico-chemical properties of the substrate, retain moisture and nutritional elements, stabilize the substrate, and prevent wind erosion, decrease mobility and dispersion of toxic elements and compounds in the environment (Maiti, 2013; Mitrović et al., 2008; Pandey, 2013, 2015; Pandey and Singh, 2014; Maiti and Maiti, 2015; Gajić and Pavlović, 2018; Gajić et al., 2016, 2018, 2019; Kostić et al., 2018). In addition, the selection of native plant species is essential for ecorestoration success because they are the best adapted to local climate promoting the natural revegetation (Kirmer and Tischew, 2010; Baasch et al., 2012; Tischew et al., 2014). Over time, native sown, and spontaneously colonized plant species on contaminated sites bind substrate with a fibrous root system, and they spread by seeds and rhizomes forming a dense vegetation cover. These plant species are stress-tolerant and they possess adaptive mechanisms to unfavorable environmental conditions, such as drought, high temperatures, toxicity, and/or deficiency of chemical elements (Pavlović et al., 2004; Djurdjević et al., 2006; Mitrović et al., 2008; Kostić et al., 2012; Pandey, 2012, 2015; Gajić et al., 2013, 2016, 2019; Pandey et al., 2014, 2015a, 2015b, 2016; Kumari et al., 2016; Randjelović et al., 2016).

1.3.1 Perennial grasses on contaminated sites

Perennial grasses belong to the family of *Poaceae* which includes 780 genera and 12,000 plant species that are widely distributed on the globe (Christenhusz and Byng, 2016). They are growing in natural grasslands, such as meadows, steppes, and prairies. Grasses present an economically important plant family, which are used in food production, industry (raw material and fuel), lawns and pasture, forage, gardening, sports, erosion control, phytoremediation, and ecorestoration (Dittberner and Olson, 1983; Longland, 2013; Maiti, 2013; Pandey and Singh, 2014; Pandey et al., 2012, 2015b, 2016; Maiti and Maiti, 2015; Engelhardt and Hawkins, 2016; Gajić et al., 2018, 2019).

Grasses have an extensive root system, large biomass, grow fast, occupy a great ground cover and as monocots show generally higher tolerance to pollutants than dicots (Prasad, 2006). Grasses from the genera *Agrostis, Agropyron, Alopecurus, Andropogon, Anthoxathum, Arrenatherum, Avena, Brachiaria, Bromus, Calamagrostis, Chloris, Cynodon, Dactylis, Digitaria, Elymus, Elytrigia, Eremochloa, Festuca, Lolium, Lygeum, Miscanthus, Panicum, Paspalum, Pennisetum, Phleum, Phragmites, Piptatherum, Poa, Setaria, Sorghum, Secale, Spartina, Stipa, Typha, Vetiveria* are often used in phytoremediation of metal(loid) and organic compounds on the contaminated sites (McCutcheon and Schnoor, 2003; Prasad, 2006; Conesa et al., 2009; Maiti, 2013; Nadgórska-Socha et al., 2013; Dželetović et al., 2013; Pandey et al., 2015b, 2016; Maiti and Maiti, 2015; Parraga-Aguado et al., 2015; Gajić et al., 2016, 2018, 2019; Andrejić et al., 2018; Rabêlo et al., 2018) (Fig. 6.2A, Fig. 6.3, Fig. 6.4).

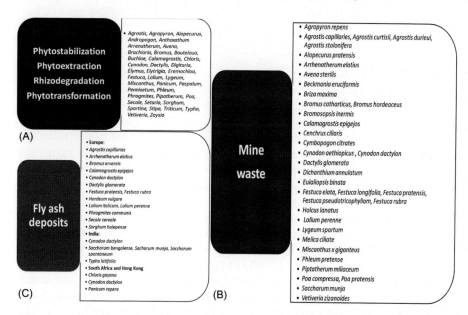

Figure 6.2 Perrenial grasses that are suitable for phytoremediation and ecorestoration of contaminated sites (A); grasses with high potential to grow and survive on mine waste (B) and fly ash deposits (C).

Figure 6.3 Mining waste in Serbia: Copper mine waste in Bor region with *Calamagrostis epigejos* (A–D); flotation tailings on the mine "Rudnik" (Pb, Cu, Zn) (E) with *Miscanthus x giganteus* (F), *Phragmites communis* (G) and *Calamagrostis epigejos* (H); flotation tailings "Stolice" (mine of antimone) after accidental flooding (I) with *Calamagrostis epigejos* (J, K) and *Phragmites communis* (L).

Perennial grass species, such as *Agropyron repens, Dactylis glomerata, Phleum pretense,* and *Setaria viridis* that are widely spread around the metallurgic plants in Kazakhstan accumulate Pb, Cu, Zn, and Cd, mainly in the roots and can be used for phytostabilization of contaminated soils (Atabayeva, 2016). *Calamagrostis epigejos* that grows around electrical, chemical, and food industries in Saransk (Russia) accumulates higher concentrations of Cu, Zn, Ni, Fe, and Mn in roots than in leaves and is also suitable for phytostabilization of contaminated lands (Bashmakov et al., 2006). According to Padmavathiamma and Li (2009), *Lolium perenne* is a good candidate for phytostabilization of Pb and Cu, *Festuca rubra* for Mn and *Poa pratensis* for Zn. In addition, *Sorghum bicolor* can be suitable for phytoextraction of Pb, Zn, and Cr in hydroponic solution (Bonfranceschi et al., 2009) whereas pot experiment showed that this grass is suitable for phytostabilization of the same elements (Abou-Shanab et al., 2007). Similarly, Yang et al. (2014) found that *Cynodon dactylon* is good for phytostabilization of Cd, Mn, Pb, and Zn while Saraswat and Rai (2009) noted that this species has high phytoextraction potential for Cd, Cr, and Ni in contaminated soil.

Grass species, such as *C. dactylon, L. perenne, Lolium multiflorum, D. glomerata, Festuca arundinaceae, Phragmites australis,* and *S. bicolor* show high potential for rhizomediation of PAHs/PCB pollutants (McCutcheon and Schnoor, 2003; Muratova

Figure 6.4 Fly ash deposits and coalmine land in Serbia: active cassette of thermal power plant "Nikola Tesla-A" (TENT-A) (A, B) and the embankment of active cassette with grass cover (C); cassette 3-years old with *Triticum aestivum* (D), *Arrhenatherum elatius* (E) and *Dactylis glomerata* (F, G); cassette 11-years old with *Sorghum halepense* (H, I), *Calamagrostis epigejos* (J, K); lignite coal mine in Kostolac, "Drmno" (L, M); coal mine tailings 3-years old with *Calamagrostis epigejos* (N) and *Sorghum halepense* (O) and coal mine tailings 22-years old with *Calamagrostis epigejos* (P), *Phragmites communis* (R) and *Sorghum halepense* (S).

et al., 2003; Parrish et al., 2005; Smith et al., 2006; Duponnois et al., 2006) whereas high potential for phytodegradation was noted in *L. multiflorum* (phenanthrene, pyrene, Gao and Zhu, 2004; Kang et al., 2010), *Sorghum drummondii* (16 PAHs, Petrová et al., 2017), *Triticum, aestivum, Penisetum* sp., and *Hordeum vulgare* (fluoranthene, pentachlorophenol 4-chloroaniline, 4-n-nonylphenols, Wilken et al., 1995; Harms, 1996; Bokern and Harms, 1997; Kolb and Harms, 2000). According to Wang and Oyaizu (2009) grasses, such as *C. dactylon, Agrostis palustris,* and *Zoysia japonica* had a high biomass and rhizospheric high polychlorinated dibenzofuran (DFB)-degraded bacteria suggesting that microbial populations are capable of degrading DFB in soils whereas *C. dactylon, L. perenne,* and *Paspalum notatum* are capable of reducing atrazine (pesticide, Karthikeyan et al., 2004). Furthermore, *L. multiflorum, Sorghum vulgare, C. dactylon,* and *Digitaria ciliaris* showed high rhizodegradation potential for petroleum hydrocarbon (PAH)-contaminated soils due to the high dehydrogenase activity (DHA) (Kaimi et al., 2006, 2007). According to Frick et al. (1999), grass species that show capability for remediation of crude oil (total petroleum hydrocarbons, TPHs) are: *Triticum aestivum, Andropogon gerardi, Bouteloua curtipendula, Bouteloua gracillis, Buchloe dactyloides, Chloris gayana,* and *Panicum virgatum.*

Biological recultivation of brown coal mine waste in Urals (Russia) by perennial grasses *A. repens, Alopecurus pratensis, Beckmania eruciformis, Bromosopsis inermis, Festuca pratensis, F. rubra, P. pretense,* and *P. pratensis* provided well-developed vegetation cover and successful rehabilitation of degraded sites (Zelensky and Sariev, 2008; Iglikov, 2016; Chibrik et al., 2018) (Fig. 6.2B). Furthermore, rehabilitation of mine sites in Ireland (Pb/Zn, Cu/pyrite) was achieved by seeding metal—tolerant grasses, such as *Agrostis stolonifera, Agrostis capillaries, F. rubra, Festuca longifolia, Holcus lanatus, L. perenne, Poa pretensis, Poa compressa, Phleum pratense* (Courtney, 2018) (Fig. 6.2B). According to Alves et al. (2018), native grass *Aristida setifolia* can be recognized as efficient plant species for reclamation of inclined slopes on uranium mine waste dump in Brazil. Successful ecorestoration of mine waste is achieved by establishment of the self-sustaining vegetation grass cover: *A. stolonifera* and *C. epigejos* in copper flotation ore in Poland (Kasowska et al., 2018); *Agrostis durieui, H. lanatus, F. rubra,* and *D. glomerata* in Pb-Zn and Hg-As mining waste in Spain (Fernandez et al., 2017); *Agrostis curtisii, Arrhenatherum elatius, Avena sterilis, Briza maxima, Bromus hordeaceus, Festuca pseudotricophylla, H. lanatus, Melica ciliate,* and *P. pretense* on mine waste in Portugal (Prasad and Freitas, 2006); *Bromus catharticus* on the abandoned mine tailings in Mexico (Santos et al., 2017); *Festuca elata* and *C. dactylon* on the degraded mine land in China (Li et al., 2007; Zhang et al., 2014); *Vetiveria zizanoides, Cymbopogon citrates, Dichanthium annulatum, Cenchrus ciliaris, Saccharum munja,* and *Eulaliopsis binata* in the coal mine land in India (Maiti, 2013) (Fig. 6.2B). According to Ellery and Walker (1985), *C. dactylon* and *Cynodon aethiopicus* showed great potential to grow on asbestos tailings as pioneer grasses due to rapid stolon production and high tolerance to harsh conditions on these sites (Fig. 6.2B). Furthermore, Grant et al. (2002) noted that *C. dactylon* is a grass species which provides the best ground cover on the abandoned three mine sites in Australia (mine site with Pb, Cu, Zn, Ag; mine site with As; mine site with asbestos) and in mulch as well as in tailings/mulch pots, showing rapid and extensive growth and high biomass.

Grasses that can grow and survive on fly ash deposits in Europe are: *A. capillaries, C. epigejos, C. dactylon, D. glomerata, F. rubra, L. perenne, Sorghum halepense, Phragmites communis, A. elatius, Lolium italicum, Secale cereale, F. pratensis, Bromus arvensis, H. vulgare, P. communis* (Hodgson and Townsend, 1973; Shaw, 1996; Djordjević-Miloradović, 1998; Pavlović et al., 2004; Djurdjević et al., 2006; Mitrović et al., 2008; Gajić et al., 2016, 2019; Gajić and Pavlović, 2018) whereas in India grasses that colonize fly ash deposits are: *C. dactylon, Saccharum bengalense, Sacharum munja, Saccharum spontaneum, Typha latifolia* (Gupta and Sinha, 2008; Maiti and Jaswal, 2008; Dwivedi et al., 2008; Pandey and Singh, 2014; Pandey et al., 2015a, 2015b, 2016; Kumari et al., 2016); *Panicum repens* can grow on fly ash landfill in Hong Kong (Chu, 2008) while *C. gayana* and *C. dactylon* can survive on fly ash deposits in South Africa (Morgenthal et al., 2001; Van Rensburg et al., 2003) (Fig. 6.2C).

2   General aspects of *F. rubra* L.

2.1   Taxonomy and geographical distribution of F. rubra L.

F. rubra L. (red fescue) is a perennial, rhizomatous coal season grass which belongs to Class *Monocots*, Order *Poales*, Family *Poaceae*, and Genus *Festuca* (Gajić, 1976; USDA, 2016) (Fig. 6.5). This grass is widely distributed circumboreal and it is native for arctic and temperate zones of Europe, Asia, and America, North Africa as well as Mexico and New Zealand (Gajić, 1976; Barkworth et al., 2007). Therefore, *F. rubra* can be found in different geographical regions around the globe: Greenland, Faroe Island, Svalbard and Jan Myen, Ireland, Great Britain, Scotland, Finland, Norway, Sweden, Denmark, Russia Federation (European Russia), Estonia, Belarus, Ukraine, Lithuania, Luxemburg, Netherlands, Belgium, Germany, Austria, Switzerland, France, Poland,

Kingdom: *Plantae*
Class: *Monocots*
Order: *Poales*
Family: *Poaceae*
Genus: *Festuca*
Species: *F. rubra L. – red fescue*

Figure 6.5   Taxonomy and morphology of *Festuca rubra* L.

Czech Republic, Slovakia, Hungary, Romania, Bulgaria, Serbia, Slovenia, Italy, Greece, Spain, Portugal, Corsica, Malta, Sicilia as well as in Alaska, Canada, United States (except of Arkansas, Louisiana, Florida, Mississipi, South Dakota) (Hitchcock, 1951; Anderson, 1959; Munz, 1973; Gajić, 1976; Lackschewitz, 1991; Stancik, 2003).

F. rubra shows great variability between the subspecies and according to Stukonis et al. (2015), this grass included 347 cultivars in the European Union database of registered plant varieties in 2014. These cultivars are phenotypically different, that is, variation of morpho-anatomical characteristics indicates great diversity of ecotypes suggesting its adaptive response to local climate and environmental conditions depending on the origin of the genotype (Dabrowska, 2011; Stukonis et al., 2015). Therefore, extensive hybridization between (sub)species of *F. rubra* leads to the polyplods (2n = 14, 21, 28, 42, 49, 56, 64, 70) that are related to the morphological variations, high adaptive capacity, and ecotype diversity (Aiken and Darbyshire, 1990; Sampoux and Huyghe, 2009).

In Serbia, this species is determined as: *F. rubra* L. 1753. Sp. Pl. ed. 1 : 74; Pančić 1884. addition to Flora of the Principality of Serbia: 241; Hayek 1933. Prodr. Fl. Pen. Balc. 3 : 284: - common name: *red fescue* (Gajić, 1976). According to Gajić (1976) several subspecies of this grass are determined and described in Serbia:

- *Festuca rubra* L. ssp. *vulgaris* Hayek op. c. : 285.
- *Festuca rubra* L., ssp. *eu-rubra* Hackel 1882. Monogr. Fest. :138.
- *Festuca rubra* L. ssp. *fallax* Thuill. 1799. Fl. Par. Ed. 2 : 50.
- *Festuca rubra* L. spp. *commutata* Gaud. 1882. Fl. Helv. 1 : 287.
- *Festuca rubra* L. spp. *caespitosa* Hackel 1878. Termèszetr. Füz. 2 : 292.

According to Preston et al. (2002) and St. John et al. (2012) subspecies of *F. rubra* in Europe and North America are as follows:

- *Festuca rubra* L. spp, *arctica* (Hack.) Govor.
- *Festuca rubra* L. spp. *arenaria* (Osbeck.) Syme.
- *Festuca rubra* L. spp. *aucta* (Krecz. and Bobr.) Hulten.
- *Festuca rubra* L. spp. *commutata* Gaud.
- *Festuca rubra* L. spp. *fallax* Thuill.
- *Festuca rubra* L. spp. *juncea* (Hack.)
- *Festuca rubra* L. spp. *litoralis* (G. Mey.)
- *Festuca rubra* L. spp. *mediana* (Pavlick) Pavlick.
- *Festuca rubra* L. spp. *planifolia* Hack.
- *Festuca rubra* L. spp. *pruinosa* (Hack.) Piper.
- *Festuca rubra* L. spp. *rubra* L.
- *Festuca rubra* L. spp. *scotica* Al-Bermani.
- *Festuca rubra* L. spp. *secunda* (J. Presl) Pavlick.
- *Festuca rubra* L. spp. *thessalica* Markgr.-Dann.
- *Festuca rubra* L. spp. *vallicola* (Rydb.) Pavlick.

2.1.1 Red List Category and conservation

F. rubra is classified as Least Concern (LC) in Red List Category by European Union (EU) (Duarte et al., 2011). However, in Norway, this grass is classified as Vulnerable

(VU) as listed in Annex I of the International Treaty on Plant Genetic Resource for Food and Agriculture (Kålås et al., 2006).

According to EURISCO (2015) 2 167 germplasm of *F. rubra* can be found in European genebanks and 875 are of wild origin. *F. rubra* has high potential for the selection of useful genes and breeding due to high diversity of ecotypes which is useful for turf industry (Stukonis et al., 2015).

2.2 Morphology and reproduction of F. rubra L.

F. rubra is a perennial grass with short rhizomes, loosely or densely tufted, intensely green or grayish-green leaf color (Fig. 6.5) (Gajić, 1976). Root depth is maximum 30 cm. Sterile stems are long and creepy, 30–90 cm tall, rigid, and smooth. Sheaths are reddish, shredding into fibers. Ligules are naked, 0.1–0.5 mm long. Basal leaves are bent with 5–7 nerves, 7–9 sclerenchima bundles, and 0.4–1.2 mm wide. Leaves are narrow as a needle, distichous, 5–15 cm long (Fig. 6.5) with 11–17 nerves and often on the upper side with hairs. Inflorescence is panicle with spikelets which are of linear shape, 6–15 cm long, and often reddish, purple to brown color with 4-6-10 flowers (5–15 cm long). Lemmas glabrous, narrow or lanceolate, 1-2-4-7-5 mm long without/ with awn 1–5 mm long (Gajić, 1976). The fruit is a caryopsis.

F. rubra reproduces effectively by seeds (sexual propagation) and spreads fast clonally by creeping rhizomes (vegetative, asexual propagation) (Eriksson, 1989). Cook (1983) noted that the clone of *F. rubra* is 220 m long and over 1000 years old. This grass does not form a stable seed bank (Roberts, 1981; Thompson, 1987). However, when the plants of *F. rubra* have established, they are long-lived and capable to endure unfavorable environmental conditions.

2.3 Ecology of F. rubra L.

F. rubra belongs to the life form of hemicryptophytes (Kojić et al., 1997). This grass is well-adapted to different ecological conditions and habitats (Fig. 6.6). *F. rubra* is capable to grow in the conditions of both light and shading (semisciophyte), warm and cold (mesothermophyte), humid and dry (submesophyte), on the soils that are rich in available nitrogen content (mesotrophyte), and can grow in neutral or slightly acidic soils (neutrophyte) (Smoliak et al., 1981; Hallsten et al., 1987; Kojić et al., 1997).

Therefore, *F. rubra* tolerates high and low temperatures, drought, flooding during spring as well as water logging due to artificial irrigation. Furthermore, it can grow on the sandy, loam, and clay soils that can be moderately rich or poor in nutrients (Smoliak et al., 1981). According to Carroll (1943) survival rate of *F. rubra* was 60%–80% at −10°C whereas at −15°C, percentage of survival was low. This grass occurs on soils with pH from 4.5 to 6.0 (Vogel, 1981), but it can grow on pH = 8 (Engelhardt and Hawkins, 2016). In addition, *F. rubra* tolerates salinity (Smoliak et al., 1981), from 3–6 dS/m (Marcum, 1999) to 6–10 dS/m (Uddin and Juraimi, 2013), and 8–12 dS/m (Butler et al., 1971, depending on subspecies and cultivars. According to Rozema et al. (1978), salt tolerance was higher in *F. rubra* spp. *litoralis* than in *F. rubra* spp. *arenaria* and *F. rubra* spp. *rubra*, and that was related to their capability to accumulate

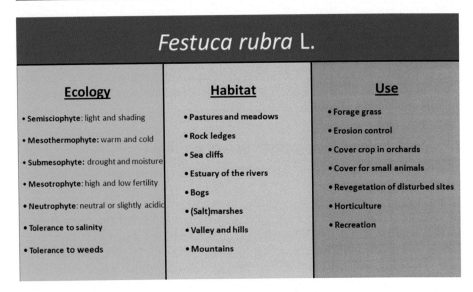

Figure 6.6 Ecology, habitat types, and use of *Festuca rubra*.

more proline and less Na$^+$ and Cl$^-$ ions than in non-tolerant ecotypes. Zhang et al. (2013) noted that *F. rubra* had high salt tolerance in comparison to the numerous turfgrass species in hydroponic system. Furthermore, *F. rubra* showed high resistance to invasion of weed and weed suppression by this grass was > 70%–80% (McKernan et al., 2001; Bertin et al., 2009).

2.3.1 Habitats and plant communities of F. rubra L.

F. rubra grows in pastures and meadows, sand dunes, rock ledges, and wetland habitats, such as sea cliffs, river banks, bogs, and (salt) marshes (Hitchcock, 1951; Voss, 1972; Hickman, 1993) (Fig. 6.6). It can be found in the grass community from the sea level to very high elevations of mountain top vegetation, such as: Alaska (396–914 m, Elliot et al., 1987), California (0–2743 m, Hickman, 1993), Colorado (2134–4115 m, Dittberner and Olson, 1983), Montana (975–1524 m, Dittberner and Olson, 1983), and Utah (1372–2835 m, Dittberner and Olson, 1983). In Serbia, *F. rubra* belongs to meadows and pasture vegetation and occurs in meadow communities of valleys and hilly-mountainous areas (1000–1600 m), on serpentine and lime substrate (Kojić et al., 1998) (Fig. 6.6).

According to EUNIS classification of habitats, which is a standard classification of European habitats developed by the European Environment Agency (http://eunis.eea. europa.eu/), *F. rubra* occurs in many habitat types (herbaceous and forest). In Europe particularly, significant populations are noted as follows (EUNIS, in press):

- British heavy metal grasslands (code E1.B11): formations, in particular of Wales and the Pennines, developed in the vicinity of former mining operations or on river gravels, with

Minuartia verna, Thlaspi caerulescens, Armeria maritima, Viola lutea, Festuca ovina s.l., *F. rubra* s.l., *Agrostis capillaris* (*Agrostis tenuis*).

- Atlantic *F. rubra-A. stolonifera* swards (code A2.5313): upper saltmarsh communities of the Atlantic.
- *F. rubra* mid-upper saltmarshes (code A2.53A).
- Mid-upper saltmarshes: sub-communities of *F. rubra* with *A. stolonifera, Juncus gerardi, Puccinellia maritima, Glaux maritima, Triglochin maritima, Armeria maritima,* and *Plantago maritima* (code A2.53B).
- Mid-upper saltmarshes: *Artemisia maritima* with *F. rubra* (code A2.539).
- *Nardus stricta* swards (code E1.71): Mesophile and xerophile *Nardus stricta*-dominated or -rich grasslands of Atlantic or sub-Atlantic lowland, collinar, and montane regions of northern Europe, middle Europe and western Iberia.
- Fenno-Scandian *Avenula pratensis-F. rubra* grasslands (code E1.7225): dry or mesophile calcareous grasslands of subarctic affinities, limited to the continental middle boreal zone of lowland Sweden and northern Finland and to the middle boreal and arcto-alpine zones of the Scandinavian mountains; dominated by *F. rubra*, with *Botrychium boreale, Botrychium lanceolatum, Botrychium lunaria, Carex brunnescens, Carex ericetorum, Cerastium alpinum, Erigeron borealis, Galium boreale, Gentiana nivalis, Gentianella amarella, Gentianella campestris, Gentianella tenella, Poa glauca, Primula scandinavica, Primula striata.*

In Serbia, *F. rubra* mainly grows in the following habitats according to EUNIS classification (Lakušić et al., 2005):

- Serpentine steppes (code E1.2B)
- Middle European *Bromus erectus* semidry grasslands (code E1.26)
- *Nardus stricta* swards (code E1.71)
- *Agrostis - Festuca* grasslands (code E1.72)
- Permanent mesotrophic pastures and aftermath-grazed meadows (code E2.1)
- Moist or wet eutrophic and mesotrophic grasslands (code E3.4)
- Moist or wet oligotrophic grasslands (code E3.5)
- Closed calciphile alpine grasslands (code E4.41).
- *Fagus* woodland (code G1.6)

According to Mucina et al. (2016), significant abundance of *F. rubra* is noted mainly in the communities of the following vegetation alliances:

- Mat-grass dry pastures in the submontane to subalpine belts of the mountain ranges of Central Europe and the Northern Balkans – *Nardo-Agrostion tenuis* Sillinger 1933 (*Nardetalia strictae* Preising 1950, *Nardetea strictae* Rivas Goday et Borja Carbonell in Rivas Goday et Mayor López 1966).
- Mesic mown meadows on mineral-rich soils in the lowland to submontane belts of temperate Europe – *Arrhenatherion elatioris* Luquet 1926 (*Arrhenatheretalia elatioris* Tx. 1931, *Molinio-Arrhenatheretea* Tx. 1937).

2.4 Multiple uses and management consideration

F. rubra has been extensively used as forage grass for grazing, erosion control, cover crop in orchards, cover for small animals (birds and mammals), revegetation of disturbed sites, cover along roadside and highways, in horticulture as ornamental, and in recreation as ski run and golf fields (Fig. 6.6) (Strausbaugh and Core, 1977; Wheeler and Hill, 1957; Dittberner and Olson, 1983; Engelhardt and Hawkins, 2016; Longland, 2013; Chen et al., 2018).

F. rubra is a palatable cool season grass commonly used in pasture and pasture management (Longland, 2013). According to Dittberner and Olson (1983), this grass is very good for grazing by cattle and horses, but less favorable for sheep. Nutrition value of leaves is relatively high despite low energy values (Smoliak et al., 1981). Digestibility of cellulose is low and this grass retains high protein content during the year. *F. rubra* for graze pastures has high yield (2–5 t/ha) with grazed height from 2.5 to 5 cm and does not have large requirements for fertilizers (Longland, 2013). However, it can decrease when it is overgrazed (Volland, 1985).

F. rubra is used to prevent erosion and stabilize slopes, hillsides, irrigation channels, and banks (Strausbaugh and Core, 1977). This grass has a fibrous root system, and higher root biomass compared to shoot. However, according to Brown et al. (2010), 83% of *F. rubra's* root biomass occurs within 8 cm of the soil and that indicates its poor water retention.

F. rubra also grows on contaminated and anthropogenically disturbed habitats, such as abandoned coal mines, fly ash deposits, along the roads and railways (Voss, 1972; Smoliak et al., 1981; Russell, 1985; Ruemmele et al., 2003; Gajić et al., 2016). *F. rubra* has moderate tolerance to traffic and it is useful as turfgrass on the roadside and highways, due to: (1) maintenance of visibility; (2) improving landscape and providing a safety zone for vehicles; (3) moderate commercial availability and cost; (4) fast establishment (high seedling vigor); (5) minimal requirements for irrigation or fertilizers; (6) high tolerance to drought, freezing and low fertility, salinity, acidity, and competition with other plant species (Beard, 1973; Engelhardt and Hawkins, 2016).

3   Ecorestoration techniques

Contaminated sites, such as roadsides, fly ash deposits, spoil heaps from coal, and ore mine tailings can be restored by human intervention or spontaneous succession of plants. Technical reclamation of degraded sites involves: (1) grading; (2) top soil covering; (3) soil amendments/ameliorative; (4) selection of plant species; (5) mulching; and (6) irrigation (Maiti, 2013). Grading implies remodeling the surface by heavy machinery, that is, benching, terracing, and creating undulating surfaces forming different microhabitats that can promote plant biodiversity. Topsoil from undisturbed land that is spreading in the area should be uniformly distributed at 10–60 cm, and it provides organic matter due to the presence of plant propagules, seeds, and microorganisms. Topsoil can also be overburden and waste rock that require amelioration with organic (animal manures and sewage sludge) and chemical amendments (gypsum, lime, inorganic fertilizers). Mulching is used for erosion control, increasing water-holding capacity, improving soil structure, aeration, and drainage as well as enhancing availability of nutrients and germination of plant seeds. Organic mulches can be straw, saw mill waste, hay, and cellulose products (bark, wood, and wood fiber) (Maiti, 2013).

Biorecultivation implies sowing commercial grass-legume mixture and planting shrubs/trees that are the best adapted to the local climate and unfavorable environmental conditions (Mitrović et al., 2008; Kostić et al., 2012; Maiti, 2013; Pandey, 2015; Gajić et al., 2016; Prach et al., 2014). Sown native plant species in grass-legume mixture act as pioneers providing dump stabilization by well-developed tuft root systems, improving physico-chemical and nutrient properties of substrate and leading to the regeneration of an ecosystem and establishment of self-sustaining vegetation on degraded sites (Maiti, 2013; Gajic´, 2014; Pandey, 2015; Gajić et al., 2016). In revegetation program, sowing time, seed sources, collection, quantity and quality, viability, and germination percentage of seeds as well as their storage and longevity are very important (Maiti, 2013).

3.1 Seed production, establishment and management of F. rubra L.

Seeds of *F. rubra* can be collected from commercial sources, such as agricultural university/institute and nursery. There are more than 3000 cultivars of *F. rubra* on the market that have different traits: high tolerance to drought, heat, winter hardiness, salinity, better turf quality, strong rhizomes, and intense color (Ruemmele et al., 2003; Engelhardt and Hawkins, 2016). In the greenhouse, *F. rubra* can be easily propagated. In the containers, seeds germinate within 15 days at 27°C and according to Baskin and Baskin (2002), 80% of transplant plants can survive. Seed production lasts only 1 or 2 years, but once established, plants of *F. rubra* effectively spread by rhizomes (St. John et al., 2012). Seedlings of *F. rubra* have a moderate growth rate to the full size and cover, and because of that it is best to sow it in the mixture of companion plants, such as *Lolium* sp., *Festuca* sp., *Trifolium* sp. *Medicago* sp., *Poa* sp. *Brassica* sp., *Triticum* sp. that act as nursery plants providing well sheltered environment for developing seedlings of *F. rubra* (St. John et al., 2012; Engelhardt and Hawkins, 2016). In the seed mixture, *F. rubra* is mixed with 20%–60% depending on application purpose (St. John et al., 2012). *F. rubra* can seed alone in around 0.30–0.35 m row spacing. This plant species has 1,350,000 seeds/kg and 900 kg/ha of seed yields (Yoder, 2000). Furthermore, *F. rubra* starts to grow in spring, it slows down in summer, and from late summer until freezing it grows vigorously. Flowering time of *F. rubra* is from May till September depending on the climate (Smoliak et al., 1981).

3.2 Biorecultivation of F. rubra L. on fly ash deposits (TENT, Serbia)

The disposal of fly ash at the TENT landfill (Obrenovac) is continuous and the biological recultivation must be carried out continuously every spring and autumn of the current year in order to reduce the effect of wind erosion. In spring, sowing works are carried out on embankments built during winter, and in autumn on those formed during summer. On the embankments that are being built, the irrigation system is installed, and sowing can be done till vegetation cover is developed. Furthermore, the seeds of the mixture of grasses and leguminoses are carried directly on the fly ash, due to large areas that should be recultivated. The seed is purchased up to 30 days prior to sowing and stored in TENT warehouse as well as mineral fertilizers.

The vegetation cover is formed from annual nursery cultures (*S. cereale, Avena sativa, Hordeum sativum, S. vulgare var Sudanese*), leguminoses (*Vicia sativa, Medicago sativa, Lotus corniculatus*) and mixes of perennial grasses (*F. rubra, D. glomerata, P. pratensis, L. multiflorum, L. perenne, P. pratense, F. pratensis*). Annual plants with their fast growth protect long-term grasses from low/high temperatures and drying, and they prevent wind erosion.

In spring (15.03–15.04) and autumn (01.09–30.09), the first grass-legume mixture consists of *V. sativa* (60 kg/ha), *S. vulgare var. sudanese* (90 kg/ha), and perennial plants (120 kg/ha): *F. rubra* (35%), *D. glomerata* (20%), *P. pratensis* (10%), *L. multiflorum* (15%), *M. sativa* (10%), and *L. corniculatus* (10%) (Table 6.1). The second grass-legume mixture consists of *V. sativa* (50 kg/ha), *H. vulgare* (100 kg/ha), perennial plants (120 kg/ha): *F. rubra* (30%), *D. glomerata* (20%), *L. multiflorum* (15%), *L. perenne* (15%), *M. sativa* (10%), and *L. corniculatus* (10%) (Table 6.1). The most favorable period for sowing is autumn, with the sowing finished by the end of September, which would avoid the adverse effects of frost on young plants.

The fertilization is carried out by mineral fertilizers NPK (15 : 15 : 15 or 8 : 16 : 24) (Table 6.1) which is most suitable in the amount of 1000 kg/ha. Further, fertilization is carried out the next season by nitrogen fertilizers (KAN) (Table 6.1) in the amount of 300 kg/ha. Irrigation or wetting of sowing areas is necessary until the vegetation cover is completed. The irrigation of the embankment is done by a stationary sprinkler system. On the flat part of fly ash deposits, there is a stationary wetting system – cannons. It is recommended that the irrigation system should stay for the next 2 years.

3.2.1 Agrotechnical technology

Sowing of grass–legume mixture can be done manually and mechanically (Table 6.1). *Hand sowing* is more convenient on slopes whereas machines are used along the up-

Table 6.1 Grass–legume mixture, fertilization, and agrotechniques applied on the fly ash deposits (TENT-A, Obrenovac, Serbia).

Perennial plants	Fertilization	Seeding	Fly ash area	Tools
Grasses	NPK	Manuall	Embankments	Hoe, roller
Festuca rubra	KAN	Machines	Flat part	Plate, cyclone,
Dactylis glomerata		Hydroseeding	Steep slope	seed drill,
Lolium multflorum				harrow, roller
Lolium perenne				Hydrogel
Poa pratensis				Hydrosseder
Sorghum vulgare var.				
sudanese				
Hordeum vulgare				
Leguminoses				
Medicago sativa				
Lotus corniculatus				
Vicia villosa				

per flat part of the fly ash landfill. Before manually sowing, it is mandatory to irrigate embankments of fly ash up to 2 h. The depth of seedbed is 10 cm and the row spacing is about 16 cm. Seeds and fertilizers are dispersed manually (10 kg NPK/acres). The seed must be covered with fly ash. After sowing it is necessary to water the sown areas. *Machine sowing* is done mechanically using agricultural machinery, such as: plate, cyclone, seed drill, harrow, and roller (Table 6.1). First, plate performs 4–6 cm holes in fly ash creating favorable conditions for the development of grass-leguminous mixtures. A cyclone is used to disperse the mineral fertilizer. Behind the cyclone, a seed driller processes fly ash inserting a mixture of grass. The next technical measures are drilling and rolling. After sowing it is necessary to water these areas. *Hydroseeding* is often used to form grass cover on steep dikes by a hydroseeder and implies mixing of seeds, fertilizers, and adhesives in order to provide fast and uniform grass cover (Table 6.1). Hydrogel is a biopolymer of organic origin which is present in the form of white crystals with different granulations. It is inserted into substrate at a depth of 5–15 cm, remaining not degraded for 4 years in the substrate, and increasing the volume by 10 times in humid conditions. The water absorption capacity of hydrogel is 100 times higher than the clay particles of the same size. It absorbs surface moisture and does not allow the loss of water from the surface of fly ash.

At the passive cassette of fly ash deposits TENT-A in Obrenovac that is 3 years old, from all seeded grass-legume plant species, only 6 remain: *F. rubra, D. glomerata, L. multiflorum, M. sativa, V. villosa,* and *L. corniculatus* with the highest abundance and plant cover of *F. rubra* (2%–100%) (Fig. 6.7). However, at the passive cassette of fly ash deposits 11 years old, only 5 plant species remain: *F. rubra, D. glomerata, M. sativa, V. villosa,* and *L. corniculatus* with the highest abundance and plant cover of *F. rubra* (5%–75%) (Gajić, 2014) (Fig. 7).

Figure 6.7 *Festuca rubra* **growing on the passive cassette of the fly ash deposits.** $PC_{3\ years\ old}$ – passive cassette 3 years old) (A,B) and $PC_{11\ years\ old}$ – passive cassette 11 years old (C,D).

4   The role of *F. rubra* L. in phytoremediation of contaminated sites

4.1   Phytoremediation potential, uptake and transport of metal(loid)s

Ability of plants to uptake, transport, and tolerate metal(loid)s on contaminated sites can be determined by bioconcentration factor (BCF) and translocation factor (TF), that is, these factors can be used to assess the phytoremediation potential and the survival strategy of plants (Baker, 1981; Yoon et al., 2006). According to Baker (1981), BCF presents the ratio of metal(loid) concentration in the root and its concentration in the soil (BCF = C_{Root}/C_{Soil}) indicating the metal(loid) accumulation in the roots and its removal from the soil. Furthermore, TF presents the ratio of the metal(loid) concentration in the leaves and its concentration in the root (TF = C_{Leaf}/C_{Root}) indicating the potential of plants to transfer metal(loid)s from root to shoot (Baker, 1981). Plants that have a combination of BCF < 1 and TF < 1, BCF < 1 and TF > 1, BCF > 1, and TF < 1 show phytostabilization potential and they are excluders, because they possess mechanisms that maintain low uptake of soil-metal contents and small shoot-metal contents (Liu et al., 2004; Yoon et al., 2006). However, plants with BCF > 1 and TF > 1 are good accumulators and they are suitable for phytoextraction (Yoon et al., 2006).

The metal(loid)s uptake and transport in plants are related to distribution and tolerance mechanisms in roots and leaves. Plant adaptive response to deficiency or toxicity of metal(loid)s is associated with the maintenance of metal(loid) homeostasis in cells which include the regulation of the transporters responsible for the uptake, efflux, translocation, and sequestration of elements in plants (Memon and Schröder, 2006). According to Arif et al. (2016) and Castro et al. (2018), ZIP/IRT family of proteins (zinc-iron regulated protein permease), NRAMP protein (natural resistance-associated macrophage protein), HM-ATPase, CDF (cation diffusion facilitator), CAX (cation exchanger), and ABC (ATP-binding cassette transporters) proteins are the main families of influx and efflux transporters in plant roots and leaves. According to Clemens (2006), the major transporter system responsible for the uptake of Fe^{2+}, Mn^{2+}, Zn^{2+}, Ca^{2+} is ZIP/IRT. Transport of metal(loid)s from roots to leaves occurs via membrane transporters, amino acids, and/or organic acids (Jabeen et al., 2009).

Plant tolerance to excess metal(loid)s can be achieved by excluding the element from further course of transport through the restricted uptake via: (1) binding the element to the cell wall; (2) preventing their entry into the cytoplasm by reducing the metal(loid) uptake (modification of ion channels or activation of efflux pump for ejection of the metal(loid) from the cell); (3) forming the chelate complexes (metal(loid) – organic acids) on the outer side of the plasma membrane (Tong et al., 2004). Accumulation of metal(loid)s in plant cell is achieved through the metal–chelate complexes, such as glutathione (M-GSH), phytochelatines (M-PCs), organic acids (M-OA), and amino acids (M-AA), as well as the compartmentalization in vacuoles and/or sequestration of metals by chaperones (proteins involved in intracellular transport of metals to the place of activation) (Tong et al., 2004; Anjum et al., 2015; Singh et al., 2016).

4.2 Phytoremediation potential of F. rubra L. grown on fly ash deposits

F. rubra growing on fly ash deposits showed a high phytostabilization potential for As, B, Cu, Zn, Mn, Mo, and Se, and can be considered as an excluder plant suitable for phytoremediation of contaminated sites (Gajić, 2014; Gajić et al., 2016). In the condition of toxicity or deficiency of some metal(loid)s, this grass possesses a different adaptive mechanism of uptake, translocation, distribution, and sequestration, preventing the uptake of high concentrations of metal(loid)s and their transport to the leaves or it activates other mechanisms that are responsible for influx of essential elements and further transport to the leaves regulating the homeostasis in the plant cells (Gajić, 2014).

Arsenic (As) is a non-essential chemical element for plants (Marschner, 1995). Availability, uptake and toxicity of As in plants depends on its concentration in soil and phosphorus content in soil (Singh and Ma, 2007). Arsenate (AsV) and arsenite (AsIII) are forms of As that plants can uptake by roots. Toxic effects of As expressed through the production of reactive oxygen species (ROS) that react with lipids, proteins, pigments, and nucleic acids leading to lipid peroxidation, plasma membrane damage, inactivation of the enzyme which decrease plant function and lead to the death of the plant cell (Singh et al., 2006). As (V) is analogous to phosphate, thus disrupts phosphate metabolism by substituting phosphate in phosphorylation reactions, such as ATP synthesis whereas As (III) reacts with sulfhydryl groups of enzymes (Meharg and Hartley-Whittaker, 2002).

Arsenic concentrations in the leaves of *F. rubra* growing on both passive cassettes at the fly ash deposits on TENT-A were toxic (Table 6.2). Generally, As accumulation in plants is relatively small due to its low availability in soil, limited root uptake, and restricted translocation from root to leaf (Wang et al., 2002). However, plants that grow on the soil contaminated with As contain high concentrations of As in tissues (Tripathi et al., 2007) which is in accordance with our results. Higher concentrations of As in root and leaves in *F. rubra* growing on the passive cassettes of fly ash deposits compared to the control site can be related to toxic As concentrations in fly ash and high available concentration of As (Table 6.2). Also, high content of As in roots can be a consequence of a generally small amount of phosphorus in fly ash, and according to Wang et al. (2002), low content of phosphorus increases the capacity of the root for arsenate (AsV) by 2.5 times indicating an increased synthesis of As-Pi transporter of the root cell plasma membrane.

In *F. rubra* growing on fly ash deposits, BCF < 1, and TF < 1 indicating a higher amount of As in roots than in leaves is probably due to sequencing, complexing, and accumulation of As at the root level. A higher concentration of As in the plant roots and smaller in the leaves was noted by Carbonell-Barrachina et al. (1998), which was in line with our results. Porter and Peterson (1977) found that in the tolerant individuals of *A. capillaris,* the largest amount of As is accumulated in the root, and only 6%–25% is transported to the leaves. Higher amounts of As in the root of *F. rubra* on fly ash deposits can be the result of increased synthesis of the arsenate reductase enzyme (AR) which reduces As (V) to As (III). However, As concentration in the leaves

Table 6.2 Total and available metal(loid) concentrations ($\mu g/g$) in fly ash deposits on TENT-A and control site, root and leaf metal(loid) concentrations, bioconcentration (BCF) and translocation (TF) factor of *Festuca rubra*.

Parameter	As	B	Cu	Mn	Zn	Mo	Se
Total concentration							
C	9.14	6.45	35.91	504.68a***b***	84.35a***b***	1.34	0.43
PC$_3$ years old	18.32a***c***	46.44a***c***	54.06a***c***	233.82c***	47.18c***	1.67a***	2.72a***cns
PC$_{11}$ years old	13.70b***	34.44b***	40.58b**	185.02	24.88	2.98b***c***	2.77b**
Range[d]	4.4–9.3	22–45	13–24	270–525	45–100	0.7–1.5	0.25–0.34
Available concentration							
C	0.143	0.292bns	3.526a***b**	22.702a***b***	4.787a***b***	0.015	0.028
PC$_3$ years old	0.329b***	1.448a***c***	1.206c***	1.472c***	0.952c***	0.022a***c***	0.052a***c**
PC$_{11}$ years old	0.410a***c*	0.304	0.647	0.785	0.484	0.014bns	0.037bns
Root							
C	4.70	8.27b*	6.18	22.16a***b***	34.42a***b***	1.05bns	2.77
PC$_3$ years old	7.59a***cns	10.67a***c***	10.11a***c***	19.00cns	23.23cns	1.49a***c***	4.76a***c**
PC$_{11}$ years old	7.11b***	7.00	8.97b**	19.25	23.08	1.04	3.96b***

Leaves						
C	3.98	3.96	8.59[a]***[b]***	38.18[a]ns[b]***	20.86[a]***[b]***	0.91	2.32
$PC_{3\ years\ old}$	6.08[a]***[c]ns	52.55[a]***[c]***	5.52[c]ns	37.01[c]***	12.70[c]*	3.25[a]***[c]***	2.57[a]*[c]ns
$PC_{11\ years\ old}$	5.39[b]***	23.65[b]***	4.94	13.71	11.69	2.24[b]***	2.60[b]*
D_R*	—	5–30	2–5	10–30	10–20	0.1–0.3	—
N_R	1–1.7	10–100	5–30	30–300	27–150	0.2–5.0	0.01–2.0
T_R	5–20	50–200	20–100	400–1000	100–400	10–50	5–30
BCF						
C	0.526	1.953	0.173	0.044	0.409	0.788	9.410
$PC_{3\ years\ old}$	0.418	0.238	0.190	0.082	0.501	0.632	1.587
$PC_{11\ years\ old}$	0.534	0.215	0.223	0.107	0.973	0.505	2.197
TF						
C	0.845	0.503	1.467	1.734	0.603	0.872	0.902
$PC_{3\ years\ old}$	0.823	5.003	0.555	1.993	0.578	2.187	0.679
$PC_{11\ years\ old}$	0.758	3.306	0.555	0.708	0.504	2.211	0.561

ANOVA, Data represent means; C, control site; *D_R, deficiency range of metal(loid)s content in leaves; n = 15; ns = not significant; N_R, normal range of metal(loid)s content in leaves; $PC_{3years\ old}$, passive cassette 3 years old; $PC_{11years\ old}$, passive cassette 11 years old; T_R, toxic range of metal(loid)s content in leaves (Kabata-Pendias, 2011).

Note: *$P < 0.05$; **$P < 0.01$; ***$P < 0.001$

[a]C-L3.

[b]C-L2.

[c]L3-L2.

[d]Range of metal(loid)s content in soils (Kabata-Pendias, 2011).

of *F. rubra* growing on both passive cassettes of fly ash deposits were toxic, which could indicate that detoxification mechanisms were not effective enough to prevent its accumulation in the leaves and potentially negative effects on the functioning of these plants. According to Zhao et al. (2003), it has not yet been proven whether the PCs-As (phytochelatin – arsenic) complex breaks down in the vacuole or the inorganic As, as a decomposition product comes out of the vacuole, which is why its accumulation in the leaves increases.

Boron as an essential element for plants plays a primary role in the biosynthesis and cell wall structure, lignification, maintenance of plasma membrane integrity, metabolism and sugar transport, and metabolism phenolic compounds (Marschner, 1995; O'Neill et al., 2004). The toxic effects of B are most expressed by binding B for the ribosome moieties of the ATP, NADH, NADPH, or RNA molecules, thereby disturbing metabolic processes (Reid et al., 2004).

Concentrations of B in the leaves of *F. rubra* were toxic on the passive cassette of fly ash deposits 3 years old, whereas on the passive cassette of fly ash deposits 11 years old were in the normal range (Table 6.2). The results of this study showed that the concentrations of B in the root and the leaves of *F. rubra* growing on the fly ash deposits were higher compared to the control site, which can be related to the high total and available concentration of B (Table 6.2). In *F. rubra* growing on both passive cassettes of fly ash deposits, BCF < 1, TF > 1 indicating that a larger amount of B is accumulated in leaves than in roots and this can be due to the activation of BOR2 and BOR4 transporters in roots, that are responsible for exporting high B concentrations in leaves when B is present in excess concentrations in soil (Takano et al., 2008).

Copper is an essential chemical element necessary for the growth and development of plants, it is a cofactor of many metalloproteins and enzymes involved in the transfer of electrons, it participates in the metabolism of the cell wall, carbohydrates, lipids, and nitrogen, regulation of oxidative stress and biogenesis molybdenum cofactors (Marschner, 1995; Yruela, 2005). Therefore, ions of Cu are cofactors of numerous enzymes: Cu/Zn superoxide dismutase (SOD), ascorbate oxidase, plastocyanins, and polyphenol oxidases. Deficiency of Cu disturbs the photosynthesis by disintegrating the chloroplast thylakoid membranes, reducing the plastocyanine biosynthesis, the amount of chlorophyll and carotenoids, and reducing the content of unsaturated fatty acids, indicating that the intracellular Cu concentration is to be strictly controlled (Yruela, 2005). Cu homeostasis in the plant cell is regulated by mechanisms which are able to maintain the optimal content for cellular processes, and they can protect the cells against the negative effects of Cu deficiency in plants.

Concentrations of Cu in the leaves of *F. rubra* growing on both passive cassettes of fly ash deposits on TENT-A were deficient and they were lower in relation to control site (Table 6.2). In addition, the total content of Cu in fly ash of passive cassettes on TENT-A was toxic whereas the available concentrations of Cu were low (Table 6.2). *F. rubra* on both passive cassettes had BCF < 1, TF < 1 indicating higher accumulation of Cu in the root than in the leaves and that can be related to the effective mechanism of Cu uptake, and limited capacity of its translocation from the root to the leaf. In response to Cu deficiency, plants activate high-affinity Cu acquisition in order to increase the absorption of Cu from the soil where Cu-responsive transcription factor SPL7 (SQUAMOSA promoter-binding protein-like7) activates ferric reductase oxidase (FRO4/5) and Cu transporters COPT1/2/6 (putative copper influx transporters) (Yamasaki et al., 2009; Bernal et al., 2012; Migocka and Malas, 2018) as well as enhance the expression of Cu chaperones (CCH) which are involved in trafficking Cu to cuproproteins and compartments in the cells (Sanceonon et al., 2003). Under Cu deprivation, cytosolic, stromal, and chloroplast Cu/Zn dismutases decrease whereas the ferric superoxide dismutase 1 (FSD1) increase and PAA2/HMA8 (ATPase) deliver Cu to the plastocyanine (Abdel-Ghany and Pilon, 2008; Tapken et al., 2012).

Manganese is an essential chemical element for plants, and has an important role in redox processes as a cofactor of many enzymes (Marschner, 1995). Thus, Mn enters the Mn-superoxide dismutase (MnSOD), catalase, phenylalanine ammonium-lyase (PAL), (Dučić and Polle, 2005). Mn builds a catalytic center of the OES (oxygen evolving complex) in PSII (Barry et al., 2006) and participates in the assimilation of nitrates, synthesis of chlorophyll, carotene, ascorbic acid, activates the lipid biosynthesis, and catalyzes the removal of O_2^- and H_2O_2 (Dučić and Polle, 2005). Deficiency of Mn in plants most often leads to high sensitivity to peroxidal stress, decreases photosynthesis and the amount of chlorophyll, reduces MnSOD activity, and increases the transpirations due to the reduction of epicuticular layers of wax (Hebbern et al., 2009).

Concentrations of Mn in the leaves of *F. rubra* on the passive cassette of fly ash deposits 11 years old were in deficiency whereas those on the passive cassette of fly ash deposits 3 years old were in normal range (Table 6.2). The concentrations of Mn in the roots and the leaves of *F. rubra* on both passive cassettes of the fly ash deposits on TENT-A were lower compared to the control site, which can be mostly associated with a significantly lower total and available amount of Mn in fly ash (Table 6.2). *F. rubra* growing on $PC_{3 \text{ years old}}$ has BCF < 1 and TF > 1 which indicates that this species accumulates a significant amount of Mn in leaves and this can be explained by the effective activation of the OsnRAMP5 protein, which was found to be responsible for the accumulation of Mn in leaves (Sasaki et al., 2012). However, *F. rubra* growing on $PC_{11 \text{ years old}}$ has BCF < 1 and TF < 1 indicating higher amount of Mn retained in the roots than in the leaves. Plants growing on fly ash deposits activate different mechanisms in the conditions of Mn deficiency, such as activation of IRT1 protein and ZIP3/4/5 transport proteins which are involved in Mn uptake (López-Millán et al., 2004; Pedas et al., 2005). Large amount of Mn in the root of *F. rubra* on the fly ash deposits could be explained by high binding capacity of Mn for the root cell walls where 83% of Mn is accumulated in the root vacuole (Pedas et al., 2005). Also, a smaller amount of Mn in the leaves can be a consequence of the reduced capacity for the uptake of Mn by root via NRAMP1 protein and reduced transport to leaves via OsNRAMP5 protein (Sasaki et al., 2012). However, under Mn deficiency NRAMP3/4 proteins transport Mn in the vacuole of leaves and according to Lanquar et al. (2010), these proteins are essential for maintaining photosynthetic activity.

Zinc is an essential chemical element for plants that has been found to act as a catalytic or structural co-factor in a large number of enzymes and regulatory proteins (Marschner, 1995). Zn is a necessary component for enzymes, such as carbonic anhydrase, Cu/Zn SOD, phospholipase, RNA, and DNA polymerase (Marschner, 1995). Zn protects the plants from oxidative stress (Cakmak, 2000) and can act as an intracellular signal molecule (Yamasaki et al., 2007). The deficiency of Zn is manifested by increased production of ROS due to the decreased activity of Cu/Zn SOD and carbonic anhydrase, inhibition of protein synthesis, damage of membrane proteins and chlorophyll, and inhibition of the photosynthesis process (Marschner, 1995; Cakmak, 2000).

Concentrations of Zn in *F. rubra* growing on both passive cassettes of the fly ash deposits on TENT-A were deficient (Table 6.2). In the roots and the leaves of *F. rubra,* concentrations of Zn were lower compared to control site which can be associated with a significantly lower total and available amount of Zn in fly ash (Table 6.2). In *F. rubra* growing on both passive cassettes of the fly ash deposits BCF < 1 and TF < 1 indicating a higher amount of Zn in the roots than in the leaves, suggesting an intense biosynthesis of phytosiderophores, which in Gramineae make it easier to uptake Zn by root in the form of the Zn-MAs complex (mugineic acid family phytosiderophores) (Marschner, 1995). Also, the activation of the ZIP1/3/4 transporters and MTP8 (metal tolerance transporter) can provide an effective input of Zn from the soil solution into the root cells (Bashir et al., 2012). A higher Zn content in the root of *F. rubra* than in the leaves can be explained by the expression of ZIF1 protein (zinc-induced facilitator), which increases the Zn accumulation in vacuole's root and it reduces translocation in leaves (Sinclair and Krämer, 2012). However, leaf transport can be reduced due to the reduced synthesis of active transporters that provide the flow of Zn from root to leaves (IRT3, OsZIP4, OsZIP5, OsZIP8, MTP3, HMA2, and HMA4) (Arrivault et al., 2006; Sinclair and Krämer, 2012).

Molybdenum is an essential chemical element for plant growth (Marschner, 1995; Mendel and Bittner, 2006). The quantities of Mo needed by plants are very small, but are essential for redox reactions of C, N, and S (Mendel, 2011). Mo becomes active in the cell only after the formation of the catalytically active site Mo-enzyme (Moco) (Mendel and Bittner, 2006; Mendel, 2011). Moco (molybdenum cofactor) represents the catalytically active site of the enzyme, which is produced by inserting Mo into the metal-binding compound protein (MPT). It is believed that a certain amount of Moco is essential for the maintenance of homeostasis Mo in the cell (Mendel and Bittner, 2006). However, reduced MTP synthesis or blocked biosynthesis of Moco leads to the loss of metabolic function, since all Mo-dependent enzymes lose their activity, which inevitably leads to the death of the plant (Mendel, 2007).

The concentrations of Mo in the leaves of *F. rubra* on both passive cassettes of fly ash deposits on TENT-A were in the normal range for plants (Table 6.2). These Mo concentrations in the leaves were higher compared to control, which can be related to high total concentrations of Mo in fly ash (Table 6.2). *F. rubra* growing on fly ash deposits had BCF < 1 and TF > 1, which means that this species had a higher amount of Mo in the leaves than in the roots. In *F. rubra*, Mo is effectively transported from roots into leaves suggesting the possible presence of MOT1 and MOT2 sulfate transporters that have been found to have an essential role in intracellular transport and the allocation of Mo between roots and leaves in plants (Fitzpatrick et al., 2008; Tejada-Jimenez et al., 2009). The concentrations of Mo in the leaves of *F. rubra* growing on fly ash deposits were optimal for normal functioning. According to Gasber et al. (2011), the amount of Mo in leaves is correlated with the total concentration of Moco and its precursor MPT, which can indicate effective regulation and activation of all Mo-enzymes.

Selenium is a chemical element that has been found not to be essential for plants (Marschner, 1995). However, Se at low and moderate concentrations has a beneficial effect, and at high concentrations, it can be toxic to plants (Hamilton, 2004). Low concentrations of Se can stimulate root and leaf growth, as well as increase the plant's antioxidant capacity (Peng et al., 2002; Hasanuzzaman et al., 2010). However, Se toxicity can reduce total pool of S, which affects the reduced synthesis of secondary S metabolites (non-protein thiols) and it also competes with S for the biochemical function of the Fe-S cluster, and increased production of ROS. According to Terry et al. (2000), toxicity of Se is manifested when it comes to the non-specific substitution of the cysteines and methionines by their Se analytes (selenocysteine, SeCys and selenometionine, SeMet). The toxic concentration of Se also depends on its form, that is, the presence of selenate (SeO_4^{2-}) and/or selenite (SeO_3^{2-}) (Terry et al., 2000). Selenite is more toxic than selenate due to its rapid conversion to Se amino acids (Zayed et al., 1998). However, other authors state that selenate is more toxic than selenite, since its uptake is much faster than selenite's (Pilon-Smits et al., 1999). Selenate is the dominant form in alkaline environments, while selenium and selenide are the dominant forms of Se in neutral or acidic environment (Kabata-Pendias, 2011).

Concentrations of Se in the leaves of *F. rubra* growing on both passive cassettes on fly ash deposits on TENT-A were in normal range (Table 6.2). Furthermore, Se concentrations in the root and leaves of *F. rubra* were higher in relation to control site, which can be associated with significantly high total concentrations of Se in fly ash (Table 6.2). *F. rubra* growing on fly ash deposits had BCF > 1, and TF < 1 indicating an accumulation of Se in the roots. Zayed et al. (1998) showed that leaf/root ratio ranges from 0.6 to 1.0 in SeMet plants and ≤ 0.5 when plants take up selenite which can suggest that selenite and SeMet are selenium forms that are probably accumulated in the roots of *F. rubra*. Uptake of selenite occurs by phosphate (Pi) and OsNIP2;1 transporters (Li et al., 2008). It has been found that selenium is retained at a higher level because it is rapidly translated to SeMe, and that only 10% of selenite is transported from root to leaves (Zayed et al., 1998; Li et al., 2008).

5   Physiological and morphological response of *F. rubra* L.

5.1   *Photosynthesis, pigments, and antioxidants of* F. rubra L. *grown on fly ash deposits*

Photosynthetic efficiency (Fv/Fm) and the ratio of Fm/Fo in *F. rubra* growing on fly ash deposits (TENT-A) were below optimal range (0.750–0.850; 5.0–6.0, Bjorkman and Demmig, 1987) (Table 6.3) indicating high sensitivity of PSII to stress and reduced vitality. Toxic concentrations of As and B, and the deficiency of Mn and Zn in the leaves of *F. rubra* decrease chlorophyll *a* fluorescence parameters, such as Fm (maximum fluorescence), Fv (variable fluorescence), and Fv/Fm indicating the photoinactivation of PSII efficiency at the donor side (Gajić, 2014; Gajić et al., 2016). In addition, reduced values of Fv indicate a decrease in non-photochemical dissipation of energy due to a lower amount of carotenoids which leads to overall decrease in

Table 6.3 Chl *a* fluorescence parameters, content of pigments (mg/g), malondialdehyde (nmol/g), phenolics (mg/g), ascorbic acid (mg/g), and total antioxidant activity (mg/mL) in *Festuca rubra* growing on control site and fly ash deposits on TENT-A.

Parameters	C	$PC_{3 \text{ years old}}$	$PC_{11 \text{ years old}}$
Chlorophyll a fluorescence			
Fo	$0.18^{ansbnscns}$	0.18	0.17
Fm	$0.87^{a***b***}$	0.39	0.57^{c**}
Fv	$0.69^{a***b***}$	0.25	0.41^{c**}
$t_{1/2}$	$163.6^{ansbnscns}$	154.87	158.80
Fv/Fm	$0.781^{a***b***}$	0.543	0.691^{c***}
Fm/Fo	$4.66^{a***b***}$	2.22	3.41^{c***}
Pigments			
Chl *a*	$3.60^{a***bns}$	2.46	3.21^{c*}
Chl *b*	$1.26^{ansbnscns}$	0.90	1.22
Chl *a + b*	4.86^{a**bns}	3.36	4.43^{c**}
Total carotenoids	$1.01^{a***b***}$	0.78^{cns}	0.87
Anthocyanins	0.501^{bns}	0.620^{a*cns}	0.563
Lipid peroxidation			
MDA	0.79	1.87^{a**c**}	1.25^{b***}
Antioxidants			
Free phenolics	7.85	12.85^{a**cns}	13.36^{b***}
Bound phenolics	17.96	26.61^{a*cns}	32.97^{b***}
Total phenolics	25.89	39.46^{a**cns}	45.74^{b***}
Ascorbic acid	0.595	$1.142^{a***cns}$	1.076^{b***}
Total antioxidant activity	$0.288a^{***bns}$	0.446	0.254^{c***}

ANOVA, Data represent means; *C*, control site; $n = 15$; *ns* = not significant; $PC_{3years\ old}$, passive cassette 3 years old; $PC_{11years\ old}$, passive cassette 11 years old.
Note: $*P < 0.05$; $**P < 0.01$; $***P < 0.001$;
[a]C-L3.
[b]C-L2.
[c]L3-L2.

photosynthetic activity of *F. rubra*. However, the changes in Fo parameter (minimal fluorescence) and $t_{1/2}$ (PQ, plastoquinone pool size of electron acceptors at the reducing side of PSII) were not significant, and it can be assumed that the photosynthetic electron transfer from OEC (oxygen evolving complex) to the PSII reaction center was effective, thus, electron transfer from primary quinone acceptor Q_A to the secondary quinone acceptor Q_B was increased, which could indicate that the acceptor side of PSII is not inhibited by metal(loid)s (Gajić, 2014). In *F. rubra* growing on the passive cassette 11 years old on fly ash deposits, photosynthetic efficiency is still below optimum, but it was higher than on the passive cassette 3 years old, indicating that this grass possesses active mechanisms for photosynthesis recovery and more functionally active PSII centers which increase tolerance of the photosynthetic apparatus in unfavorable fly ash conditions (Gajić, 2014).

In the leaves of *F. rubra* growing on fly ash deposits, the content of chlorophylls (Chl *a*, Chl *a + b*) was lower in relation to control site (Table 6.3) which indicates that this photosynthetic pigments show high sensitivity to stress that prevail on fly ash, such as drought, intense light, radiation, as well as toxic concentrations of As and B, but also deficiency of Cu, Mn, and Zn in leaves (Gajić, 2014). The values of ratio of Chl *a/b* were in normal range for plants (2.5–3.5, Lichtenthaler and Buschmann, 2001) which can indicate that *F. rubra* possesses high functional potential for maintaining physiological activity.

During the season, *F. rubra* on the fly ash deposits maintained green color and large content of Chl *a*, Chl *a* + *b*, and Chl *a/b* and that could be potentially explained by the existence of a *stay-green* mutation of *sid* gene (Gajić, 2014). According to Hauck et al. (1997), in *stay-green* mutants the amount of Chl *a*, Chl *b*, Tot Carot, and Chl *a/b* ratio increase or remain stable over the season. This mutation was first described by Thomas (1987) in *F. pratensis*, whereby at the end of the life cycle, leaves of the mutant line Bf993 were green, while in the wild type, they were completely yellow. The type C - *stay green* mutation in *F. pratensis* is classified into a non-functional "cosmetic" category, where chlorophyll is retained and photosynthetic activity decreases (Thomas and Howarth, 2000; Armstead et al., 2007). During the aging of leaves, there is no degradation of chlorophylls, thus, there is damage in the catabolic pathway of chlorophyll, wherein the membrane proteins of the thylakoid remain stable, such as the LHCPII (light harvesting protein complex of photosystem II) and D1 protein of reaction center of PSII (Thomas, 1987; Thomas et al., 1989; Sakuraba et al., 2012; Kusaba et al., 2013). In the normal process of leaf aging, LHCP proteins are not strongly attached to chlorophyll, which makes them suitable for the action of proteolytic enzymes. The activity of enzymes chlorophyllase, Mg dehelatase, and red chlorophyll catabolit reductase (RCCR) remains unchanged, while the activity of phaeophorbide *a* oxygenase (PaO) in the mutants is blocked, indicating that the mutation is localized to the PaO gene responsible for the loss of green color in the old leaves (Thomas et al., 1989; Roca et al., 2004; Ren et al., 2007). The same authors have found that the block at the PaO enzymes lead to the accumulation of highly polar green components, such as chlorophyllide *a* and phaeophorbide *a* and plant leaves stay green.

The concentration of *total carotenoids* in the leaves of *F. rubra* growing on fly ash deposits was lower compared to control site (Table 6.3) and can be a consequence of toxic concentrations of As and B as well as deficiency of Cu, Zn, and Mn (Gajic´, 2014). The low content of carotenoids in *F. rubra* indicates reduced photoprotection of PSII, thus, the content of β-carotene, lutein, and the content of zeaxanthin probably decrease leading to the dissociation of the LHCP complex from the PSII reaction center, reduction of dissipation of excited energy in the form of heat and inefficient quenching of triplet chlorophyll ($^3Chl^*$) and single oxygen ($^1O_2^*$) and all that together decreases the photosynthetic activity.

The content of *malondialdehyde (MDA)* in the leaves of *F. rubra* growing on fly ash deposits was higher compared to control site (Table 6.3) and indicates oxidative stress. MDA is considered as indicator of the production of free radicals and lipid peroxidation of membranes (Weber et al., 2004). Schmid-Siegert et al. (2012) observed that in the cells of chloroplast, the highest amount of MDA originated from TFAs (trienoic fatty acid, α-linoleic acid, 75%) (Falcone et al., 2004) and it leads to decrease of photochemical efficiency (Vollenweider et al., 2000), which is in line with our results.

Large amount of MDA in the leaves of *F. rubra* can be a result of high temperature, drought, toxic As concentration and B, as well as the Mn and Zn deficiencies in the leaves.

The increased amount of *anthocyanins* in the leaves of *F. rubra* growing on the passive cassette 3 years old in comparison to control site (Table 6.3) can be explained by toxic concentrations of As and B, and the deficiency of Zn. Anthocyanins protect the photosynthetic tissue from photoxidation through a reduced amount of available light and a lower amount of excited energy, reducing the generation of O_2^- and photoinhibition of PSII (Merzlyak and Chivkunova, 2000; Quina et al., 2009). According to Das et al. (2011), anthocyanins are responsible for the redox state of PQ pool in the photosynthetic electron transport. Furthermore, anthocyanins inhibit lipid peroxidation, that is, C3G (cyanidine 3-O-beta-D-glucoside) and its aglycone may reduce the formation of MDA in the liposomal system where it can directly quench the O_2^-, OH and H_2O_2 (Tsuda et al., 1996; Yamasaki et al., 1996).

In the leaves of *F. rubra* growing on fly ash deposits, the content of *phenolics (free, bound, and total phenolics)* was higher compared to the control site (Table 6.3), and that can be a stress response to the harsh conditions on fly ash, such as high temperature, drought, As, and B toxicity, as well as Cu, Mn, and Zn deficiency. The highly soluble fractions of phenolics (free phenolics) can act as effective antioxidants whereas increased content of bound phenolics points out to the enhanced polymerization of phenolics to the lignin. The antioxidant properties of phenolics are reflected in their ability to: (1) directly remove ROS (O_2^-, OH, H_2O_2, $^1O_2$, peroxyl radical) by donating electron or H atoms, (2) chelating the ion of metal(loid), (3) changing the kinetics of peroxidation, by modifying the lipid packing and reducing the fluidity of the membranes, (4) stabilizing and delocalizing an unpaired electron phenoxyl radical (Rice-Evans et al., 1996; Arora et al., 2000; Majer et al., 2014). According to Santos et al. (2011), the balance between elements, such as Cu, Zn, and Mn can affect the competition between different classes of phenolic compounds.

The content of *ascorbic acid (AsA)* in the leaves of *F. rubra* growing on the fly ash deposits was higher than on the control site (Table 6.3) point out to the physiological significance of ascorbate as an antioxidant, photoprotectant, regulator of the process of photosynthesis, and lignification of cell walls (De Tullio et al., 1999; Smirnoff, 2005). Therefore, increased amount of ascorbate may have an important role in the accumulation of anthocyanins and flavonoids, since it is a co-factor of enzymes involved in their biosynthesis (Page et al., 2012). Furthermore, ascorbate can alleviate the inactivation of the PSII reaction centers by acting as an alternative donor of the electron (Mano et al., 1997). Ascorbate prevents damage of D1 proteins by donating electrons to Tyr_Z^+ and P_{680}^+ (Tóth et al., 2011). In order to maintain the electron transport from AsA to PSII, the ascorbate regeneration system is very important, and after electron donation, oxidized ascorbate is rapidly regenerated through the Halliwell-Asada cycle to its reduced form (Potters et al., 2002).

Antioxidant system is a protection system that removes or prevents the formation of ROS in conditions of oxidative stress and maintains the metabolic and structural integrity of organisms (Gill and Tuteja, 2010). DPPH (1,1-diphenyl-2-picrylhydrazyl) is a free radical which is used to evaluate the total antioxidant activity of compounds that have the ability to give H and/or remove free radicals (Brand-Williams et al., 1995; Molyneux, 2004). The lower *DPPH antioxidant activity* in *F. rubra* growing on passive cassette 3 years old compared to control site (Table 6.3) suggests that in spite of high content of some components of the antioxidant system (free and bound phenolics, anthocyanins, and ascorbate), it was not sufficient to reduce membrane damage (production of MDA products), and functional molecules (chlorophylls and carotenoids), as well as damage of PSII activity due to toxic concentrations of As and deficiency of Cu. However, in *F. rubra* growing on the passive cassette 11 years old, the total antioxidant DPPH activity was high which coincides with relatively high free and bound phenolics, anthocyanins, ascorbate, and photosynthetic efficiency indicating that this grass over time developed adaptive mechanisms to overcome oxidative stress. The maintenance of redox homeostasis and regulation of cellular antioxidant potential are necessary for plant's survival in the environmental stress conditions (Gill and Tuteja, 2010).

5.2 Leaf morphology of F. rubra L.

Morphological symptoms of leaf damage are expressed in the form of chlorosis and necrosis (Kabata-Pendias, 2011; McCauley et al., 2009). Concentration of chemical elements in the soil and tissue of plants, as well as the determination of physiological and biochemical changes (photosynthesis, amount of chlorophyll, and anthocyanins) are necessary in the determination and early diagnosis of morphological symptoms of metal(loid) deficiency or toxicity.

F. rubra at the control site in May had green leaves whereas in July they were light green and in August leaves were dark green to grayish green (Fig. 6.8A–C). Leaves of *F. rubra* growing on the passive cassette 3 and 11 years old were green in May whereas in July and August, leaves were green, yellow, and red and there were some completely dry plants, that is, symptoms of leaf morphological damages in the form of chlorosis and necrosis were observed (Fig. 6.8D–W). *F. rubra* growing on the passive cassette 3 years old was exposed to high available As and B content, as well as small soluble amounts of Cu and Zn, which coincided with toxic concentrations of As and B in leaves, as well as with deficiency of Cu and Zn. This perennial grass on the passive cassette 11 years old was additionally exposed to the deficiency of Mn. Low vitality of *F. rubra* may indicate that this species was additionally exposed to adverse environmental conditions, such as drought and high temperatures in July and August.

Symptoms of leaf damage in *F. rubra* growing on the passive cassette 3 years old were expressed in the form of necrosis of red, brown, and black color from the top of the leaves to the base and could be related to the toxic concentrations of As and B, but also with the lack of Cu and Zn (Kabata-Pendias, 2011). Chlorosis which is yellow color in the middle of the leaves could occur due to Cu and Zn deficiency and high B concentrations (McCauley et al., 2009). Some leaves were with white tips which may be the result of Zn deficiency (Kabata-Pendias, 2011). Morphological damage of leaves in *F. rubra* growing on the passive cassette 11 years old was in the form of chlorosis light green and yellow color and necrosis black and brown color and all together

Figure 6.8 *Festuca rubra* growing on the control site in May (A), July (B) and August (C); *F. rubra* growing on the fly ash deposits on TENT-A: $PC_{3\ years\ old}$ – passive cassette 3 years old in May (D–F), July (G–I), August (J–M); $PC_{11\ years\ old}$ – passive cassette 11 years old in May (N–P), July (R–T), August (U–W).

may be the result of the synergistic effect of drought, toxic concentrations of As and B, and deficiency of Cu, Zn, and Mn in the leaves (Kabata-Pendias, 2011).

5.3 SEM analysis of leaf surface structure of F. rubra L.

Scanning electron microscopy (SEM) is used to observe the micro-morphology of the surface structure of plant leaves. Furthermore, the shape, size, and chemical composition of fly ash particles can be determined by SEM microscopy using the EDS program (energy-dispersive spectrometry). In general, particle deposition on the surface of plant leaves depends on climatic conditions (direction and speed of the wind, relative air humidity, precipitation, temperature), chemical properties of the elements, particle size, and morphological characteristics of leaves (Beckett et al., 1998; Jamil et al., 2009; Sawidis et al., 2011). Heavy rain and intense wind can wash deposited particles from the surface of the leaves. Periods with increased air humidity, dew, fog, low precipitation, and low intensity winds favor the retention and efficient binding of particles to the leaf surface (Schönherr et al., 2005). Deposited particles are more easily dissolved and metal(loid)s are more easily uptaken through the cuticle or stoma into the tissues leading to the functional damage. Also, smooth leaves with a thick layer of epicuticular wax retain a smaller amount of particles on the surface, unlike hairy leaves or rough surfaces of leaves (Sawidis et al., 2011).

The surface of *F. rubra's* leaves on the control site is uneven. The epidermis is built from the long cells, arranged in arrays, in such a way that one series consists of the protruding, curly, and narrow epidermal cells, and the second series builds wide and less protruding epidermal cells (Fig. 6.9A). On the surface of the wide cells of epidermis, silicate bodies of circular shape are present (Fig. 6.9A, B). Also, on the leaf surface a large number of very fine particles that were arranged in the dent between the cells were detected. The particle diameter of the irregular shape ranged from 7.81–10.3 µm (Fig. 6.9B). SEM (EDS) analysis of the deposited particles on the leaf surface showed chemical elements, such as C, O, Mg, Al, Si, K, Ca, Ti, and Fe (Fig. 6.10A).

The leaf surface of *F. rubra* growing on the passive cassette 3 years old was not damaged, although more particles were detected on its surface than at the control site (Fig. 6.9C). Deposited particles were of spherical or irregular shape (cenospheres and plerospheres of fly ash particles). The diameter of deposited particles ranged from 18.0 to 68.0 µm (Fig. 6.9D). The spectral analysis of the chemical composition of the particles showed the presence of the same chemical elements such as on the control site, except for Mg and Si (Fig. 6.10B).

The leaf surface of *F. rubra* growing on the passive cassette 11 years old was without damage (Fig. 6.9E). A small number of deposited particles were of spherical or irregular shape, and their diameter ranged from 4.84 to 21.1 µm (Fig. 6.9F). SEM (EDS) analysis of the deposited particles on the leaf surface shows the following chemical elements: O, Mg, Al, Si, K, Ca, Ti, and Fe (Fig. 6.10C).

Generally, the results showed that the damage of leaf cuticle and epidermis in *F. rubra* growing on fly ash deposits was not large, although a certain amount of fly ash particles was deposited on the leaf surfaces. Slightly protruding and elongated epidermal cells retain particles on the surface of leaves. However, deposited particles

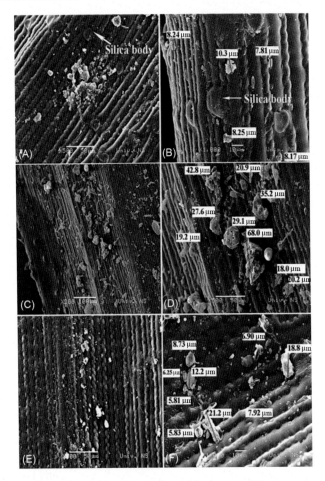

Figure 6.9 SEM micrograph of the upper surface of the leaves of *Festuca rubra* growing on the control site (×500) (A) and fly ash deposits on TENT-A: $PC_{3\text{ years old}}$ − passive cassette 3 years old (×200) (C) and $PC_{11\text{ years old}}$ − passive cassette 11 years old (×200) (E); the diameter of the deposited particles on the upper surface of the *F. rubra* growing on the control site (×1000) (B) and fly ash deposits on TENT-A: $PC_{3\text{ years old}}$ − passive cassette 3 years old (×500) (D) and $PC_{11\text{ years old}}$ − passive cassette 11 years old (×1000) (F).

were easily washed away by rainfall or wind, causing no damage to epidermis due to smooth cuticle. The particles deposited on the surface of leaves can change plant physiology by increasing the temperature of the leaf, passing through the stoma or mechanically closing them completely or partially, leading to disorders in the photosynthetic and water regime of plants and finally to the occurrence of necrosis of leaves (Farmer, 2002). Deposited particles may reduce the absorption of CO_2 that decrease the photosynthetic activity, and they can also block stoma preventing their closure, increasing the uptake of some toxic ions into tissues which leads to water stress (Farmer, 2002).

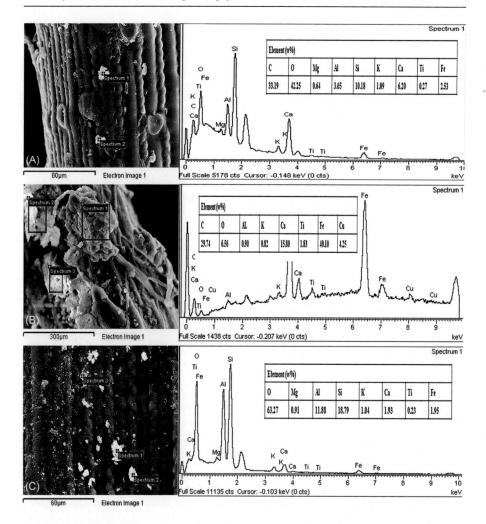

Figure 6.10 SEM (EDS) spectral analysis of the chemical composition of the deposited particles on the upper surface of *Festuca rubra* leaves on the control site (A) and fly ash deposits on TENT-A: $PC_{3\ years\ old}$ – passive cassette 3 years old (B) and $PC_{11\ years\ old}$ – passive cassette 11 years old (C).

SEM analysis of the leaves of *F. rubra* showed that this perennial grass forms silicate bodies on its surface, which is characteristic of the genus *Festuca* (Ortunez and Cano-Ruiz, 2013). According to Zhang et al. (2013), Si is deposited in the cell walls of the silica cells of the epidermis of the leaves. In cell walls of silica cells, Si is combined with Ca and complexes of carbohydrates—lignin (Inaga and Okasaka, 1995). Silicate bodies are thought to be fully formed when the leaves are fully developed (Zhang et al., 2013). Since silicon increases the resistance of plants to various abiotic and biotic stress factors in the environment (Kim et al., 2014), it can be assumed that silica bodies may contribute to the tolerance of *F. rubra* to harsh conditions prevailing on fly ash.

6 Conclusion and future outlook

The main anthropogenic sources of environmental pollution are industrial, urban, and agricultural activities, waste disposal and accidents that lead to land contamination with metal(loid)s and organic pollutants. Soil pollution reduces fertility, decreases growth, and yields of plants, and negatively affects the human health. Sustainable phytomanagement of polluted sites involves assessment of contaminated sites and pollutant characteristics as well as the selection of the best remediation option with low risk/emission cost in order to protect environment and human beings.

Ecorestoration and phytoremediation are eco-friendly, low-cost and green technologies that use plants to clean up contaminated sites and initiate the recovery of degraded ecosystems. Establishment of self-sustaining vegetation cover on the contaminated sites is achieved by sowing selected plant species or plants can spontaneously colonize the polluted area. Perennial grasses are suitable candidates for phytoremediation and ecorestoration of industrial and urban sites as well as coal and ore mine land and fly ash deposits: *Agrostis, Arrhenatherum, Bromus, Clamagrostis, Chloris, Cynodon, Cymbopogon, Dactylis, Festuca, Lolium, Miscanthus, Panicum, Paspalum, Poa, Saccharum, Sorghum, Typha, Vetiveria.*

F. rubra (red fescue) is a perennial grass that belongs to the familiy of *Poaceae*. This plant species has wide geographical distribution around the globe, great phenotypical variability and a number of cultivars. *F. rubra* is a rhizomatous plant with green leaves with needle shape and it can spread very fast by rhizomes or by seeds. This grass is tolerant to different ecological conditions and it can be characterized as semisciophyte, mesotermophyte, submesophyte, mesotrophyte, neutrophyte with high tolerance to salinity. Due to its high ecological plasticity, *F. rubra* can be found in different habitat types, such as pastures and meadows, rock clifs, rivers coastline, bogs and saltmarshers, and growing in grasslands on different elevation in the submontane, subalpine, and mountain regions. It is extensively used as forage grass in pastures, erosion control of banks, hillsides and irrigation channels, revegetation of highways, mine land, and fly ash deposits and it is also used as ornamental and for recreation.

F. rubra is part of the commercial grass-legume mixture of plants that are used in biorecultivation of contaminated sites where it is acting as a pioneer in improving physico-chemical and nutrient properties of substrate and stabilizing the soil/mine spoil/fly ash by a well-developed root system. This grass can be sown by hand, machine, and hydroseeder. *F. rubra* shows a high phytoremediation potential for Pb/Zn/Cu/pyrite and Pb/Zn/Hg/As from the mining waste. Furthermore, this grass has a high potential for phytoremediation of fly ash deposits. It is considered as a suitable plant for phytostabilization of As (BCF < 1, TF < 1), B (BCF < 1, TF > 1), Cu (BCF < 1, TF < 1), Zn (BCF < 1, TF < 1), Mn (BCF < 1, TF < 1), Mo (BCF < 1, TF > 1), and Se (BCF > 1, TF < 1) indicating that *F. rubra* is an excluder that retains a great amount of metal(loid)s in root than in leaves. This grass has effective mechanisms of metal(loid) uptake where the element is in deficiency and limited capacity for uptake and translocation of excess amount of elements from root to leaves which is achieved by specific transporters, amino acids, and organic acids.

Ecophysiological response of *F. rubra* growing on fly ash deposits shows reduced photosynthetic efficiency, photopigments content, high oxidative stress, and increased content of phenolics, ascorbate, and total antioxidant activity indicating high adaptive potential for surviving harsh conditions prevailing on fly ash. Visible leaf damage was in the form of chloroses and necroses due to sinergistic effect of drought, high temperatures, excess amount of As and B and deficiency of Cu, Zn, and Mn in leaves. SEM (EDS) analysis of leaf surface shows that deposited particles of fly ash consist of following chemical elements: C, O, Mg, Al, Si, Fe, K, Ca. Despite deposited particles on the leaf surface of *F. rubra*, damage of cuticle and epidermis is not large probably due to smooth cuticle and presence of silica body.

Great effort has been made to show that *F. rubra* possesses a high potential for phytostabilization of metal(loid)s and ecophysiological adaptation to unfavorable environmental conditions. However, further experiments are required to elucidate mechanisms of acquisition, uptake, distribution, translocation, and sequestration of metal(loid) and organic pollutants in roots and shoots of *F. rubra* growing on different types of contaminated land. There is a need to determine how this grass in polluted environment can stimulate fungal and bacterial activity by root excudates. Transgenic studies can be important for a better understanding of the connection between novel genes and adaptive response of *F. rubra* to contaminants. Therefore, genetic engineering and breeding of *F. rubra* together with its molecular biology, physiology, biochemistry, metabolomes, and proteomes can improve the knowledge about the selection of phytoremediation technologies supporting better land-use management. Taken together, utilization of *F. rubra* in phytomanagement practice offers great prospects in developing policy frameworks to promote cleanup of polluted sites worldwide.

Acknowledgments

This work was supported by the Ministry of Education, Science and Technology Development of Serbia (Grant No. 173018). This chapter is a part of the doctoral thesis of Gordana Gajić. Special thanks to Dr. Branka Stevanović, full time professor of the Biological Faculty, University of Belgrade, Department of Plant Ecology and Phytogeography for supervising and support. Authors gratefully acknowledge Dr. Anka Dinić, principal research fellow and Dr. Snežana Jarić, senior research associate from the Department of Ecology, Institute for Biological Research "Siniša Stanković," University of Belgrade for determining plant species on the fly ash deposits. Authors are particularly grateful to Dr. Vladan Djordjević, research assistant from the Faculty of Biology, University of Belgrade, Department of Plant Ecology and Phytogeography for determining habitat types and plant communities of *Festuca rubra*. Authors gratefully acknowledge Dr. Lola Djurdjević, principal research fellow from the Department of Ecology, Institute for Biological Research "Siniša Stanković," University of Belgrade for biochemical research. We also thank Dr. Olga Kostić, research associate from the Department of Ecology, Institute for Biological Research "Siniša Stanković," University of Belgrade for assistance on the field.

References

Abdel-Ghany, S.E., Pilon, M., 2008. MicroRNA-mediated systemic down-regulation of copper protein expression in response to low copper availability in Arabidopsis. J. Biol. Chem. 283, 15932–15945.

Abou-Shanab, R., Ghanem, N., Ghanem, K., Al-Kolaibe, A., 2007. Phytoremediation potential of crop and wild plants for multi-metal contaminated sites. Res. J. Agric. Biol. Sci. 3 (5), 370–376.

Aiken, S.G., Darbyshire, S.J., 1990. *Fescue* Grasses of Canada. Canadian Government Publishing Centre, Ottawa.

Alves, L. de J., Nunes, F.C., Prasada, M.N.V., Mangabeira, P.A.O., Gross, E., Loureiro, D.M., Medrado, H.H.S., Bomfim, P.S.E., 2018. Uranium Mine Waste Phytostabilization With Native Plants – A Case Study From Brazil. In: Prasad, M.N.V., Favas, P.J.C. Maiti, S.K. (Eds.), Bio-Geotechnologies for Mine Site Rehabilitation, Elsevier, Amsterdam, Netherlands, pp. 299–333.

Anderson, J.P., 1959. Flora of Alaska and Adjacent Parts of Canada. Iowa State University Press, Ames, IA, pp. 543.

Andrejić, G., Gajić, G., Prica, M., Dželetović, Ž., Rakić, T., 2018. Zinc accumulation, photosynthetic gas exchange, and chlorophyll *a* fluorescence in Zn-stressed *Miscanthus* × *giganteus* plants. Photosynthetica 56, 1249–1258.

Anjum, N.A., Hasanuzzaman, M., Hossain, M.A., Thangavel, P., Roychoudhury, A., Gill, S.S., Rodrigo, M.A., Adam, V., Fujita, M., Kizek, R., Duarte, A.C., Pereira, E., Ahmad, I., 2015. Jacks of metal/metalloid chelation trade in plants–an overview. Front. Plant Sci. 6, 192.

Arif, N., Yadav, V., Singh, S., Singh, S., Ahmad, P., Mishra, R.K., Sharma, S., Tripathi, D.K., Dubey, N.K., Chauhan, D.K., 2016. Influence of high and low levels of plant-beneficial heavy metal ions on plant growth and development. Front. Environ. Sci. 4, 69.

Armstead, I., Donnison, I., Aubry, S., Harper, J., Hörtensteiner, S., James, C., Mani, J., Moffet, M., Ougham, H., Roberts, L., Thomas, A., Weeden, N., Thomas, H., King, I., 2007. Cross–species identification of mendel's I locus. Science 315, 73.

Arora, A., Byrem, T.M., Nair, M.G., Strasburg, G.M., 2000. Modulation of liposomal membrane fluidity by flavonoids and isoflavonoids. Arch. Biochem. Biophys. 373, 102–109.

Arrivault, S., Senger, T., Kramer, U., 2006. The *Arabidopsis* metal tolerance protein AtMTP3 maintains metal homeostasis by mediating Zn exclusion from the shoot under Fe deficiency and Zn oversupply. Plant J. 46, 861–879.

Atabayeva, S., 2016. Heavy metals accumulation ability of wild grass species from industrial areas of Kazakhstan. In: Ansari, A.A., Gill, S.S., Gill, R., Lanza, G.R., Newman, L., (Eds.), Phytoremediation. In: Management of Environmental Contaminants, Vol. 3, Springer, Switzerland, pp. 157–208.

Baasch, A., Kirmer, A., Tischew, S., 2012. Nine years of vegetation development in a postmining site: effects of spontaneous and assisted site recovery. J. Appl. Ecol. 49, 251–260.

Baker, A.J.M., 1981. Accumulators and excluders–strategies in the response of plants to heavy metals. J. Plant Nutr. 3 (1–4), 643–654.

Barkworth, M., Anderson, L., Capels, K., Long, S., Piep, M., 2007. Manual of Grasses for North America. Intermountain Herbarium and Utah State University Press, Logan, UT, USA, p. 627.

Barry, B.A., Cooper, I.B., De Riso, A., Brewer, S.H., Vu, D., M., Dyer, R.B., 2006. Time-resolved vibrational spectroscopy detects protein-based intermediates in the photosynthetic oxygen-evolving cycle. Proc. Natl. Acad. Sci. USA 103, 7288–7291.

Bashir, K., Ishimaru, Y., Nishizawa, N.K., 2012. Molecular mechanisms of zinc uptake and translocation in rice. Plant Soil 36, 189–201.

Bashmakov, D.I., Lukatkin, A.S., Prasad, M.N.V., 2006. Temperate weeds in Rusia: Sentinels for monitoring trace element pollution and possible application in phytoremediation. In: Prasad, M.N.V., Sajwan, K.S., Naidu, R. (Eds.), Trace Elements, in the Environment (Bio-geochemistry, Biotechnology, and Bioremediation), CRC Taylor and Francis, Boca Raton, FL, pp. 439–450.

Baskin, C., Baskin, J., 2002. Propagation protocol for production of container *Festuca rubra* L. plants. University of Kentucky, Lexington, Kentucky. In: Native Plat Network. University of Idaho, College of Natural Resources, Forest Research Nursery, Moscow, ID. Available from: http://www.nativeplantnetwork.org.

Beard, J., 1973. Turfgrass: Science and Culture. Prentice Hall PTR, Englewood Cliffs, NJ, p. 658.

Beckett, K.P., Freer-Smith, P.H., Taylor, G., 1998. Urban woodlands: their role in reducing the effects of particulate pollution. Environ. Pollut. 99, 347–360.

Bernal., M., Casero, D., Singh, V., Wilson, G.T., Grande, A., Yang, H., Dodani, S,C., Pellegreni, M., Huijser, P., Conolly, E.L., Merchant, S.S., Kramer, U., 2012. Transcriptome sequencing identifies SPL7-regulated copper acquisitionngenes FRO4/FRO5 and the copper dependence of iron homeostasis in Arabidopsis. Plant Cell 24, 738–761.

Bertin, C., Senesac, A., Rossi, F., DiTommaso, A., Weston, L., 2009. Evaluation of selected fine-leaf fescue cultivars for their turfgrass quality and weed suppressive ability in field settings. Horttechnology 19 (3), 660–668.

Bjorkman, O., Demmig, B., 1987. Photon yield of O_2 evolution and chlorophyll fluorescence characteristics at 77 K among vascular plant of divers origins. Planta 170, 489–504.

Boivin, J.P., Ricour, J., 2005. Sites et Sols Pollues-Out-ils juridiques, techniques et financiers de la remise en etat des sites pollues. Editions Le Moniteur, Paris, France.

Bokern, M., Harms, H., 1997. Toxicity and metabolism of 4-n-nonxlphenol in cell suspension cultures of different plant species. Environ. Sci. Technol. 31 (7), 1849–1854.

Bonfranceschi, B.A., Flocco, C.G., Donati, E.R., 2009. Study of the heavy metal phytoextraction capacity of two forage species growing in an hydroponic environment. J. Hazard. Mater. 165 (1–3), 366–371.

Brand-Williams, W., Cuvelier, M.E., Berset, C., 1995. Use of a free radical method to evaluate antioxidant activity. Lebensm. Wiss. Tecnol. 28, 25–30.

Brevik, E.C., Burgess, L.C., 2013. Soils and Human Health. CRC Press, Boca Raton.

Brown, R., Percivalle, C., Narkiewicz, S., DeCuollo, S., 2010. Relative rooting depths of native grasses and amenity grasses with potential for use on roadsides in New England. Hort-Science 45 (3), 393–400.

Burken, J.G., 2003. Uptake and metabolism of organic compounds: green liver model. In: Mc-Cutcheon, S.C., Schnoor, J.L. (Eds.), Phytoremediation. In: Transformation and Control of Contaminants, John Wiley and Sons, Inc, Publications, Hoboken, New Jersey, pp. 59–84.

Butler, J., Fults, J., Sanks, G., 1971. Review of grasses for saline and alkali areas. Int. Turfgrass Soc. 2, 551–556.

Cakmak, I., 2000. Possible roles in zinc in protecting plant cells from damage by reactive oxygen species. New Phytol. 146, 185–205.

Carbonell-Barrachina, A.A., Aarabi, M.A., De Laune, R.D., Gambrell, R.P., Patrich, W.H., 1998. The influence of arsenic chemical form and concentration on *Spartina patens* and *Spartina alterniflora* growth and tissue arsenic concentration. Plant Soil 198, 33–43.

Carroll, J., 1943. Effects of drought, temperature and nitrogen on turf grasses. Plant Physiol. 18 (1), 19–36.

Castro, P.H., Lilay, G.H., Assunção, A.G.L., 2018. Regulation of micronutrient homeostasis and deficiency response in plants. In: Hossain, M.A., Kamiya, T., Burritt, D.J., Phan Tran, L-S., Fujiwara, T. (Eds.), Plant Micronutrient Use Efficiency. Molecular and Genomics Perspectives in Crop Plants, Academic Press, Elsevier, Inc., London, United Kingdom, pp. 1–15.

Chen, Y., Pettersen, T., Kvalbein, A., Aamild, T.S., 2018. Playing quality, growth rate, thatch, accumulation and tolerance to moss and annual bluegrass invasion as influenced by irrigation strategies on red fescue putting greens. J. Agro. Crop Sci. 204, 185–195.

Chibrik, T.S., Lukina, N.V., Filimonova, E.I., Glazyrina, M.A., Rakov, E.A., Prasad, M.N.V., 2018. Establishment of phytocenosis on brown coal mine waste in Urals Sverdlovsk and Chelyabinsk regions, Russia – drivers, constraints, and trade-offs. In: Prasad, M.N.V., Favas, P.J.C., Maiti, S.K. (Eds.), Bio-Geotechnologies for Mine Site Rehabilitation, Elsevier, Amsterdam, Netherlands, pp. 529–546.

Christenhusz, M.J.M., Byng, J.W., 2016. The number of known plants species in the world and its annual increase. Phytotaxa 261 (3), 201–217.

Chu, M., 2008. Natural revegetation of coal fly ash in a highly saline disposal lagoon in Hong Kong. Appl. Veg. Sci. 11, 297–306.

Clemens, S., 2006. Toxic metal accumulation, response to exposure and mechanisms of tolerance in plants. Biochimie 88, 1707–1719.

Colombano, S., Merly, C., Gaboriau, H., 2009. Risk management and decision making in remediation option selection. Land Degrad. Reclamat. 17 (3–4), 345–353.

Commission Proposal COM, 2006. 232 of 22 September 2006 for a Directive of the European Parliament and of the Council establishing a framework for the protection of soil and amending Directive 2004/35/EC.

Conesa, H.M., Moradi, A.B., Robinson, B.H., Kühne, G., Lehmann, E., Schulin, R., 2009. Response of native grasses and *Cicer arietinum* to soil polluted with mining wastes: implications for the management of land adjacent to mine sites. Environ. Exp. Bot. 65, 198–204.

Cook, R.E., 1983. Clonal plant populations. Am. Sci. 71, 244–253.

Courtney, R., 2018. Irish mine sites rehabilitation – a case study. In: Prasad, M.N.V., Favas, P.J.C., Maiti, S.K. (Eds.), Bio-Geotechnologies for Mine Site Rehabilitation, Elsevier, Amsterdam, Netherlands, pp. 439–456.

Dabrowska, A., 2011. Variability of morphological and anatomical traits in natural populations of *Festuca rubra* and *F. nigrescens*. Acta Agrobot. 64 (4), 159–170.

Das, P.K., Geul, B., Choi, S.-B., Yoo, S.-D., Park, Y., 2011. Photosynthesis-dependent anthocyanin pigmentation in *Arabidopsis*. Plant Signal Behav. 6 (1), 23–25.

De Tullio, M.C., Paciola, C., Dalla Vechia, F., Rascio, N., D'Emerico, S., De Gara, L., Liso, R., Arrigoni, O., 1999. Changes in onion root development induced by the inhibition of peptidyl prolyl hydroxylase and influence of ascorbate system on cell division and elongation. Planta 109, 424–434.

Dimpka, C.O., Merten, D., Svatos, A., Büchel, G., E. Kothe, E., 2009. Sideropohores mediates reduced and increased uptake of cadmium by *Streptomyces tendae* F4 and sunflower (*Helianthus annus*), respectively. J. Appl. Microbiol. 107, 1687–1696.

Dittberner, P.L., Olson, M.R., 1983. The Plant Information Network (PIN) Data Base: Colorado, Montana, North Dakota, Utah, and Wyoming. FWS/OBS-83/86. US Department of the Interior, Fish and Wildlife Service, Washington, DC.

Djordjević-Miloradović, J., 1998. Population dynamics of plants in the primary succession of the vegetation on the waste ash deposits of the thermoelectric power station Kostolac. PhD thesis, Faculty of Biology, University of Belgrade, Belgrade, 458 pp. (in Serbian).

Djurdjević, L., Mitrović, M., Pavlović, P., Gajić, G., Kostić, O., 2006. Phenolic acids as bioindicators of fly ash deposit revegetation. Arch. Environ. Contam. Toxicol. 50, 488–495.

Duarte, M.C., Holubec, V., Uzundzhalieva, K., Kell, S.P., Vögel, R., Economou, G., Vörösváry, G., 2011. *Festuca rubra*. IUCN Red List of Threatened Species 2011: eT176558A7266236.

Dučić, T., Polle, A., 2005. Transport and detoxification of manganese and copper in plants. Braz. J. Plant Physiol. 17 (1), 103–112.

Duponnois, R., Kisa, M., Assigbetse, K., Prin, Y., Thioulouse, J., Issartel, M., Moulin, P., Lepage, M., 2006. Fluorescent pseudomonas occurring in *Macrotermes subhyalinus* mound structure decrease Cd toxicity and improve its accumulation in sorghum plants. Sci. Total Environ. 370, 391–400.

Dwivedi, S., Srivastava, S., Mishra, S., Dixit, B., Kumar, A., Tripathi, K.R.D., 2008. Screening of native plants and algae growing on fly-ash affected areas near National Thermal Power Corporation, Tanda, Uttar Pradesh, India for accumulation of toxic heavy metals. J. Hazard. Mater. 158, 359–365.

Dželetović, Ž., Mihailović, N., Živanović, I., 2013. Prospects of using bioenergy crop *Miscanthus* × *giganteus* in Serbia. In: Méndez-Vilas, A. (Ed.), Materials and Processes for Energy: Communicating Current Research and Technological Developments. Formatex Research Center, Badajoz, Spain, pp. 360–370.

Ellery, K.S., Walker, B.H., 1985. Growth xharacteristics of selected plant species on asbestos tailings from Msauli Mine, eastern Transval. S. Afr. Tydskr. Plankt. 52 (3), 201–206.

Elliot, C.L., McKendrick, J.D., Helm, D., 1987. Plant biomass, cover, and survival of species, used for stripmine reclamation in south-central Alaska, USA. Arct. Alp. Res. 19 (4), 572–577.

Engelhardt, K.A.M., Hawkins, K., 2016. Identification of low growing, salt tolerant turfgrass species suitable for use along highway right of way. Research Report. Maryland Department of transportation state highway administration. University of Maryland Center for Environmental Science, Appalachian Laboratory. Frostburg, Maryland.

Eriksson, O., 1989. Seedling dynamics and life histories in clonal plants. Oikos 55, 231–238.

EUNIS classification of habitats. Standard classification of European habitats. European Environment Agency. Available from: http://eunis.eea.europa.eu/.

EURISCO, 2015. EURISCO Catalogue release 1.1.6.

European Commission, 2013. Soil Contamination: Impacts on Human Health. Science for Environmental Policy. Report produced for the European Commission DG Environment, September 2013. Science Communication Unit, University of the West of England, Bristol.

Falcone, D.L., Ogas, J.P., Somerville, C.R., 2004. Regulation of membrane fatty acid composition by temperature in mutants of *Arabidopsis* with alterations in membrane lipid composition. BMC Plant Biol. 4, 1–15.

Farmer, A., 2002. Effects of particulates. In: Bell, J.N.B., Treshow, M. (Eds.), Air Pollution and Plant Life. John Wiley and Sons Ltd., Chichester, England, pp. 187–199.

Fernandez, S., Poschenrieder, C., Marceno, C., Gallego, J.R., Jimenez-Gamez, D., Bueno, A., Afif, E., 2017. Phytoremediation capability of native plant species living on Pb-Zn and Hg-As mining wastes in the Cantabrian range, north Spain. J. Geochem. Explor. 174, 10–20.

Fitzpatrick, K.L., Tyerman, S.D., Kaiser, B.N., 2008. Molybdate transporter through the plant sulfate transporter SHST1. FEPBS Lett. 582, 1508–1513.

Frick, C.M., Farrell, R.E., Germida, J.J., 1999. Assessment of phytoremediation as an in-situ technique for cleaning oil-contaminated sites. Petroleum Technology Alliance Canada, Calgary, Canada.

Gajić, G., 2014. Ecophysiological adatations of selected species of herbaceous plants at the fly ash landfill of the thermal power plant 'Nikola Tesla –A' in Obrenovac, PhD thesis, Faculty of Biology, University of Belgrade, Belgrade, 406 pp. (in Serbian).

Gajić, G., Djurdjević, L., Kostić, O., Jarić, S., Mitrović, M., Pavlović, P., 2018. Ecological potential of plants for phytoremediation and ecorestoration of fly ash deposites and mine wastes. Front. Environ. Sci. 6, 124. doi: 10.3389/fenvs.2018.00124.

Gajić, G., Djurdjević, L., Kostić, O., Jarić, S., Mitrović, M., Stevanović, B., Pavlović, P., 2016. Assessment of the phytoremediation potential and an adaptive response of *Festuca rubra* L. Sown on fly ash deposits: native grass has a pivotal role in ecorestoration management. Ecol. Eng. 93, 250–261.

Gajić, G., Mitrović, M., Pavlović, P., 2019. Ecorestoration of fly ash deposits by native plant species at thermal power stations in Serbia. In: Pandey, V.C., Bauddh, K. (Eds.), Phytomanagement of Polluted Sites. Market Opportunities in Sustainable Phytoremediation. Elsevier, Amsterdam, pp. 113–177.

Gajić, G., Pavlović, P., 2018. The role of vascular plants in the phytoremediation of fly ash deposits. In: Mathichenkov, V. (Ed.), Phytoremediation. Methods, Management and Assessment. Nova Science Publishers, Inc., New York, pp. 151–236.

Gajić, G., Pavlović, P., Kostić, O., Jarić, S., Djurdjević, L., Pavlović, D., Mitrović, M., 2013. Ecophysiological and biochemical traits of three herbaceous plants growing of the disposed coal combustion fly ash of different weathering stage. Arch. Biol. Sci. 65 (1), 1651–1667.

Gajić, M., 1976. Genus *Festuca*. In: Josifović, M. (Ed.), Flora of SR Serbia. Serbia Acad. Sci. and Art. Belgrade, Serbia (in Serbian), pp. 415–442.

Gao, Y., Zhu, L., 2004. Plant uptake, accumulation and translocation of phenanthrene and pyrene in soils. Chemosphere 55, 1169–1178.

Gasber, A., Klaumann, A., Trentmann, O., Trampczynska, A., Clemens, S., Schneider, S., Sauer, N., Feifer, I., Bittner, F., Mendel, R.R., Neuhaus, H.E., 2011. Identification of an *Arabidopsis* solute carrier critical for itracellular transport and interorgan allocation of molbdate. Plant Biol. 13, 710–718.

Gill, S.S., Tuteja, N., 2010. Reactive oxgen species and antioxidant mashinery in abiotic stress tolerance in crop plants. Plant Physiol. Biochem. 48, 909–930.

Grant, C.D., Campbell, C.J., Charnock, N.R., 2002. Selection of species suitable for derelict mine site rehabilitation in New South Wales, Australia. Water Air Soil Pollut. 139, 215–235.

Gruiz, K., 2009. Integrated and efficient assessment of contaminated sites. Land Degradation Reclam. 17 (3–4), 371–384.

Gupta, A.K., Sinha, S., 2008. Decontamination and/or revegetation of fly ash dykes through naturally growing plants. J. Hazard. Mater. 153, 1078–1087.

Hallsten, G.P., Skinner, Q.D., Beetle, A.A., 1987. Grasses of Wyoming, third ed. Research Journal 202. Laramie, W.Y. University of Wyoming, Agricutural Experiment Station.

Hamilton, S.J., 2004. Review of selenium toxicity in the aquatic food chain. Sci. Total Environ. 326, 1–31.

Harms, H., 1996. Bioaccumulation and metabolic fate of sewage sludge derived organic xenobiotics in plants. Sci Total Environ. 185, 83–92.

Hasanuzzaman, M., Hossain, M.A., Fujita, M., 2010. Selenium in higher plants: physiological role, antioxidant metabolism and abiotic stress tolerance. J. Plant Sci. 5, 354–375.

Hauck, B., Gay, A.P., Macduff, J., Griffiths, C.M., Thomas, H., 1997. Leaf senescence in a non-yellowing mutant of *Festuca pratensis*: implications of the stay-green mutation for photosynthesis, growth and nitrogen nutrition. Plant Cell Environ. 20, 1007–1018.

Hebbern, C., Laursen, K.H., Ladegaard, A.H., Schmidt, S.B., Pedas, P., Bruhn, D., Schjoerring, J.K., Wulfsohn, D., Husted, S., 2009. Latent manganese deficiency increases transpiration in barley (*Hordeum vulgare*). Physiol. Plantarum 135, 307–316.

Hickman, J.C., 1993. The Jepson Manual: Higher Plants of California. University of California Press, Berkley, CA.

Hitchcock, A.S., 1951. Manual of the grasses of the United States. Misc. Publ. No. 200. Washington, DC: US Department of Agriculture, Agricultural Research Administration. Second edition revised by Agnes Chase in two volumes. Dover Publications, Inc. New York.

Hodgson, D.R., Townsend, W.N., 1973. The amelioration and revegetation of pulverized fuel ash. In: Chadwick, M.J., Goodman, G.T. (Eds.), Ecology and Reclamation of Devastated Land. Ecology and Reclamation of Devastated Land, Gordon and Breach, London, pp. 247–250.

Iglikov, A., 2016. The development of artificial phytocenosis in environmental construction in the far north. Proc. Eng. 165, 800–805.

Inaga, S., Okasaka, A., 1995. Calcium and silicon binding compounds in cell walls of rice shoots. Soil Sci. Plant Nutr. 41, 103–110.

Jabeen, R., Ahmad, A., Iqbal, M., 2009. Phytoremediation of heavy metals: physiological and molecular mechanisms. Bot. Rev. 75, 339–364.

Jamil, S., Abhilash, P.C., Singh, A., Singh, N., Behl, H.M., 2009. Fly ash trapping and metal accumulating capacity of plants: Implication for green belt around thermal power plants. Landsc. Urban Plan. 92, 136–147.

Kabata-Pendias, A., 2011. Trace Elements in Soils and Plants, fourth ed. CRC Press LLC, Boca Raton, London, New York, Washington.

Kaimi, E., Mukaidani, T., Miyoshi, S., Tamaki, M., 2006. Reygrass enhancement of biodegradation in diesel-contaminated soil. Environ. Exp. Bot. 55, 110–119.

Kaimi, E., Mukaidani, T., Tamaki, M., 2007. Screening of twelve plant species for phytoremediation of petroleum hydrocarbon-contaminated soil. Plant Prod. Sci. 10 (2), 211–218.

Kala, S., 2014. Rhizoremediation: a promising rhizosphere technology. J. Eviron. Sci Toxicol. Food Technol. 8 (8), 23–27.

Kålås, J.A., Viken, Å., Bakken, T., 2006. Norsk Rødliste 2006 – 2006 Norwegian Red List. Artsdatabanken.

Kang, F., Chen, D., Gao, Y., Zhang, Y., 2010. Distribution of polyclic aromatic hydrocarbons in subcellular root tissues of ryegrass (*Lolium multiflorum* L). BMC Plant Biol. 10, 210.

Karthikeyan, R., Davis, L.C., Erickson, L.E., Al-Khatib, K., Kulakow, P.A., Barnes, P.L., Hutchinson, S.L., Nurzhanova, A.A., 2004. Potential for plant –based remediation of pesticide-contaminated soil and water using nontarget plants such as trees, shrubs and grasses. Crit. Rev. Plant Sci. 23 (1), 91–104.

Kasowska, D., Gediga, K., Spiak, Z., 2018. Heavy metal and nutrient uptake in plants colonizing post-flotation copper tailings. Environ. Sci. Pollut. Res. 25, 824–835.

Khan, M.U., Sessitsch, A., Harris, M., Fatima, K., Imran, A., Arslan, M., Shabir, G., Khan, Q., Afzal, M., 2014. Cr-resistant rhizo- and endophytic bacteria associated with *Prosopis juliflora* and their potential as phytoremediation enhancing agents in metal-degraded soils. Front. Plant Sci. 5, 755.

Kidd, P., Barcelo, J., Bernal MP., Navari-Izo-F., Poschenrieder, C., Shilev, S., Clemente, R., Monterroso, C., 2009. Trace element behaviour at the root-soil interface: implications in phytoremediation. Environ. Exp. Bot. 67, 243–259.

Kim, Y-H., Khan, A.L., Kim, D-H., Lee, S-Y., Kim, K-M., Waqas, M., Jung, H-Y., Shin, J-H., Kim, J-G., Lee, I-J., 2014. Silicon mitigates heavy metal stress by regulating P-type heavy metal ATPases, *Oryza sativa* low silicon genes, and endogenous phytohormones. BMC Plant Biol. 14, 1–13.

Kirmer, A., Tischew, S., 2010. Near-natural restoration strategies in post-mining landscapes. In: Müller, N., Werner, P., Kelcey, J.G. (Eds.), Urban Biodiversity, Design. Blackwell Publishing, Oxford, pp. 539–555.

Kojić, M., Popović, R., Karadžić, B., 1997. Vascular Plants of Serbia as Indicators of Habitats. Agricultural Research Institute "Serbia", Institute for Biological Research "Siniša Stanković", Belgrade.

Kojić, M., Popović, R., Karadžić, B., 1998. Syntaxonomy overview of the vegetation of Serbia. Institute for Biological Research "Siniša Stanković", Belgrade (in Serbian).

Kolb, M., Harms, H., 2000. Metabolism of fluoranthene in different plant cell cultures and intact plants. Environ. Toxicol. Chem. 19, 1304–1310.

Kostić, O., Jarić, S., Gajić, G., Pavlović, D., Pavlović, M., Mitrović, M., Pavlović, P., 2018. Pedological properties and ecological implications of substrates derived 3 and 11 years after the revegetation of lignite fly ash disposal sites in Serbia. Catena 163, 78–88.

Kostić, O., Mitrović, M., Knežević, M., Jarić, S., Gajić, G., Djurdjević, L., Pavlović, P., 2012. The potential of four woody species for the revegetation of fly ash deposits from the 'Nikola Tesla –a' thermoelectric plant (Obenovac, Serbia). Arch. Biol. Sci. 64 (1), 145–158.

Kumari, A., Lal, B., Rai, U.N., 2016. Assessment of native plant species for phytoremediation of heavy metals growing in the vicinity of NTPC sites, Kahagon, India. Int. J. Phytoremediat. 18 (6), 592–597.

Kusaba, M., Tanaka, A., Tanaka, R., 2013. Stay-green plants: what do they tell us about the molecular mechanism of leaf senescence. Photosynth. Res. 117, 221–234.

Lackschewitz, K., 1991. Vascular plants of west-central Montana – indetification guidebook. Gen. Tech. Rep. INT-227. Ogden, UT: US Department of Agriculture, Forest Service, Intermountain Research Station.

Lakušić, D., Blaženčić, J., Ranđelović, V., Butorac, B., Vukojičić, S., Zlatković, B., Jovanović, S., Šinžar-Sekulić, J., Žukovec, D., Ćalić, I., Pavićević, D., 2005. Habitats of Serbia—Handbook with descriptions and basic data. In: Lakušić, D. (Ed.), Habitats of Serbia, Results of the project "Harmonization of the National Nomenclature in Habitat Classification with the Standards of the International Community". Institute of Botany and Botanical Garden "Jevremovac", Faculty of Biology, University of Belgrade, Ministry of Science and Environmental Protection of the Republic of Serbia, Belgrade, Serbia. http://habitat. bio.bg.ac.rs/.

Lanquar, V., Ramos, M.S., Lelievre, F., Barbier-Brygoo, H., Krieger-Liszkay, A., Krämer, U., Thomine, S., 2010. Export of vacuolar manganese by AtNRAMP3 and AtNRAMP4 is required for optimal photosynthesis and growth under manganese deficiency. Plant Physiol. 152, 1986–1999.

Li, H.F., McGrath, S.P., Zhao, F.J., 2008. Selenium uptake, translocation and speciation in wheart supplied with selenate or selenite. New Phytol. 178, 92–102.

Li, M.S., Luo, Y.P., Su, Z.Y., 2007. Heavy metal concentrations in soils and plant accumulation in a restored manganese mineland in Guangxi, South China. Environ. Pollut. 147, 168–175.

Lichtenthaler, H.K., Buschmann, C., 2001. Chlorophylls and Carotenoids: Measurement and Characterization by UV-VIS Spectroscopy. In: Wrolstad, R.E., Acree, T.E., An, H., Decker, E.A., Penner, M.H., Reid, D.S., Schwartz, S.J., Shoemaker, C.F., Sporns, P. (Eds.), Current Protocols in Food Analytical Chemistry (CPFA) F4.3.1-F4.3.8. John Wiley and Sons, New York.

Liu, J., Li, K., Xu, J., Zhang, Z., Ma, T., Lu, X., Yang, J., Zhu, Q., 2004. Lead toxicity, uptake, and translocation in different rice cultivars. Plant Sci. 165, 793–802.

Longland, A.C., 2013. Pastures and pasture managements. In: Geor, R.J., Harris, P.A., Coenen, M. (Eds.), Equine Applied and Clinical Nutition. Health, Welfare and Perfomance. Elsevier, London.

López-Millán, A.F., Ellis, D.R., Grusak, M.A., 2004. Identification and characterization of several new members of the ZIP family of metal ion transporters in *Medicago truncatula*. Plant Mol. Biol. 4, 583–596.

Ma, Y., Prasad, MNV., Rajkumar, M., H. Freitas, H., 2011. Plant growth promoting rhizobacteria and endophytes accelerate phytoremediation of metalliferous soils. Biotech. Adv. 29, 248–258.

Maiti, S.K., 2013. Ecorestoration of the Coalmine Degraded Lands. Springer, New Delhi, India.

Maiti, S.K., Jaswal, S., 2008. Bioaccumulation and translocation of metals in the natural vegetation growing on the fly ash dumps: a field study from Santaldih thermal power plant, West Bengal, India. Environ. Monit. Assess. 136, 355–370.

Maiti, S.K., Maiti, D., 2015. Ecological restoration of waste dumps by topsoil blanketing, coirmatting and seeding with grass-legume mixture. Ecol. Eng. 77, 74–84.

Majer, P., Neugart, S., Krumbein, A., Schreiner, M., Hideg, E., 2014. Singlet oxygen scavenging by leaf flavonoids contributes to sunlight acclimation in *Tilia platyphyllos*. Environ. Exp. Bot. 100, 1–9.

Mano, S., Yamaguchi, K., Hayashi, M., Nishimura, M., 1997. Stromal and thylakoid-bound ascorbate peroxidases are produced by alternative splicing in pumpkin. FEBS Lett. 413, 21–26.

Marcum, K., 1999. Salinity tolerance mechanisms of grasses in the subfamily Chloridoideae. Crop Sci. 39 (4), 1153–1160.

Marschner, H., 1995. Mineral Nutrition of Higher Plants. Academic Press, London.

McCauley, A., Jones, C., Jacobson, J., 2009. Plant Nutrient Functions and Defficiency and Toxicity Symptoms. Nutrient Management Module No.9. A self course from the MSU Extension Service Continuing Education Series. Montana State University, extension.

McCutcheon, S.C., Schnoor, J.L., 2003. Phytoremediation. Transformation and Control of Contaminants. John Wiley and Sons, Inc, Publications, Hoboken, New Jersey.

McKernan, D., Ross, J., Tompkins, D., 2001. Evaluation of grasses grown under low maintenance conditions. Int. Turfgrass Soc. 9, 25–32.

Meharg, A.A., Hartley-Whittaker, J., 2002. Arsenic uptake and metabolism in arsenic resistant and nonresistant plant species. New Phytol. 154, 29–43.

Memon, A.R., Schröder, P., 2006. Implications of metal accumulation mechanisms to phytoremediation. Environ. Sci. Pollut. Res. 16, 162–175.

Mendel, R.R., 2007. Biology of the molybdenum cofactor. J. Exp. Bot. 58 (9), 2289–2296.

Mendel, R.R., 2011. Cell biology of molybdenum in plants. Plant Cell Rep. 30, 1787–1797.

Mendel, R.R., Bittner, F., 2006. Cell biology of molybdenum. Biochim. Biophys. Acta 1763, 621–635.

Merzlyak, M.N., Chivkunova, O.B., 2000. Light stress induced pigment changes and evidence for anthocyanin photoprotection in apple fruit. J. Photochem. Photobiol. B Biology (B) 55, 154–162.

Migocka, M., Malas, K., 2018. Plant responses to copper: Molecular and regulatory mechanisms of copper uptake, distribution and accumulation in plants. In: Hossain, M.A., Kamiya, T., Burritt, D.J., Phan Tran, L-S., Fujiwara, T. (Eds.), Plant Micronutrient Use Efficiency. Molecular and Genomics Perspectives in Crop Plants, Academic Press, Elsevier, Inc. London, United Kingdom, pp 71–86.

Mitrović, M., Pavlović, P., Lakušić, D., Stevanović, B., Djurdjević, L., Kostić, O., Gajić, G., 2008. The potencial of *Festuca rubra* and *Calamagrostis epigejos* for the revegetation on fly ash deposits. Sci. Tot. Environ. 72, 1090–1101.

Molyneux, P., 2004. The use of the stable free radical diphenylpicryl-hydrazyl (DPPH) for estimating antioxidant activity. Songklanakarin J. Sci. Technol. 26, 211–219.

Morgenthal, T.L., Cilliers, S.S., Kellner, K., van Hamburg, H., Michael, M.D., 2001. The vegetation of fly ash disposal sites at Hendrina Power Station II: floristic composition. S. Afr. J. Bot. 67, 520–532.

Mucina, L., Bültmann, H., Dierßen, K., Theurillat, J.-P., Raus, T., Čarni, A., Šumberová, K., Willner, W., Dengler, J., Gavilán García, R., Chytrý, M., Hájek, M., Di Pietro, R., Iakushenko, D., Pallas, J., Daniëls, F. J. A., Bergmeier, E., Santos Guerra, A., Ermakov, N., Valachovič, M., Schaminée, J. H. J., Lysenko, T., Didukh, Ya. P., Pignatti, S., Rodwell, J. S., Capelo, J., Weber, H. E., Solomeshch, A., Dimopoulos, P., Aguiar, C., Freitag, H., Hennekens, S. M., Tichý, L., 2016. Vegetation of Europe: hierarchical floristic classification system of plant, lichen, and algal communities. Appl. Veg. Sci. 19, 3–264.

Munz, P.A., 1973. A California Flora and Supplement. University of California Press, Berkley, CA.

Muratova, A., Hübner, Th., Narula, N., Wand, H., Turkovskaya, O., Kuschik, P., Jahn, R., Merbach, W., 2003. Rhizosphere microflora of plants used for the phytoremediation of bitumen-contaminated soil. Microbiol. Res. 158, 151–161.

Nadgórska-Socha, A., Ptasiński, B., Kitam, A., 2013. Heavy metal bioaccumulation and antioxidative responses in *Cardaminopsis arenosa* and *Plantago lanceolata* leaves from metalliferous and non-metalliferous sites: a field study. Ecotoxicology 22, 1422–1434.

Nathanail, J., Bardos, P., Nathanail, P., 2002. Contaminated Land Management Ready Reference. EPP Publications/Land Quality Press, London, United Kingdom.

O'Neill, M.A., Ishii, T., Albersheim, P., Darvil, A.G., 2004. Rhamnogalacturonan II: structure and function of a borate cross-linked cell wall pectic polysaccharide. Annu. Rev. Plant Biol. 55, 109–139.

Ortunez, E., Cano-Ruiz, J., 2013. Epidermal micromorphology of the genus *Festuca* L. subgenus *Festuca*(*Poaceae*). Plant Syst. Evol. 299 (8), 1471–1483.

Padmavathiamma, P.K., Li, L.Y., 2009. Phytoremediation of metal-contaminated soil in temperate humid regions of British Columbia, Canada. Int. J. Phytoremediat. 11 (6), 575–590.

Page, M., Sultana, N., Paszkiewicz, K., Florance, H., Smirnoff, N., 2012. The influence of ascorbate on anthocyanin accumulation during high light in *Arabidopsis thaliana*: further evidence for redox control of anthocyanin synthesis. Plant Cell Environ. 35, 388–404.

Pandey, V.C., 2012. Invasive species based efficient green technology for phytoremediation of fly ash deposits. J. Geochem. Explor. 123, 13–18.

Pandey, V.C., 2013. Suitability of *Ricinus communis* L. cultivation for phytoremediation of fly ash disposal sites. Ecol. Eng. 57, 336–341.

Pandey, V.C., 2015. Assisted phytoremediation of fly ash dumps through naturally colonized plants. Ecol. Eng. 82, 1–5.

Pandey, V.C., Bajpai, O., Sinhg, N., 2016. Plant regeneration potential in fly ash ecosystem. Urban For. Urban Gree. 15, 40–44.

Pandey, V.C., Pandey, D.N., Singh, N., 2015a. Sustainable phytoremediation based on naturally colonizing and economically valuable plants. J. Clean Prod. 86, 37–39.

Pandey, V.C., Prakash, P., Bajpai, O., Kumar, A., Sing, N., 2015b. Phytodiversity on fly ash deposits: evaluation of naturally colonized species for sustainable phytorestoration. Environ. Sci. Pollut. Res. 22 (4), 2776–2787.

Pandey, V.C., Singh, K., Singh, R.P., Singh, B., 2012. Naturally growing *Saccharum munja* L. on the fly ash lagoons: a potential ecological engineer for the revegetation and stabilization. Ecol. Eng. 40, 95–99.

Pandey, V.C., Singh, N., 2014. Fast green capping on coal fly ash basins through ecological engineering. Ecol. Eng. 73, 671–675.

Pandey, V.C., Singh, N., Singh, R.P., Singh, D.P., 2014. Rhizomediation potential of spontaneous grown *Typha latifolia* on fly ash basins: study from the field. Ecol. Eng. 71, 722–727.

Parraga-Aguado, I., Gonzáles-Alcaraz, M.N., Schulin, R., Conesa, H.M., 2015. The potential use of *Piptatherum miliaceum* for the phytomanagement of mine tailings in semiarid areas: role of soil fertility and plant competition. J. Environ. Manage. 158, 74–84.

Parrish, Z.D., Banks, M.K., Schwab, A.P., 2005. Effect of root death and decay on dissipation of polycyclic aromatic hydrocarbons in the rhizosphere of yellow sweet clover and tall fescue. J. Environ, Manage. 34, 207–216.

Pavlović, P., Mitrović, M., Djurdjević, L., 2004. An ecophysiological study of plants growing on the fly ash deposits from the Nikola Tesla – A thermal power station in Serbia. Environ. Manage. 33, 654–663.

Pedas, P., Hebbern, C.A., Schjoerring, J.K., Holm, P.E., Husted, S., 2005. Differential capacity for high-affinity manganese uptake contributes to differences between barley genotypyes in tolerance to low manganese availability. Plant Physiol. 139, 1411–1420.

Peng, X.L., Liu, Y.Y., Luo, S.G., 2002. Effects of selenium on lipid peroxidation and oxidizing ability of rice roots under ferrous stress. J. N. E. Agric. Univ. 19, 9–15.

Petrová, Š., Rezek, J., Soudek, P., Vaněk, T., 2017. Preliminary study of phytoremediation of brownfield soil contaminatedby PAHs. Sci. Total Environ. 599-600, 572–580.

Pilon-Smits, E., 2005. Phytoremediation. Annu. Rev. Plant. Biol. 56, 15–39.

Pilon-Smits, E.A.H., de Souza, M.P., Hong, G., Amini, A., Bravo, R.C., 1999. Selenium volatilization and accumulation by twenty aquatic plant species. J. Environ. Qual. 28, 1011–1017.

Pilon-Smits, E., LeDuc, D.L., 2009. Phytoremediation of selenium using transgenic plants. Curr. Opin. Biotechnol. 20, 207–212.

Porter, E.K., Peterson, P.J., 1977. Arsenic tolerance in grasses growing on mine waste. Environ. Pollut. 14, 255–265.

Potters, G., De Gara, L., Asard, H., Horemans, N., 2002. Ascorbate and glutathione: guardians of the cell cycle, paterns in crime? Plant Physiol. Biochem. 40, 537–548.

Prach, K., Rehounkova, K., Lencova, K., Jirova, A., Konvalinkova, P., Mudrak, O., Student, V., Vanecek, Z., Tichy, L., Petrık, P., Smilauer, P., Pysek, P., 2014. Vegetation succession in restoration of disturbed sites in Central Europe: the direction of succession and species richness across 19 seres. Appl. Veg. Sci. 17, 193–200.

Prasad, M.N.V., 2006. Stabilization, remediation, and integrated management of metal-contaminated ecosystems by grasses (Poaceae). In: Prasad, M.N.V., Sajwan, K.S., Naidu, R. (Eds.), Trace Elements in the Environment (Biogeochemistry, Biotechnology, and Bioremediation). CRC Taylor and Francis, Boca Raton, pp. 405–424.

Prasad, M.N.V., Freitas, H., 2006. Metal-tolerant plants: biodiversity prospecting for phytoremediation technology. In: Prasad, M.N.V., Sajwan, K.S., Naidu, R. (Eds.), Trace Elements in the Environment (Biogeochemistry, Biotechnology, and Bioremediation). CRC Taylor and Francis, Boca Raton, p. 483.

Preston, C.D., Pearman, D.A., Dines, T.D., 2002. New Atlas of the British and Irish Flora. Oxford University Press, Oxford.

Quina, F.H., Moreira, P.F., Vautier-Giongo, C., Rettori, D., Rodrigues, R.F., Freitas, A.A., Silva, P.F., Macanita, A.L., 2009. Photochemistry of anthocyanins and their biological role in plant tissues. Pure Appl. Chem. 81 (9), 1687–1694.

Rabêlo, F.H.S., Borgo, L., Lavres, J., 2018. The use of forage grasses for the phytoremediation of heavy metals: plant tolerance mechanisms, classifications, and new prospects. In: Mathichenkov, V. (Ed.), Phytoremediation. Methods, Management and Assessment. Nova Science Publishers, Inc, New York, pp. 59–102.

Randjelović, D., Gajić, G., Mutić, J., Pavlović, P., Mihailović, N., Jovanović, S., 2016. Ecological potential of *Epilobium dodonaei* Vill. for restoration of metalliferous mine waste. Ecol. Eng. 95, 800–810.

Ranieri, E., Bombardelli, F., Gikas, P., Chiaia, B., 2016. Soil pollution, prevention and remediation. Appl. Environ. Soil Sci. 16, Article ID 9415175.

Raskin, I., Smith, R.D., Salt, D.E., 1997. Phytoremediation of metals: using plants to remove pollutants from the environment. Curr. Opin. Biotechnol. 8, 221–226.

Reeves, R.D., Baker, A.J.M., 2000. Metal-accumulating plants. In: Raskin, I., Ensley, B.D. (Eds.), Phytoremediation of Toxic Metals: Using Plants to Clean Up the Environment. John Wiley and Sons, Inc, Publications, New York, pp. 193–229.

Reid, R.J., Haynes, J.E., Post, A., Stangoulis, J.C.R., Graham, R.D., 2004. A critical analysis of the boron toxicity in plants. Plant Cell Environ. 27, 1405–1414.

Ren, G., An, K., Liao, Y., Zhou, X., Cao, Y., Zhao, H., Ge, X., Kuai, B., 2007. Identification of a novel chloroplast protein AtNYE1 regulating chlorophyll degradation during leaf senescence in *Arabidopsis*. Plant Physiol. 144, 1429–1441.

RESCUE, 2005. Best Practice Guidance for Sustainable Brownfield Regeneration. Available from: www.rescue-europe.com.

Rice-Evans, C.A., Miller, J.M., Paganga, G., 1996. Structure-antioxidant activity relationship of flavonoids and phenolic acids. Free Radic. Biol. Med. 20, 933–956.

Roberts, H.A., 1981. Seed banks in soils. Appl. Biol. 5, 1–55.

Roca, M., James, C., Pružinska, A., Hörtensteiner, S., Thomas, H., Ougham, H., 2004. Analysis of the chlorophyll catabolism pathway in leaves of an introgression senescence mutant of *Lolium termulentum*. Phytochemistry 65, 1231–1238.

Rozema, J., Rozema-Dijust, E., Freijsen, A.H.J., Huber, J.J.L., 1978. Population differentiation within *Festuca rubra* L. with regard to soil salinity and soil water. Oecologia 34, 329–341.

Ruemmele, B., Wipoff, J., Brilman, L., Hignight, K., 2003. Fine-leaved Festuca species. In: Casler, M.D., Duncan, R.R. (Eds.), Turfgrass Biology. Genetics and Breeding. John Wiley & Sons, Hoboken, NJ, pp. 129–174.

Russell, W.B., 1985. Vascular flora of abandoned coal-mined land, Rocky Mountain Foothills, Alberta. Can. Field-nat. 99 (4), 503–516.

Sakuraba, Y., Schelbert, S., Park, S.-Y., Han, S.-H., Lee, B.-D., Andres, C.B., Kessler, F., Hörtensteiner, S., Paek, N.-C., 2012. Stay-green and chlorophyll catabolic enzymes interact at light-harvesting complex II for chlorophyll detoxification during leaf senescence in *Arabidopsis*. Plant Cell 24, 507–518.

Salt, D.E., Smith, R.D., Raskin, I., 1998. Phytoremediation. Annu. Rev. Plant Biol. 49, 643–668.

Sampoux, J.P., Huyghe, C., 2009. Contribution of ploidy level variation and adaptive trait diversity to the environmental distribution of taxa in the 'fine-leaved' lineage (genus *Festuca* subg. Festuca). J. Biogeogr. 36, 1978–1993.

Sanceonon, V., Puig, S., Mira, H., Thiele, D.J., Peñarrubia, L., 2003. Identification of copper transporter family in *Arabidopsis thaliana*. Plant. Mol. Biol. 51, 577–587.

Santos, A.E., Cruz-Ortega, R., Meza-Figueroa, D., Romero, F.M., Sanchez-Escalante, J.J., Maier, R.M., Neilson, J.W., Alcarez, L.D., Freaner, F.E.M., 2017. Plants from the abandoned Nacozari mine tailings: evaluation of their phytostabilization potential. PerrJ 5, e3280. doi: 10.7717/peerj.3280.

Santos, R.M., Fortes, G.A.C., Ferri, P.H., Santos, S.C., 2011. Influence of foliar nutrients on phenol levels in leaves of *Eugenia uniflora*. Braz. J. Pharmacog. 21 (4), 581–586.

Saraswat, S., Rai, J.P.N., 2009. Phytoextraction potential of six plant species grown in multi-metal contaminated soil. Chem. Ecol. 25 (1), 1–11.

Sasaki, A., Yamaji, N., Yokosho, K., Ma, J.F., 2012. Nramp5 is a major transporter responsible for manganese and cadmium uptake in rice. Plant Cell 24, 2155–2167.

Sawidis, T., Metentzoglou, E., Mitrakas, M., Vasara, E., 2011. A Study of chromium, copper, and lead distribution from lignite fuels using cultivated and non-cultivated plants as biological monitors. Water Air Soil Poll. 220, 339–352.

Schmid-Siegert, E., Loscos, J., Farmer, E.E., 2012. Inducible malondialdehyde pools i zones of cell proliferation and developing tissues in Arabidopsis. J. Biol. Chem. 287 (12), 8954–8962.

Schönherr, J., Fernandez, V., Schreiber, L., 2005. Rates of cuticular penetration of chelated Fe(III): role of humidity, concentration, adjuvants, temperature and type of chelate. J. Agric. Food Chem. 53, 4484–4492.

SER, 2002. The SER Primer on Ecological Restoration. Society for Ecological Restoration and Policy Working Group.

Shaw, P.J.A., 1996. Role of seedbank substrates in the revegetation of fly ash and gypsum in the United Kingdom. Restor. Ecol. 4 (1), 61–70.

Sinclair, S.A., Krämer, U., 2012. The zinc homeostasis network of land plants. Biochim. Biophys. Acta 1823, 1553–1567.

Singh, N., Ma, L.Q., 2007. Assessing plants for phytoremediation of arsenic –contaminated soils. In: Willey, N. (Ed.), Phytoremediation: Methods and Reviews. Humana Press Inc, Totowa, NJ, pp. 319–347.

Singh, N., Ma, L.Q., Srivastava, M., Rathinasabapathi, B., 2006. Metabolic adaptations to arsenic-induced oxidative stress in *Pteris vittata* L. and *Pteris ensiformis* L. Plant Sci. 170, 274–282.

Singh, S., Parihar, P., Singh, R., Singh, V.P., Prasad, S.M., 2016. Heavy metal tolerance in plants: role of transcriptomics, proteomics, metabolomics, and ionomics. Front. Plant Sci. 6, 1143.

Smirnoff, N., 2005. Ascorbate, tocopherol and carotenoids: metabolism, pathway engineering and function. In: Smirnoff, N. (Ed.), Antioxidants and Reactive Oxgen Species in Plants. Blackwell Publishing Ltd., Oxford, UK, pp. 353–386.

Smith, M.J., Flowers, T.H., Duncan, H.J., Alder, J., 2006. Effects of polycyclic aromatic hydrocarbons on germination and subsequent growth of grasses and legumes in freshly contaminated soil and soil with aged PAHs residues. Environ. Pollut. 131, 519–525.

Smoliak, S., Penney, D., Harper, A.M., Horricks, J.S., 1981. Alberta forage manual. Alberta Agriculture, Print Media Branch, Edmonton, B.A.

Spira, Y., Edwards, D., Henstock, J., Gaboriau, H., Merly, C., Müller, D., Birke, V., Duijne, H., 2007. EURODEMO-improving the uptake of efficient soil and groundwater remediation technologies. Land Degrad. Reclamat. 17 (3–4), 685–692.

St. John, L., Tilley, D., Hunt, P., Wright, S., 2012. Plant Guide for Red Fescue (*Festuca rubra*) USDA – Natural Resources Conservation Service, Plant Materials Center, Aberden, Idaho 83210.

Stancik, D., 2003. Las species del gènero *Festuca* (*Poaceae*) en Colombia. Darwiniana, nueva serie 41(1–4), 93-153 (in Spanish).

Strausbaugh, P.D., Core, E.L., 1977. Flora of West Virginia. In: Morgantown, W.V. (Ed.), *Flora of West Virginia*, second ed. Seneka Books, Inc., Morgantown, West VA, USA, p. 1079.

Stukonis, V., Juzenas, S., Cesevičiene, J., Norkevičiene, E., 2015. Assessment of morpho-anatomical traits of red fescue (*Festuca rubra* L.) germplasm differing in origin. Zemdirbyste – Agriculture 102 (4), 437–442.

Takano, J., Miwa, K., Fujiwara, T., 2008. Boron transport mechanisms: collaboration of channels and transporters. Trends Plant Sci. 13 (8), 451–457.

Tapken, W., Ravet, K., Pilon, M., 2012. Plastocyanin controls the stabilization of the thylakoid Cu-transporting P-type ATPase PAA2/HMA8 in response to low copper in *Arabidopsis*. J. Biol. Chem. 287, 18544–18550.

Tejada-Jimenez, M., Galvan, A., Fernandez, E., Llamas, A., 2009. Homeostasis of the micronutrients Ni, Mo, and Cl with specific biochemical functions. Curr. Opin. Plant Biol. 12, 358–363.

Terry, N.A.M., Zayed, M.P., de Souza, M.P., Tarun, A.S., 2000. Selenium in higher plants. Annu. Rev. Plant Physiol. Plant Mol. Biol. 51, 401–432.

Thomas, H., 1987. Sid: a Mendelian locus controlling thylakoid membrane disassembly in senescing leaves of *Festuca pratensis*. Theor. Appl. Genet. 73, 551–555.

Thomas, H., Bortlik, K., Rentsch, D., Schellenberg, M., Matile, P., 1989. Catabolism of chlorophyll in vivo: significance of polar chlorophyll catabolites in a non-yellowing senescence mutant of *Festuca pratensis* Huds. New Phytol. 111, 3–8.

Thomas, H., Howarth, C.J., 2000. Five ways to stay green. J. Exp. Bot. 51, 329–337.

Thompson, K., 1987. Seeds and seed banks. New Phytol. 106, 23–34.

Tischew, S., Baasch, A., Grunert, H., Kirmer, A., 2014. How to develop native plant communities in heavily altered ecosystems: examples from large-scale surface mining in Germany. Appl. Veg. Sci. 17, 288–301.

Tokala, R.K., Strap, J.L., Jung, C.M., Crawford, D.L., Salove, H., Deobald, L.A., Bailey, F.J., Morra, M.J., 2002. Novel plant– microbe rhizosphere interaction involving *S. lydicus* WYEC108 and the pea plant (*Pisum sativum*). Appl. Environ. Microbiol. 68, 2161–2171.

Tong, Y.P., Kneer, R., Zhu, Y.G., 2004. Vacuolar compartmentalization: a second–generation approach to engineering plants for phytoremediation. Trends Plant Sci. 9 (1), 8–9.

Tóth, S.Z., Nagy, V., Puthur, J.T., Kovacs, L., Garab, G., 2011. The physiological role of ascorbate as photosystem II electron donor: protection against photoinactivation in heat-stressed leaves. Plant Physiol. 156, 382–392.

Tripathi, R.D., Srivastava, S., Mishra, S., Singh, N., Tuli, R., Gupta, D.K., Maathius, F.J.M., 2007. Arsenic hazards: strategies for tolerance and remediation by plants. Trends Biotechnol. 25 (4), 158–165.

Tsuda, T., Shiga, K., Ohshima, K., Kawakishi, S., Osawa, T., 1996. Inhibition of lipid peroxidation and the active oxygen radical scavenging effect of anthocyanin pigments isolated from *Phaseolus vulgaris* L. Biochem. Pharmacol. 52, 1033–1039.

Uddin, M., Juraimi, A., 2013. Salinity tolerance turfgrass: history and prospects. Sci. World J. 13, 6.

US Department of Agriculture, Natural Resources Conservation Service, 2016. USDA PLANTS Database.

Van Rensburg, L., Morgenthal, T.L., Van Hamburg, H., Michael, M.D., 2003. A comparative analysis of the vegetation and topsoil cover nutrient status between two similarly rehabilitated ash disposal sites. Environmentalist 23, 285–295.

Vogel, W.G., 1981. A guide for revegetatating coal minesoils in the eastern United States. Gen. Tech. Rep. NE-68. US Department of Agriculture, Forest Service, Northeastern Forest Experiment Station, Broomall, PA.

Volland, L., 1985. Guidelines for forage resource evaluation within central Oregon Pumice Zone. R6-Ecol-177-1985. US, Department of Agriculture, Forest Service, Pacific Northwest Region, Portland, OR.

Vollenweider, S., Weber, H., Stolz, S., Chetalat, A., Farmer, E.E., 2000. Fatty acid ketodienes and fatty acid ketotrienes: Michael addition acceptors that accumulate in wounded and deseased *Arabidopsis* leaves. Plant J. 24, 467–476.

Voss, E.G., 1972. Michigan flora. Part I. Gymnosperms and monocots. University of Michigan Herbarium; Cranbrook Institute of Science, Ann Arbor; Bloomfield Hills, MI.

Wang, J., Zhao, F.-J., Meharg, A.A., Raab, A., Feldmann, J., McGrath, S.P., 2002. Mechanisms of arsenic hyperaccumulation in *Pteris vittata*. Uptake kinetiks, interactios with phosphate, and arsenic speciation. Plant Physiol. 130, 1552–1561.

Wang, Y., Oyaizu, H., 2009. Evaluation of the phytoremediation potential of four plant species for dibenzofuran-contaminated soil. J. Hazard. Mater. 168, 760–764.

Weber, H., Chetelat, A., Reymond, P., Farmer, E.E., 2004. Selective and powerful stress gene expression in *Arabidopsis* in response to malondialdehyde. Plant J. 37, 877–888.

Wheeler, W.A., Hill, D.D., 1957. Grassland Seeds. D. Van Nostrand Company, Inc, Princeton, NJ.

WHO, 2010. Fact sheet N°225. Dioxins and their effects on human health. World Health Organization, Geneva.

WHO, 2013. Ten chemicals of major health concern. Available from: www.who.int/ipcs/assessment/public_health/chemicals_phc/en/index.html.

WHO/UNECE, 2006. Health risk of heavy metals from long-range transboundary air pollution. Draft of May 2006. World Health Organisation Regional Office for Europe and Geneva: United Nations Economic Commission for Europe (UNECE), Copenhagen.

Wilken, A., Bock, C., Bokern, M., Harms, H., 1995. Metabolism of different PCB congeners by plant cell cultures. Environ. Toxicol. Chem. 14, 2017–2022.

Yamasaki, H., Abdel-Ghany, S.E., Cohu, C.M., Kobayashi, Y., Shikanai, T., Pilon, M., 2009. SQUAMOSA promoter binding protein-like7 is a central regulator for copper homeostasis in *Arabidopsis*. Plant Cell 21, 347–361.

Yamasaki, H., Uefuji, H., Sakihama, Y., 1996. Bleaching of the red anthocyanin induced by superoxide radical. Arch. Biochem. Biophys. 332, 183–186.

Yamasaki, S., Sakata-Sogawa, K., Hasegawa, A., Suzuki, T., Kabu, K., Sato, E., Kurosaki, T., Yamashita, S., Tokunaga, M., Nishid, K., Hirano, T., 2007. Zinc is a novel intracellular second messenger. J. Cell Biol. 177, 637–645.

Yang, S., Liang, S., Yi, L., Xu, B., Cao, J., Guo, Y., Zhou, Y., 2014. Heavy metal accumulation and phytostabilization potential of dominant plant species growing on manganese mine tailings. Front. Env. Sci. Eng. 8 (3), 394–404.

Yoder, C., 2000. Creeping Red Fescue Seed Production on the Peace River Region. Agri-Facts. Alberta Agriculture, Food and Rural Development.

Yoon, J., Cao, X., Zhou, Q., Ma, L.Q., 2006. Accumulation of Pb, Cu, and Zn in native plants growing on a contaminated Florida site. Sci. Total Environ. 368 (2), 456–464.

Yruela, I., 2005. Copper in plants. Braz. J. Plant Physiol. 17, 145–156.

Zayed, A., Lytle, C.M., Terry, N., 1998. Accumulation and volatilization of different chemical species of selenium by plants. Planta 206, 284–292.

Zelensky, V.M., Sariev, A.H., 2008. Perennial grasses for land reclamation in the subarctic tundra of Taimyr. Siberian Bull. Agric. Sci. 2, 40–46.

Zhang, Q., Zuk, A., Rue, K., 2013. Salinity tolerance of nine fine fescue cultivars compared to other cool-season turfgrasses. Sci. Hort. 159, 67–71.

Zhang, Y., Yang, J., Wu, H., Shi, C., Zhang, C., Li, D., Feng, M., 2014. Dynamic changes in soil and vegetation during varying ecological - recovery conditions of abandoned mines in Beijing. Ecol. Eng. 73, 676–683.

Zhao, F.J., Wang, J.R., Barker, J.H.A., Schat, H., Bleeker, P.M., McGraph, S.P., 2003. The role of phytochelatins in arsenic tolerance in the hyperaccumulator *Pteris vittata*. New Phytol. 159, 403–410.

Reed canary grass (*Phalaris arundinacea* L.): coupling phytoremediation with biofuel production

7

Vimal Chandra Pandey[a,], Ambuj Mishra[b], Sudhish Kumar Shukla[c], D.P. Singh[a]*
[a]Department of Environmental Science, Babasaheb Bhimrao Ambedkar University, Lucknow, Uttar Pradesh, India; [b]School of Environmental Sciences, Jawaharlal Nehru University, New Delhi, India; [c]Department of Chemistry, Manav Rachna University, Faridabad, Haryana, India
[*]Corresponding author

1 Introduction

Phytoremediation technique has gained much popularity world-wide as a vital, simple, cost-effective, widely acceptable, eco-friendly, compatible, sustainable, and reliable. Therefore, it is a promising technology and applicable in large areas when using socioeconomically and ecologically valuable native plants, that is helpful in the generation of income from produced phytoproducts from contaminated sites (Pandey et al., 2009, 2015a, 2016a; Pandey and Bajpai, 2019). Five different types of strategies are used for phytoremediation of contaminated sites with metal(loid)s, such as phytostabilization, phytoextraction/phytoaccumulation, rhizofiltration, phytodegradation, and phytovolatilization (Pandey and Bajpai, 2019). The usefulness of energy crops in remediation of a wide-ranging waste dumpsites and contaminated sites has been explored in detail for socio-economic and ecological benefits toward nature sustainability (Pandey et al., 2012a; Pandey, 2017; Pandey and Souza-Alonso, 2019). The suitability of aromatic crops in phytoremediation of a variety of waste dumpsites and polluted sites has been proved for multiple benefits (Pandey and Singh, 2015; Verma et al., 2014; Pandey et al., 2019). Several potential plant species (i.e., *Typha latifolia, Azolla caroliniana, Ricinus communis, Thelypteris dentata, Saccharum munja, Saccharum spontaneum, Ipomoea carnea,* and *Ziziphus mauritiana*) have been evaluated for phytoremediation of fly ash dumpsites, and suggested for remediation and management of newly fly ash dumpsites (Pandey, 2012a, 2012b, 2013; Pandey et al., 2012b, 2014, 2015b, 2015c; Kumari et al., 2013; Pandey and Mishra, 2018). In addition, assisted phytoremediation of fly ash dumps by naturally colonized native plants through ecological engineering can be used for rehabilitation programs with multiple benefits (Pandey and Singh, 2012; Pandey, 2015). Phytoremediation of red mud deposit sites has also been suggested by naturally growing plants (Mishra et al., 2017; Mishra and Pandey, 2012). The success of phytoremediation of contaminated sites can further be improved through the use of microbes via harnessing plant–microbe interactions (Praveen et al., 2019).

Phytoremediation Potential of Perennial Grasses. http://dx.doi.org/10.1016/B978-0-12-817732-7.00007-9

Recently, perennial grasses have been identified as potential asset for quick remediation of polluted sites with toxic metals due to its fast growing ability. Reed canary grass (*Phalaris arundinacea* L.) is a perennial grass spread broadly in both wet and terrestrial habitats for instance wetlands, lakeshores, wet meadows, riverbanks, and floodplains (Lavergne and Molofsky, 2004). This species can survive under different environmental stress conditions, such as drought, flooding, grazing, freezing, contaminated sites, degraded lands. Numerous studies showed the capacity of reed canary grass for phytoremediation of trace metal bioaccumulation (Samecka-Cymerman and Kempers, 2001; Korzeniowska and Stanislawska-Glubiak, 2019; Polechońska and Klink, 2014b; Dzantor et al., 2000; Deng et al., 2004; Rosikon et al., 2015). The potential use of this perennial grass in remediation of polluted sites is discussed with biofuel production potential in view of some of its other features, for instance, high tolerance to site conditions, cosmopolitan range, and the potential capability to accumulate certain trace metals (Lavergne and Molofsky, 2004). The aims of this chapter are to reveal potential characteristics of reed canary grass for phytoremediation of a wide-ranging problematic soil under different environmental stress conditions along with biomass production potential for biofuel production.

2 Origin and geographical distribution

P. arundinacea L. is a circumboreal (occurs in Boreal region) species and occurs in temperate region of five continents (Galatowitsch et al., 1999; Waggy and Melissa, 2010) in which it is known as native to temperate parts of Europe, Asia (Waggy and Melissa, 2010; Ali et al., 2010), and North America (Barkworth et al., 2007) and introduced to New Zealand and South America (Waggy and Melissa, 2010). In Indian sub-continent, it is known to be native to Afghanistan, Pakistan, Kashmir, Himachal Pradesh, Meghalaya, and West Bengal of India (Waggy and Melissa, 2010; Ali et al., 2010; Lansdown, 2014). The early native origin of the reed canary grass is not certainly known and can't be correctly predicted (Galatowitsch et al., 1999; Waggy and Melissa, 2010). Cultivation of the grass began in early 1749 in Sweden and spread across the Europe extensively (Waggy and Melissa, 2010). Introduction of these European cultivars in North American continent has been reported in 1930 at first (Alway, 1931) and cultivated widely throughout the continent (Galatowitsch et al., 1999; Maurer et al., 2003; Waggy and Melissa, 2010). Lavoie and Dufresne (2005) postulated the evidences for native origin of the grass in Quebec (Canadian province) remote areas with study of herbaria records collected in 19th century. Several studies based on DNA marker indicate the clear distinction between invasive and non-invasive strains of reed canary grass (Jakubowski et al., 2011; Jensen et al., 2018). Invasive strains occur throughout North America are either non-native or hybrid between European cultivars and native North American cultivars, while non-invasive strains are native to the region (Lavergne and Molofsky, 2007; Dore and McNeill, 1980) which may no longer occur in North America (Waggy and Melissa, 2010).

3 Ecology

Grassland comprises approximately 20% of the global vegetation. Occurrence of grasses can be seen almost in every habitat around the globe, apart from rain forest and dense thickets. Grasses spread widely as exhibit more than one breeding system; thus its availability may exploit the new habitat insuring its own survival. The genus *Phalaris* is distributed across the world except Antarctica and Greenland (Anderson, 1961); and *P. arundinacea* is one among the species of the genus *Phalaris*. It occurs in cold and wet habitat with high nutrient (Barnes, 1999; Anderson, 2012), and dominates over the native vegetation to become invasive (Anderson, 1961). It can be found near marshes, wet meadows, prairies, swamp, riparian area, peat land, and flood plains (Waggy and Melissa, 2010; Lansdown, 2014). The grass can also be seen as monoculture in uplands oak savannah, where it grows slowly (Meisel et al., 2002). In wetlands, it grows thicker and appears with maximum vegetation cover (McCain and Christy, 2005). It shows prominent association with graminoids in mixed habitats (Muldavin et al., 2000). In riparian areas, reed canary grass may occur in mixed or dominant in under story of woody plants (McCain and Christy, 2005; Narumalani et al., 2009). Reed canary grass grows in thick, dense monotypic colonies sometimes bear few other plant species and provides shelter or nesting for few wetland animals (Anderson, 2012). It is a light demanding grass showed comparatively decreased dominance and biomass in shaded condition (Maurer and Zedler, 2002; Kim et al., 2006). The invasive *P. arundinacea* is severely aggressive and competitive to eliminate native species for space and nutrient; thus often a threat to the native biodiversity (Anderson, 2012). In fact invasive strain of the grass is threat to its native species itself (Lavergne and Molofsky, 2007; Waggy and Melissa, 2010; Anderson, 2012). As result of its productiveness and rapid growth, this species presents a threat to many wetland ecosystems. It exhibits excellent frost tolerance and can manage to grow easily in flood plains or poorly drained soil.

4 Botanical description

P. aundinacea is one of the most aggressive and wide spread species among the species of the genus *Phalaris* (Anderson, 2012). Main stem of this grass can grow up to 2 m tall and girth may extend up to 1.25 cm with sometimes variegated leaf blades (some garden varieties). With usually streaked spikelets (green and purple), panicle may grow 30 cm tall (Waggy and Melissa, 2010). Flourishing occurs from late May to August. Glumes are scabrous and glabrous (Fernald, 1950). Seed also occurs in reed canary grass. Generally, it is a vigorous, high yielding, perennial, and sod-forming grass. Roots grow thick and vigorously but shallow to produce vegetative stems by extending horizontally, causes clubbed growth of new plants (Anderson, 2012). On chromosomal basis, there are two races of this grass, such as tetraploid and hexaploid (Jakubowski et al., 2011).

5 Propagation

Mainly two types of propagation occur in reed canary grass, that is, through vegetative propagation (within an established population) and by means of seed (propagation within an established population as well as new colonization).

Vegetative propagation (within an established population)—*P. aundinacea* is a sod-forming grass. Creeping rhizomes grow very shallow and parallel to ground as well as forms stolons and generates new vegetative shoots. Rhizomes are below the ground stem, often growing few meters in reed canary grass, forms new shoots breaking through the crust of soil. Stolons are also the stem basically remains at surface of soil give rise to new shoots and root at nodes. Thus this vegetative growth turns the patch denser within an established population as well as extends area of the establishment.

Seed (propagation within an established population as well as new colonization)— Reed canary grass produces huge amount of seeds which also varies within and among the populations (Leck and Simpson, 1994; Coops and Van der Velde, 1995). The latitudes, where day light is for longer time, seed production is higher with the fact that longer the day light, higher the flowering (Allard and Evans, 1941). The grass prominently appears to be cross pollinated (Fryxell, 1957), occurs through wind (Merigliano and Lesica, 1998). It has been noticed a buoyancy in the seeds for early few days form fall so that it may also provide dispersal through water as often grow in wet areas. It also forms soil seed banks; seeds buried shallower or deeper than 22 cm are less viable than buried at 22 cm of depth (Goss, 1924; Toole and Brown, 1946). Viability of the soil seed bank is more than the floating seeds (Comes et al., 1978). Moreover, for the case seeds submerged in water, variable cases came into notice; in few, the seeds lost germination capability if submerged more than a couple of years (Comes et al., 1978) while in other case it germinated even after 3 years of being submerged (Beule, 1979). Thus, seed dispersal is the method with propagation take place within an established population as well as it extends it colonization to the new sites.

6 Main features of reed canary grass in relation to phytoremediation

Reed canary grass has wide-ranging stress tolerant potential, such as drought, flooding, grazing, and freezing. It is broadly adapted temperate grass and has also capacity to grow healthy in dry and wet areas. It can survive with a wide pH range from 4.9 to 8.2 (Carlson et al., 1996). Its drought tolerant capacity facilitates harvesting and makes it comparatively more productive in the summer season than the cold season (Carlson et al., 1996). The use of flexibility of reed canary grass for animal feed and fuel production during phytoremediation of degraded/polluted soils is an opportunity in multiple geographies with carbon sink and biodiversity benefits. Reed canary grass has been reported as a potential bioaccumulator for trace elements (Deng et al., 2004; Samecka-Cymerman and Kempers, 2001; Abreu et al., 2012; Rosikon et al., 2015). Some important features of reed canary grass regarding phytoremediation are given in the following subsections.

6.1 A good phytoextractor

A lots of research proved that *P. arundenacea* is an effective crop for phytoremediation (Dzantor et al., 2000; Deng et al., 2004; Polechońska and Klink, 2014b; Rosikon et al., 2015; Korzeniowska and Stanislawska-Glubiak, 2019). Several investigations revealed that it has potential to remediate the heavy metals (Samecka-Cymerman and Kempers, 2001; Deng et al., 2004; Abreu et al., 2012; Polechońska and Klink, 2014b; Lord, 2015; Rosikon et al., 2015; Korzeniowska and Stanislawska-Glubiak, 2019), organic pollutants (Dzantor et al., 2000), and radioactive elements (Lasat and Kochian, 1997; Rickard and Price, 1990). Besides, the reed canary grass is a good phytoextracter of macronutrient N, P, Na, Mg, etc. (Polechońska and Klink, 2014a), therefore can be used in the eutrophic lakes to extract the nutrient.

6.2 A good phytostabilizer

P. arundinacea has been demonstrated as a good phytostabilizer with low translocation factor (ratio of metal concentration in above ground parts with metal concentration in roots), shows low motilities of the metal content in above ground parts of the grass which is helpful to hinder the biomagnification of heavy metals and its spread in other systems, as well as confirms high tolerance of the species by limiting translocation of heavy metal in its fragile parts (Deng et al., 2004). Optimum phytoremediation of organic pollutants like PCB, TNT, pyrene, etc. has also been reported with the reed canary grass (Dzantor et al., 2000).

6.3 A good bioindicator

This grass has properties like wide-spread occurrence around the globe, facile identification, easy harvesting with vigorous growth, and high accumulation strength, which are the suitable characteristics for an effective bioindicator (Samecka-Cymerman and Kempers, 1996). Moreover, Polechońska and Klink, (2014b) reported the significant and positive correlations between the content of Co and Zn in the substrate and the levels of these elements in the plant parts of the reed canary grass which indicate potential application of *P. arundinacea* in the biomonitoring of environmental contamination with these metals. It has been appeared as an interesting species for phytostabilization of contaminated soil, particularly for Co and Zn, due to limited mobility and translocation of metal content, absorbed by roots (Polechońska and Klink, 2014b). However, accumulation of heavy metals put adverse effect on the accumulation of above and below ground biomass of the grass (Korzeniowska and Stanislawska-Glubiak, 2019).

6.4 Tolerance to wide-ranging stress conditions

Reed canary grass tolerates a wide range of stress conditions (i.e., drought, flooding, grazing, freezing, pH range from 4.9 to 8.2, marginal lands, and polluted sites). As such, it can be used as adaptive agricultural practices for phytoremediation of polluted sites, pasture, straw or bedding for livestock, pulp and paper, hay or silage production, soil conservation, biomaterials, biomass production for bioenergy, feedstock to industrial biotechnology (Jensen et al., 2018).

6.5 Biomass yield

Korzeniowska and Stanislawska-Glubiak (2019) assessed and found that *P. arundinacea* has high tolerance for Ni metal. This tolerance was tested in terms of Tolerance Index (TI), which is the percentage ratio of yield of the biomass in the contaminated soil and in the control.

$$TI = \frac{\text{mean biomass yield in contaminated soil/treatment}}{\text{biomass yield in control}} \times 100$$

Biomass of the above ground parts of the grass decreased in the contaminated soil. While this decrease in the biomass yield, was for very high concentrations only. Infact in the next growing season, the reduction of the biomass yield observed significantly smaller than the previous one. This might be due to the growth of the roots, deeper than the contamination (Korzeniowska and Stanislawska-Glubiak, 2019). Lord (2015) also reported it as tolerant species with better biomass yield for zinc, copper, nickel, and cadmium.

6.6 Rate of photosynthesis

Chen et al. (2009) and Shafeeq et al. (2012) suggested that accumulation of heavy metal like Ni inhibits photosynthesis by damaging stomatal aperture which disturbs the photosynthetic rate. Korzeniowska and Stanislawska-Glubiak (2019) also observed the decrease in photosynthetic rate of reed canary grass at very high concentrations of Ni only.

6.7 Easy propagation, establishment, and biomass production

Reed canary grass can be propagated easily by seed or vegetative method. Reed canary grass is inexpensive to establish and fits current farming practice, providing flexibility and minimum risk to farmers to grow to gain confidence in farming for energy. Therefore, the low cost of establishment and faster rates of return on financial investment make it a potential candidate to grow on a wide-ranging stress condition for bioenergy (Jensen et al., 2018). This grass has ability to produce good biomass from late summer until early spring thereby producing biomass earlier in the year than other energy grasses, thus reducing storage requirements for end users (Jensen et al., 2018).

7 Multiple uses of reed canary grass

7.1 Biomass production

P. arundenacea is a severely aggressive and sod-forming grass with vigorous growth and can be grown in marginal lands (Jensen et al., 2018). Although native to the temperate zone, it can be grown to several other habitats, multiple topologies and adapted to variety of stress like flooding, drought, etc. It gives more biomass production in summers when other grasses exhibits very less production. Moreover, being cost-effective with huge biomass production, the earlier indicated properties turn it to be an efficient bioenergy crop with emergence of new cultivars (Jensen et al., 2018). Efforts produce new and efficient cultivars in Northern America and Northern parts of Europe reflected its potential to increase its yield in terms of biomass production (Casler et al., 2009; Jensen et al., 2018). Several trials and modifications to improve the breed for greater biomass production (tall shoot, multiple nodes, heavy straw quantity, etc.) and efficient combustion (minimized lignifications, decreased silicon, chlorine, and potassium levels with lodging resistance) are also being made since long (Casler et al., 2009; Boateng et al., 2008). In the United Kingdom, Bs5321, the highest biomass yielding synthetic cultivar was derived from four local accessions (Jensen et al., 2018). Such kind of more efforts are being made in Northern Europe to produce more efficient cultivars with the research project "Optimization of Reed Canary Grass as a Native European Energy Crop" by setting up the collaboration between the United Kingdom and Sweden (Jensen et al., 2018).

7.2 Carbon sequestration

Grasses sink most of the carbon in the roots, prevents the stock from wild file, and adapted to drought conditions, performs sequestration even in extreme stress. Dass et al. (2018) performed a set of modeling experiment and reported superior resilience of grass lands C sink than forest in response to changes in climate change in 21st century. The research quotes "the resilience of grassland to rising temperatures, drought and fire, coupled with the preferential banking of C to below ground sinks, helps to preserve sequestered terrestrial C and prevent it from re-entering the atmosphere." Yang et al. (2019) also suggested the role and importance of grasslands to restore agriculturally degraded and abandoned lands to scrub atmospheric carbon dioxide and sequester it to soil organic matter. By such a means grasses increases the importance of such degraded and abandoned lands in carbon sequestration and productivity. Perennial grasses, which persists in land for a long time and contribute its parts and residues to add organic matter in the soil, remains for a very long time in the soil. Peat lands are the best examples for it which is the largest soil C sinks among all.

P. arundenacea is perennial and deep and rooted grass with vigorous growth, and serves all the purposes indicated earlier for presenting itself as efficient and potential crop for removal of atmospheric carbon and to sequester it in soil organic carbon for long time. Several research confirms the efficient ability of reed canary grass for C sequestration potential (Bills, 2009; Shurpali et al., 2008; Mander et al., 2012; Don et al., 2012; Xiong and Kätterer, 2010). It observed that reed canary grass exhibit high carbon accumulation capabilities in below ground parts with vigorous and dense root networks, ads in soil carbon heavily (Xiong and Kätterer, 2010) with lower respiratory carbon losses in the soil (Shurpali et al., 2008). Cultivation of the grass as a potential bioenergy crop at peat extraction sites also showed low GHG emission of nitrous oxide and methane, and converted it to net carbon sink from net carbon source (Ní Choncubhair et al., 2017). Thus cropping of reed canary grass in degraded and abandoned land can be the best land use change practice to cultivate it as bioenergy crop, sequestering atmospheric carbon and enhancing soil organic matter and promoting restoration of the land.

7.3 Pulp and paper

The pulp industry with non-woody fibers, mainly based on straw is growing extensively in European and Nordic countries since more than a decade (Finell, 2003). Development of more advance technologies for reed canary grass has also being made to develop high quality paper since long (Pavilainen, 1998). Swedish Agro-Fiber Project (1987–91) and The Finish Agro-Fiber Project (1993–95) are some of the examples for developing cost effective and eco-friendly grasses to replace woody species (particularly birch) for production of better quality of paper pulp and reed canary grass has been proved one among the best species for the purpose (Finell, 2003; Paavilainen and Tulppala, 1996). In the EU Reed Canary Grass Project (1995–99), *P. arundenacea* has been investigated with the objectives to develop it as an economically and environmentally competitive industrial crop for combined production of high quality chemical pulp and bio-fuel (Finell, 2003; Andersson and Lindvall, 1997). Paavilainen and Torgilsson (1994) reported reed canary grass as a better crop producing almost double pulp than temperate woody species birch, annually. The reed canary grass has been used for both the paper grades, such as fine paper which is high quality writing and printing paper (Finell, 2003; Paavilainen and tulppala, 1996) and white-top liner paper with good printing surface (Finell, 2003).

8 Conclusions and future prospects

In conclusion, reed canary grass is a viable bio-energy perennial grass which has potential ability to grow on a variety of degraded lands to meet our societal needs. In this regard, a number of studies have been done to evaluate the usefulness of reed canary grass to the phytoremediation of the contaminated soil along with biomass production. Being tolerant in nature in order to a wide-ranging of environmental stress conditions (i.e., drought, flooding, grazing, and freezing, etc.) it can be used in future changing environment.

Acknowledgments

Financial assistance given to Dr. V.C. Pandey under the Scientist's Pool Scheme (Pool No. 13 (8931-A)/2017) by the Council of Scientific and Industrial Research, Government of India is gratefully acknowledged. The authors are thankful to the reviewers for their valuable comments. The authors declare no conflicts of interest.

References

Abreu, M.M., Santos, E.S., Magalhães, M.C.F., Fernandes, E., 2012. Trace elements tolerance, accumulation and translocation in *Cistus populifolius*, *Cistus salviifolius* and their hybrid growing in polymetallic contaminated mine areas. J. Geochem. Explor. 123, 52–60.

Ali, S. I., Qaiser, M., et al., 2010. Flora of Pakistan, [Online]. Islamabad: Pakistan Agricultural Research Council; Karachi, Pakistan: University of Karachi; St. Louis, MO: Missouri Botanical Garden. In: eFloras. St. Louis, MO: Missouri Botanical Garden; Cambridge, MA: Harvard University Herbaria (Producers). Available from: http://www.efloras.org/flora_page.aspx?flora_id=5; http://www.mobot.org/MOBOT/research/pakistan/welcome.shtml

Allard, H.A., Evans, M.W., 1941. Growth and flowering of some tame and wild grasses in response to different photoperiods. J. Agric. Res. 62(4), 193–228.

Alway, F.J., 1931. Early trials and use of reed canary grass as a forage plant. J. Am. Soc. Agron. 23 (1), 64–66.

Anderson, D.E., 1961. Taxonomy and distribution of the genus *Phalaris*. IOWA State J. Sci. 36, 1–96.

Anderson, H., 2012. Invasive Reed Canary Grass (*Phalaris arundinacea* subsp. arundinacea). Best Management Practices in Ontario. Ontario Invasive Plant Council, Peterborough, ON.

Andersson, B., Lindvall, E., 1997. Use of biomass from reed canary grass (*Phalaris arundinacea*) as raw material for production of paper pulp and fuel. In: Proceedings of the XVIII International Grassland Congress, Canada, Vol. 1, No. 3.

Barkworth, M.E., Capels, K.M., Long, S., Anderton, L. K., Piep, M.B. (Eds.), 2007. Flora of North America north of Mexico. Volume 24: Magnoliophyta: Commelinidae (in part): Poaceae, part 1. Oxford University Press, New York, p. 911.

Barnes, W.J., 1999. The rapid growth of a population of reed canary grass (*Phalaris arundinacea* L.) and its impact on some river bottom herbs. J. Torrey Bot. Soc. 126 (2), 133–138.

Beule, J. D., 1979. Control and management of cattails in southeastern Wisconsin wetlands. Tech. Bull No. 112. Department of Natural Resources, Madison, WI, p. 40.

Bills, J.S., 2009. Invasive reed canary grass (*Phalaris arundinacea*) and carbon sequestration in a wetland complex. Doctoral dissertation.

Boateng, A.A., Weimer, P.J., Jung, H.G., Lamb, J.F., 2008. Response of thermochemical and biochemical conversion processes to lignin concentration in alfalfa stems. Energy Fuels 22, 2810–2815.

Carlson, I.T., Oram, R.N., Surprenant, J., 1996. Reed canarygrass and other Phalaris species. In: Moser, L.E., Buxton, D.R., Casler, M.D. (Eds.), Cool-Season Forage Grasses. Agronomy 34, 569–604.

Casler, M.D., Phillips, M.M., Krohn, A.L., 2009. DNA polymorphisms reveal geographic races of reed canarygrass. Crop Sci. 45 (6), 2139–2148, doi: 10.2135/cropsci2009.02.0055.

Chen, C., Huang, D., Liu, J., 2009. Functions and toxicity of nickel in plants: recent advances and future prospects. Clean Soil Air Water 37(4–5), 304–313.

Comes, R.D., Bruns, V.F., Kelley, A.D., 1978. Longevity of certain weed and crop seeds in fresh water. Weed Sci. 26 (4), 336–344.

Coops, H., Van der Velde, G., 1995. Seed dispersal, germination and seedling growth of six helophyte species in relation to water-level zonation. Freshwater Biol. 34 (1), 13–20.

Dass, P., Houlton, B.Z., Wang, Y., Warlind, D., 2018. Grasslands may be more reliable carbon sinks than forests in California. Environ. Res. Lett. 13 (7), 074027.

Deng, H., Ye, Z.H., Wong, M.H., 2004. Accumulation of lead, zinc, copper and cadmium by 12 wetland plant species thriving in metal-contaminated sites in China. Environ. Pollut. 132 (1), 29–40.

Don, A., Osborne, B., Hastings, A., et al., 2012. Land-use change to bioenergy production in Europe: implications for the greenhouse gas balance and soil carbon. Glo-bal Change Biol. Bioenergy 4, 372–391.

Dore, W.G., McNeill, J., 1980. Grasses of Ontario. Monograph No. 26. Agriculture Canada, Research Branch, Ottawa, ON, p. 566.

Dzantor, E.K., Chekol, T., Vough, L.R., 2000. Feasibility of using forage grasses and legumes for phytoremediation of organic pollutants. J. Environ. Sci. Health Part A 35 (9), 1645–1661.

Finell, M., 2003. The use of reed canary-grass (*Phalaris arundinacea*) as a short fibre raw material for the pulp and paper industry, Vol. 424.

Fryxell, P.A., 1957. Mode of reproduction of higher plants. Bot. Rev. 23, 135–233.

Galatowitsch, S.M., Anderson, N.O., Ascher, P.D., 1999. Invasiveness in wetland plants in temperate North America. Wetlands 19 (4), 733–755.

Goss, W.L., 1924. The vitality of buried seeds. J. Agric. Res. 29 (7), 349–362.

Jakubowski, A., Jackson, R., Johnson, R., Hu, J., Casler, M., 2011. Genetic diversity and population structure of Eurasian populations of reed canary grass: cytotypes, cultivars, and interspecific hybrids. Crop Pasture Sci. 62, 982–991.

Jensen, E.F., Casler, M.D., Farrar, K., Finnan, J.M., Lord, R., Palmborg, C., Valentine, J., Donnison, L.S., 2018. Reed canary grass: from production to eEnd use. In: Alexopoulou, E. (Ed.), Perennial Grasses for Bioenergy and Bioproducts. Academic Press, London, UK, pp. 153–173.

Kim, K.D., Ewing, K., Giblin, D.E., 2006. Controlling *Phalaris arundinacea* (reed canary grass) with live willow stakes: a density-dependent response. Ecol. Eng. 27 (3), 219–227.

Korzeniowska, J., Stanislawska-Glubiak, E., 2019. Phytoremediation potential of *Phalaris arundinacea, Salix viminalis* and *Zea mays* for nickel-contaminated soils. Int. J. Environ. Sci. Technol. 16, 1999–2008.

Kumari, A., Pandey, V.C., Rai, U.N., 2013. Feasibility of fern *Thelypteris dentata* for revegetation of coal fly ash landfills. J. Geochem. Explor. 128, 147–152.

Lansdown, R.V., 2014. *Phalaris arundinacea. The IUCN Red List of Threatened Species*2014: e.T164064A1021826. Available from: http://dx.doi.org/10.2305/IUCN.UK.2014 1.RLTS. T164064A1021826.en.

Lasat, M.M., Kochian, L.V., 1997. Potential for phytoextraction of 137 Cs from a contaminated soil. Plant Soil 195 (1), 99–106.

Lavergne, S., Molofsky, J., 2004. Reed canary grass (*Phalaris arundinacea*) as a biological model in the study of plant invasions. Crit. Rev. Plant Sci. 23 (5), 415–429.

Lavergne, S., Molofsky, J., 2007. Increased genetic variation and evolutionary potential drive the success of an invasive grass. Proc. Nat. Acad. Sci. 104 (10), 3883–3888.

Lavoie, C., Dufresne, C., 2005. The spread of reed canary grass (*Phalaris arundinacea*) in Quebec: a spatio-temporal perspective. Ecoscience 12 (3), 366–375.

Leck, M.A., Simpson, R.L., 1994. Tidal freshwater wetland zonation: seed and seedling dynamics. Aquat. Bot. 47, 61–75.

Lord, R.A., 2015. Reed canary grass (*Phalaris arundinacea*) outperforms *Miscanthus* or willow on marginal soils, brownfield and non-agricultural sites for local, sustainable energy crop production. Biomass Bioenerg. 78, 110–125.

Mander, Ü., Järveoja, J., Maddison, M., Soosaar, K., Aavola, R., Ostonen, I., Salm, J.-O., 2012. Reed canary grass cultivation mitigates greenhouse gas emissions from abandoned peat extraction areas. Global Change Biol. Bioenergy 4 (4), 462–474.

Maurer, D.A., Lindig-Cisneros, R., Werner, K.J., Kercher, S., Miller, R., Zedler, J.B., 2003. The replacement of wetland vegetation by reed canarygrass (*Phalaris arundinacea*). Ecol. Restor. 21 (2), 116–119.

Maurer, D.A., Zedler, J.B., 2002. Differential invasion of a wetland grass explained by tests of nutrients and light availability on establishment and clonal growth. Oecologia 131 (2), 279–288.

McCain, C., Christy, J.A., 2005. Field guide to riparian plant communities in northwestern Oregon. Tech. Pap. R6-NR-ECOL-TP-01-05. U.S. Department of Agriculture, Forest Service, Pacific Northwest Region, Portland, OR, p. 357.

Meisel, J., Trushenski, N., Weiher, E., 2002. A gradient analysis of oak savanna community composition in western Wisconsin. J. Torrey Bot. Soc. 129 (2), 115–124.

Merigliano, M.F., Lesica, P., 1998. The native status of reed canarygrass (*Phalaris arundinacea* L.) in the inland Northwest, U.S.A. Nat. Areas J. 18 (3), 223–230.

Mishra, T., Pandey, V.C., 2012. Phytoremediation of red mud deposits through natural succession. In: Pandey, V.C., Bauddh, K. (Eds.), Phytomanagement of Polluted Sites. Elsevier, Amsterdam, Netherlands, pp. 409–424.

Mishra, T., Pandey, V.C., Singh, P., Singh, N.B., Singh, N., 2017. Assessment of phytoremediation potential of native grass species growing on red mud deposits. J. Geochem. Explor. 182, 206–209.

Muldavin, E., Durkin, P., Bradley, M., Stuever, M., Mehlhop, P., 2000. Handbook of Wetland Vegetation Communities of New Mexico. Volume 1: Classification and Community Descriptions. University of New Mexico, Biology Department; New Mexico Natural Heritage Program, Albuquerque, NM, p. 172.

Narumalani, S., Mishra, D.R., Wilson, R., Reece, P., Kohler, A., 2009. Detecting and mapping four invasive species along the floodplain of North Platte River, Nebraska. Weed Technol. 23, 99–107.

Ní Choncubhair, O.N., Osborne, B., Finnan, J.M., Lanigan, G.J., 2017. Comparative assessment of ecosystem C exchange in Miscanthus and reed canary grass during early Establishment. GCB Bioenergy 9, 280–298, doi: 10.1111/gcbb.12343.

Paavilainen, L., 1998. Modern non-wood pulp mill—process concepts and economic aspects. North American Non-Wood Fiber Symposium. TAPPI Press, Atlanta, GA, USA.

Paavilainen, L., Torgilsson, R., 1994. Reed canary-grass—a new Nordic paper making fibre. Proceeding from Tappi Pulping Conference, November 6–10, San Diego, CA, USA. Tappi Press, Atlanta, GA, USA.

Paavilainen, L., Tulppala, J., 1996. Fine paper from reed canary-grass. Proceedings of Uses for Non-Wood Fibres—Commercial and Practical Issues for Papermaking, October 29–30, Pira International, Peterborough, UK, Leatherhead, UK.

Pandey, V.C., Abhilash, P.C., Singh, N., 2009. The Indian perspective of utilizing fly ash in phytoremediation, phytomanagement and biomass production. J. Environ. Manage. 90, 2943–2958.

Pandey, V.C., Bajpal, O., Singh, N., 2016a. Energy crops in sustainable phytoremediation. Renew. Sust. Energ. Rev. 54, 58–73.

Pandey, V.C., Pandey, D.N., Singh, N., 2015a. Sustainable phytoremediation based on naturally colonizing and economically valuable plants. J. Clean Prod. 86, 37–39.

Pandey, V.C., Prakash, P., Bajpai, O., Kumar, A., Singh, N., 2015b. Phytodiversity on fly ash deposits: evaluation of naturally colonized species for sustainable phytorestoration. Environ. Sci. Pollut. Res. 22, 2776–2787.

Pandey, V.C., Singh, B., 2012. Rehabilitation of coal fly ash basins: current need to use ecological engineering. Ecol. Eng. 49, 190–192.

Pandey, V.C., Singh, N., Singh, R.P., Singh, D.P., 2014. Rhizoremediation potential of spontaneously grown *Typha latifolia* on fly ash basins: study from the field. Ecol. Eng. 71, 722–727.

Pandey, V.C., 2012a. Phytoremediation of heavy metals from fly ash pond by *Azolla caroliniana*. Ecotoxicol. Environ. Saf. 82, 8–12.

Pandey, V.C., 2013. Suitability of *Ricinus communis* L. cultivation for phytoremediation of fly ash disposal sites. Ecol. Eng. 57, 336–341.

Pandey, V.C., 2015. Assisted phytoremediation of fly ash dumps through naturally colonized plants. Ecol. Eng. 82, 1–5.

Pandey, V.C., 2017. Managing waste dumpsites through energy plantations. In: Bauddh, K., Singh, B., Korstad, J. (Eds.), Phytoremediation Potential of Bioenergy Plants. Springer, Singapore, pp. 371–386.

Pandey, V.C., 2012b. Invasive species based efficient green technology for phytoremediation of fly ash deposits. J. Geochem. Explor. 123, 13–18.

Pandey, V.C., Bajpai, O., 2019. Phytoremediation: from theory toward practice. In: Pandey, V.C., Bauddh, K. (Eds.), Phytomanagement of Polluted Sites. Elsevier, Amsterdam, Netherlands, pp. 1–49.

Pandey, V.C., Bajpai, O., Pandey, D.N., Singh, N., 2015c. *Saccharum spontaneum*: an underutilized tall grass for revegetation and restoration programs. Genet. Resour. Crop Evol. 62 (3), 443–450.

Pandey, V.C., Mishra, T., 2018. Assessment of *Ziziphus mauritiana* grown on fly ash dumps: prospects for phytoremediation but concerns with the use of edible fruit. *Int. J. Phytoremediation* 20 (12), 1250–1256. Available from: http://dx.doi.org/10.1080/15226514.2016 .1267703.

Pandey, V.C., Rai, A., Korstad, J., 2019. Aromatic crops in phytoremediation: from contaminated to waste dumpsites. In: Pandey, V.C., Bauddh, K. (Eds.), Phytomanagement of Polluted Sites. Elsevier, Amsterdam, Netherlands, pp. 255–275.

Pandey, V.C., Singh, K., Singh, J.S., Kumar, A., Singh, B., Singh, R.P., 2012a. Jatropha curcas: a potential biofuel plant for sustainable environmental development. Renew. Sust. Energ. Rev. 16, 2870–2883.

Pandey, V.C., Singh, K., Singh, R.P., Singh, B., 2012b. Naturally growing *Saccharum munja* on the fly ash lagoons: a potential ecological engineer for the revegetation and stabilization. Ecol. Eng. 40, 95–99.

Pandey, V.C., Singh, N., 2015. Aromatic plants versus arsenic hazards in soils. J. Geochem. Explor. 157, 77–80.

Pandey, V.C., Souza-Alonso, P., 2019. Market opportunities in sustainable phytoremediation. In: Pandey, V.C., Bauddh, K. (Eds.), Phytomanagement of Polluted Sites. Elsevier, Amsterdam, Netherlands, pp. 51–82.

Polechońska, L., Klink, A., 2014a. Accumulation and distribution of macroelements in the organs of *Phalaris arundinacea* L.: implication for phytoremediation. J. Environ. Sci. Health Part A 49 (12), 1385–1391.

Polechońska, L., Klink, A., 2014b. Trace metal bioindication and phytoremediation potentialities of *Phalaris arundinacea* L. (reed canary grass). J. Geochem. Explor. 146, 27–33.

Praveen, A., Pandey, V.C., Marwa, N., Singh, D.P., 2019. Rhizoremediation of polluted sites: harnessing plant–microbe interactions. In: Pandey, V.C., Bauddh, K. (Eds.), Phytomanagement of Polluted Sites. Elsevier, Amsterdam, Netherlands, pp. 389–407.

Rickard, W.H., Price, K.R., 1990. Strontium-90 in Canada Goose eggshells and reed canary grass from the Columbia River, Washington. Environ. Monitor. Assess. 14 (1), 71–76.

Rosikon, K., Fijalkowski, K., Kacprzak, M., 2015. Phytoremediation potential of selected energetic plants (*Miscanthus giganteus* L and *Phalaris arundinacea* L) in dependence on fertilization. J. Environ. Sci. Eng. A 10, 2162–5298.

Samecka-Cymerman, A., Kempers, A.J., 1996. Bioaccumulation of heavy metals by aquatic macrophytes around Wrocław, Poland. Ecotoxicol. Environ. Safety 35 (3), 242–247.

Samecka-Cymerman, A., Kempers, A.J., 2001. Concentrations of heavy metals and plant nutrients in water, sediments and aquatic macrophytes of anthropogenic lakes (former open cut brown coal mines) differing in stage of acidification. Sci. Total Environ. 281 (1–3), 87–98.

Shafeeq, A., Butt, Z.A., Muhammad, S., 2012. Response of nickel pollution on physiological and biochemical attributes of wheat (*Triticum aestivum* L.) var. Bhakar-02. Pak. J. Bot. 44 (1), 111–116.

Shurpali, N.J., Hyvönen, N.P., Huttunen, J.T., Biasi, C., Nykänen, H., Pekkarinen, N., Martikainen, P.J., 2008. Bare soil and reed canary grass ecosystem respiration in peat extraction sites in eastern Finland. Tellus 60B, 200–209.

Toole, E.H., Brown, E., 1946. Final results of the Duvel buried seed experiment. J. Agric. Res. 72, 201–210.

Verma, S.K., Singh, K., Gupta, A.K., Pandey, V.C., Trivedi, P., Verma, R.K., Patra, D.D., 2014. Aromatic grasses for phytomanagement of coal fly ash hazards. Ecol. Eng. 73, 425–428.

Waggy, Melissa, A., 2010. *Phalaris arundinacea*. In: Fire Effects Information System. U.S. Department of Agriculture, Forest Service, Rocky Mountain Research Station, Fire Sciences Laboratory.

Xiong, S., Kätterer, T., 2010. Carbon-allocation dynamics in reed canary grass as affected by soil type and fertilization rates in northern Sweden. Acta Agri. Scandinavica Sec. B Soil Plant Sci. 60, 24–32.

Yang, Y., Tilman, D., Furey, G., Lehman, C., 2019. Soil carbon sequestration accelerated by restoration of grassland biodiversity. Natur. Comm. 10 (1), 718.

Switchgrass—an asset for phytoremediation and bioenergy production

8

Divya Patel[a], Vimal Chandra Pandey[b,*]
[a]Department of Biotechnology, Sant Gadge Baba Amravati University, Amravati, Maharashtra, India; [b]Department of Environmental Science, Babasaheb Bhimrao Ambedkar University, Lucknow, Uttar Pradesh, India
*Corresponding author

1 Introduction

Initially, scientist focused on switchgrass (*Panicum virgatum* L.) as a livestock. It was considered as same till 1936 when L.C. Newell, who was an agronomist with the Bureau of Plant Industry, USDA, in Lincoln, Nebraska, started working with switchgrass and other grasses to re-vegetate large areas devastated by the drought of 1930s in the central Great Plains and Midwest. Since that time, establishment and management practices have been developed and refined, genetic resources have been evaluated, seed production has been improved, and a wealth of information has been made available to producers.

Switchgrass was chosen by the US Department of Energy (DOE) under Herbaceous Energy Crops Program (HECP) as the "model" bioenergy crop for further research (Parrish and Fike, 2005; Wright and Turhollow, 2010) and the primary objective of the HECP initiated in 1984 was to develop data and information regarding commercially viable systems for producing herbaceous biomass for fuels and energy feedstock. Further objectives using herbaceous plants for biofuel included under the HECP were: (1) to achieve the primary goal while minimizing adverse environmental effects, (2) to increase the production of biomass for energy without significantly reducing food production; (3) to produce fuels or energy feedstock rather than chemicals; (4) to have the greatest possible impact on total biomass energy use (highlighting the lignocellulosic crops) (Wright and Turhollow, 2010). The sticking features found in switch grass by HECP are given in Fig. 8.1.

Apart from the earlier-mentioned reasons of perennial grasses, like switchgrass, is desirable bioenergy feedstock for several reasons. They can be combusted for direct heat and electricity (Lewandowski et al., 2003; Sanderson et al., 2006), or as lignocellulosic feedstock for the production of bioethanol (Ragauskas et al., 2006; Schubert, 2006). Bioenergy production from perennial grasses does not necessarily displace food crop production, as with corn or soybean for bioethanol and biodiesel production (Tilman et al., 2006). In comparison to annual biofuel crops (i.e., corn, wheat, soybean), the perennial nature of grasses results in greater energy return, as less intensive labor, equipment, and fossil fuel energy are required each production year (Parrish and

Phytoremediation Potential of Perennial Grasses. http://dx.doi.org/10.1016/B978-0-12-817732-7.00008-0

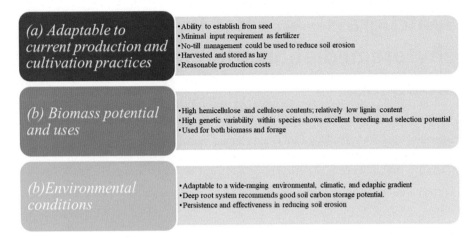

Figure 8.1 The important features of switchgrass related to adaptive agricultural practices, biomass potential and uses as well as environmental conditions.

Fike, 2005). Moreover, perennial grasses such as switchgrass utilize the C4 photosynthetic pathway, and these crops usually have higher nutrient, water, and solar radiation efficiencies than C3 plants (Lewandowski et al., 2003; Parrish and Fike, 2005).

Perennial crops generally have lower N fertilizer demands (Parrish and Fike, 2005). As such, crop harvest after senescence (either late autumn/early winter or early spring) usually results in a better quality biofuel as the feedstock will have a low ash (Si), mineral (mainly N, Cl, K), and water contents and therefore release less pollution when combusted (Lewandowski et al., 2003; Parrish and Fike, 2005). Perennial grasses have the potential for soil carbon storage; as much as 64% of the plant can be comprised of lignin because of their perennial nature and extensive root systems (Frank et al., 2004; Parrish and Fike, 2005). Perennial grasses can also be adapted to marginal lands and are of potential use in phytoremediation strategies on contaminated soils (Tilman et al., 2006). This chapter focuses on exploring the multi potential switchgrass with special attention to its phytoremediation properties and its use as bioenergy crop.

2   General aspect of switchgrass

Ecology—Switchgrass is a warm-season perennial grass native to North America. Its habitat ranges from the Atlantic Coast to Nevada, United States, and latitudinally from Central America to the prairies of southern Canada, with a northern adaptation limit of about 51°N (Parrish and Fike, 2005). However, it can be cultivated in other regions of the world, such as Northern Europe (Elbersen et al., 2001). For ornamental purposes, the companion plants of switchgrass are tall sedums, boltonia, asters, Russian sage, and mums. Fig. 8.2 represents Switchgrass as the collection of energy plants on the field of Poltava State Agrarian Academy (PSAA), Ukraine. Mixed grass species plantings have been noticed some advantages for conservation, remediation,

Figure 8.2 Switchgrass as the collection of energy plants on the field of Poltava State Agrarian Academy (PSAA), Ukraine.
Photo credits: Marina Galitskaya, PSAA.

and biomass purposes as well as may be beneficial for bioenergy drives as well. The practice of high-diversity mixtures with low-input has been suggested to produce the sustainable development of bioenergy on degraded land (Tilman et al., 2006). For this, grass species selection is an important criterion for long-term success and must have similar growth characteristics, biomass quality characteristics, seed vigor, restoration potential, maturity dates, easy harvesting, and tolerance to a wide-ranging environmental stress conditions.

The common warm-season grass species as the dominant species of tall grass prairie in North America are *Andropogon gerardii, Sorghastrum nutans, Panicum virgatum,* and *Schizachyrium scoparium.* These grass species also are common in mixtures for set-aside plantings within the Central United States. Moreover, a positive connection has been established between plant diversity and soil heterogeneity for grassland ecosystems. Companion crops have been used successfully to establish perennial grasses especially when herbicide options are limited. The benefits of using companion crops are reduction in soil erosion and weed populations during switchgrass establishment. While the main drawback of using companion crops is that in certain regions, an establishment year switchgrass harvest is not feasible because of insufficient biomass. Switchgrass has been fruitfully established with the mixture of maize and sorghum-sudangrass (Cossar and Baldwin, 2004; Hintz et al., 1998). *Triticum aestivum, Avena sativa,* and other annual cool-season grasses managed as a hay crop have been effectively used to establish perennial grasses such as switchgrass.

Morphological description—Switchgrass (*P. virgatum* L.) belongs to *Poaceae* family. Switchgrass is a C4 warm-season perennial grass. It looks like a bunchgrass and it spreads slowly by seeds and rhizomes. It has erect stems with a height that could range 0.5–2.7 m. Open panicles of 15–50 cm long are developed on the top of the tillers at inflorescence. Its deep root system can be up to 3 m in depth (Alexopoulou et al., 2018). Leaves are 30–90 cm long with wide-ranging color gray–green or blue–green. It has leaf blade with 6–12 mm wide, flat, elongate, and distinctly veined. It contains leaf sheath with round, red purplish, and glabrous. Auricle is absent, while ligule is present as 1.5–3.5 mm long, pilose, ciliate, and membranous. Flower is often up to 60 cm long with wide-ranging color gray-blue, soft purple, yellowish, and pinkish red. Seeds are 3–6-mm-long and up to 1.5-mm-wide. Its spikelet has two flowered including one fertile floret and other sterile floret. Two ecotypes of switchgrass are recognized; one is lowland varieties which are commonly found in wetter environments of lower latitudes, another is upland varieties are predominantly found in drier more mesic regions of mid- to northern latitudes (Elbersen et al., 2001; Parrish and Fike, 2005). Ploidy levels vary between ecotypes, as lowland types being generally tetraploid ($2n = 4x = 36$), while most upland types are octoploid ($2n = 8x = 72$), including cultivars, such as "Cave-In-Rock" and "Pathfinder" (Parrish and Fike, 2005). The high ploidy level of switchgrass explains its wide adaptation to a variety of environments. Thus, switchgrass displays cultivar x environment variability, such that difference is often observed for the same cultivars grown under different environmental conditions (Parrish and Fike, 2005).

Origin and Geographic distribution—Switchgrass is native to the tall grass plain of North America. The first appearance of this grass was estimated 2 million years ago (Parrish et al., 2012). Nearly 450 species relatively heterogeneous comes under the *Panicum* genus. Switchgrass is considered a New World species; it occurs naturally in Canada southward into the United States and Mexico (Jefferson et al., 2002). Switchgrass radiated and adapted across major portions of the North American continent (Huang et al., 2011). When Europeans arrived in the New World, switchgrass distribution ranged from central to eastern North America. Initially, switchgrass was of interest as a member of prairie ecosystems and slowly became popular as potential forage crop and then for other uses. Switchgrass history as a real crop planted or studied counts dates back only a few decades. It began to emerge from the anonymity of being "just" a prairie grass in the 1940s. In the 1980s, a large number of studies were reported and although the majority of them dealt with its forage value and breeding, there were a few reports that dealt with reclamation, erosion control, and diseases. Also in the 1980s, switchgrass was identified as a candidate energy crop for the United States by the US DOE (Wright and Turhollow, 2010; Wright, 2007). Besides the distribution of switchgrass in United States, Canada, Europe, and China, a number of studies on switchgrass for bioenergy have been conducted in other areas of the globe, namely, South and Central America (Argentina, Colombia, Mexico, and Venezuela), Australia, Asia (Korea, Japan, and Pakistan), and Africa (Sudan) (Parrish et al., 2012; Alexopoulou et al., 2018). Fig. 8.3 depicts the research works on switchgrass in the world.

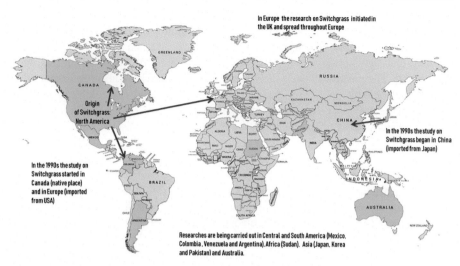

Figure 8.3 The research works on switchgrass in the world.
Modified from Alexopoulou et al., 2018.

Plant propagation and regeneration—Switchgrass spreads slowly by seeds and rhizomes. There are two principal ecotypes of switchgrass: upland and lowland. Their origin is reflected by their names: the upland ecotype was originally found in upland habitats, often characterized by droughty soils, while the lowland ecotype originated along riverine habitats and flood plains. There is a difference between lowland switchgrass as this is taller and has longer and wider leaf blades, fewer tillers per plant, and larger stem diameter and is later in heading and flowering than upland switchgrass. Apart from that a bluish waxy bloom on leaf sheaths and blades is typically associated with the lowland ecotype. Lowland varieties of switchgrass (including Alamo, Kanlow, and Summer) are pseudotetraploid.

3 Multiple uses

Nowadays, switchgrass has been adopted as an energy crop from Canada (Samson et al., 2005). For the production of advance biofuel, switchgrass is considered one of the most important energy crops in Europe, despite the fact that it is a non-native species. In case of China where switchgrass was imported in the 1990s from Japan together with other fodder species, among which only switchgrass has survived (Ichizen et al., 2005). Only recently has switchgrass been considered as a candidate crop for cellulosic bioenergy feedstock in northern China and considerable research has been published since 2010. Its adaptability and productivity have been tested on arid and semiarid marginal areas of China (Yue et al., 2017). Bransby et al. (1998) stated that switchgrass could reduce CO_2 emissions and improve soil quality by carbon sequestration. Apart from the large amount of above-ground biomass, switchgrass has an extensive and deep root system that is beneficial for increasing soil carbon storage. It is reported that it is being at least 50% more effective in water use compared to cool-season grasses (Stout et al., 1988). Switchgrass can also be used to improve surface water quality. Lee et al. (1998) compared switchgrass filter strips to cool-season grass filter strips and reported that switchgrass was more effective in removing phosphorous and nitrogen from runoff. The positive impacts of switchgrass on wildlife have also been reported as it provides a suitable habitat for grassland birds that are rapidly lessening in numbers (Murray et al., 2003). The appropriate typical harvesting season for switchgrass is late summer or the fall. As the breeding season for most of the grassland birds species get over by this time and there is minimal disturbance to nesting birds (Roth et al., 2005). A balance between harvested and non-harvested switchgrass fields to preserve species richness of grassland birds was proposed by Murray and Best (2003).

Another aspect of multiple applications of switchgrass is an alternate option for pulping raw material. Due to environmental issues such as forest preservation and reduction of CO_2 emissions, the alternative of raw material for pulping application have changed from wood to non-wood resources. Perennial grasses like switchgrass can be harvested without annual reestablishment and also they have low lignin content which makes them an attractive raw material for pulping. Study for using switchgrass as an attractive crop and alternative for local pulp industry was conducted by Fox et al. (1999) in which they concluded this by a regional economic evaluation for eastern Ontario and western Quebec. Assessment of suitability of switchgrass for pulping industry and producing paper of writing and printing quality was made by Ververis et al. (2004) which can also be mixed with conventional woody sources for making paper with a wider range of applications.

4   Limiting factors

The limiting factor of using switchgrass is related to the combustion of its biomass because of the presence of alkali metals that react to form sulfates, chlorides, silicate, and hydroxides that contribute to slag formation and fouling of combustion systems (Dayton et al., 1995). For the alkali metals, leaching of switchgrass with water was proposed by Dayton et al. (1999) so that the removal of alkali metals and chlorine prior to combustion can be done. A limiting growth factor of switchgrass cultivation is the air temperature (Madakadze et al., 1998). The satisfied basal temperature for sowing is 10°C (Sanderson and Wolf, 1995). Switchgrass is a short-day crop and therefore its phenology is influenced by day length (Giannoulis et al., 2016).

5   Phytoremediation

Switchgrass is often referred to marginal agricultural areas with minimal inputs and yields the highest biomass among native grasses in North America. It was planted for phytoremediation along stream banks and riparian lands of petroleum contaminants (Euliss et al., 2008) and also used as a filter strip to prevent soil nutrient loss in to runoff water (Sanderson et al., 2012). Well- known agronomic characteristics of cultivating switchgrass, such as fertilization, irrigation, soil pH optima, plant density, biomass field-drying, and harvest schedule encourages switchgrass to be used for the phytoextraction of metals. Earlier studies suggested that switchgrass was used for the phytoremediation of 137Cs- and 90Sr-contaminated soil (Entry and Watrud, 1998) and of chromium (Cr)-contaminated soil (Shahandeh and Hossner, 2000). Alamo is the better cultivar of switchgrass suited for the phytoextraction of Cd from contaminated soil suggested by Chen et al. (2012). However, in previous studies reduction of biomass on the performance of phytoextraction involving switchgrass for high metal concentrations in plants was not determined. Despite this, it has received little attention as a potential agent for the phytoremediation of lead contaminated soils (Salt et al. 1998; Gleeson, 2007). Better understanding of the ability of switchgrass to remove Cd, Cr, and Zn by phytoextraction from contaminated sites, the relationship between plant metal content and biomass yield is expressed in different models to predict the amount of metals switchgrass can extract. Some of those models for phytoextraction from switchgrass were applied by Chen et al. (2012) in their study, which showed linear and exponential decay models are more suitable for presenting the relationship between plant cadmium and dry weight and approached 40 and 34 mg Cd per pot, respectively. The log normal model was superior in predicting the relationship between plant Cr and dry weight. In addition, the exponential decay and log normal models were better than the linear model in predicting the relationship between plant Zn and dry weight. Switchgrass is frequently studied for metal removal in natural conditions for soil polluted by chromium. The results showed that switchgrass has high tolerance to Cr. The concentration of Cr in shoots and roots all increased with the increasing Cr concentration in soil. The highest accumulation of Cr in plants reached 1543.8 mg/pot. The studies conducted by Li et al. (2011) implies that the root of switchgrass has great ability to uptake Cr from polluted soil and it is speculated that switchgrass has great potential for phytoremediation of Cr-polluted soil.

Switchgrass is used for various phytoremediation works; one of the researches is associated with the removal of atrazine from agricultural lands. Atrazine is widely used herbicide in agriculture. Phytoremediation of these kinds of chemicals can provide a great solution to the non-point source contamination from its runoff. Study conducted by Murphy and Coats (2011) characterizes the ability of switchgrass to take up atrazine from soils, estimated the amount of biodegradation occurring in the plant, and the amount of degradation occurring in the rhizosphere. Results demonstrated that leaf biomass is capable of detoxifying atrazine because of metabolites present in leaf material (Murphy and Coats, 2011). To overcome the major challenge in phytoextraction, that is, to increase plants' removal rates of metals from contaminated soils. Juang et al. (2011) developed a phytoextraction model, by coupling an energy-based toxicity model and a saturable Michaelis–Menten type accumulation model. This model was designed to predict copper (Cu) removal by switchgrass grown hydroponically in different exposure levels. Results of their study indicated that the accumulation capacity exists in the plant for Cu. In addition to this, the results suggested that under a lower dissolved concentration, Cu removal is increased more efficiently as the exposure duration increases. This approach can offer a first approximation in predicting the phytoextraction time required for plant species to eliminate a specific metal from polluted areas, which is vital in assessing the economic costs for remediation purposes. To see the combine effects of Pb-phytoextraction by two grass species (i.e., switchgrass and corn) grown in Pb-contaminated soil or sand-culture with supplemented Pb given combined applications of an arbuscular mycorrhizal fungi (AMF) suppressant (benomyl), chelating agent (EDTA), and Fe (foliar- or soil-application) by Johnson et al., (2015). Plants in the foliar-Fe experiment showed similar shoot concentration of Pb regardless of chemical treatment. However, they showed significant increase in Pb-translocation. Similarly, plants treated with the combined application of root-Fe and EDTA (FE) had significantly the highest Pb-translocation. In this study, total phytoextraction of Pb was increased significantly when plants were treated with a combined application of benomyl, root-Fe, and EDTA (BFE) in comparison to those amendments alone. Corn and switchgrass were capable of extracting Pb from the soil. Research suggested that experiments would need additional manipulation to allow for effective phytoextraction. Translocation of metals was significantly influenced by foliar-Fe or root-Fe in combination with EDTA application which could potentially be explored further in chemically induced phytoextraction.

Bioremediation potential of a high biomass yielding grass, that is, *P. virgatum*, along with plant associated microbes (AMF and *Azospirillum*), was tested against lead and cadmium in pot trials by Arora et al. (2016). Growth parameters and bioremediation potential of AMF and *Azospirillum* against different concentrations of Pb and Cd were compared which showed that AM fungi and *Azospirillum* increased the root length, branches, surface area, and root and shoot biomass. The lower value of bioconcentration factor and translocation index even at highest concentration of Pb and Cd, revealed the capability of switchgrass of accumulating high concentration of Pb and Cd in the roots, while preventing the translocation of Pb and Cd to aerial biomass. Phytoremediation makes use of plants which accompanied by microorganisms to clean up soils and sediments contaminated with inorganic and organic pollutants. In various studies, switchgrass was used to test for its efficiency for phytoremediation. It was tested for removal of three specific polychlorinated biphenyl (PCB) congeners (PCB 52, 77 and 153) in soil microcosms. The congeners were chosen for their ubiquity, toxicity, and recalcitrance by Liang et al. (2014). The improved PCB removal in switchgrass-treated soils explained by phytoextraction processes and enhanced microbial activity in the rhizosphere also bio-augmentation with *Burkholderia xenovorans* LB400 was performed and suggested to further enhance aerobic PCB degradation. The combination of phytoremediation and bio-augmentation offers an efficient and environmental-friendly strategy to eliminate recalcitrant PCB congeners and remediate PCB contamination in the environment. Switchgrass was also studied to determine the effect of fly ash (FA) on various cropping system because of its increasing amendments in agriculture lands. The experiments were conducted by to assess mutagenic response of a mutant strain of *Salmonella enterica serovar Typhimurium* (TA100) to varying concentrations of FA-water extracts, determine oxidative stress in switchgrass at varying levels of FA-soil admixtures, and evaluate mycorrhiza-mediated modulation of oxidative stress responses of FA-grown switchgrass (Lushola et al., 2017).

6 Bioenergy production

Switchgrass is a perennial grass, native to much of the Great Plains, and has been identified as a significant potential bioenergy crop based on research conducted across 31 locations over several years in the late 1980s and early 1990s. In 1991, under the Biofuels Feedstock Development Program, switchgrass was selected as a model bioenergy crop majorly due to its steadily high productivity, large range of adaptability, and low water and nutrient requirements. These lower requirements allow switchgrass to be grown even on marginal lands. Traditionally switchgrass has been used for forage production, soil conservation, and as ornamental grass (Gupta and Demirbas 2010). The viability of producing switchgrass as a bioenergy feedstock in the Great Plains has been the topic of much exploration (Perlack et al., 2005; Mapemba and Epplin, 2004; Epplin et al., 2007; Bangsund et al., 2008; Perrin et al., 2008). Switchgrass has been identified as a model herbaceous energy crop for the Unites States. Several constraints such as economic production of switchgrass for biomass feedstock including reliable establishment practices to ensure productive stands in the seeding year, efficient use of fertilizers, and more efficient methods to convert lignocellulose to biofuels are still remaining. The balance between the existing higher costs of biofuels and fossil fuels may be achieved by precisely evaluating environmental benefits related to the perennial grasses such as reduced runoff and erosion and comparatively reduced losses of soil nutrients and organic matter, increased incorporation of soil carbon and reduced use of agricultural chemicals. Atmospheric CO_2 may also lessen by the use of warm-season perennial grasses in bioenergy cropping systems.

There are several benefits of using switchgrass as bioenergy crop, such as the requirement of low maintenance after its formation phase, being non-invasive, and suitability to wide-ranging soils, including marginal lands not as productive for high-value crops, such as corn or soybeans (Wright, 2007). The harvesting, transporting, and storing switchgrass is related to well-established hay production practices (Wright, 2007), although long-term biomass storage may reduce ethanol yields (Rigdon et al., 2013). Despite this, it has received little attention as a potential agent for the phytoremediation of lead contaminated soils (Salt et al. 1998, Gleeson, 2007). The annual production of switchgrass has been reported to range from 6 to 9 ton/acre/year (Lewandowski et al., 2003). Biomass dry matter partitioning and growth characteristics are important selection criteria for energy crops. Though switchgrass is a promising bioenergy feedstock, but industrial-scale production may lead to negative environmental effects. In a study conducted by Balsamo et al. (2015), potential consequence such as long-term exposure to fine particulate matter (PM2.5) and total air quality impacts of switchgrass production has been identified in which the work points to the human health damage from air pollution as a potentially large social cost from switchgrass production and suggests means of mitigating that impact via strategic geographical deployment and management. The study of fundamentals of switchgrass cultivation technology is very important because the yield of the crop, weather conditions, and variety properties significantly depend on land treatment. The optimal switchgrass seeding term for the conditions of Ukraine has been established. Experimentally proved that switchgrass with a width between rows of 45 cm forms the highest yield of above-ground vegetative mass (wet and dry) with minimum water consumption and forms maximum yield of energy intensive biofuel.

7 Carbon sequestration

Switchgrass could reduce CO_2 emissions and improve soil quality by carbon sequestration (Bransby et al., 1998). Gebhart et al. (1994) showed that perennial grasses like switchgrass can store 1.1 Mg of carbon per hectare annually in the upper 1 m of the soil on conservation reserve program (CRP) lands. However, McLaughlin et al. (2002) showed that switchgrass grown on bioenergy research plots could add 1.7 Mg of carbon per hectare annually, which is higher than CRP estimate reported by Gebhart et al. (1994). Bransby et al. (1998) specified that these studies on CRP lands are probably not representative of switchgrass grown for energy production since the importance would shift from conservation to maximizing yields. A 33% increase in soil carbon in the 0–5 cm layer and a slight decrease in the 5–15 cm and 15–30 cm layers 7 years after planting was reported by Sanderson (2008). A systematic study made by Ma et al. (2000) of carbon dynamics following the establishment of switchgrass showed that over a 2-year period, the top 15 cm of sandy loam soil revealed a 122% increase in carbon mineralization, a 168% increase in microbial biomass carbon, and a 116% increase in net carbon turnover. Wu et al. (2006) assessed that the net amount of CO_2 sequestered would be around 48.5 kg per dry metric ton of switchgrass. Frank et al. (2004) studied Sunburst and Dacotah switchgrass cultivars and noted that the net system carbon gain doubled over a 3-year period. Combined with the zero net carbon exchange as a result of burning bioethanol from switchgrass, addition of soil carbon results in the overall reduction of atmospheric release of CO_2 (Lynd et al., 1991). However, such gains as a result of carbon sequestration are not assured. Bransby et al. (1998) observed that switchgrass will provide net gains in carbon sequestration when grown on soil. McLaughlin et al. (2002) also estimated smaller gains in soil carbon sequestration following conversion of pastures to switchgrass and noted that significant gains can be achieved for highly degraded soils in warm climates.

8 Physiological adaptation

Physiological adaptations against increased air temperature in climate can drive plant growth and productivity (Morgan et al., 2004), with dominant plant species influencing the community structure, dynamics, invisibility, and overall ecosystem function (Emery and Gross, 2007). Grass species responses to changed environments differ based on species-specific responses, including differential sensitivity to changes in temperature or water availability (Nippert et al., 2009). Thus, future productivity of dominant grass species recognized for biofuel cultivation (i.e., switchgrass) will reflect leaf-level and whole-plant responses to specific environmental drivers as well as the magnitude of the environmental change (Fay et al., 2011). Switchgrass is a perennial C4 grass species adapted to a wide-range of environmental conditions (i.e., temperature and precipitation) across North America (Parrish and Fike, 2005). Historically, switchgrass was valued in the use of forage, but over 20 years most of researches have focused to evaluate its suitability for biofuel production (Wright and Turhollow, 2010). Currently, switchgrass application in phytoremediation of contaminated sites with biofuel production is a growing interest to meet societal needs without displacing food productivity. Switchgrass shows large genetic variability and diversity (Das et al., 2004). These specific traits facilitate its suitability for cultivation in a wide-ranging of environmental stress conditions and geographic ranges across North America (Parrish and Fike, 2005), although this range will likely increase and shift north under future climate change conditions (Barney and DiTomaso, 2010). Some important features of switchgrass for physiological adaptation are the presence of C4 pathway, water and nutrient efficiency, water stress potential (Alexopoulou et al., 2018).

9 Conclusion and future perspectives

Switchgrass is a hopeful feedstock for bioethanol production, thermal energy conversion, and pulping applications. The encouraging environmental benefits associated with switchgrass include improvement of wildlife diversity, improvement of soil and water quality, reduced pesticide use, and carbon sequestration. Presently, switchgrass has been proved that it is adaptable grass species under changing environment for multiple benefits from socio-economic to ecological importance. In future, switchgrass could be used for biofuel production along with phytoremediation and utilization of polluted sites.

Acknowledgment

Financial assistance given to Dr. V.C. Pandey under the Scientist's Pool Scheme (Pool No. 13 (8931-A)/2017) by the Council of Scientific and Industrial Research, Government of India is gratefully acknowledged. The authors declare no conflicts of interest.

References

Alexopoulou, E., Monti, A., Elbersen, H.W., Zegada-Lizarazu, W., Millioni, D., Scordia, D., Zanetti, F., Papazoglou, E.G., Christou, M., 2018. Switchgrass: from production to end use. In: Alexopoulou, E. (Ed.), Perennial Grasses for Bioenergy and Bioproducts. Academic Press, London, United Kingdom, pp. 61–105.
Arora, K., Sharma, S., Monti, A., 2016. Bio-remediation of Pb and Cd polluted soils by switchgrass: a case study in India. Int. J. Phytoremediation 18 (7), 704–709.

Balsamo, R.A., Kelly, W.J., Satrio, J.A., Ruiz-Felix, M.N., Fetterman, M., Wynn, R., Hagel, K., 2015. Utilization of grasses for potential biofuel production and phytoremediation of heavy metal contaminated soils. Int. J. Phytoremediation 17 (5), 448–455.

Bangsund, D.A., DeVuyst, E.A., Leistritz, F.L., 2008. Evaluation of Breakeven Farmgate Switchgrass Prices in South Central North Dakota. Agribusiness and Applied Economics Report No. 632, North Dakota State University, Agribusiness and Applied Economics.

Barney, J.N., DiTomaso, J.M., 2010. Bioclimatic predictions of habitat suitability for the biofuel switchgrass in North America under current and future climate change scenarios. Biomass Bioenerg. 34, 124–133.

Bransby, D.I., McLaughlin, S.B., Parrish, D.J., 1998. A review of carbon and nitrogen balances in switchgrass grown for energy. Biomass Bioenerg. 14 (4), 379–384.

Chen, B.C., Lai, H.Y., Juang, K.W., 2012. Model evaluation of plant metal content and biomass yield for the phytoextraction of heavymetals by switchgrass. Ecotoxicol. Environ. Saf. 80, 393–400.

Cossar, R.D., Baldwin, B.S., 2004. Establishment of switchgrass with sorghum sudangrass. In: Randall, J., Burns, J.C. (Eds.), In: Proceedings of Third Eastern Native Grass Symposium. Omnipress, Chapel Hill, NC.

Das, M.K., Fuentes, R.G., Taliaferro, C.M., 2004. Genetic variability and trait relationships in switchgrass. Crop Sci. 44, 443–448.

Dayton, D.C., French, R.J., Milne, T.A., 1995. Direct observation of alkali vapor release during biomass combustion and gasification. 1. Application of molecular-beam mass-spectrometry to switchgrass combustion. Energ. Fuels 9 (5), 855–865.

Dayton, D.C., Jenkins, B.M., Turn, S.Q., Bakker, R.R., Williams, R.B., Belle- Oudry, D., Hill, L.M., 1999. Release of inorganic constituents from leached biomass during thermal conversion. Energ. Fuels 13 (4), 860–870.

Elbersen, H.W., Christian, D.G., El Bassam, N., Bacher, W., Sauerbeck, G., Aleopoulou, E., Sharma, N., Piscioneri, I., de Visser, P., van den Berg, D., 2001. Switchgrass variety choice in Europe. Aspects Appl. Biol. 65, 21–28.

Emery, S.M., Gross, K.L., 2007. Dominant species identity, not community evenness, regulates invasion in experimental grassland communities. Ecology 88, 954–964.

Entry, J.A., Watrud, L.S., 1998. Potential remediation of 137Cs and 90Sr contaminated soil by accumulation in Alamo switchgrass. Water Air Soil Pollut. 104, 339–352.

Epplin, F.M., Clark, C.D., Roberts, R.K., Hwang, S., 2007. Challenges to the development of a dedicated energy crop. Am. J. Agric. Econom. 89 (5), 1296–1302.

Euliss, K., Ho, C.H., Schwab, A.P., Rock, S., Banks, M.K., 2008. Greenhouse and field assessment of phytoremediation for petroleum contaminants in a riparian zone. Bioresour. Technol. 99, 1961–1971.

Fay, P.A., Blair, J.M., Smith, M.D., Nippert, J.B., Carlisle, J.D., Knapp, A.K., 2011. Relative effects of precipitation variability and warming on tallgrass prairie ecosystem function. Biogeosciences 8, 3053–3068.

Fox, G., Girouard, P., Syaukat, Y., 1999. An economic analysis of the financial viability of switchgrass as a raw material for pulp production in eastern Ontario. Biomass Bioenergy 16 (1), 1–12.

Frank, A.B., Berdahl, J.D., Hanson, J.D., Liebig, M.A., Johnson, H.A., 2004. Biomass and carbon partitioning in switchgrass. Crop Sci. 44, 1391–1396.

Gebhart, D.L., Johnson, H.B., Mayeux, H.S., Polley, H.W., 1994. The CRP increases soil organic-carbon. J. Soil Water Conserv. 49 (5), 488–492.

Giannoulis, K.D., Karyotis, T., Sakellariou-Makrantonaki, M., Bastiaans, L., Struik, P.C., Danalatos, N.G., 2016. Switchgrass biomass partitioning and growth characteristics under different management practices. NJAS Wageningen J. Life Sci. 78, 61–67.

Gleeson, A.N., 2007. Phytoextraction of Lead From Contaminated Soil By *Panicum Virgatum* L. (Switchgrass) And Associated Growth Responses. A Thesis submitted to the Department of Biology, The degree of Master of Science, Queen's University, Kingston, Ontario, Canada.

Gupta, R.B., Demirbas, A., 2010. Gasoline, Diesel and Ethanol From Grasses and Plants. Cambridge University Press, England, United Kingdom, pp. 6–10.

Hintz, R.L., Harmoney, K.R., Moore, K.J., George, J.R., Brummer, E.C., 1998. Establishment of switchgrass and big bluestem in corn with atrazine. Agron. J. 90, 591–596.

Huang, S., Zalapa, J.E., Jakubowski, A.R., Price, D.L., Acharya, A., Wei, Y., Brummer, E.C., Kaeppler, S.M., Casler, M.D., 2011. Post-glacial evolution of *Panicum virgatum*: centers of diversity and gene pools releaved by SSR markers and cpDNA sequences. Genetica 139, 933–948.

Ichizen, N., Takashashi, H., Nishio, T., Liu, G.B., Huang, J., 2005. Impacts of switchgrass (*Panicum virgatum* L.) planting on soil erosion in the hills of the Loess Plateau in China. Weed Biol. Manag. 5, 31–34.

Jefferson, P.G., McCaughey, W.P., May, K., Woosaree, J., McFarlane, L., Wright, S.M.B., 2002. Performance of American native grass cultivars in the Canadian prairie provinces. Native Plants J. 3, 24–33.

Johnson, D.M., Deocampo, D.M., El-Mayas, H., Greipsson, S., 2015. Induced phytoextraction of lead through chemical manipulation of switchgrass and corn; role of iron supplement. Int. J. Phytoremediation 17 (12), 1192–1203.

Juang, K.W., Lai, H.Y., Chen, B.C., 2011. Coupling bioaccumulation and phytotoxicity to predict copper removal by switchgrass grown hydroponically. Ecotoxicology 20, 827–835.

Lee, K.H., Isenhart, T.M., Schultz, R.C., Mickelson, S.K., 1998. Nutrient and sediment removal by switchgrass and cool-season grass filter strips in central Iowa, USA. Agrofor. Syst. 44 (2–3), 121–132.

Lewandowski, I., Scurlock, J.M.O., Lindvall, E., Christou, M., 2003. The development and current status of perennial rhizomatous grasses as energy crops in the US and Europe. Biomass Bioenerg. 25, 335–361.

Li, C., Wang, Q.H., Xiao, B., Li, Y.F., 2011. Phytoremediation potential of switchgrass (*Panicum virgatum* L.) for Cr-polluted soil. 2011 International Symposium on Water Resource and Environmental Protection, 20-22 May 2011, INSPEC Accession Number: 12070656, DOI: 10.1109/ISWREP.2011.5893582.

Liang, Y., Meggo, R., Hu, D., Schnoor, J.L., Mattes, T.E., 2014. Enhanced polychlorinated biphenyl removal in a switchgrass rhizosphere by bioaugmentation with *Burkholderia xenovorans* LB400. Ecol. Eng. 71, 215–222.

Lushola, M., Awoyemi, E., Dzantor, K., 2017. Toxicity of coal fly ash (CFA) and toxicological response of switchgrass in mycorrhiza-mediated CFA-soil admixtures. Ecotoxicol. Environ. Saf. 144, 438–444.

Lynd, L.R., Cushman, J.H., Nichols, R.J., Wyman, C.E., 1991. Fuel ethanol from cellulosic biomass. Science 251, 1318–1323.

Ma, Z., Wood, C.W., Bransby, D.I., 2000. Impacts of soil management on root characteristics of switchgrass. Biomass Bioenerg. 18, 105–112.

Madakadze, I.C., Coulman, B.E., Stewart, K., Peterson, P., Samson, R., Smith, D.L., 1998. Phenology and tiller characteristics of big bluestem and switchgrass cultivarsin a short growing season area. Agron. J. 90, 489–495.

Mapemba, L., Epplin, F.M., 2004. Lignocellulosic Biomass Harvest and Delivery Cost. Paper presented at the Southern Agricultural Economics Association Annual Meetings, February 14–18. Tulsa, OK.

McLaughlin, S.B., Ugarte, D.G.D.L., Garten, C.T., Lynd, L.R., Sanderson, M.A., Tolbert, V.R., Wolf, D.D., 2002. High-value renewable energy from prairie grasses. Environ. Sci. Technol. 36 (10), 2122–2129.

Morgan, J.A., Pataki, D.E., Körner, C., Clark, H., Del Grosso, S.J., Grünzweig, J.M., Knapp, A.K., Mosier, A.R., Newton, P.C.D., Niklaus, P.A., Nippert, J.B., Nowak, R.S., Parton, W.J., Polley, H.W., Shaw, M.R., 2004. Water relations in grassland and desert ecosystems exposed to elevated atmospheric CO_2. Oecologia 140, 11–25.

Murphy, I.J., Coats, J.R., 2011. The capacity of switchgrass (*Panicum virgatum*) to degrade atrazine in a phytoremediation setting. Environ. Toxicol. Chem. 30 (3), 715–722.

Murray, L.D., Best, L.B., 2003. Short-term bird response to harvesting switchgrass for biomass in Iowa. J. Wildl. Manage. 67 (3), 611–621.

Murray, L.D., Best, L.B., Jacobsen, T.J., Braster, M.L., 2003. Potential effects on grassland birds of converting marginal cropland to switchgrass biomass production. Biomass Bioenerg. 25 (2), 167–175.

Nippert, J.B., Fay, P.A., Carlisle, J.D., Knapp, A.K., Smith, M.D., 2009. Ecophysiological responses of two dominant grasses to altered temperature and precipitation regimes. Acta Oecol. 35, 400–408.

Parrish, D., Casler, M.D., Monti, A., 2012. The evolution of switchgrass as an energy crop (Chapter 1) in the book. In: Switchgrass: A Valuable Biomass Crop for Energy. Springer-Verlag, London.

Parrish, D.J., Fike, J.H., 2005. The biology and agronomy of switchgrass for biofuels. Crit. Rev. Plant Sci. 24, 423–459.

Perlack, R.D., Wright, L.L., Turhollow, A.F., Graham, R. L., Stokes, B.J., Erbach, D.C., 2005. Biomass as Feedstock for a Bioenergy and Bioproducts Industry: The Technical Feasibility of a Billion-Ton Annual Supply. Oak Ridge National Laboratory, Oak Ridge, TN.

Perrin, R., Vogel, K., Schmer, M., Mitchell, R., 2008. Farm-scale production cost of switchgrass for biomass. Bioenerg. Res. 1 (1), 91–97.

Ragauskas, A.J., Williams, C.K., Davison, B.H., Britovsek, G., Cairney, J., Eckert, C.A., Frederick, Jr., W.J., Hallett, J.P., Leak, D.J., Liotta, C.L., Mielenz, J.R., Murphy, R., Templer, R., Tschaplinski, T., 2006. The path forward for biofuels and biomaterials. Science 311, 484–489.

Rigdon, A.R., Jumpponen, A., Vadlani, P.V., Maier, D.E., 2013. Impact of various storage conditions on enzymatic activity, biomass components and conversion to ethanol yields from sorghum biomass used as a bioenergy crop. Bioresour. Technol. 132, 269–275.

Roth, A.M., Sample, D.W., Ribic, C.A., Paine, L., Undersander, D.J., Bartelt, G.A., 2005. Grassland bird response to harvesting switchgrass as a biomass energy crop. Biomass Bioenerg. 28 (5), 490–498.

Salt, D.E., Smith, R.D., Raskin, I., 1998. An annual review of plant physiology and plant molecular biology. Phytoremediation 49, 643–668.

Samson, R., Mani, S., Boddey, R., et al., 2005. The potential of C4 perennial grasses for developing a global Bioheat industry. Crit. Rev. Plant Sci. 24, 461–495.

Sanderson, M.A., 2008. Upland switchgrass yield, nutritive value, and soil carbon changes under grazing and clipping. Agron. J. 100 (3), 510–516.

Sanderson, M.A., Adler, P.R., Boateng, A.A., Casler, M.D., Sarath, G., 2006. Switchgrass as a biofuels feedstock in the USA. Can. J. Plant Sci. 86, 1315–1325.

Sanderson, M.A., Schmer, M., Owens, V., Keyser, P., Elbersen, W., 2012. Crop management of switchgrass. In: Monti, A. (Ed.), Switchgrass. Green Energy and Technology. Springer, London, pp. 87–112.

Sanderson, M.A., Wolf, D.D., 1995. Morphological development of switchgrass in diverse environments. Agron. J. 87, 908–915.

Schubert, C., 2006. Can biofuels finally take center stage? Nat. Biotechnol. 24, 777–784.

Shahandeh, H., Hossner, L.R., 2000. Plant screening for chromium phytoremediation. Int. J. Phytorem. 2, 31–51.

Stout, W.L., Jung, G.A., Shaffer, J.A., 1988. Effects of soil and nitrogen on water-use efficiency of tall fescue and switchgrass under humid conditions. Soil Sci. Soc. Am. J. 52 (2), 429–434.

Tilman, D., Hill, J., Lehman, C., 2006. Carbon-negative biofuels from low-input high-diversity grassland biomass. Science 314, 1598–1600.

Ververis, C., Georghiou, K., Christodoulakis, N., Santas, P., Santas, R., 2004. Fiber dimensions, lignin and cellulose content of various plant materials and their suitability for paper production. Ind. Crops Products 19 (3), 245–254.

Wright, L., 2007. Historical Perspective on How and Why Switchgrass was Selected as a "Model" High-Potential Energy Crop. Available from: https://energy.gov/sites/prod/files/2014/04/f14/ornl_switchgrass.pdf.

Wright, L., Turhollow, A., 2010. Switchgrass selection as a "model" bioenergy crop: a history of the progress. Biomass Bioenerg. 34, 851–868.

Wu, M., Wu, Y., Wang, M., 2006. Energy and emission benefits of alternative transportation liquid fuels derived from switchgrass: a fuel life cycle assessment. Biotechnol. Prog. 22 (4), 1012–1024.

Yue, Y., Hou, X., Fan, X., Zhu, Y., Zhao, C., Wu, J., 2017. Biomass yield components of 12 switchgrass cultivars grown in Northern China. Biomass Bioenerg. 102, 44–51.

Cymbopogon flexuosus—an essential oil-bearing aromatic grass for phytoremediation

Vimal Chandra Pandey[a,*], Apurva Rai[b], Anuradha Kumari[c], D.P. Singh[a]
[a]Department of Environmental Science, Babasaheb Bhimrao Ambedkar University, Lucknow, Uttar Pradesh, India; [b]Plant Ecology and Environmental Science Division, National Botanical Research Institute, Lucknow, Uttar Pradesh, India; [c]School of Environmental Sciences, Jawaharlal Nehru University, New Delhi, India
*Corresponding author

1   Introduction

In recent years, phytoremediation has emerged as one of the most viable approach of bioremediation methods for the restoration of the heavily disturbed and degraded sites (Pandey et al., 2015a; Pandey and Bajpai, 2019). It is an eco-friendly, cost-effective, economically feasible, and sustainable initiative for reclamation and decontamination of polluted sites, such as metal contaminated soil, fly ash dumps, coal mine spoil, red mud dumps, chromite-asbestos mine, copper mine tailings, etc. (Pandey, 2017). Several plants have been identified for the phytoremediation, phytorestoration, and rhizoremediation of fly ash dumpsites and contaminated sites (Pandey, 2012a,b, 2013, 2015; Pandey et al., 2012, 2014, 2015b,c). Conventional approaches, such as physical and chemical are developed to remediate heavy metal contaminated sites but phytoremediation techniques are considered to have huge potential to remediate heavy metal contaminated sites with multiple benefits from ecological to socio-economic importance (Pandey and Souza-Alonso, 2019). It can stabilize the heavy-metal contaminated sites, resulted due to mineral extraction from the mines and other industrial activities. When studied for metal stress tolerance capacity, lemongrasses showed distinguish property of proline accumulation in shoot and root under high level of chromium (VI) concentration in the soil. The proline accumulation due to Cr^{+6} induced oxidative stresses in lemongrass indicates that it can be recommended for phytostabilization of chromium (VI) polluted sites (Patra et al., 2018). There are different types of phytoremediation techniques known, such as phytovolatilization, rhizofiltration, phytodegradation, phytoextraction, and phytostabilization (Pandey and Bajpai, 2019). However, from utility point of view, the two most prominent aspects of phytoremediation which are having scientific justifications are phytoextraction and phytostabilization. According to previous studies, aromatic grasses are metal accumulator or metal stress tolerant plants and are efficient for the phytoremediation of metalliferous soils (Verma et al., 2014; Pandey and Singh, 2015; Pandey et al., 2019a; Pandey et al., 2019b). The metal accumulation techniques involve uptake of metals and translocation from soil

Phytoremediation Potential of Perennial Grasses. http://dx.doi.org/10.1016/B978-0-12-817732-7.00009-2

to the various aerial/harvested parts of the plant. The lemongrass (*Cymbopogon flexuosus*) is an excellent example of metal stress tolerant perennial aromatic grass with rapid growth, high biomass, high essential bio-oil producing capacity, and extensive root system. It has a huge potential for revegetation, reclamation, carbon sequestration, eco-restoration through phytoremediation, and is considered economically viable due to wide use in pharmaceuticals, perfumery, aromapathy, industries, and sectors (Table 9.1). Moreover, lemongrass can be a suitable species to grow as cover crop in poor and alkaline soils which can stabilize steep slopes and can be useful in reclamation of mining and industrial dumping sites or waste lands (Pandey et al., 2019a). The chapter deals with the origin, geographic distribution, ecology, propagation, and various uses of lemongrass like phytoremediation of mine dumps, essential oil production, carbon sequestration, and socio-economic development.

2 Ecology

Cymbopogon citratus and *C. flexuosus* are the two most common cultivars of lemongrass. Both cultivars are aromatic, perennial, rapidly growing, and essential oil producing grasses of family *Poaceae*, sub family *Panicoidaea,* and genus *Cymbopogon*. According to the reports, *C. citratus* grows best at an elevation of up to 750 m while *C. flexuosus* up

Table 9.1 An indicative list of potential attributes and uses of lemongrass (*Cymbopogon flexuosus*) for environmental and socio-economic development.

S. no.	Multiple uses and attributes of lemongrass	References
1	Essential-oil production	Singh et al. (1996); Singh (2001); Pandey et al. (2003); Zheljazkov et al. (2011)
2	Soil carbon sequestration	Singh et al. (2014)
3	Soil quality improvement of degraded soil	Pankaj et al. (2017); Singh (2001)
4	Controlling runoff and soil erosion	Thomas et al. (2012)
5	Wasteland reclamation	Kiran et al. (2009); Pandey et al. (2019a)
6	Phytoremediation of fly ash deposits	Maiti and Prasad (2017); Panda et al. (2018); Verma et al. (2014)
7	Restoration of waste dumpsites	Maiti and Kumar (2016); Das and Maiti (2009); Kumar and Maiti (2015)
8	Phytoremediation of heavy metals contaminated soils	Patra et al. (2018); Gautam et al. (2017); Pandey and Singh (2015); Pandey et al. (2019a); Pandey et al. (2019b)
9	Vermicomposting	Puttanna and PrakasaRao (2003)
10	Medicinal properties against pathogens	Chandrashekar and Joshi (2006); Anaruma et al. (2010); Kumar et al. (2009); Pandey et al. (2003)
11	Food industry, cosmetics, and perfumery	Sarkic and Stappen (2018)

to 2200 m above sea level. The essential part of the plant is stem/stalk and leaves/foliage. Cultivation of crop is dependent on good precipitation, quality of soil condition, and its pH variability. The cultivation of crop with good oil content requires a temperature of 10°C–33°C, pH (5.0–8.4), and precipitation 700–3000 mm followed by insolation (sunshine). However, the lemongrass plant is quite sensitive toward the winter conditions and intolerant to snow or frost (Lemongrass, 2012). Moreover, lemongrass is adapted to grow well in wide range of soil; it is efficiently cultivated on sandy to clay loam soils in the pH range less acidic to less alkaline. Essential oil is extracted through the process of steam distillation from the fresh plant material parts like leaves and stalk. The essential oil contents of lemongrass depend on the altitude of the region and basicity of the soil, the more is the basic soil and lower the altitude, the higher will be the citral content of the lemongrass oil. Commercially, high citral content oil producing cultivar is in demand. It is an important CO_2 sequestering, metal tolerant and metal accumulator plant species. In the regions of scarce rainfall, the crop should be cultivated/grown by following various irrigation methods. The lemongrass plant is quite sensitive toward the cold climatic conditions and intolerant to snow or frost (Lemongrass, 2012). Fig. 9.1 represents the growing nature of *C. flexuosus* on degraded land.

3 Origin and distribution

The origin of Lemongrass (*C. flexuosus*) is found to be in the tropical and sub-tropical climatic regions of the world. It is extensively cultivated around the world, countries such as India, Indonesia, Madagascar, Japan, Somalia, Brazil, Cuba, Ecuador, Singapore, China, France, Haiti, Puerto Rico, Mexico, Guatemala, Honduras, Salvador, and Thailand, for its essential oil enrich with terpene compounds (Soto-Ortiz

Figure 9.1 *Cymbopogon flexuosus* growing on degraded land.

et al., 2002; Anaruma et al., 2010; Ganjewala and Luthra, 2010). The lemongrass plant derived its' name due to presence of lemon scented aroma in their leaves. It is perennial in nature, multi-harvest aromatic grass found both in non-irrigated and irrigated status of land. But a significant proportion is harvested from the natural habitats like banks of water bodies or mixed forests ecosystem (Singh and Sharma, 2001; Weiss, 1997). There are two important species/cultivar of *Cymbopogon* which produces essential oil, *C. flexuosus* (Steud.) Wats, (syn *Andropogonnardus* var. flexuosus Hack; *A. flexuosus* Nees), also called as East Indian lemongrass, France Indian verbena, Cochin grass, and Malabar grass. The species is widely distributed over and native to sub tropics of India, Sri Lanka, Thailand, and Myanmar. Another one is *Cymbopogon citratus* (DC.) Stapf., commonly known as Lemon grass, Citronella grass, Madagascar lemongrass and West Indian lemongrass is native to South India, Indonesia, Madagascar, and Sri Lanka.

4 Botanical description

C. flexuosus grows rapidly up to height of 150 cm and has lemon-scented aroma. It produces ovules and has distinct dark green leaves/foliage. Whereas, *C. citrates* also has the characteristic lemon-scented aroma, and stem height reaches upto 100 m but it does not produce seed/ovule and has bluish-green foliage (Lemongrass, 2012). These two species of *Cymbopogon* produces many bulbous stalks that increase clump diameter with plant's maturity (Lemongrass, 2012). *Cymbopogon* sp. produces extensive root system and network of rootlets that facilitate binding of soil into it.

5 Propagation

The lemongrass plant effectively reproduces by vegetative propagation through splitting the clumps into slips and it is preferable to obtain high grade and yield of an essential oil. The clump's top portion should be cut off within 20–25 cm of the root and the latter part should be split into slips. However, the bottom brown sheath should be separated for the exposure of young roots. *C. flexuosus* also be effectively propagated through seeds (Lemongrass, 2012). A healthy *C. flexuosus* plant yields seeds on an average of about 100–200 g. For its seed germination, 5–6 days are required along with desirable temperature and moisture level. The propagation through transplantation of seedlings requires about 60 days post germination. It is economical to raise seedlings in nursery whenever the water stress condition prevails/exists. For cultivation of lemongrass, it is recommended to transplant the seedlings raised in the nursery in spite of sowing seeds directly in the field.

6 Important aspects in relation to phytoremediation

The lemongrass plant being perennial, rapid and fast growing, metal tolerant, hyper-accumulator, extensive root system, and luxuriant biomass producing can play an important role in phytoremediation. Some important aspects of lemongrass with respect to phytoremediation are the retarded antioxidant enzyme activity under high fly ash concentrations which implies that lemongrass planted in such area encounter negative impact on photosynthesis. Another study on chromium (VI) toxicity reveals efficiency of lemongrass plant to stabilize highly chromium (VI) contaminated sites. Cr^{+6} are a stable oxidation state of chromium in natural conditions and are highly toxic form (Panda and Patra, 1997; Panda et al., 2003; Mohanty and Patra, 2011; Das et al., 2014). Studies have shown that an increasing trend in chromium (VI) accumulation with increase in concentration. Consequently, there is decreasing trend in protein and chlorophyll content of lemongrass (Patra et al., 2018). The elevated chromium accumulation in lemongrass accelerated the proline level and anti-oxidant enzyme activity which implies that lemongrass has efficient damage control activity in chromium (VI) stress condition. The analysis of various physicochemical properties of the soil amended with different concentration of Cr^{+6} revealed that maximum amount of organic carbon was present in 100 mg/kg of Cr^{+6} contaminated soil. The lemongrass showed growth inhibition at very high Cr^{+6} concentration levels along with increased concentration of available N, P, and K in soil. Whereas, treatment of 10 mg/kg of Cr^{+6} showed better growth in both root and shoot due to the fact that chromium affects Fe availability. Higher chromium concentration in lemongrass increased proline content in them that signifies that lemongrass plant can cope up with Cr toxicity by higher proline accumulation. Fig. 9.2 shows important features and properties of lemongrass (*C. flexuosus*) related to its phytoremediation potential.

7 Multiple uses of lemongrass

7.1 Phytoremediation

Phytoremediation of chromite-asbestos mine—The lemongrass plant with combination of different organic manure amendments in soil has been used for the massive biomass production and growth with maximum metal tolerance potential. The studies have suggested that application of treatments in combination of 90% mine waste,

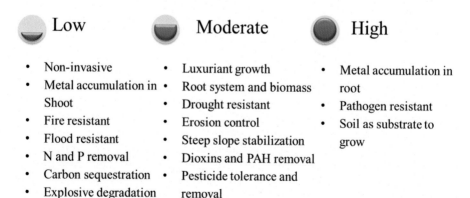

Figure 9.2 **Important aspects of *Cymbopogon flexuosus* related to phytoremediation.**

2.5% chicken manure, 2.5% farmyard manure, and 5% garden soil (w/w %) leads to luxuriant biomass production and growth of lemongrass. An observation of the given amendment shows low Ni and Cr accumulation in roots and decrease in translocation to shoots. Moreover, lemongrass can efficiently accumulate 82.01% of Cr in roots and 44.63% in shoots (Kumar and Maiti, 2015). Results from the pot scale study indicate that total biomass of *C. citratus* was highest in all treatment combinations except in the control (un-amended mine waste). Moreover, for the efficient phytostabilization of abandoned chromite-asbestos mine waste sites, lemongrass can be cultivated efficiently with combination of different organic manure amendments/ treatments. Consequently, it will enhance the ecological, commercial, and esthetic value of the phytostabilized sites (Kumar and Maiti, 2015).

Phytostabilization of copper mine tailings—The metal mine tailings left unmanaged after extraction of minerals are highly prone to erosion and cause extensive damage to the surrounding environment by leaching of toxic heavy metals. Cultivation of cover crops like lemongrass, a potential heavy metal accumulator plant species can be used to stabilize, reduce erosion, and immobilize the toxic metals from the abandoned toxic mine tailings. However, phytostabilization technique will require effective cultivation methods, such as application of chicken manure or soil- manure mixture to increase the plant growth manifold. The study suggests that application of about 2.5% of chicken manure in the tailings can increase biomass production of the lemongrass up to 10 times compared to unamended/pure tailings (Das et al., 2014). Moreover, comparative study on metal accumulation by same species of lemongrass planted in different amended tailings reveals that lemongrass can accumulate more Cu and Ni in roots when cultivated on pure tailings whereas, shoot accumulates higher proportion of Cu, Ni, and Pb in amended tailings (Das et al., 2014). In spite of chemical limitations and stress; lemongrass can grow on toxic copper tailings with restricted/stunted growth but application of chicken manure and soil as an amendment could enhance the growth both in biomass and number of tillers in manifold. Chicken manure amendment alone could lead to increase in metal accumulation by lemongrass in shoots but reverse trend observed in roots. However, it is important to note that mixture of soil and chicken manure decrease metal accumulation except for Pb, in both shoot and root but lead to massive increase in biomass production of lemongrass. In an another study for reclamation of toxic copper mine tailings by cultivation of cover crops like lemongrass, it was found that addition of manures and soil will drastically improve the nutrient status like enhancement of organic carbon, exchangeable K, and available N & P which consequently enhance the supply of nutrients to plant and improve the physical status of the tailing. Lemongrass can efficiently accumulate heavy metals, such as Mn, Zn, and Pb from the tailings when cultivated in 5% chicken manure amended field as it produce significantly higher shoot and luxuriant plant biomass (Das and Maiti, 2009). Thus, the application of Cu tailings and chicken manure as amendments will enhance the biomass production and consequently lead to more production of essential bio oil. These amendments in soil will also increase the organic matter concentration in tailings which increase the moisture content of tailings and facilitate stabilization of unstable toxic Cu mine tailings (Das and Maiti, 2009).

Reclamation and re-vegetation of fly ash disposal sites—Globally India is the third largest fly ash producing country (Ram et al., 2008). Fly ash deposited sites pose severe environmental problems near the coal-based thermal power plants. Therefore, it is urgent need for reclamation and restoration of fly ash dumpsites. Fly ash dump sites are rich in Mn, Cu, Zn, Cd, As, Cr, Ni, and Pb, when released in environment as leachates lead to contamination of ground water and soil (Pandey et al., 2009, 2011; Pandey and Singh, 2010). The lemongrass plant is an early colonizer species it can potentially survive in initial stress condition of the fly ash dumpsites. Cultivation of cover crops like lemongrass on these sites can potentially reduce the reclamation phase duration and enhance the establishment of other native grasses. Previous studies suggest that fly ash dump sites require certain amount of amendments for the improvement in growth and biomass of *C. citratus*. Addition of farmyard manure as an amendment enhances the growth of lemongrass high root:shoot biomass ratio. However, fly ash already contains sufficient amount of organic nutrients which can promote plant growth but adding small quantity of easily available NPK supplements along with farmyard manure or garden soil can further enhance the growth of economically valuable aromatic grasses. Consequently, massive deep fibrous root system of *C. citratus* will develop with high root to shoot biomass ratio and stabilize the upper surface/layer of fly ash by holding the loose materials of fly ash. Further, for reclamation of fly ash dump sites stress tolerant, efficient biomass producing and rapidly growing plant species are recommended that will help in the development of fast vegetation cover (Maiti and Prasad, 2017). Whereas, when studied for fly ash tolerance potential, lemongrass cultivated in 25% fly ash amendment showed increase in plant growth in terms of root, shoot, and total biomass (11%) and has significant metal tolerance index. Addition to this, no significant change observed in stomatal conductance, leaf photosynthetic rate, and photosystem II activity (Panda et al., 2018). Verma et al. (2014) and Pandey et al. (2019a) suggested that lemon grass has potential to grow on fly ash dumpsites.

Ecorestoration of over burden dumping site rich in sulfur content—Abandoned mine sites emerges as severe challenge for the protection of environment from degradation. One such study on mine dumping site containing high sulfur over burden near Assam-India suggests that lemongrass plant has potential to withstand against macro and micro nutrient deficient soil, very low pH, and elevated level of heavy metal and trace metal concentrations. The plantation of variety plant species along with lemongrass have resulted in primary and secondary ecological succession that will lead to eco-restoration of the mine over burden (Dowarah et al., 2009). Mining activities has severe depleting impact on ecology and environment of that area. Consequently, this lead to reduction in biotic diversity, changes in soil horizons, alteration in structure, and functioning of ecosystems. As a result, these alterations may influence the trophic level interactions, water, and nutrient dynamics (Matson et al., 1997; Almas et al., 2004; Ghose, 2004). The findings of experimental study on mine over burden dumping site suggests that plantation of herbaceous monocot species with grasses like lemongrass can accelerate the ecological processes leading to long term sustainable eco-restoration process. Lemongrass help in covering the exposed area rapidly with their quick growth and fibrous root system which enable them withstand and resists the stressful conditions of mine over burden and make them most stress tolerant plant species. The plant cover can establish and survive on the mine over burden site despite of low biological activity, absence of true soil characteristics, and highly acidic pH showing potential for building self-sustaining ecosystem through ecological succession (Dowarah et al., 2009).

Lemongrass-legume cover establishment on degraded land for stabilization—The industrial and mining wastes generated by an integrated sponge iron plant possess serious threat as the waste contains hazardous substances. To encounter the problem of land pollution due to these waste dumps an environment friendly and inexpensive eco-restoration technique should be encouraged to follow. Covering the wastes with topsoil and re-vegetating with grass-legume mixtures are ecologically sound approaches to stabilize these wastes dumping sites that will serve multipurpose benefits, such as improve soil nutrient contents, restore the soil conditions prior to the damage, prevent soil erosion, and stabilize the wastes (Maiti and Maiti, 2014). Nitrogen plays a crucial role in eco-restoration of degraded land (Kendle and Bradshaw, 1992,) so the legume-grass mixture will restore soil fertility in a natural way without use of heavy doses of conventional fertilizers. The legume-grass mixtures create a nitrogen pool in soil by reducing the loss of nutrients through leaching. Furthermore, grasses used for these mixtures are rapid growing, provides luxuriant biomass, and can survive on toxic waste materials. Moreover, these grasses are stress tolerant species which can survive in adverse pH conditions, heavy metal pollution or poor nutrient availability conditions. The extensive root system of grasses will provide mechanical support to the loose soil particles and restricts soil erosion. When dried the plant biomass act as mulch and it perhaps conserve moisture. Moreover the legumes are draught resistant, rapid growing, and perennial species that will facilitate soil nitrogen enrichments, soil moisture conservation, accelerate organic matter contents of soil and create a nitrogen rich soil (Maiti and Maiti, 2014), enhance soil formation processes and improve soil nitrogen deficient status of mine spoil (Maiti, 2013).

Phytoremediation of bauxite residue also referred as red mud—Red mud is highly alkaline and saline in nature. Phytoremediation of these heavily polluted red mud dump sites can be done by plantation of lemongrass with combination of cow dung manure and sewage sludge as amendments. The experimental study on the effects of red mud on soil properties, plant growth performance, and metal accumulation in the lemongrass plant revealed that lemongrass is a potential metal tolerant plant species and it is having more than 100% metal tolerance index. The other two parameters used in the study for phytoextraction potential of lemongrass are translocation factor and bio-concentration factor, which showed lemongrass is a potential phytostabilizer of Fe, Mn, and Cu in roots whereas, it efficiently translocate Al, Zn, Cd, Pb, Ni, Cr, and As from roots to shoot. Moreover, bio-concentration factor is less than 1 which implies that lemongrass act as a potential metal excluder for all the metals studied (Gautam et al., 2017).

7.2   Essential oil production

The lemongrass plant produces luxuriant biomass that contains essential bio-oil which may have potential applications in bio-fuel generation. The can be extracted by the steam distillation technique. However, study revealed that essential oil content in lemongrass biomass decreases as the plant grow older and biomass exhibits linear increase with increasing N rates (Singh, 2001) and can reach to optimal at 100 kg/ha of N (Singh et al., 1997; Rao et al., 1998). Lemongrass produces high citral content oil, which has pleasant smell and refreshing effect. The oil has antibacterial, antifungal, and antiviral properties, which made it economically important in medicinal and pharmaceutical industries (Anaruma et al., 2010; Guynot et al., 2003; Pandey et al., 2003; Kumar et al., 2007, 2009; Inouye et al., 2001; Pattnaik et al., 1996). In addition to this, lemongrass oil is also used as perfumery, food, and eco-friendly pesticides. Due to high-value essential oil content, lemongrass may prove to a rich source for bio-fuel. Literature studies report that lemongrass oil chemical composition varies considerably. Kulkarni et al. (1997) found 25.9% geranylacetate, 13.1% geraniol, and 11.2% citronellyl acetate; Pandey et al. (2003) found 43.8% geranial, 18.9% citral, 5.27% geranyl acetate, and 3.66% trans-geraniol. Chandrashekhar and Joshi (2006) reported 6% limonene, 8% geranyl acetate, 13% geraniol, and 61% citral or neral. However, recent experimental report on composition of lemongrass oil by Zheljazkov et al. (2011) suggests that there is variation as a function of N and S treatments, harvest and geographical location. The geranial varies from 25%–53%, citral 20%–45%, caryophyllene oxide 1.3%–7.2%, and t-caryophyllene 0.3%–2.2%. Furthermore, it is reported that lemongrass can be a potential/considerable/potent option for biofuel production owing to its' luxuriant biomass production and multi harvest property.

7.3   Carbon sequestration

The drastic rise in CO_2 level in the atmosphere is a main cause of the global warming. Therefore, it is pressing need to sequester atmospheric carbon dioxide in potential sinks. For this purpose, plant system can play a major role in land based carbon sequestration by acting as a sink. However, agricultural land and forests can play crucial role in the strategy designed for carbon sequestration but it should be a priority that such practices should avoid conflict between economic interest of the farmer and environmental concerns. Lemongrass (*C. flexuosus*) is an economically viable cash crop due to its essential oil contents, luxuriant biomass production, extensive root system, and massive growth, and can be consider as a potential carbon sequestering aromatic grass. Carbon sequestration in plant biomass and soil for the reduction of rising level of atmospheric CO_2 has gained importance due to its mitigation potential against the drastic increase in CO_2 concentration in the atmosphere. Study suggests that lemongrass possess moderate carbon sequestration capacity when compared to vetiver grass (Singh et al., 2014). According to experimental result, lemongrass can sequester 5.38 Mg/ha/year in biomass and 3.08 Mg/ha/year in soil (Singh et al., 2011) that indicates it as a potential candidate for carbon sequestration. The lemongrass cultivation has significant economic and environmental benefits, as it is most extensively cultivated for the essential oil production from the luxuriant biomass production. Consequently, it stabilizes the derelict lands, improves soil conditions by acting as metal hyperaccumulator, restores mine spoils, and dumps sites, etc. Moreover, the waste biomass generated after the essential oil extraction can potentially employed in the degraded land area for improving soil carbon content.

8 Medicinal use

Lemongrass (*C. flexuosus*) oil has stronger antimicrobial properties (Pandey et al., 2003). Therefore, it is extensively used in pharmaceutical and cosmetic industries. Combination of lemongrass oil and mint oil can exhibit broad range of antibacterial, antifungal, and antiviral properties against some selected strains of bacteria, viruses, and fungi (Chao and Young, 2000). According to study on chemical composition and antimycin activity of lemongrass oil against few fungal species suggests that 100% concentrated oil may have maximum anti-fungal property against few tested fungal species, such as *T. mentagrophytes* and *F. oxysporum*. Moreover, it is significant to note that lemongrass oil show maximum fungitoxic effect against both the fungal species (Pandey et al., 2003). The major constituents of lemongrass oil (*C. flexuosus*) are citral (43.80%), trans-geraniol (3.66%), z-citral (18.93%), and geranyl acetate (5.27%) (Pandey et al., 2003).

9 Other commercial uses

Lemongrass oil can be used in food and flavoring industries. The fresh plant parts like stem and leaves are widely used in preparation of oriental cuisines. Whereas, dried leaves are used to make tea. In industries, lemongrass oil is used in making candles, waxes, and insect repellents. It serves as source of vitamin A and E and constituent of organic pesticides. In cosmetic industries, lemongrass oil is an ingredient for soaps, detergents, tissue toner, facial treatments, and fragrance agent. In therapeutic and pharmaceuticals, lemongrass is a diuretic, antiseptic, tonic, and stimulant. It is used to cure stomach-ache, headaches, fevers, diarrhea, flu, muscular pain, poor circulation, and slack tissue. The lemongrass oil has antiseptic property and used in the treatment of acne and athlete's foot disease. In veterinary, it is used as ingredients of shampoos for cat and dog, it act as repellant for ticks, lice, and fleas.

10 Socio-economic development

The plantation of lemongrass can provide potential alternative source of income to the farmers as they provide essential oil which serve as ingredients for perfumery and food industry and medicine or pharmaceutical uses. The small and marginal farmers can derive good amount of financial benefits as the cultivation of lemongrass is very economical, provide significant turnover and require minimal care for its growth and development. Moreover, in terms of economics, lemongrass can produce fresh biomass upto 27.7 Mg/ha/year, essential oil around 0.8% and will give turnover/net returns of 84550 ₹ (USD 1219)/ha/year (Singh et al., 2011). However, India is one of the largest lemongrass oil producing country in the world, around 300–350 tons annually from which it exports about 80% of its oil production to the developed countries, such as France, Germany, Japan, United Kingdom, China, Spain, United States, and Italy

(Lal et al., 2013). According to one study, it was estimated that 20–25 tons of fresh lemongrass per acre can produce 70–75 kg of essential oil per annum, and the profit gain to farmers will be ranged in 80%–120% (Janhit Foundation, 2014).

11 Implementation strategies

Incentives to the marginal farmers for lemongrass cultivation are an important strategy. Cultivation of lemongrass for environmental benefits like restoration of derelict land, re-vegetation of heavily disturbed dump sites, and mine spoils may be included in the National Development Programs on watershed development, afforestation of degraded forests etc. Provision of providing subsidies to farmers for plantation of lemongrass can enhance the essential oil production and certainly it will give economic benefits to marginal and small farmers. Consequently, it will accelerate the carbon sequestration in plant biomass and soil accomplishing dual targets environmental and economic benefits. Furthermore, massive biomass producing lemongrass plant will restrict the harmful and excess emission of CO_2 from biomass burning after essential oil extraction, when converted into vermicompost following standard procedure (Puttanna and PrakasaRao, 2003). Incorporation of this vermicompost into the degraded soil will enhance the carbon sequestration in the soil and limits the excess CO_2 release into atmosphere. Therefore, lemongrass cultivation can potentially and efficiently incorporate in land management practices toward the utilization of degraded land. Its systematic and large scale plantation will provide economic benefits to small and marginal farmers resulting into significant environmental benefits.

12 Conclusion and future prospects

A holistic approach is required for eco-restoration, reclamation, re-vegetation, and management of heavily disturbed sites by anthropogenic activities, such as copper mine tailing dumps, fly ash disposal sites, chromite-asbestos mine wastes. Phytoremediation techniques can be efficiently implemented with planning to achieve the goal of remediation and management of waste dumped sites. The various studies done on lemongrass emphasized that it can grow vigorously in nutrient deficient conditions and get established within a short range of time period, the massive fibrous root network of lemongrass can hold loose soil particles and restricts soil erosion, the grass–legume mixture application could be a viable method for eco-restoration by re-vegetation approach on wide-ranging industrial dumpsites.

References

Almas, A.R., Bakken, L.R., Mulder, J., 2004. Changes in tolerance of soil microbial communities in Zn and Cd contaminated soils. Soil Biol. Bioch. 36, 805–813.

Anaruma, N.D., Schmidt, F.L., Duarte, M.C.T., Figueira, G.M., Delarmelina, C., Benato, E.A., Sartoratto, A., 2010. Control of *Colletotrichum gloeosporioides*(Penz.) Sacc. in yellow passion fruit using *Cymbopogon citratus*essential oil. Braz. J. Microbiol. 41, 66–73.

Chandrashekar, K.S., Joshi, A.B., 2006. Chemical composition and anthelmintic activity of essential oils of three *Cymbopogon* species of South Canara, India. J. Saudi Chem. Soc. 10, 109–111.

Chao, S.C., Young, D.C., 2000. Screening of inhibitory activity of essential oils on selected bacteria, fungi and viruses. J. Essent. Oil Res. 12, 639–649.

Das, M., Maiti, S.K., 2009. Growth of *Cymbopogon citratus* and *Vetiveria zizanioides* on Cu mine tailings amended with chicken manure and manure-soil mixtures: a pot scale study. Int. J. Phytorem. 11 (8), 651–663.

Das, S., Mishra, J., Das, S.K., Pandey, S., Rao, D.S., Chakraborty, A., Sudarshan, M., Das, N., Thato, H., 2014. Investigation on mechanism of Cr (VI) reduction and removal by *Bacillus amyloliquefaciens*, a novel chromate tolerant bacterium isolated from chromite minesoil. Chemosphere 96, 112–121.

Dowarah, J., Boruah, H.D., Gogoi, J., Pathak, N., Saikia, N., Handique, A.K., 2009. Eco-restoration of a high-sulphur coal mine overburden dumping site in northeast India: a case study. J. Earth Syst. Sci. 118 (5), 597–608.

Dowarah, J., Boruah, H.D., Gogoi, J., Pathak, N., Saikia, N., Handique, A.K., 2009. Eco-restoration of a high-sulphur coal mine overburden dumping site in northeast India: a case study. J. Earth Syst. Sci. 118 (5), 597–608.

Lemongrass, 2012. Essential Oil Crops, Production Guidelines For Lemongrass. Published by Directorate communication Services Department of agriculture, forestry and fisheries private Bag X144, pretoria, 0001 South Africa.

Ganjewala, D., Luthra, R., 2010. Essential oil biosynthesis and regulation in the genus *Cymbopogon*. Nat. Prod. Commun. 5, 163–172.

Gautam, M., Pandey, D., Agrawal, M., 2017. Phytoremediation of metals using lemongrass (*Cymbopogon citratus* (DC) Stapf.) grown under different levels of red mud in soil amended with biowastes. Int. J. Phytorem. 19 (6), 555–562.

Ghose, M.K., 2004. Effect of opencast mining on soil fertility. J. Sci. Indust. Res. 63, 1006–1009.

Guynot, M.E., Ramos, A.J., Seto, L., Purroy, P., Sanchis, V., Marin, S., 2003. Antifungal activity of volatile compounds generated by essential oils against fungi commonly causing deterioration of bakery products. J. Appl. Microbiol. 94, 893–899.

Inouye, S., Takizawa, T., Yamaguchi, H., 2001. Antibacterial activity of essential oils and their major constituents against respiratory tract pathogens by gaseous contact. J. Antimicrob. Chemother. 47, 565–573.

Janhit Foundation, 2014. Factsheet on lemon grass. www.janhitfoundation.

Kendle, A.D., Bradshaw, A.D., 1992. The role of soil nitrogen in the growth of trees on derelict land. Arboric. J. 16 (2), 103–122.

Kiran, Kudesia, R., Rani, M., Pal, A., 2009. Reclaiming degraded land in India through the cultivation of medicinal plants. Bot. Res. Int. 2(3), 174–181.

Kulkarni, R.N., Mallavarapu, G.R., Baskaran, K., Ramesh, S., Kumar, S., 1997. Essential oil composition of a citronella-like variant of lemongrass. J. Essen. Oil Res. 9, 393–395.

Kumar, A., Maiti, S.K., 2015. Effect of organic manures on the growth of Cymbopogon citratus and Chrysopogon zizanioides for the phytoremediation of chromite-asbestos mine waste: a pot scale experiment. Int. J. Phytorem. 17, 437–447.

Kumar, A., Shukla, R., Singh, P., Dubey, N.K., 2009. Biodeterioration of some herbal raw materials by storage fungi and aflatoxin and assessment of *Cymbopogon flexuosus* essential oil and its components as antifungal. Int. Biodeterior. Biodegradation 63, 712–716.

Kumar, R., Dubey, N.K., Tiwari, O.P., Tripathi, Y.B., Sinha, K.K., 2007. Evaluation of some essential oils as botanical fungi toxicants for the protection of stored food commodities from fungal infestation. J. Sci. Food Agric. 87, 1737–1742.

Lal, K., Yadav, R.K., Kaur, R., Bundela, D.S., Khan, M.I., Chaudhary, M., Meena, R.L., Dar, S.R., Singh, G., 2013. Productivity, essential oil yield, and heavy metal accumulation in lemon grass (*Cymbopogon flexuosus*) under varied wastewater-groundwater irrigation regimes. Ind. Crop. Prod. 45, 270–278.

Maiti, D., Prasad, B., 2017. Studies on colonisation of fly ash disposal sites using invasive species and aromatic grasses. J. Environ. Eng. Landscape Manage. 25 (3), 251–263.

Maiti, S.K., 2013. Establishment of grass and legume cover. In: Ecorestoration of the Coalmine Degraded Lands. Springer, India, pp. 151–161.

Maiti, S. K., Kumar, A., 2016. Energy plantations, medicinal and aromatic plants on contaminated soil. In: Bioremediation and Bioeconomy. Elsevier, The Netherlands, pp. 29–47.

Maiti, D., Maiti, S.K., 2014. Ecorestoration of waste dump by the establishment of grass-legume cover. Int. J. Sci. Technol. Res. 3 (3), 37–41.

Matson, P.A., Parton, W.J., Powere, A.G., Swift, M.J., 1997. Agricultural intensification and ecosystem properties. Science 277, 504–509.

Mohanty, M., Patra, H.K., 2011. Attenuation of chromium toxicity in mine waste water using water hyacinth. J. Stress Physiol. Biochem. 7, 335–346.

Panda, D., Panda, D., Padhan, B., Biswas, M., 2018. Growth and physiological response of lemongrass (*Cymbopogon citratus* (D.C.) Stapf.) under different levels of fly ash-amended soil. Int. J. Phytorem. 20, 538–544.

Panda, S.K., Chaudhury, I., Khan, M.H., 2003. Heavy metal induces lipid peroxidation and affects antioxidants in wheat leaves. Biol. Plant. 46, 289–294.

Panda, S.K., Patra, H.K., 1997. Physiology of chromium toxicity in plants—a review. Plant Physiogl. Biochem. 4, 10–17.

Pandey, A.K., Rai, M.K., Acharya, D., 2003. Chemical composition and antimycotic activity of the essential oils of corn mint (*Mentha arvensis*) and lemon grass (*Cymbopogon flexuosus*) against human pathogenic fungi. Pharmaceut. Biol. 41, 421–425.

Verma, S.K., Singh, K., Gupta, A.K., Pandey, V.C., Trivedi, P., Verma, R.K., Patra, D.D., 2014. Aromatic grasses for phytomanagement of coal fly ash hazards. Ecol. Eng. 73, 425–428.

Pandey, V.C., Singh, N., 2015. Aromatic plants versus arsenic hazards in soils. J. Geochem. Explor. 157, 77–80.

Pandey, V.C., Rai, A., Korstad, J., 2019a. Aromatic crops in phytoremediation: from contaminated to waste dumpsites. In: Pandey, V.C., Bauddh, K. (Eds.), Phytomanagement of Polluted Sites. Elsevier, Amsterdam, Netherlands, pp. 255–275.

Pandey, V.C., Singh, K., Singh, R.P., Singh, B., 2012. Naturally growing *Saccharum munja* on the fly ash lagoons: a potential ecological engineer for the revegetation and stabilization. Ecol. Eng. 40, 95–99.

Pandey, V.C., Abhilash, V.C., Singh, N., 2009. The Indian perspective of utilizing fly ash in phytoremediation, phytomanagement and biomass production. J. Environ. Manage. 90, 2943–2958.

Pandey, V.C., Bajpai, O., 2019. Phytoremediation: from theory toward practice. In: Pandey, V.C., Bauddh, K. (Eds.), Phytomanagement of Polluted Sites. Elsevier, Amsterdam, Netherlands, pp. 1–49.

Pandey, V.C., Bajpai, O., Pandey, D.N., Singh, N., 2015c. *Saccharum spontaneum*: an underutilized tall grass for revegetation and restoration programs. Genet. Resour. Crop Evol. 62 (3), 443–450.

Pandey, V.C., 2017. Managing waste dumpsites through energy plantations. In: Bauddh, K., Singh, B., Korstad, J. (Eds.), Phytoremediation Potential of Bioenergy Plants. Springer, Singapore, pp. 371–386.

Pandey, V.C., 2013. Suitability of *Ricinus communis* L. cultivation for phytoremediation of fly ash disposal sites. Ecol. Eng. 57, 336–341.

Pandey, V.C., Souza-Alonso, P., 2019. Market opportunities in sustainable phytoremediation. In: Pandey, V.C., Bauddh, K. (Eds.), Phytomanagement of Polluted Sites. Amsterdam, Netherlands, Elsevier, pp. 51–82.

Pandey, V.C., Pandey, D.N., Singh, N., 2015a. Sustainable phytoremediation based on naturally colonizing and economically valuable plants. J. Clean. Prod. 86, 37–39.

Pandey, V.C., 2015. Assisted phytoremediation of fly ash dumps through naturally colonized plants. Ecol. Eng. 82, 1–5.

Pandey, V.C., Prakash, P., Bajpai, O., Kumar, A., Singh, N., 2015b. Phytodiversity on fly ash deposits: evaluation of naturally colonized species for sustainable phytorestoration. Environ. Sci. Pollut. Res. 22, 2776–2787.

Pandey, V.C., 2012a. Phytoremediation of heavy metals from fly ash pond by *Azolla caroliniana*. Ecotoxicol. Environ. Saf. 82, 8–12.

Pandey, V.C., Singh, N., Singh, R.P., Singh, D.P., 2014. Rhizoremediation potential of spontaneously grown *Typha latifolia* on fly ash basins: study from the field. Ecol. Eng. 71, 722–727.

Pandey, V.C., 2012b. Invasive species based efficient green technology for phytoremediation of fly ash deposits. J. Geochem. Explor. 123, 13–18.

Pandey, V.C., Singh, N., 2010. Impact of fly ash incorporation in soil systems. Agric. Ecosyst. Environ. 136, 16–27.

Pandey, V.C., Singh, J.S., Singh, R.P., Singh, N., Yunus, M., 2011. Arsenic hazards in coal fly ash and its fate in Indian scenario. Resour. Conserv. Recy. 55, 819–835.

Pandey, J., Verma, R.K., Singh, S., 2019a. Screening of most potential candidate among different lemongrass varieties for phytoremediation of tannery sludge contaminated sites. Int. J. Phytorem. 21 (6), 600–609. doi: 10.1080/15226514.2018.1540538.

Pankaj, U., Verma, S.K., Semwal, M., Verma, R.K., 2017. Assessment of natural mycorrhizal colonization and soil fertility status of lemongrass [(*Cymbopogon flexuosus*, Nees ex Steud) W. Watson] crop in subtropical India. J. Appl. Res. Med. Aromatic Plants 5, 41–46.

Patra, D.K., Pradhan, C., Patra, H.K., 2018. An in situ study of growth of lemongrass *Cymbopogon flexuosus* (Nees ex Steud. W. Watson on varying concentration of Chromium (Cr^{+6}) on soil and its bioaccumulation: perspectives on phytoremediation potential and phytostabilisation of chromium toxicity. Chemosphere 193, 793–799.

Pattnaik, S., Subramanyam, V.R., Kole, C., 1996. Antibacterial and antifungal activity of ten essential oils in vitro. Microbios 86, 237–246.

Puttanna, K., PrakasaRao, E.V.S., 2003. Vermicomposting distillation residues of some aromatic crops. Fertil. News 48 (2), 67–68.

Ram, L.C., Jha, S.K., Tripathi, R.C., Masto, R.E., Selvi, V.A., 2008. Remediation of fly ash landfills through plantation. Remediation 18, 71–90, http://dx.doi.org/10.1002/rem.20184.

Rao, B.R.R., Chand, S., Bhattacharya, A.K., Kaul, P.N., Singh, C.P., Singh, K., 1998. Response of lemongrass (*Cymbopogon flexuosus*) cultivars to spacings and NPK fertilizers under irrigated and rainfed conditions in semi-arid tropics. J. Med. Aromatic Plant Sci. 20, 407–412.

Sarkic, A., Stappen, I., 2018. Essential oils and their single compounds in cosmetics—a critical review. Cosmetics 5, 11. doi: 10.3390/cosmetics5010011.

Singh, M., 2001. Long-term studies on yield, quality and soil fertility of lemongrass (*Cymbopogon flexuosus*) in relation to nitrogen application. J. Hortic. Sci. Biotechnol. 76, 180–182.

Singh, M., Sharma, S., 2001. Influence of irrigation and nitrogen on herbage and oil yield of palmarosa (*Cymbopogon martinii*) under semi-arid tropical conditions. Eur. J. Agron. 14, 157–159.

Singh, M., Shivaraj, B., Sridhara, S., 1996. Effect of plant spacing and nitrogen levels on growth, herb and oil yields of lemongrass (*Cymbopogon flexuosus*(Steud.) Wats. var. Cauvery). J. Agron. Crop Sci. 177, 101–105.

Singh, M., Guleria, N., Rao, E.P., Goswami, P., 2011. A strategy for sustainable carbon sequestration using Vetiver (*Vetiveria zizanioides* (L.)): a quantitative assessment over India. *Project document CM PD-1101, CSIR Centre for Mathematical Modelling and Computer Simulation, India.*

Singh, M., Guleria, N., Rao, E.V.P., Goswami, P., 2014. Efficient C sequestration and benefits of medicinal vetiver cropping in tropical regions. Agron. Sustain. Develop. 34 (3), 603–607.

Singh, M., Rao, R.S.G., Ramesh, S., 1997. Irrigation and nitrogen requirement of lemongrass [*Cymbopogon flexuosus*(Steud) Wats] on a red sandy loam soil under semiarid tropical conditions. J. Essential Oil Res. 9, 569–574.

Soto-Ortiz, R., Vega-Marrero, G., Tamajon-Navarro, A.L., 2002. Technical instruction of *Cymbopogon citratus* (DC) Stapf (lemongrass). Cuban J. Med. Plant 7 (2), 89–95.

Thomas, T.P., Sankar, S., Unni, K.K., 2012. A field study to evaluate the efficacy of lemon grass in controlling runoff and soil erosion. Final report of the project KFRI 543/08. ISSN 0970-8103.

Weiss, E.A., 1997. Lemongrass. In: Weiss, E.A. (Ed.), Essential Oil Crops. Cambridge University Press, Cambridge, pp. 86–103.

Zheljazkov, V.D., Cantrell, C.L., Astatkie, T., Cannon, J.B., 2011. Lemongrass productivity, oil content, and composition as a function of nitrogen, sulfur, and harvest time. Agron. J. 103 (3), 805–812.

Saccharum spp.—potential role in ecorestoration and biomass production[#]

Vimal Chandra Pandey[a,*], Ashutosh Kumar Singh[b]
[a]Department of Environmental Science, Babasaheb Bhimrao Ambedkar University, Lucknow, Uttar Pradesh, India; [b]CAS Key Laboratory of Tropical Forest Ecology, Xishuangbanna Tropical Botanical Garden, Chinese Academy of Sciences, Menglun, Yunan, China
*Corresponding author

1 Introduction

Growing population and infrastructural development resulted increase in wide range of waste dumps such as coal mine spoils, fly ash landfills, uranium tailings, red mud deposits, sponge iron waste deposits, chromite–asbestos mine waste dumps is a serious concern worldwide (Pandey and Singh, 2012; Kumar and Maiti, 2015; Verma and Verma, 2017). These anthropogenic waste dumps led land degradation affect biodiversity of nearby area; deteriorate the environmental quality; and causes severe disturbance in the cultural landscape (Bryan et al., 2012; Smith et al., 2016; Bhuiyan et al., 2018). These waste dumps are hazardous for humans and other being owing to the presence of numerous potentially toxic metals or metalloids such as Cd, Cr, As, Pb, Hg, etc. above their environmental threshold limits (Pandey et al., 2011; Nayak et al., 2015; Lata et al., 2015). These toxicant contaminates the surrounding environment (create air, soil, and water pollution) and causes several life threatening diseases including cancer, cardiac disease, genetic disorders, mesothelioma, pneumoconiosis, and serious problems in human being living near the dumpsites (Pandey et al., 2009, 2011). Thus, there is a strong public pressure and need of present and future prospect for the restoration and management of waste dump sites in environmental and economically friendly manner.

In this context, restoration of waste dump through revegetation (phytoremediation) is considered as cost-effective, popularly accepted, eco-friendly, large scale applicable, sustainable, and ecorestoration approach, which could remove contaminants from waste dumps and simultaneously improves edaphic properties (e.g., C, N, and P nutrients) (Pandey et al. 2015a; Wang et al., 2017; Burges et al., 2018; Kumar et al., 2018). Though, the efficiency of phytoremediation or ecorestoration approach is limited by several interacting factors such as plant functional traits (e.g., tree, shrub, and herb), ability of plants to uptake contaminants, ability to survive on waste land, nutrients availability on waste land, and low bioavailability of the contaminants (Sheoran

[#] V.C. Pandey and A.K. Singh contributed equally.

Phytoremediation Potential of Perennial Grasses. http://dx.doi.org/10.1016/B978-0-12-817732-7.00010-9

et al., 2016; Sarwar et al., 2017). Plenty of multiutility plants have been subjected to characterization of their potential for sustainable phytoremediation on specific dumps. Researchers have identified numerous plants which have the ability to survive on numerous waste dumps and remediate contaminants (phytostabilize or phytoaccumulate) and restore waste land comparable to adjacent undisturbed land (Pandey et al. 2015b; Sheoran et al., 2016; Kumar et al., 2018). Moreover, some plants can naturally grow on such waste dumps with inherent ability of remediation of contaminants and restoration of waste land (Rau et al., 2009; Pandey et al., 2012; Fernández et al., 2017). Such naturally growing plants have an edge over those plants which have been screened under controlled laboratory conditions, owing to their proven self-sustaining ability, natural adoptability to present environment and their native distribution. Such plants require very less care if they are revegetated over waste dumps.

In addition, few plants group proved a great potential in view of their superlative adaptation at the abandoned sites and their multiple uses. *Saccharum* spp. (a genus of perennial grasses) is a group of native colonizers which grow well on a variety of abandoned sites including coal mine spoils, coal fly ash dump, rock mined land, red mud dumps, iron waste dumps, and copper and uranium tailings (Singh and Soni, 2010; Sharma and Mishra, 2011; Hansda et al., 2017; Mishra et al., 2017; Kumar et al., 2018). *Saccharum* spp. have good root network which allow them to utilize resources more efficiently and survive under poor resource condition. By growing on waste land, these species perform phytoremediation of contaminants and improve soil characteristics of the waste land sites (Chandra et al., 2018). The biomass of *Saccharum* spp. are very promising in improving rural economy owing to their application as a stable timber material (roofing houses, cattle sheds, baskets, etc.), paper and pulp production, and as a medicine to cure multiple disease such as asthma, jaundice, leprosy, eye disease, piles, sexual weaknesses (Kumar et al., 2010; Slathia and Paul, 2012; Dalal et al., 2013; Jahan et al., 2016). These species may be also used as fodder during juvenile stage (Pandey et al., 2012). Moreover, *Saccharum* spp. are also useful in bioethanol production, attesting their multiutility potential for improving rural economy particularly in countries (i.e., Indian subcontinent region) where agriculture is a dominant industry.

Among *Saccharum* spp., *Saccharum spontaneum* and *Saccharum munja* are the wild species and the two most common species which are predominantly distributed across the waste dumps. Thus the sections that follow, we discuss about the *Saccharum* spp. with primary focus on *S. spontaneum* and *S. munja*. We discuss the ecology of species, geographic distribution, morphological description, multiple uses, and the suitability of the species for ecorestoration. We will also highlight the suitability of these species for the restoration and stabilization of fly ash (FA) dumps.

2 Ecology

Saccharum spp. are robustly growing tropical perennial weed at the wasteland sites of Indian subcontinent. These species belong to C4 grass (Poaceae family) with deep root network, which can grow up to 4 m height (Pandey et al., 2015a). Five species have been reported from *Saccharum* (*S. officinarum*, *S. spontaneum*, S. barberi,

S. sinensi, and *S. munja*), of which *S. spontaneum* L. and *S. munja* are considered as wild species. Moreover, the *S. spontaneum* and *S. munja* are the two most common species in terms of their potential to survive under adverse climatic condition, and their natural distribution on diverse wasteland sites (Dowarah et al., 2009; Pandey et al., 2015b; Chandra et al., 2018). The ecology of these species are as follows:

S. munja, an abundantly growing perennial wild grass, believed to be native of India. In northern plains of India, it is generally pronounced as sarapat, munj, sara, sarkanda, and kana (local names) (Vasudevan et al., 1984). It has extensive and deep root network which allows it to grow on rocky habitats or less weathered soil. Moreover, its dense root network facilitates soil formation through binding of soil particles and forms tall thick clumps with high biomass tufts. It can grow to a wide range of moisture condition, from arid/semiarid region to moist region. The height of *S. munja* varied between 0.5 and 2 m, with stunned growth observed under arid condition (north-west desert areas of India) and full-flourished growth can be seen in moist areas of Indo-Gangetic plains. The flowering season of this perennial grass occurs during October to December in Indian subcontinent (Vasudevan et al., 1984; Pandey et al., 2012).

S. spontaneum, an abundantly growing perennial wild grass, believed to be native of India. In northern plains of India, it is generally pronounced as "Kans" and "Kansa" (Hindi name) and Tharu tribes of Terai region (India and Nepal) pronounce as "Jhaksi" (local name) (Dangol, 2005). It commonly grows along the banks of water bodies (river, lakes, and ponds), roadsides, railway tracks, alluvial plains, damp depressions, and swamps (Holm et al., 1997). It has extensive and deep root network which allows it to grow at the base of the Himalayan range in India, Nepal, China, and Bhutan. *S. spontaneum* occurs at an altitude ranging from 0 to 1800 msl (meter above sea level) (Holm et al., 1997). It has some allelopathic effects on crop through release of certain aromatic compound from root (Amritphale and Mall, 1978). The grass lands of *S. spontaneous* are an important habitat for the Indian chital (*Axis axis*) and rhinoceros (*Rhinoceros unicornis* L.) in the Himalayan Terai and Duar region. The flowering season of *S. spontaneous* occurs during October to December (post rainy season) in the Indian subcontinent. *Saccharum* species like *S. munja* and *S. spontaneum* grows naturally as a colonizer in the fly ash dumping sites (Fig. 10.1A,B) (Pandey et al., 2012; 2015a).

3 Morphological description

The morphological descriptions of some (wild) *Saccharum* spp. are as follows:

S. munja, is a tall (up to 10–12 feet) densely tufted terrestrial grass, with a creeping root stock. Stem is thick, ascending, caespitose, tufted, nodes hairy, internodes solid or spongy, stem with tall inflorescence. Leaves mostly cauline, conspicuously 2-ranked, sheathing at base, leaf sheath, and blade differentiated. Leaf blades linear, 2–10 mm wide, mostly flat, scabrous, and roughened. The inflorescence of *S. munja* is solitary, with 1 spike, fascicle, glomerule, head, or cluster per stem or culm, inflorescence has numerous branches (>10), inflorescence branches are deciduous and falling intact. The flowers are bisexual, spikelets pedicellate, sessile or subsessile, dorsally

Figure 10.1 *Saccharum* species naturally colonizing in the fly ash dumping sites —
(A) *Saccharum munja* and (B) *Saccharum spontaneum*. Photo credits: V.C. Pandey.

compressed or terete, spikelet less than 3 mm wide, and with 1 fertile floret, spikelets
paired at rachis nodes, all spikelets are similar and fertile, spikelets are in paired units,
1 sessile, 1 pedicellate, spikelets bisexual and spikelets conspicuously hairy. Glumes
are present in *S. munja*, glumes 2 also present, glumes equal or subequal, glumes

either equal or longer than adjacent lemma, glume surface hairy, lemmas thin, lemma 3 nerved, lemma apex acute or acuminate, lemma straight, callus or base of lemma clearly hairy, stamens 3, styles 2-fid, deeply 2-branched, stigmas 2, fruit is caryopsis.

S. spontaneum is a tall (up to 20 feet) densely tufted terrestrial grass, with a creeping, tufted and rhizomatous rootstock. Stem is woody, 10–15 feet in length, erect habit, fistula below (=Culm), polished, robust; internodes solid; node 5–10, waxy, tall inflorescence. Leaves blades linear, green with white midrib, long hairs at base, margins finely serrated and prickly, involute, ligule 2–8 mm long, ovate, brown, membranous, glabrous, apex accuminate, base simple or tapering to the white midrib, scabrid to serrate along margins; sheath longer than internode. Inflorescence plumose panicles; panicle 15–60 cm long, peduncle hirsute above; white, axis silky pilose, open, ovate, dense; racemes 3–17 cm long; rachis internodes filiform; spikelets homomorphic, lanceolate, reddish-brown, paired (one sessile and the other pedicelled), pilose with long silky hairs. Fertile spikelets are sessile, lanceolate, dorsally compressed, 2 in the cluster, subequal; pedicels filiform, ciliate. Glumes are similar, membranous above, chartaceous below, ovate-lanceolate to elliptic, subcoriaceous to coriaceous, acuminate at apex, ciliate along margins, 2-keeled, ovate-lanceolate, coriaceous, acute at apex, ciliate along margins, mucronate, without keels. Florets basal sterile and upper fertile; sterile florets barren, without significant palea, lanceolate, hyaline, 0-veined, without midvein, without lateral veins, acute at apex; fertile floret bisexual, linear, hyaline; linear-lanceolate, hyaline; Palea absent or minute. Flower lodicules, cuneate, ciliate. Stamens are 3; anthers yellow or reddish. Ovary is oblong; stigma white. Flowering and fruiting occurs from June to September. Flowers begins just onset of rainy season and takes 1–2 months to produce seeds.

4 Geographic distribution

There are five *Saccharum* spp. have been reported till date, of which four occurs in India. Mukherjee (1957) reported that these species are primarily centered in India, Malaysia, and China. Panje (1970) further documented that these *Saccharum* spp. are distributed throughout the tropical countries of Asia, Africa, America, and Australia. These previous studies are also emphasized that *Saccharum* complex originated in the Indo-Burma-Chinese region and spread to adjoining areas. For instance, the biotypes of *S. spontaneum* have been recorded from Turkestan, Uzbekistan, Afghanistan, and Iran extending to the northern and eastern regions of Africa (Anon, 1972).

5 Propagation

Saccharum spp. propagated mainly vegetative means through creeping rhizomes and stem cuttings. However, propagation through seeds is also common in few species such as *S. munja* and *S. spontaneum*. These species commonly produces huge amount of seeds (up to 12,800 seeds/plant in *S. spontaneum*) and mainly disperse through wind (Pancho, 1964; Datta and Banerjee, 1973). Seeds are light weighted and comprised of callus hairs or spikelets, which facilitate its movement through air. Sometimes seeds become entangled in a woolly mass, which leads to its longer dispersal through air (Sharma and Tiagi, 1979). The germination of seeds takes place below the ground (hypogeal type). There is no dormancy period in seeds of *Saccharum* spp., they start germinating after 24 h. The best germination occurs at temperature between 15°C–25°C, and early rainy season (July–August) is considered as the best period of germination in Indian subcontinent. The seeds mortality is also very high among these species due to the adverse conditions of surrounding habitat, and germination percentage is merely 30%. The vegetative regeneration takes place through rhizomes and stem fragments (Artschwager, 1942), where stem displays good regeneration potential even after drying. The *Saccharum* spp. (*S. spontaneum* and *S. munja*) cultivated around the barren lands as hedges and along the water canals to prevent soil erosion (Bhandari, 1990; Sastry and Kavathekar, 1990).

6 Multiple uses

Saccharum spp. provides multiutility from a long time in many countries. Their traditional and modern application can be grouped into few categories (Fig. 10.2).

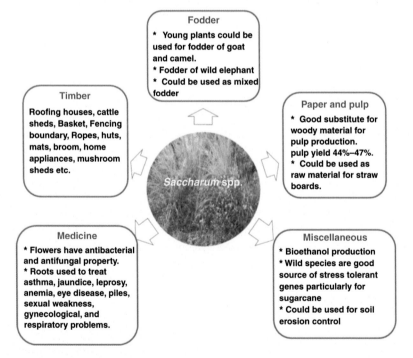

Figure 10.2 Multiple applications of biomass of *Saccharum* spp. as a source to improve rural livelihood and economy.

Timber—Leaves of *Saccharum* spp. (particularly *S. spontaneum* and *S. munja*) are an excellent thatching material and used by local people for roofing of houses, cattle sheds, dry fodder, storing baskets, fencing of boundary, and thatching for off season vegetation cultivation (Slathia and Paul, 2012). The thatched leaves of *Saccharum* spp. are also very stable and thereby used for making ropes, baskets, mats, broom, huts, etc. to support their livelihood (Wapakala, 1966; Pandey et al., 2012). The timber material (leaves and culms) produced by *Saccharum* spp. is frequently used for seasonal roofing of the mushroom sheds which acts as insulator in mushroom cultivation (Slathia and Paul, 2012). The timbers of *Saccharum* spp. are also a demanding material in the making of decorative huts at the picnic spots. Home appliances such as tables, chairs, and stools prepared from culms of *Saccharum* spp. are believed that resistant against white ants (Vasudevan et al., 1984). These materials are therefore considered as a good substitute for appliances made of plastic materials.

Fodder—*Saccharum* spp. are naturally present in arid and semiarid areas of India. Therefore, attention was first directed to its possible utilization as a cattle feed. However, mature or dry leaves of *Saccharum* spp. as such are rarely consumed by cattle (Vasudevan et al., 1984). In contrast, Thakur (1984) reported that during juvenile growth stage they might be used as fodder for goats and camels. Kehar (1944) experimentally proved that the leaves of *S. munja* might be used as a mixed fodder ingredient along with other leafy fodder. Slathia and Paul (2012) and Singh et al. (2017) also advocated the possibility of application of *Saccharum* spp. as a prominent source of fodder. Joshi and Singh (2008) studied the feeding behavior of Asian elephant's (*Elephas maximus*) in Rajaji National Park area (Uttarakhand, India) between years 1999 and 2006. They found that *S. spontaneum* and *S. munja* are among few choice species of shrub which have been preferred by elephant particularly when they move across the slope from top to low-lying drier areas.

Paper and pulp—*Saccharum* spp. is a good choice for nonwood raw materials for pulp production. For instance, Jahan et al. (2006, 2016) reported that pulp yield from *S. spontaneum* is very high and the kappa number is very low, which is an indication of good quality pulp. Similarly, Bhat and Jaspal (1959) reported that a good quality paper (pulp yield of 43.7%–46.7%) appropriate for writing and printing can be obtained from *S. munja*. Vasudevan et al. (1984) documented that *S. munja* can be used as raw material for production of straw boards.

Medicine—Medical applications of *Saccharum* spp. have been well documented. Ripa et al. (2009) performed an experiment to explore the in vitro antimicrobial activity of chloroform extracted from flowers of *S. spontaneum*. The chloroform extract was shown significant potential of antibacterial and antifungal activity against multiple strains. The roots of *Saccharum* spp. (both *S. spontaneum* and *S. munja*) are important constituent of "Trinpanchmool" (an Ayurvedic drug prepared using roots of five different grasses), used to treat various ailments like asthma, jaundice, galactogogue, diuretic, gout, leprosy, anemia, eye disease, and also used as aphrodisiac (Jayalakshmi et al., 2011). Moreover, the roots of *S. spontaneum* and *S. munja* alone can be used to treat dyspepsia, burning sensation, piles, sexual weakness, gynecological, and respiratory problems (Kumar et al., 2010; Dalal et al., 2013; Saha et al., 2014). Some tribal people of India also use fresh juice of *S. spontaneum* stem to treat various types of mental illness. Besides India, medical applications of *Saccharum* spp. are common in other countries like Philippines and Indonesia (Pancho and Obien, 1983).

Miscellaneous application—Chandel et al. (2011) reported that *Saccharum* spp. could be a good source of bioethanol. They had successfully converted cellulose obtained from *S. spontaneum* to ethanol using yeast (*Pichia stipitis* NCIM3498) as a biocatalyst. The wild *Saccharum* spp. (*S. spontaneum* and *S. munja*) is considered as a valuable genetic resource because it contains various stress tolerant genes particularly for sugarcane (*Saccharum officinarum* L.) (Anon 1972). From ecological point of view, these species are very effective to cope with soil erosion problem owing to their extensive root network (Pandey et al., 2015b).

7   Role of *Saccharum* spp. in ecological restoration of waste land

Saccharum spp. are having inherent ability to colonize wastelands and marginal lands (Pandey et al., 2012, 2015a; Awasthi et al., 2017; Mandal et al., 2017). These species can survive under adverse climatic conditions, produce high biomass, and have good root system, which provides them ecological compatibility to colonize wasteland and restore disturbed soil. For instance, Mandal et al. (2017) performed a meta-analysis including the use of grasses to reduce runoff and soil erosion. They suggested that *S. munja* is a promising species to effectively control runoff and soil erosion in highlands of Himalayan region, and such grasses can reverse the process of land degradation by improving fertility of soil. Pandey (2015) documented that *S. spontaneum* and *S. munja* can naturally colonize coal fly ash dumped wasteland and perform ecological restoration without any external input. Pandey et al. (2015a) highlighted that *Saccharum* spp. can grow on various waste-dumps where other plants are unable to grow. Moreover, *Saccharum* spp. may have a competitive edge over other species because these species utilize soil resources (e.g., nutrients and water) more efficiently than any other species in a degraded land (Craven et al., 2009; Park et al., 2010). For instance, Das et al. (2013) investigated vegetation composition across mine-spoiled dump of West Bengal, India. They found that *S. spontaneum* dominate among the spoil-vegetation with a very high level of prevalence (80%). Similarly, *S. munja* has been identified for eco-restoration of morrum mine abundant sites in north India (Sharma and Mishra, 2011). Dowarah et al. (2009) identified *S. spontaneum* for eco-restoration of a high-sulfur coal mine overburden dumping site in northeast India. *S. spontaneum* has been identified in eco-restoration of degraded land following volcanic eruption (Mount Pinatubo) in Philippines (Marler and del Moral, 2011). Mishra et al. (2017) surveyed a red mud deposited sites in central India, they found that *Saccharum bengalense* Retz (considered as synonym of *S. munja*) was the most dominant vegetation across the red mud deposits. They also showed that *S. bengalense* significantly improved soil nutrients and phytostabilized toxic ions under the rhizosphere. *S. spontaneum* has been identified in eco-restoration of sulfur-rich coal mine spoiled sites in northeast India (Dowarah et al. 2009). *Saccharum* spp. can be also used for restoration of uranium tailings (waste produced during uranium mining). For instance, Singh and Soni (2010) demonstrated that *S. spontaneum* is capable of eco-restoration of uranium tailing in eastern India. They identified that *S. spontaneum* has ability of uranium mining tailing pond restoration on the basis of its root penetration and radioactive ion binding capacity. *S. spontaneum* is a species of choice for the eco-restoration rock phosphate mined abundant land, owing to its ability to stabilize or hyperaccumulate toxic metals ions (e.g., As) (Chandra et al., 2018). It can be also used for eco-restoration of Cu tailings, produced during mining, smelting, and processing of copper ores (Hansda et al., 2017). Similarly, *S. spontaneum* was reported for eco-restoration of sponge iron solid waste dumps (Kullu and Behera, 2011) and for revegetation and biostabilization of coal mine overburden dump slope (Chaulya et al., 1999).

Saccharum spp. could be a choice of species to improve soil carbon sequestration. For instance, Singh et al. (2016) measured soil organic carbon content under the different long-term land use systems in Indo-Gangetic plains. They found that soil organic carbon content remained unchanged between the tree dominant land use and *S. munja*. Similarly, *S. spontaneum* has been suggested to be used for revegetation of abundant land to improve soil carbon stock (Arif et al., 2017). However, *Saccharum* spp. is also considered as invasive species in some countries (e.g., *S. spontaneum* in Panama) (Wishnie et al., 2007), therefore, prevention of invasion is an important issue regarding biodiversity loss. Thus, *Saccharum* spp. for revegetation or eco-restoration should be used with caution. These species if become invasive then their control and management can be achieved using invasive species management techniques (Doren et al. 2009; Wald et al., 2018). Fig. 10.3 represents ecorestoration potential of *Saccharum* spp. to a wide-ranging waste lands.

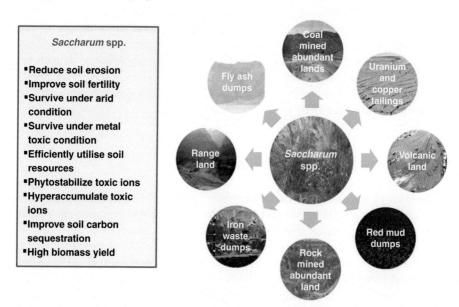

Figure 10.3 Ecorestoration potential of *Saccharum* spp. to a range of waste lands. Square box describes ecorestoration benefits of *Saccharum* spp. plantations on waste lands.

8   Role of *Saccharum* spp. in ecological restoration of fly ash dumps

The huge amount of FA produced as by-product during combustion of coal in a thermal power plant, while its disposal as FA dumps is considered as serious environmental concern across the world (Pandey et al., 2009; Pandey and Singh, 2012; Madhumita et al., 2018). Characteristically FA is a nutrient rich material, with the both micro- and macro-nutrients present in abundant quantity (Pandey and Singh, 2010; Ram and Masto 2014; Ribeiro et al., 2018). FA can be used alone (Jala and Goyal, 2006) or in combination of other inorganic and organic amendments to improve crop productivity (Ram and Masto, 2014; Lim et al., 2016; Hong et al., 2018). For instance, 10% FA and ~90% sand (v/v) has been recommended as a suitable rooting media for the propagation of *Leucaena leucocephala* (Pandey and Kumar, 2013). However, FA is a source of several toxic substances such as toxic metals (e.g., arsenic), radioactive nuclei, organic pollutant, and its pH were also not suitable to be used as a plant growth medium alone (Pandey et al., 2011; Yao et al., 2015). Therefore, remediation of FA dumps is an urgent need across the world. In this context, remediation through plants (phytoremediation) is emerged as a holistic and sustainable approach, which has been also found effective for the remediation of FA dumps (Ram et al., 2008; Pandey et al., 2009; Madhumita et al., 2018). *Saccharum* spp. could be used for remediation of FA dumps (Rau et al., 2009; Pandey et al., 2012; Pandey, 2015). Despite adverse condition for plant growth on FA dump, these species can naturally colonize FA dumps owing to their extensive root network and ability to tolerate under stress condition (Pandey et al., 2012; Verma et al., 2014; Malhotra et al., 2017). *Saccharum* spp. colonization enhances bacterial indole acetic acid (IAA) production through improving NO_3-N and PO_4-P enrichment in rhizosphere (Malhotra et al., 2017). Increased IAA production in turn improves plant perpetuation on FA dumps. Because *Saccharum* spp. rarely used to feed cattle, therefore, if any toxic elements accumulated in plant parts, it would be very less probability of their biomagnification through food chain. This dwindling consumption by grazing animals also helps self-perpetuation of *Saccharum* spp. over FA dumps (Pandey et al., 2012). *Saccharum* spp. has a vital and inherent potential to restore fly ash dumpsites (Fig. 10.3).

9   Biomass and bioenergy production

In the draft of the European OPTIMA project (G.A. 289642), perennial grasses at their wild or near-wild state and well-adapted to Mediterranean semi-arid environment are collected from riparian areas and tested in field trial conditions in order to assess their appropriateness as energy crops. Out of numerous perennial grasses explored, a grass from *Saccharum* genus looks the promising one (Cosentino et al., 2015). Grasses are well identified as suitable candidates for bioenergy crops in Thailand (Banka et al. 2015) due to lignocellulosic biomass production, having good adaptability to biotic stresses (i.e., grazing, diseases, and other pests) and edaphic stresses (i.e., arid soils, mine spoils, salt-affected soils, fly ash deposits, red mud sites, heavy metal polluted sites, etc.). *Saccharum* genus has approximately 40 species, mostly native to South-Eastern Asia (Clayton and Renvoize, 1986). Among *Saccharum* genus, *Saccharum spontaneum* L. is an herbaceous and rhizomatous perennial grass with high polymorphism, robust, and resistant to the physiopaties (Pignatti, 1982). The biomass of *S. spontaneum* can be used for butanol production because it is rich in carbohydrate and fiber content. As a source for bioenergy, it has ability for dry matter production. It is also called the waste land weeds. Maximum aboveground biomass dry matter (37.86 Mg/ha) can be achieved by using 1150 mm of water via the crop (Cosentino et al., 2015). In a 1-year-old plantation, establishment percentage and total (aboveground–belowground) biomass production of *S. spontaneum* on fly ash dumps have been reported as 87% and 950 g per plant fresh biomass (Pandey and Singh, 2014; Cosentino et al. 2015), respectively. This indicates that high biomass production ability of *S. spontaneum* under hostile conditions at fly ash dumpsite.

10 Conclusion

Saccharum spp. are comprises of wild native perennial grasses which can naturally colonize extreme habitats of a range of waste land. These species can utilize soil resources more efficiently and sustained to a large range of moisture regimes, from arid to humid environment. These species produces high biomass and have dense root network which allow them to hold soil particles from runoff losses. *Saccharum* spp. can phytostabilize and hyperaccumulate toxic metal ions (e.g., As). However, these toxic metal ions less likely to enter in food chain through biomagnification, because mature *Saccharum* spp. are rarely consumed by grazing animals. Along with ecological functions, *Saccharum* spp. also improves rural economy. Their timber materials (leaves and culm) are used for roofing of houses, cattle and mushroom sheds, fencing of boundary, making of baskets, and multiple home appliances. *Saccharum* spp. are useful in pulp and paper industry and can be used as a substitute of woody trees. The flowers and roots of *Saccharum* spp. contain medicinal properties and are used to treat asthma, jaundice, leprosy, anemia, eye disease, sexual weaknesses, gynecological, and respiratory problems. *Saccharum* spp. during juvenile stage may be also used to feed goat and other cattle. The biomass of *Saccharum* spp. can be used to produce bioethanol. Despite wide scale utility of *Saccharum* spp., the plantation should be done with caution owing to their rapid propagation and invasion may threaten native taxa.

Acknowledgments

The authors acknowledge Director, CSIR-NBRI for providing indispensable conveniences and his incessant encouragement. AKS is thankful to University Grants Commission, Govt of India for PhD Fellowship (UGCJRF/SRF). Financial assistance given to Dr. V.C. Pandey under the Scientist's Pool Scheme (Pool No. 13 (8931-A)/2017) by the Council of Scientific and Industrial Research, Government of India is gratefully acknowledged. The authors declare no conflicts of interest.

References

Amritphale, D., Mall, L.P., 1978. Allelopathic influence of *Saccharum spontaneum* L. on the growth of three varieties of wheat [India]. Sci. Cult. 44, 28–30.

Anon, 1972. The Wealth of India. Raw Materials. New Delhi, India: CSIR Publications & Information Directorate, Vol. IX.

Arif, M., Shahzad, M.K., Elzaki, E.E.A., Hussain, A., Zhang, B., Yukun, C., 2017. Biomass and carbon stocks estimation in Chichawatni irrigated plantation in Pakistan. Int. J. Agric. Biol. 19, 1339–1349.

Artschwager, E., 1942. Comparative Analysis Of The Vegetative Characteristics Of Some Variants Of *Saccharum Spontaneum*. United States Department Of Agriculture, Washington, DC.

Awasthi, A., Singh, K., Singh, R.P., 2017. A concept of diverse perennial cropping systems for integrated bioenergy production and ecological restoration of marginal lands in India. Ecol. Eng. 105, 58–65.

Banka, A., Komolwanich, T., Wongkasemjit, S., 2015. Potential Thai grasses for bioethanol production. Cellulose 22, 9–29.

Bhandari, M.M., 1990. Flora of the Indian desert. Revised edition. Jodhpur MPS Repros 435p.-illus., col. illus., map. ISBN 8185304130 En Icones, Keys. Geog 6.

Bhat, R.V., Jaspal, N.S., 1959. Chemical pulp & writing & printing papers from Kana grass (S. munja Roxb.). Indian For. 85, 241–248.

Bhuiyan, M.A., Jabeen, M., Zaman, K., Khan, A., Ahmad, J., Hishan, S.S., 2018. The impact of climate change and energy resources on biodiversity loss: evidence from a panel of selected Asian countries. Renew. Energy 117, 324–340.

Bryan, Jr., A.L., Hopkins, W.A., Parikh, J.H., Jackson, B.P., Unrine, J.M., 2012. Coal fly ash basins as an attractive nuisance to birds: parental provisioning exposes nestlings to harmful trace elements. Environ. Pollut. 161, 170–177.

Burges, A., Alkorta, I., Epelde, L., Garbisu, C., 2018. From phytoremediation of soil contaminants to phytomanagement of ecosystem services in metal contaminated sites. Int. J. Phytoremediation 20, 384–397.

Chandel, A.K., Singh, O.V., Rao, L.V., Chandrasekhar, G., Narasu, M.L., 2011. Bioconversion of novel substrate *Saccharum spontaneum*, a weedy material, into ethanol by *Pichia stipitis* NCIM3498. Bioresour. Technol. 102, 1709–1714.

Chandra, R., Kumar, V., Tripathi, S., Sharma, P., 2018. Heavy metal phytoextraction potential of native weeds and grasses from endocrine-disrupting chemicals rich complex distillery sludge and their histological observations during in-situ phytoremediation. Ecol. Eng. 111, 143–156.

Chaulya, S.K., Singh, R.S., Chakraborty, M.K., Dhar, B.B., 1999. Numerical modelling of biostabilisation for a coal mine overburden dump slope. Ecol. Modell. 114, 275–286.

Clayton, W.D., Renvoize, S.A., 1986. Genera Graminum, grasses of the world. Kew Bull Additional Series XIII.

Craven, D., Hall, J., Verjans, J., 2009. Impacts of herbicide application and mechanical cleanings on growth and mortality of two timber species in *Saccharum spontaneum* grasslands of the Panama Canal Watershed. Restor. Ecol. 17, 751–761.

Cosentino, S.L., Copani, V., Testa, G., Scordia, D., 2015. *Saccharum spontaneum* L. ssp. *Aegyptiacum* (Willd.) Hack. A potential perennial grass for biomass production in marginal land in semi-arid Mediterranean environment. Ind. Crops Prod. 75, 93–102, http://dx.doi.org/10.1016/j.indcrop.2015.04.043.

Dalal, P.K., Tripathi, A., Gupta, S.K., 2013. Vajikarana: treatment of sexual dysfunctions based on Indian concepts. Indian J. Psychiatry 55, S273.

Dangol, D.R., 2005. Species composition, distribution, life forms and folk nomenclature of forest and common land plants of Western Chitwan. Nepal. J. Inst. Agric. Anim. Sci. 26, 93.

Das, M., Dey, S., Mukherjee, A., 2013. Floral succession in the open cast mining sites of Ramnagore Colliery, Burdwan District, West Bengal. Indian J. Sci. Res. 4, 125.

Datta, S., Banerjee, A., 1973. Weight and number of weed seeds. Proceedings of the 4th Asian-Pacific Weed Science Society Conference. Asian Pacific Weed Science Society 1, 87–91.

Doren, R.F., Trexler, J.C., Gottlieb, A.D., Harwell, M.C., 2009. Ecological indicators for system-wide assessment of the greater everglades ecosystem restoration program. Ecol. Indic. 9, S2–S16.

Dowarah, J., Boruah, H.P.D., Gogoi, J., Pathak, N., Saikia, N., Handique, A.K., 2009. Eco-restoration of a high-sulphur coal mine overburden dumping site in northeast India: a case study. J. Earth Syst. Sci. 118, 597–608.

Fernández, S., Poschenrieder, C., Marcenò, C., Gallego, J.R., Jiménez-Gámez, D., Bueno, A., Afif, E., 2017. Phytoremediation capability of native plant species living on Pb-Zn and Hg-As mining wastes in the Cantabrian range, north of Spain. J. Geochem. Explor. 174, 10–20.

Hansda, A., Kumar, V., Anshumali, 2017. Cu-resistant Kocuria sp. CRB15: a potential PGPR isolated from the dry tailing of Rakha copper mine. 3 Biotech, 7(2), 132.

Holm, L.G., Holm, L., Holm, E., Pancho, J.V., Herberger, J.P., 1997. World Weeds: Natural Histories And Distribution. John Wiley & Sons.

Hong, C., Su, Y., Lu, S., 2018. Phosphorus availability changes in acidic soils amended with biochar, fly ash, and lime determined by diffusive gradients in thin films (DGT) technique. Environ. Sci. Pollut. Res. 25, 30547–30556.

Jahan, M.S., Akter, T., Nayeem, J., Samaddar, P.R., Moniruzzaman, M., 2016. Potassium hydroxide pulping of *Saccharum spontaneum* (kash). J-FOR-Journal Sci. Technol. For. Prod. Process. 6, 46–53.

Jahan, M.S., Chowdhury, D.A.N., Islam, M.K., 2006. NS-AQ pulping of Kash (*Saccharum spontaneum*). IPPTA 18, 69.

Jala, S., Goyal, D., 2006. Fly ash as a soil ameliorant for improving crop production—a review. Bioresour. Technol. 97, 1136–1147.

Jayalakshmi, S., Ghosh, A.K., Shrivastava, B., Singla, R.K., 2011. Preliminary investigation of antipyretic activity of Trinpanchmool extracts. Int. J. Phytomedicine 3, 147–150.

Joshi, R., Singh, R., 2008. Feeding behaviour of wild Asian elephants (*Elephas maximus*) in the Rajaji National Park. J. Am. Sci. 4, 34–48.

Kehar, N.D., 1944. Investigations on famine rations for livestock. 1. Munj and molasses as famine rations. Indian J. Vet. Sci. 14, 40–48.

Kullu, B., Behera, N., 2011. Vegetational succession on different age series sponge iron solid waste dumps with respect to top soil application. Res. J. Environ. Earth Sci. 3, 38–45.

Kumar, C.A.S., Varadharajan, R., Muthumani, P., Meera, R., Devi, P., Kameswari, B., 2010. Psychopharmacological studies on the stem of *Saccharum spontaneum*. Int. J. PharmTech Res. 2 (1), 319–321.

Kumar, A., Maiti, S.K., 2015. Assessment of potentially toxic heavy metal contamination in agricultural fields, sediment, and water from an abandoned chromite-asbestos mine waste of Roro hill, Chaibasa, India. Environ. Earth Sci. 74, 2617–2633.

Kumar, S., Singh, A.K., Ghosh, P., 2018. Distribution of soil organic carbon and glomalin related soil protein in reclaimed coal mine-land chronosequence under tropical condition. Sc. Total Environ. 625, 1341–1350. doi:10.1016/j.scitotenv.2018.01.061

Lata, S., Singh, P.K., Samadder, S.R., 2015. Regeneration of adsorbents and recovery of heavy metals: a review. Int. J. Environ. Sci. Technol. 12, 1461–1478.

Lim, S.-S., Lee, D.-S., Kwak, J.-H., Park, H.-J., Kim, H.-Y., Choi, W.-J., 2016. Fly ash and zeolite amendments increase soil nutrient retention but decrease paddy rice growth in a low fertility soil. J. Soils Sediments 16, 756–766.

Madhumita, R.O.Y., Roychowdhury, R., Mukherjee, P., 2018. Remediation of fly ash dumpsites through bioenergy crop plantation and generation: a review. Pedosphere 28, 561–580.

Malhotra, S., Mishra, V., Karmakar, S., Sharma, R.S., 2017. Environmental predictors of indole acetic acid producing rhizobacteria at fly ash dumps: nature-based solution for sustainable restoration. Front. Environ. Sci. 5, 59.

Mandal, D., Srivastava, P., Giri, N., Kaushal, R., Cerda, A., Alam, N.M., 2017. Reversing land degradation through grasses: a systematic meta-analysis in the Indian tropics. Solid Earth 8, 217–233.

Marler, T.E., del Moral, R., 2011. Primary succession along an elevation gradient 15 years after the eruption of Mount Pinatubo, Luzon, Philippines. Pacific Sci. 65, 157–173.

Mishra, T., Pandey, V.C., Singh, P., Singh, N.B., Singh, N., 2017. Assessment of phytoremediation potential of native grass species growing on red mud deposits. J. Geochem. Explor. 182, 206–209.

Mukherjee, S.K., 1957. Origin and distribution of *Saccharum*. Bot. Gaz. 119, 55–61.

Nayak, A.K., Raja, R., Rao, K.S., Shukla, A.K., Mohanty, S., Shahid, M., Tripathi, R., Panda, B.B., Bhattacharyya, P., Kumar, A., 2015. Effect of fly ash application on soil microbial response and heavy metal accumulation in soil and rice plant. Ecotoxicol. Environ. Saf. 114, 257–262.

Pancho, J.V., 1964. Seed sizes and production capacities in common weed species of the rice fields of the Philippines. Philipp. Agric 48, 307–316.

Pancho, J., Obien, S., 1983. Manual of Weeds of Tobacco Farms in the Philippines. Quezon City, Philippines: New Mercury Printing Press. Panje R, 1970. The evolution of a weed. PANS 16, 590–595.

Pandey, V.C., 2015. Assisted phytoremediation of fly ash dumps through naturally colonized plants. Ecol. Eng. 82, 1–5.

Pandey, V.C., Singh, N., 2010. Impact of fly ash incorporation in soil systems. Agric. Ecosyst. Environ. 136, 16–27.

Pandey, V.C., Kumar, A., 2013. *Leucaena leucocephala*: an underutilized plant for pulp and paper production. Genet. Res. Crop Evol. 60, 1165–1171.

Pandey, V.C., Singh, N., 2014. Fast green capping on coal fly ash basins through ecological engineering. Ecol. Eng. 73, 671–675.

Pandey, V.C., Abhilash, P.C., Singh, N., 2009. The Indian perspective of utilizing fly ash in phytoremediation, phytomanagement and biomass production. J. Environ. Manage. 90, 2943–2958.

Pandey, V.C., Bajpai, O., Pandey, D.N., Singh, N., 2015a. *Saccharum spontaneum*: an underutilized tall grass for revegetation and restoration programs. Genet. Resour. Crop Evol. 62, 443–450.

Pandey, V.C., Prakash, P., Bajpai, O., Kumar, A., Singh, N., 2015b. Phytodiversity on fly ash deposits: evaluation of naturally colonized species for sustainable phytorestoration. Environ. Sci. Pollut. Res. 22, 2276–2287.

Pandey, V.C., Singh, B., 2012. Rehabilitation of coal fly ash basins: current need to use ecological engineering. Ecol. Eng. 49, 190–192.

Pandey, V.C., Singh, J.S., Singh, R.P., Singh, N., Yunus, M., 2011. Arsenic hazards in coal fly ash and its fate in Indian scenario. Resour. Conserv. Recycl. 55, 819–835.

Pandey, V.C., Singh, K., Singh, R.P., Singh, B., 2012. Naturally growing *Saccharum munja* on the fly ash lagoons: a potential ecological engineer for the revegetation and stabilization. Ecol. Eng. 40, 95–99.

Panje, R.R., 1970. The evolution of a weed. PANS Pest Artic. News Summ. 16, 590–595.

Park, A., Friesen, P., Serrud, A.A.S., 2010. Comparative water fluxes through leaf litter of tropical plantation trees and the invasive grass *Saccharum spontaneum* in the Republic of Panama. J. Hydrol. 383, 167–178.

Pignatti, S., 1982. Flora d'Italia. Edagricole, Bologna, Italia.

Ram, L.C., Jha, S.K., Tripathi, R.C., Masto, R.E., Selvi, V.A., 2008. Remediation of fly ash landfills through plantation. Remediat. J. J. Environ. Cleanup Costs, Technol. Tech. 18, 71–90.

Ram, L.C., Masto, R.E., 2014. Fly ash for soil amelioration: a review on the influence of ash blending with inorganic and organic amendments. Earth Sci. Rev. 128, 52–74.

Rau, N., Mishra, V., Sharma, M., Das, M.K., Ahaluwalia, K., Sharma, R.S., 2009. Evaluation of functional diversity in rhizobacterial taxa of a wild grass (*Saccharum ravennae*) colonizing abandoned fly ash dumps in Delhi urban ecosystem. Soil Biol. Biochem. 41, 813–821.

Ribeiro, J.P., Tarelho, L., Gomes, A.P., 2018. Incorporation of biomass fly ash and biological sludge in the soil: effects along the soil profile and in the leachate water. J. Soils Sediments 18, 2023–2031.

Ripa, F.A., Haque, M., Imran-Ul-Haque, M., 2009. In vitro antimicrobial, cytotoxic and antioxidant activity of flower extract of *Saccharum spontaneum* Linn. Eur. J. Sci. Res. 30, 478–483.

Saha, M.R., Sarker, D. De, Sen, A., 2014. Ethnoveterinary practices among the tribal community of Malda district of West Bengal, India. Indian J. Traditional Knowledge 13(2), 359–366.

Sarwar, N., Imran, M., Shaheen, M.R., Ishaque, W., Kamran, M.A., Matloob, A., Rehim, A., Hussain, S., 2017. Phytoremediation strategies for soils contaminated with heavy metals: modifications and future perspectives. Chemosphere 171, 710–721.

Sastry, T.C.S., Kavathekar, K.Y., 1990. Plants for reclamation of wastelands. Pbl. CSIR, New Delhi, India 360–362 p.

Sharma, M., Mishra, V., 2011. Functionally diverse rhizobacteria of *Saccharum munja* (a native wild grass) colonizing abandoned morrum mine in Aravalli hills (Delhi). Plant Soil, 341(1–2), 447–459. doi:10.1007/s11104-010-0657-y

Sharma, S., Tiagi, B., 1979. Flora of north-east Rajasthan. New Delhi, Kalyani Publ. xx, 540p.-keys. En Geog 6.

Sheoran, V., Sheoran, A.S., Poonia, P., 2016. Factors affecting phytoextraction: a review. Pedosphere 26, 148–166.

Singh, A.K., Rai, A., Singh, N., 2016. Effect of long term land use systems on fractions of glomalin and soil organic carbon in the Indo-Gangetic plain. Geoderma 277, 41–50. doi: 10.1016/j.geoderma.2016.05.004.

Singh, J., Singh, A., Laxmi, V., 2017. Appraisal of non timber forest products in morni and Raipur Rani ranges of Shivalik hills, India. Int. J. Usuf. Mngt 18, 2–13.

Singh, L., Soni, P., 2010. Binding capacity and root penetration of seven species selected for revegetation of uranium tailings at Jaduguda in Jharkhand, India. Curr. Sci. 99, 507–513.

Slathia, P., Paul, N., 2012. Traditional practices for sustainable livelihood in Kandi Belt of Jammu. Indian J. Tradit. Knowl. 11, 548–552.

Smith, P., House, J.I., Bustamante, M., Sobocká, J., Harper, R., Pan, G., West, P.C., Clark, J.M., Adhya, T., Rumpel, C., 2016. Global change pressures on soils from land use and management. Glob. Chang. Biol. 22, 1008–1028.

Thakur, C., 1984. Weed Science. New Delhi, India: Metropolitan Book Co.(P) Ltd.

Vasudevan, P., Gujral, G.S., Madan, M., 1984. *Saccharum munja* Roxb., an underexploited weed. Biomass 4, 143–149.

Verma, P., Verma, R.K., 2017. Species diversity of *Arbuscular Mycorrhizal* (AM) Fungi in Dalli-Rajhara iron mine overburden dump of Chhattisgarh (Central India). Int. J. Curr. Microbiol. App. Sci 6, 2766–2781.

Verma, S.K., Singh, K., Gupta, A.K., Pandey, V.C., Trivedi, P., Verma, R.K., Patra, D.D., 2014. Aromatic grasses for phytomanagement of coal fly ash hazards. Ecol. Eng. 73, 425–428.

Wald, D.M., Nelson, K.A., Gawel, A.M., Rogers, H.S., 2018. The role of trust in public attitudes toward invasive species management on guam: A case study. J. Environ. Manage. 229, 133–144. https://doi.org/10.1016/j.jenvman.2018.06.047.

Wang, L., Ji, B., Hu, Y., Liu, R., Sun, W., 2017. A review on in situ phytoremediation of mine tailings. Chemosphere 184, 594–600.

Wapakala, W., 1966. A note on the persistence of mulch grasses. Kenya Coffee 31, 111–112.

Wishnie, M.H., Dent, D.H., Mariscal, E., Deago, J., Cedeno, N., Ibarra, D., Condit, R., Ashton, P.M.S., 2007. Initial performance and reforestation potential of 24 tropical tree species planted across a precipitation gradient in the Republic of Panama. For. Ecol. Manage. 243, 39–49.

Yao, Z.T., Ji, X.S., Sarker, P.K., Tang, J.H., Ge, L.Q., Xia, M.S., Xi, Y.Q., 2015. A comprehensive review on the applications of coal fly ash. Earth Sci. Rev. 141, 105–121.

Bermuda grass –its role in ecological restoration and biomass production

Vimal Chandra Pandey[a,*], Jitendra Ahirwal[b]
[a]Department of Environmental Science, Babasaheb Bhimrao Ambedkar University, Lucknow, Uttar Pradesh, India; [b]Department of Environmental Science and Engineering, Indian Institute of Technology (Indian School of Mines), Dhanbad, Jharkhand, India
*Corresponding author

1 Introduction

World's natural ecosystem and its ecological integrity have been significantly degraded by the anthropogenic activities. These heavily degraded areas need human intervention which leads to regeneration of site condition prior to disturbance. Historically, these degraded lands were often left unmanaged or rarely restored by developing vegetation cover. Plantation on degraded sites was carried out to restore site conditions, however, use to exotic tree species to accelerate restoration process were commonly reported. Selection of appropriate tree species used for restoration is very much important to develop a self-sustaining ecosystem. The growth of tree species and amelioration of site conditions along the age of plantation are crucial parameters to measure restoration success. Plant species take time to grow under adverse conditions which may delay the ecosystem development. Therefore, recent management practices focus on the broader aspect of restoration which includes soil amendments, selection of tree species, development of native vegetation cover, the introduction of grass and legumes species, biomass production, conserve and enhance biodiversity, and resilient community structure on degraded areas.

Remarkable attention is given around the world on various grass species due to their high root density, metal accumulation, biomass production, and ecological restoration. The application of grass species as cover crops enhances the soil quality by improving its physicochemical and biological characteristic. Higher root density absorbed water along the soil profiles which promotes the overall soil functions. Moreover, below ground function increases the soil microbial activity, physical attributes of soils and stabilizes the surface layer. Metal accumulation potential of native or invasive grass species was highly explored for the decontamination of the polluted sites (Shu et al., 2002; Maiti et al., 2016). Development of perennial grasses produces higher biomass which is subsequently used for forage and organic matter supply to soils. In view of the above, grasses have been considered as indispensable cover crops in terms of biomass production and ecological restoration. Bermuda grass (*Cynodon dactylon*) is a warm season perennial creeping grasses commonly planted for lawn cover, beautification, nutrition, fodder, golf courses, ground cover, and restoration of

Phytoremediation Potential of Perennial Grasses. http://dx.doi.org/10.1016/B978-0-12-817732-7.00011-0

industrial waste lands. The present chapter describes the ecology, geographical distribution, morphology, propagation, abiotic stress tolerance, and multiple uses, such as soil conservation, biomass production, phytoremediation potential, and other ecological values of the Bermuda grass.

2 Origin, geographical distribution, and occurrence

The origin of Bermuda grass was believed in North African countries and now found worldwide in tropical and subtropical countries. The name Bermuda grass is derived from its dominance as an exotic species in Bermuda. It is mainly distributed in Asia, United States of America, islands of the Pacific Ocean, Australia, and temperate regions of Europe. Bermuda grass is very much sensitive to low temperatures; therefore, its growth was found lowest in European countries. It has the ability to naturally colonize in agriculture fields and can act as competitive crop or weed which reduces the crop yield to a certain level (Harlan et al., 1970; Horowitz, 1996; Kalita et al., 1999). It is recorded as an invasive species in many Asian, Australia, and European countries, such as Indonesia, Singapore, Mexico, Costa Rica, Chile, Colombia, Uruguay, Argentina, and Brazil. Bermuda grass is listed as the most important weed worldwide only after *Cyperus rotundus* (Holm et al., 1977, 1979). In many subtropical countries, it is well known as a weed in both annual and perennial crops and in pastures, fallows, and waste areas. Fig. 11.1 represents the growth, occurrence, and natural colonization of Bermuda grass in agricultural field, with other weed species and in the wasteland, respectively.

3 Ecology

Bermuda grass is commonly known as warm-season grass found in tropical and subtropical countries. It is drought resistant, salt tolerant, highly versatile, and easily regenerates from seed and grass plugs (Kamal Uddin et al., 2012). Some common names of the Bermuda grass are doob, duba, bahama grass, devil grass, quick grass, star grass, grama bermuda, chiendent dactyle, herbes-des-Bermudes, and grama. Bermuda grass is a perennial warm-season grass thus it grows at a faster rate during the spring to hot summer season every year. It is geographically distributed between 45°N and 45°S, and probably further south outside its native range (Harlan et al., 1970). It also grows from sea level over much of this latitudinal range to about 4000 m above sea level in the Himalayas. It can grow in a wide temperature range where the annual temperature was 6°C–28°C. However, best growth was observed where the mean daily temperatures above 24°C or diurnal temperature ranged between 17°C and 35°C (De Wet and Harlan, 1970; Bogdan, 1977). Its seed germinates rapidly in a different type of soils like broad pH (4.5–8.5), salinity tolerance (1–10 dS/m), and a different texture from clay to sandy soils. It cannot be destroyed easily after germination due to its persistent and aggressive nature. Thus, subsequent tilling and removal of roots is generally preferred to eradicate Bermuda grass (Guglielmini and Satorre, 2004).

Fig. 11.1 (A) Growth of the Bermuda grass in agriculture field, (B) occurrence of Bermuda grass with the other weed species, (C) natural colonization in the wasteland and growth of inflorescence, and (D) luxuriant growth of Bermuda grass in lawn.
Photo courtesy: J. Ahirwal.

Bermuda grass can tolerate a wide range of moisture regime and long-term drought. Most of the roots are growing within the top surface and it can reach up to 1–2 m that makes it much tolerant to critical environmental conditions. Very drought tolerant due to rhizome survival through drought-induced dormancy over periods of up to 7 months and can tolerate up to several weeks of deep flooding. The luxuriant growth of Bermuda grass can observe where average annual rainfall ranged between 625 and 1750 mm. Bermuda grass is not shade tolerant and yields decrease rapidly with increasing shade. It usually dies out under medium to dense shade. It will stand severe fires due to the extensive rhizome development in most varieties and cultivars.

4 Morphology and propagation

C. dactylon is a robust and stoloniferous perennial grass (10–50 cm tall, rarely to 90 cm), with underground rhizomes and on the ground runners. The runners spread horizontally in shape of flattened or cylindrical and contain nodes with internodes of about 10 cm length. The stems are slightly flattened, often tinged purple in color. The flowering stem contains leaves in two rows. Its leaves consist of a sheath, and

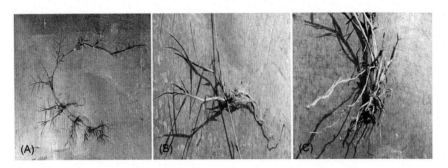

Fig. 11.2 Naturally growth of the Bermuda grass (A) steam growth up to 27 cm, (B) stolen and rhizome part of the grass and (C) root growth up to 6–8 cm.
Photo courtesy: J. Ahirwal.

leaf blades that are green to blue-green, flat, linear-lanceolate, and scabrous at margins. Both surfaces of leaf and outer surface of sheath are glabrous or with short or long hairs. The ligule is very little but with a noticeable fringe of white hairs with no auricles. Sheaths are small, roundish, with sparse hairiness. The inflorescence is supported on a culm up to 25 cm high and bears of a single whorl of 3–7 narrow racemes, each 3–8 cm long. Inflorescence is raceme-like panicle, pilosus at base, and with 3–6 rarely to 9 spikes. Spikes are glabrous, elongated, green to pale green, or green with purple spots, yellowish at maturity. Glumes are 1-nerved. Lemma silky pubescent on the keel. Palea glabrous. No awns. Finger-like spikes containing one seed. Caryopses are sub-elliptical or ovoid, yellow to reddish, brownish, compressed, about 1.5 mm long; 3–4.5 million seeds kg^{-1} (Kissmann, 1991).

Bermuda grass reproduces through seeds, runners, and rhizomes. It spreads by both above-ground stems known as stolons and below-ground stems called rhizomes (Fig. 11.2). Eco-physiological and genetic traits found in seeds and stolons helps to propagate in many ecosystems and alter ecosystem functions (D'Antonio and Vitousek, 1992). *C. dactylon* is very drought tolerant by feature of rhizome survival through drought-induced dormancy over periods of up to 7 months. After dormancy, it has the ability to easily re-sprout by stolons or rooted runners. Plants also recover quickly after fire and can tolerate at least several weeks of deep flooding (Cook et al., 2005). Rhizome biomass exhibits an annual cyclic pattern and, as with any perennial weed, low temperatures reduce biomass and viability is lost as a consequence of the consumption of materials due to respiration and maintenance.

5   Abiotic stress tolerance of Bermuda grass

Plant growth is often challenged by abiotic factors, such as low or excess water availability, wide-ranging temperatures from frost to humid conditions, photoperiod, salinity, and mineral toxicity. Abiotic stresses resulted in dramatic changes in plant metabolism, growth, and its ecosystem functions. Drought stress in plants increases the production of reactive oxygen species, such as superoxide, singlet oxygen, hydroxyl, and hydrogen peroxide molecules which can cause a negative impact on proteins, lipids, carbohydrates, and nucleic acid synthesis (Smirnoff, 1993; Mahajan and Tuteja, 2005). This drought stress can be compensated by increasing the antioxidant enzymes activity (Bian and Jiang, 2009). The study of the application of nitric oxide on the Bermuda grass after 7 or 9 days increases the antioxidant enzymes activity thus its capacity to drought tolerance (Boogar et al., 2014). The proteomic analysis to understand the molecular mechanism of Bermuda grass response to drought stress reported drought-responsive proteins which involved in photosynthesis, glycolysis, N-metabolism, tricarboxylic acid, and redox pathways in both leaves and stems of two Bermuda grass varieties including drought sensitive variety (Yukon) and drought tolerant variety (Tifgreen) (Shi et al., 2014). The study of differential proteomic responses to drought stress (15 days) in hybrid and common Bermuda grass showed superior drought tolerance in hybrid Bermuda grass mainly due to the less severe decline proteins level involved in photosynthesis and a greater increase in the antioxidant defense proteins (Zhao et al., 2010). Flooding can severely affect physiological integration, metabolism of carbohydrates, and reactive oxygen species. Flooding-induced responses including decreased root biomass, accumulation of soluble sugar and starch, as well as increased activity of superoxide dismutase and ascorbate peroxidase in apical ramets (Li et al., 2014).

Bermuda grass tolerates a wide range of temperatures, especially very high temperatures in near-desert conditions to very low during the frost. Bermuda grass was mainly dominant in tropical countries where the temperature remains high over the year. It is very much sensitive to low temperatures thus found very often in cold conditions. The annual growth-form becomes dormant and turns brown when the temperatures fall below freezing during the night or average temperatures are below $10°C$ (Cook et al., 2005). Investigation of gene profile of 4-year-old cold-acclimated versus non-acclimated Bermuda grass revealed that AP2, NAC, and WRKY family members of these upregulated genes are associated with cold stress (Zhu et al., 2015). The phytohormone ethylene is associated with plant response to cold stress (Hu et al., 2017). Effect of the atmospheric temperature on the dormancy of Bermuda grass was studied under different temperature and photoperiods (Esmaili and Salehi, 2012). Results of the study showed that decreasing temperature and photoperiod decreased fresh and dry biomass, shoot height, tiller density, leaf area, chlorophyll, and relative water contents and concluded that problem of dormancy may be could be solved by increasing the photoperiod.

Development of Bermuda grass in salt affected soils may cause the salinity stress resulted in decreased grass growth. Salinity stress potentially reduces turf quality, leaf relative water content, photosynthetic rate, transpiration rate, stomatal conductance, and cellular membrane stability (Yu et al., 2015). Study of the Bermuda grass tolerance to salinity stress showed that increased CO_2 concentration may counteract the negative effects of salinity stress (Yu et al., 2015). Exogenous application of H_2S donor showed improved salt, osmotic and freezing stress tolerances in Bermuda grass, which were further proved by a decrease in electrolyte leakage and increased survival rate under stress conditions (Shi et al., 2013). Greenhouse gas study evaluating the impact of salinity and deficit irrigation on physiological and biochemical changes in Bermuda grass reported an increase in ion leakage, soluble sugars, and a decrease in leaf relative water content, photosynthetic rate and antioxidant enzyme activity with increasing levels of both stresses (Manuchehri and Salehi, 2014). Bermuda grass growing in high salinity areas showed high salt tolerance due to its restricted uptake of Na^+ along with the increased uptake of K^+ and Ca^{2+} in the roots as well as shoot due to its higher photosynthetic rate and accumulation of organic osmotica under saline conditions (Hameed and Ashraf, 2008).

Plant absorbs soil nutrients form soil in response to concentration gradient. Plants accumulate several trace elements, such as Cd and Pb which do not have specific biological functions and can be proven harmful to plants even at low concentrations (Albornoz et al., 2016). Tan et al. (2017) investigate the effects of waterlogging and cadmium on ecophysiological responses and metal bio-accumulation in Bermuda grass under four simulated stress condition; normal condition, waterlogging, cadmium or cadmium-and-waterlogging treatments, respectively. Authors found that Bermuda grass was a cadmium hyperaccumulator and stress tolerant to waterlogging, cadmium toxicity or both.Wang et al. (2009) studied the effects of copper accumulation and resistance in two biotypes of Bermuda grass and reported that the growth is unaffected under low level of Cu concentration (<100 mg kg^{-1}), however, with the increase in Cu concentration Bermuda grass growth was decreased. Authors also observed that Bermuda grass can tolerate 500 mg kg^{-1} of Cu in the polluted area.

6   Multiple uses

Bermuda grass is widely distributed as a perennial and annual warm season turf and forage grass across the tropical, subtropical, and temperate countries. Ability to stand under adverse environmental conditions due to its reproductive system and stress resistance ability make it suitable choice for multiples uses (Table 11.1). Some of the major uses of Bermuda grass like land restoration, biomass production, carbon sequestration; green capping and bioenergy production are briefly described herein.

6.1   Wasteland's restoration

Wastelands are ecologically unstable lands which are low in productivity and severely affected by soil erosion, stress conditions, and hostile environmental conditions. These areas can be abandoned mine lands, mine tailing disposal site, deforested areas, overgrazing land, and barren lands. Poor land use practices resulted in loss of soil quality and land productivity. It is estimated that biomass production in wasteland reduced by 80% of its actual potential. To restore the productivity and functions of the wasteland, it is important to restore its vegetation and soil cover similar to pre-disturbance conditions.

Bermuda grass has higher potential to grow on severely sodic lands and can be used for the rehabilitation of degraded lands (Singh et al., 2013). Assessment of the

Table 11.1 List of multiple attributes/use of Bermuda grass (*Cynodon dactylon*) toward ecosystem services.

Title	Country	Main use	Authors
Phytoremediation of crude oil-contaminated soil using *C. dactylon* (L.) Pers.	India	Rehabilitation	Basumatary and Bordoloi (2016)
Evaluating bermudagrass (*C. dactylon*), seashore paspalum (*Paspalum vaginatum*), and weeping love grass (*Eragrostis curvula*), as a vegetative cap for industrial brin landform stabilization and phytoremediation	USA	Vegetation cap	Fontenot et al. (2015)
Antidiabetic and antidiarrhoeal potentials of ethanolic extracts of aerial parts of *C. dactylon* Pers		Medicinal value	Rahman et al. (2015)
C. dactylon: an efficient perennial grass to revegetate sodic lands	India	Revegetation	Singh et al. (2013)
Total and available soil carbon fraction under the perennial grass *C. dactylon* (L.) Pers and the bioenergy crop *Arundo donax* L.	USA	Carbon sequestration	Sarkhot et al. (2012)
Biomass productivity and nutrient availability of *C. dactylon* (L.) Pres growing on soils of different sodicity stress	India	Biomass production	Shukla et al. (2011)
Comparing repeated forage Bermuda grass harvest data to single, accumulated bioenergy feedstock harvests		Bioenergy production	Muir et al. (2010)
Accumulation of As, Pb, Zn, Cd, and Cu and arbuscular mycorrhizal status in populations of *C. dactylon* grown on metal-contaminated soils	China	Phytoremediation	Wu et al. (2010)
Lead, zinc and copper accumulation and tolerance in populations of *Paspalum distichum* and *C. dactylon*	China	Metal accumulation	Shu et al. (2002)
Use of aquatic and terrestrial plants for removing phosphorus from dairy wastewater	USA	Nutrient removal	DeBusk et al. (1995)

growth and its impact of soil properties were carried out on the sodic land in India and found that decreases in soil sodicity and increases in soil nutrients, microbial biomass, and enzyme activities in rhizosphere soil of Bermuda grass. Authors also suggested that it can be used to restore and recuperate biological activities wasteland. Blanketing of fly ash lagoons with topsoil significantly decreased the metal contamination to nearby sites and facilitates the natural colonization of pioneer autochthonous species such as Bermuda grass.

Growth of Bermuda grass on fly ash lagoon recuperates the site conditions long with the toxic metal accumulation. On the other side, it may also act as transferring link of heavy metal to the food chain (Maiti et al., 2016). Ecological restoration of highway slope by application of straw-mat and seeding with grass–legume mixture including Bermuda grass in China showed significant improvements in soil physicochemical properties, such as moisture, temperature, water holding capacity that enhance vegetation recovery, and stabilization of highway slope (Yuan et al., 2016).

6.2   Phytoremediation of toxic elements

Soil and water pollution caused by the mobility and solubility of toxic metals caused harmful effect on living being, human health, and the environment. Phytoremediation is a cost-effective technology widely used for decontamination of metal pollutant sites (Shu et al., 2002; Singh and Tripathi, 2007; Wu et al., 2010; Yuan et al., 2016; Kumar et al., 2017). Researchers often use the stress tolerant invasive species which can grow under stress conditions (Pandey, 2012; Maiti et al., 2016). Metal accumulation potential of Bermuda grass naturally colonized on mine tailing of Guangdong Province, China was investigated and found higher tolerance to lead, zinc, and copper metal. Authors also recommend this species as a potential candidate for revegetation of wastelands contaminated sites (Shu et al., 2002). Bermuda grass growing at different metal contaminated sites of China was investigated and significantly higher metal (As, Pb, Zn, Cd, and Cu) concentration in shoot was reported (Wu et al., 2010). The growth of grass along with the arbuscular mycorrhizal fungi in contaminated sites showed its phytorestoration potential of metal contaminated sites. The study of heavy metals accumulation by native herbaceous plant in an antimony mine of China reported considerable potential for As and Sb stabilization by the Bermuda grass which can be supported by higher bioaccumulation factor of 2.02 and 6.62, respectively (Xue et al., 2014). Investigation to reuse of heavy metal-accumulating Bermuda grass in remediation of contaminated water showed that plant residue can remove up to 97% for Cu. In addition, the facile acid-treatment to desorb accumulated heavy metals in Bermuda grass found 90% recovery of Zn, Pb, and Cu (Ma and Gao, 2014).

Grass–legume cover including Bermuda grass showed tremendous potential of metal accumulation and may act as a barrier for metal transport from chromite-asbestos waste to nearby agricultural or uncontaminated soils (Kumar et al., 2017). Lead and zinc concentration in soil and plant samples of Bermuda grass naturally colonized in industrial soils of Argentina was found 472% more for Pb and 288% more for Zn compared to control soils that showed higher metal reservoirs and can act as indicator of metal contamination (Albornoz et al., 2016). Bermuda grass growing in toxic levels of 16,492 mg kg^{-1} of Cr near tanning sludge storage sites, accumulate moderately high concentrations of Cr which make it good bioindicators of heavy metal pollution (Yuan et al., 2016).

Heavy mental contamination around the thermal power plants due to release of fly ash and its mix up with water body pose hazardous effect on aquatic animals, humans, and nearby agriculture land. The screening of naturally colonized plant species on fly ash dump for phytoremediation showed higher enrichment ratio of metal in

Fig. 11.3 Natural colonization and luxuriant growth of Bermuda grass on fly ash deposited sites.
Photo courtesy: V.C. Pandey.

plants where Bermuda grass accumulate 90.28 mg kg^{-1} in shoot and proved efficient plant for phytoremediation of metal contaminated site (Kisku et al., 2018). Fig. 11.3 represents natural colonization of Bermuda grass on fly ash deposited sites which reflects its capability toward restoration programs of fly ash dumps. Assessment of heavy metal phytoextraction potential of native weeds and grasses from endocrine-disrupting chemicals rich complex distillery sludge showed higher accumulation of 438 mg kg^{-1}Fe, 147 mg kg^{-1}Zn, 289 mg kg^{-1}Cu, 35 mg kg^{-1}Mn, and 23 mg kg^{-1}Pb in their root, shoot and leaves of Bermuda grass which indicated high accumulation and translocation capabilities (Chandra et al., 2018).

Estimation of metal translocation and bioaccumulation in Bermuda grass growing at the reclaimed opencast mine showed metal excluding properties in shoots with low TF (<1 for Fe, Ni, and Pb), high BCF (>1 for Zn and Pb), and low bioaccumulation coefficient (<1 for Fe, Mn, Ni, Cu, and Cd) that suggests its potential use phytoremediation of fly ash lagoons (Kumar et al., 2015). Phytoremediation potential of the Bermuda grass growing on different metal contaminated sites is given in Table 11.2. Uptake of As and Hg by Bermuda grass growing on amended soils showed 15, 25, and 200 mg kg^{-1} As and 1.4–37.6, 0.4–1.2, and 80–800 mg kg^{-1} Hg in leaves, stems and roots, respectively, can pose a serious hazard to foraging animals (Weaver et al., 1984).

Table 11.2 Phytoremediation potential of the Bermuda grass (*Cynodon dactylon*) growing on metal contaminated sites.

Study site	Plant part	Metal concentration (mg kg^{-1})								Authors
		As	Pb	Zn	Cd	Cu	Mn	Ni	Cr	
Distillery sludge contaminated site, India	Root; Shoot	na	10; 10	64; 52	na	253; 8.7	8.8; na	na	na	Chandra et al. (2018)
Mine contaminated site, India	Root; Shoot	na	12; na	54; na	na -	8; na	30; na	35; na	40; na	Kumar et al. (2015)
Fly ash lagoon, India	Root; Shoot	na	2.5; 3.3	67; 87	0.76; 1.55	9.8; 12	21; 33	11; 8.4	na	Maiti et al. (2016)
Top soil, India	Root; Shoot	na	1.5; 0.4	87; 80	1.2; 1.5	13; 12	64; 35	17; 9	na	Maiti et al. (2016)
Tanning sludge storage site, China	Root; Shoot	na	na; 0.9	na; 201	na; 0.03	na; 5.27	na	na	na; 96	Yuan et al. (2016)
Vegetation cap, Louisiana, US	Leaf	na	na	71	0.2	5.1	79	na	2.8	Fontenot et al. (2015)
Waste water treatment plant, Greece	Root; Shoot	na	na; 14	na; 90	na	na; 18	na; 110	na; 1.0	na; 30	Suchkova et al. (2014)
Metal contaminated site, China	Root; Shoot	373; 17	5.2; 46	na	2.3; 1.9	na	na	na	na	Xue et al. 2014
Metal contaminated site, Hong Kong, China	Root; Shoot	na; 94	na; 417	na; 698	na; 5.8	na	na	na	na	Wu et al. (2010)
Fly ash lagoon, India	Whole plant	na	9.0	25.3	na	65.8	17.0	10.5	na	Maiti and Nandhini (2006)
Reclaimed mine land, Spain	Root; Shoot	na; 10	na; 26	na; 352	na	na; 22	na	na	na	Del Río et al. (2002)
Mine tailing China	Root; Shoot	na	644; 351	1015; 688	na	45; 22	na	na	na	Shu et al. (2002)

na, Not analyzed.

6.3 Soil conservation

Bermuda grass with good adaptation to flooding and the ability of clonal propagation is one of the few perennial grasses to adapt to the changed environment (Chen et al., 2010). It is reported that soil-root system of Bermuda grass has capability for soil conservation and riverbank support (Chen et al., 2015). Therefore, it is pressing need to restore the Bermuda grass community for taking full benefit of its soil-root system in soil conservation, slope protection, and riverbank stability against soil erosion. Singh et al. (2013) suggested that Bermuda grass can be used to restore and increase the biological activities of sodic lands and to aid the further vegetation cover establishment. Recently, it is also observed that Bermuda grass root system can defense the rain erosion of slope surface soil, restrain the soil displacement, and prevent water and soil erosion of slope (Ji and Yang, 2019).

6.4 Biomass and biofuel

Bermuda grass has ability to grow on a wide range of substrates such as soil, wasteland, fly ash dump, red mud soil, mine spoil, etc. and produce good biomass on such hostile conditions. Bermuda grass grown on sodic soils produces approximately 134 g total biomass m^{-2} (roots contribute about 40%) (Shukla et al., 2011) and 158 g m^{-2} (involvement of roots was 78 g m^{-2}) (Singh et al., 2013). Xu et al. (2011) reviewed Bermuda grass as feedstock for biofuel production and found as a promising feedstock for fuel to produce ethanol in the Southern United States. Bermuda grass cultivar, growing conditions, and management practices affect biomass yield. A number of significant amounts of research on the conversion of Bermuda grass to ethanol have been done. The major challenge of biomass conversion of Bermuda grass comes from the recalcitrance of lignocellulose. A range of chemical, physicochemical, and biological pre-treatment techniques have been explored for the improvement of the digestibility of Bermuda grass with positive results stated. The following enzymatic hydrolysis and fermentation steps have also been broadly studied and efficiently enhanced. It is likely that the progress of genetic engineering knowledge for the grass and fermenting organisms has the potential to significantly increase the economic viability of Bermuda grass-based fuel ethanol production systems. Other energy applications of Bermuda grass include anaerobic digestion for biogas generation and pyrolysis for syngas production. Bermuda grass discussed in this chapter is a promising candidate to play an important role in phytoremediation programs with biomass production. The success of Bermuda grass as a phytoremediator grass is not only dependent on their biomass yield and efficiency in using remediation, but also on their willingness for improvements through breeding efforts or genetic engineering.

6.5 Green capping

Thermal power plants, metal mining industries, distillery and wastewater treatment plants have been recognized as major sources of environmental pollutants. Thus, proper management and disposal of industrial waste is a major environmental issue.

Presence of toxic metals mainly As, Cr, Cd, Pb, and Hg along with airborne particles can pose serious health issue. Therefore, remediation of waste disposal sites and stabilization of site conditions to combat pollution is currently needed. Development of green cap to combat dispersion and mobility of pollutants are widely recognized. Species used for the green capping should have perennial nature, good soil binder with extensive root system, self-propagation, and tolerance to high temperature and drought conditions (Pandey and Singh, 2014).

Bermuda grass was evaluated for a vegetative cover over a brine solid waste surface impoundment in Louisiana (Fontenot et al., 2015). Authors also opined that germination from seed and sod, higher nutrient accumulation capacity, fast growth and tolerant to hostile condition makes Bermuda grass a suitable candidate for stabilizing and covering brine impacted areas. Bermuda grass naturally colonized on fly ash lagoon may be suitable to develop green turf, which prevents airborne erosion of FA from the site and provides easily available and natural palatable fodder for cattle (Kumar et al., 2015).

6.6 Carbon sequestration

Climate change is a global issue. Increased emission of CO_2 into the atmosphere resulted in a rise in global ambient temperature and caused global warming. Development of vegetation cover sequesters atmospheric CO_2 and can significantly reduce the global carbon concentration. In addition, urban greenery may combat global warming by storing large amount of CO_2 in vegetation and soils. Urban greenery system largely includes Bermuda grass which can potentially contribute to carbon sequestration (Kong et al., 2014). However, extensive management like mowing and chemical application of the home lawns turf grasses contribute to the CO_2 emission in to the atmosphere. With increase age of revegetation, the C emitted to the atmosphere through home lawn decreased substantially, which can turn home lawns from sinks to sources of atmospheric C.

Estimation of net carbon sequestration potential and emissions in home lawn turf grasses of the Albuquerque, New Mexico, USA, reported mean SOC sequestration rate of 2.3 Mg C ha^{-1} year^{-1} and mean potential SOC sink capacity of 58.0 Mg C ha^{-1} (Selhorst and Lal, 2012). The study of carbon budgeting in golf course soils of Central Ohio, USA reported that farmland soils converted to golf course turf grasses can sequestered 3.55 Mg C ha^{-1} year^{-1}in fairways and 2.64 Mg C ha^{-1} year^{-1}in rough areas (Selhorst and Lal, 2011). Soil carbon sequestration in turf grass system of the urban landscapes in New Zealand was analyzed in the topsoil (0–0.25 m) of different ages (5, 9, 20, 30, 40 years) in a golf course and reported SOC sequestrated at a rate of 69 ± 8 g m^{-2} year^{-1} over a 40-year period (Huh et al., 2008). Sarkhot et al. (2012) studied the total and available soil carbon fractions under the Bermuda grass growing near the agriculture land and found 37.2 Mg C ha^{-1} in the top 50 cm depth.

6.7 Hindu rituals

In Hindu rituals, Bermuda grass (Durva) plays an important role. It is mainly used by the priest during the prayer to offer it to the Lord Ganesha and Lord Shiva (Fig. 11.4).

Fig. 11.4 Use of Bermuda grass during the (A) Puja and (B) to offer water droplet on Lord Shiva as per the Hindu rituals.
Photo courtesy: J. Ahirwal.

It is considered the second most important blessed plant after Tulsi (*Ocimum tenuiflorum*). Mostly, the shoot of the Bermuda grass is used in ritual to offer water droplets on the idol. According to the Hindu mythology, Bermuda grass consists of three blades which represent the three principles of the primal Shiva, the primal Shakti, and the primal Ganesha. In Hinduism, most people are agreed to that those suffering from negative energies feel positive when coming in the contact of Bermuda grass.

6.8 Medicine

Ayurveda is the predominant source of plant drugs for many incurable diseases. Bermuda grass is well known and traditionally used to treat epistaxis, hematuria, inflamed tumors, cuts, wounds, cystitis, nephritis, and skin infection. It also exhibits diuretic, antidiarrheal, and antiseptic properties. The antimicrobial activity of its extract and the alkaloids against human pathogenic bacteria and dermatophytes was also reported (Raman et al., 2002). The plant is used as hemeostatic and wound healing agent from the ethnopharmacological point of view. The study of aqueous extract of Bermuda grass reported its potential to wound healing activity in animals and probability in human also (Biswas et al., 2017). Murali et al. (2015) obtained luteolin and apigenin rich fraction from the ethanolic extract of Bermuda grass and reported that it can be used as a potential therapeutic agent against Chikungunya virus.

Investigation of ethanolic extracts of Bermuda grass aerial parts in Wistar rats model showed antidiabetic and antidiarrheal agents (Rahman et al., 2015). Bermuda grass is a potential source of metabolites such as flavonoids, alkaloids, glycosides, and β-sitosterol and has been traditionally used to treat urinary tract, microbial infections, and dysentery (Annapurna et al., 2013). Authors study its extracts for glycemic control and found that it inhibits diabetic neuropathy thus enabling a possibility to treat diabetes. The development of collagen-silica wound heal scaffold incorporated with the extract of Bermuda grass showed accelerated healing with enhanced collagen deposition and also proposed as a biomaterial for tissue engineering (Perumal et al., 2018). Babu et al. (2018) conduct a green approach to synthesize silica nanoparticles from Bermuda grass and reported that it is a green source for production of potential bio antimicrobial nanoparticle.

7 Conclusion

Bermuda grass [*C. dactylon* (L.) Pers.] is a promising perennial grass which can naturally colonize extreme habitats of a range of waste land. Bermuda grass can utilize soil resources more efficiently and sustained to a wide range of moisture regimes with good biomass production capability. It produces high biomass and has dense root system which is beneficial for soil conservation and helpful in soil erosion. Bermuda grass can phytostabilize the heavy metals. However, the toxic metals can enter in food chain through biomagnification and may be toxic to grazing animals due to its palatable nature. Along with ecological aspects, Bermuda grass also restores degraded land and improves rural economy. The Bermuda grass is used as fodder material to feed goat, cow, buffalo, and other cattle. The leaves, stems, and roots of Bermuda grass contain medicinal properties and are used to treat epistaxis, hematuria, inflamed tumors, cuts, wounds, cystitis, nephritis, and skin infection. The biomass of Bermuda grass can be used to produce bioethanol. Despite the multiple uses of Bermuda grass, its use in phytoremediation programs is pressing need due to rapid propagation, tolerant nature against hostile conditions, dense root system, and growing nature in wide range of lands as well as its wide occurrence.

Acknowledgments

Financial assistance given to Dr. V.C. Pandey under the Scientist's Pool Scheme (Pool No. 13 (8931-A)/2017) by the Council of Scientific and Industrial Research, Government of India is gratefully acknowledged. The authors declare no conflict of interest.

References

Albornoz, C.B., Larsen, K., Landa, R., Quiroga, M.A., Najle, R., Marcovecchio, J., 2016. Lead and zinc determinations in *Festuca arundinacea* and *Cynodon dactylon* collected from contaminated soils in Tandil (Buenos Aires Province, Argentina). Environ. Earth Sci. 75 (9), 742.

Annapurna, H.V., Apoorva, B., Ravichandran, N., Arun, K.P., Brindha, P., Swaminathan, S., Vijayalakshmi, M., Nagarajan, A., 2013. Isolation and in silico evaluation of antidiabetic molecules of *Cynodon dactylon* (L.). J. Mol. Graph. Model. 39, 87–97.

Babu, R.H., Yugandhar, P., Savithramma, N., 2018. Synthesis, characterization and antimicrobial studies of bio silica nanoparticles prepared from *Cynodon dactylon* L.: a green approach. Bull. Mater. Sci. 41 (3), 65.

Basumatary, B., Bordoloi, S., 2016. Phytoremediation of crude oil-contaminated soil using *Cynodon dactylon* (L.) Pers. In: Ansari, A., Gill, S., Gill, R., Lanza, G., Newman, L. (Eds.), Phytoremediation. Springer, Cham, Switzerland, pp. 41–51.

Bian, S., Jiang, Y., 2009. Reactive oxygen species, antioxidant enzyme activities and gene expression patterns in leaves and roots of Kentucky bluegrass in response to drought stress and recovery. Sci. Hor. 120 (2), 264–270.

Biswas, T.K., Pandit, S., Chakrabarti, S., Banerjee, S., Poyra, N., Seal, T., 2017. Evaluation of *Cynodon dactylon* for wound healing activity. J. Ethnopharmacol. 197, 128–137.

Bogdan, A.V., 1977. Tropical Pasture and Fodder Plants (Grasses and Legumes). Longman, London and New York, pp. 92–98.

Boogar, A.R., Salehi, H., Jowkar, A., 2014. Exogenous nitric oxide alleviates oxidative damage in turfgrasses under drought stress. S. Afr. J. Bot. 92, 78–82.

Cook, B., Pengelly, B., Brown, S., Donnelly, J., Eagle, D., Franco, A., Hanson, J., Mullen, B., Partridge, I., Peters, M., Schultze-Kraft, R., 2005. Tropical Forages: An Interactive Selection Tool. CSIRO, DPI&F (Qld), CIAT and ILRI, Brisbane, Australia.

Chandra, R., Kumar, V., Tripathi, S., Sharma, P., 2018. Heavy metal phytoextraction potential of native weeds and grasses from endocrine-disrupting chemicals rich complex distillery sludge and their histological observations during in-situ phytoremediation. Ecol. Eng. 111, 143–156.

Chen, F., Huang, Y., Fan, D., Xie, Z. 2010. Ecological response of vegetative propagule of *Cynodon dactylon* to simulated summer flooding. Guihaia 30, 488–492.

Chen, F., Zhang, J., Zhang, M., Wang, J., 2015. Effect of *Cynodon dactylon* community on the conservation and reinforcement of riparian shallow soil in the Three Gorges Reservoir area. Ecol. Process. 4, 3. DOI 10.1186/s13717-014-0029-2.

D'Antonio, C.M., Vitousek, P.M., 1992. Biological invasions by exotic grasses, the grass/fire cycle, and global change. Annu. Rev. Ecol. Evol. Syst. 23 (1), 63–87.

De Wet, J.M.J., Harlan, J.R., 1970. Biosystematics of *Cynodon* L.C. rich. (*Gramineae*). Taxon 19, 565–569.

DeBusk, T.A., Peterson, J.E., Reddy, K.R., 1995. Use of aquatic and terrestrial plants for removing phosphorus from dairy wastewaters. Ecol. Eng. 5 (2–3), 371–390.

Del Río, M., Font, R., Almela, C., Vélez, D., Montoro, R., Bailon, A.D.H., 2002. Heavy metals and arsenic uptake by wild vegetation in the Guadiamar river area after the toxic spill of the Aznalcóllar mine. J. Biotechnol. 98 (1), 125–137.

Esmaili, S., Salehi, H., 2012. Effects of temperature and photoperiod on postponing bermudagrass (*Cynodon dactylon* [L.] Pers.) turf dormancy. J. Plant Physiol. 169 (9), 851–858.

Fontenot, D., Bush, E., Beasley, J., Fontenot, K., 2015. Evaluating Bermudagrass (*Cynodon dactylon*), seashore Paspalum (*Paspalum vaginatum*), and weeping love grass (*Eragrostis curvula*), as a vegetative cap for industrial brine landform stabilization and phytoremediation. J. Plant Nutr. 38 (2), 237–245.

Guglielmini, A.C., Satorre, E.H., 2004. The effect of non-inversion tillage and light availability on dispersal and spatial growth of *Cynodon dactylon*. Weed Res. 44 (5), 366–374.

Hameed, M, Ashraf, M., 2008. Physiological and biochemical adaptations of Cynodon dactylon (L.) Pers. from the Salt Range (Pakistan) to salinity stress. Flora-Morphology, Distribution. Functional Ecology of Plants 203 (8), 683–694.

Harlan, J.R., De Wet, J.M.J., Rawal, K.M., 1970. Geographic distribution of the species of *Cynodon* LC Rich (*Gramineae*). East Afr. Agric. For. J. 36 (2), 220–226.

Holm, L.G., Pancho, J.V., Herberger, J.P., Plucknett, D.L., 1979. A Geographical Atlas of World Weeds. John Wiley and Sons, New York, USA.

Horowitz, M., 1996. Bermudagrass (*Cynodon dactylon*): a history of the weed and its control in Israel. Phytoparasitica 24 (4), 305–320.

Holm, L.G., Plucknett, D.L., Pancho, J.V., Herberger, J.P., 1977. The World's Worst Weeds. Distribution and Biology. University Press of Hawaii, Honolulu, Hawaii, USA.

Hu, Z., Liu, A., Bi, A., Amombo, E., Gitau, M.M., Huang, X., Chen, L., Fu, J., 2017. Identification of differentially expressed proteins in bermudagrass response to cold stress in the presence of ethylene. Environ. Exp. Bot. 139, 67–78.

Huh, K.Y., Deurer, M., Sivakumaran, S., McAuliffe, K., Bolan, N.S., 2008. Carbon sequestration in urban landscapes: the example of a turfgrass system in New Zealand. Soil Res. 46 (7), 610–616.

Ji, X.-L., Yang, P., 2019. The effect of bermuda grass root morphology on the displacement of slope. MATEC Web of Conferences 275, 03004. Available from: https://doi.org/10.1051/matecconf/201927503004.

Kalita, D., Choudhury, H., Dey, S.C., 1999. Assessment of allelopathic potential of some common upland rice weed species on morpho-physiological properties of rice (*Oryza sativa* L.) plant. Crop Res. 17 (1), 41–45.

Kamal Uddin, M., Juraimi, A.S., Ismail, M.R., Rahim, M.A., Radziah, O., 2012. Physiological and growth responses of six turfgrass species relative to salinity tolerance. Sci. World J. 10, 1–10.

Kisku, G.C., Kumar, V., Sahu, P., Kumar, P., Kumar, N., 2018. Characterization of coal fly ash and use of plants growing in ash pond for phytoremediation of metals from contaminated agricultural land. Int. J. Phytoremediation 20 (4), 330–337.

Kissmann, K.G., 1991. Plantas Dominantes e Nocivas. São Paulo: TOMOI; BASF. pp. 608.

Kong, L., Shi, Z., Chu, L.M., 2014. Carbon emission and sequestration of urban turfgrass systems in Hong Kong. Sci. Total Environ. 473–474, 132–138.

Kumar, A., Ahirwal, J., Maiti, S.K., Das, R., 2015. An assessment of metal in fly ash and their translocation and bioaccumulation in perennial grasses growing at the reclaimed opencast mines. Int. J. Environ. Res. 9 (3), 1089–1096.

Kumar, A., Maiti, S.K., Prasad, M.N.V., Singh, R.S., 2017. Grasses and legumes facilitate phytoremediation of metalliferous soils in the vicinity of an abandoned chromite–asbestos mine. J. Soil. Sediment. 17 (5), 1358–1368.

Li, Z.J., Fan, D.Y., Chen, F.Q., Yuan, Q.Y., Chow, W.S., Xie, Z.Q., 2015. Physiological integration enhanced the tolerance of *Cynodon dactylon* to flooding. Plant Biol. 17 (2), 459–465.

Ma, D., Gao, H., 2014. Reuse of heavy metal-accumulating *Cynondon dactylon* in remediation of water contaminated by heavy metals. Front. Env. Sci. Eng. 8 (6), 952–959.

Mahajan, S., Tuteja, N., 2005. Cold, salinity and drought stresses: an overview. Arch. Biochem. Biophys. 444 (2), 139–158.

Maiti, S.K., Kumar, A., Ahirwal, J., Das, R., 2016. Comparative study on bioaccumulation and translocation of metals in bermuda grass (*Cynodon dactylon*) naturally growing on fly ash lagoon and topsoil. Appl. Ecol. Env. Res. 14 (1), 1–12.

Maiti, S.K., Nandhini, S., 2006. Bioavailability of metals in fly ash and their bioaccumulation in naturally occurring vegetation: a pilot scale study. Environ. Monit. Assess. 116 (1–3), 263–273.

Manuchehri, R., Salehi, H., 2014. Physiological and biochemical changes of common bermudagrass (*Cynodon dactylon* [L.] Pers.) under combined salinity and deficit irrigation stresses. S. Afr. J. Bot. 92, 83–88.

Muir, J.P., Lambert, B.D., Greenwood, A., Lee, A., Riojas, A., 2010. Comparing repeated forage bermudagrass harvest data to single, accumulated bioenergy feedstock harvests. Bioresour. Technol. 101 (1), 200–206.

Muir, J.P., Lambert, B.D., Greenwood, A., Lee, A., Riojas, A., 2010. Comparing repeated forage bermudagrass harvest data to single, accumulated bioenergy feedstock harvests. Bioresour. Technol. 101, 200–206.

Murali, K.S., Sivasubramanian, S., Vincent, S., Murugan, S.B., Giridaran, B., Dinesh, S., Gunasekaran, P., Krishnasamy, K., Sathishkumar, R., 2015. Anti—chikungunya activity of luteolin and apigenin rich fraction from *Cynodon dactylon*. Asian Pac. J. Trop. Med. 8 (5), 352–358.

Pandey, V.C., 2012. Invasive species based efficient green technology for phytoremediation of fly ash deposits. J. Geochem. Explor. 123, 13–18.

Pandey, V.C., Singh, N., 2014. Fast green capping on coal fly ash basins through ecological engineering. Ecol. Eng. 73, 671–675.

Perumal, R.K., Gopinath, A., Thangam, R., Perumal, S., Masilamani, D., Ramadass, S.K., Madhan, B., 2018. Collagen-silica bio-composite enriched with *Cynodon dactylon* extract for tissue repair and regeneration. Mat. Sci. Eng. C 92, 297–306.

Rahman, M.S., Akter, R., Mazumdar, S., Islam, F., Mouri, N.J., Nandi, N.C., Mahmud, A.S.M., 2015. Antidiabetic and antidiarrhoeal potentials of ethanolic extracts of aerial parts of *Cynodon dactylon* Pers. Asian Pac. J. Trop. Biomed. 5 (8), 658–662.

Raman, N., Radha, A., Balasubramanian, K., Raghunathan, R., Priyadarshini, R., 2002. Bioactive molecules from *Cynodon dactylon* of Indian biodiversity. In: Şener, B. (Ed.), Biodiversity. Springer, Boston, MA.

Sarkhot, D.V., Grunwald, S., Ge, Y., Morgan, C.L.S., 2012. Total and available soil carbon fractions under the perennial grass *Cynodon dactylon* (L.) Pers and the bioenergy crop *Arundo donax* L. Biomass Bioenerg. 41, 122–130.

Selhorst, A., Lal, R., 2012. Net carbon sequestration potential and emissions in home lawn turfgrasses of the United States. Environ. Manage. 51 (1), 198–208.

Selhorst, A.L., Lal, R., 2011. Carbon budgeting in golf course soils of Central Ohio. Urban Ecosyst. 14 (4), 771–781.

Shi, H., Ye, T., Chan, Z., 2013. Exogenous application of hydrogen sulfide donor sodium hydrosulfide enhanced multiple abiotic stress tolerance in bermudagrass (*Cynodon dactylon* (L). Pers.). Plant Physiol. Biochem. 71, 226–234.

Shi, H., Ye, T., Chan, Z., 2014. Comparative proteomic responses of two bermudagrass (*Cynodon dactylon* (L). Pers.) varieties contrasting in drought stress resistance. Plant Physiol. Biochem. 82, 218–228.

Shu, W.S., Ye, Z.H., Lan, C.Y., Zhang, Z.Q., Wong, M.H., 2002. Lead, zinc and copper accumulation and tolerance in populations of *Paspalum distichum* and *Cynodon dactylon*. Environ. Pollut. 120 (2), 445–453.

Shu, W.S., Ye, Z.H., Lan, C.Y., Zhang, Z.Q., Wong, M.H., 2002. Lead, zinc and copper accumulation and tolerance in populations of *Paspalum distichum* and *Cynodon dactylon*. Environ. Pollut. 120 (2), 445–453.

Shukla, S.K., Singh, K., Singh, B., Gautam, N.N., 2011. Biomass productivity and nutrient availability of *Cynodon dactylon* (L.) Pers. growing on soils of different sodicity stress. Biomass Bioenerg. 35 (8), 3440–3447.

Singh, K., Pandey, V.C., Singh, R.P., 2013. *Cynodon dactylon*: an efficient perennial grass to revegetate sodic lands. Ecol. Eng. 54, 32–38.

Smirnoff, N., 1993. The role of active oxygen in the response of plants to water deficit and desiccation. New Phytol. 125 (1), 27–58.

Suchkova, N., Tsiripidis, I., Alifragkis, D., Ganoulis, J., Darakas, E., Sawidis, T., 2014. Assessment of phytoremediation potential of native plants during the reclamation of an area affected by sewage sludge. Ecol. Eng. 69, 160–169.

Tan, S., Dong, F., Yang, Y., Zeng, Q., Chen, B., Jiang, L., 2017. Effects of waterlogging and cadmium on ecophysiological responses and metal bio-accumulation in Bermuda grass (*Cynodon dactylon*). Environ. Earth Sci. 76 (20), 719.

Wang, Y., Zhang, L., Yao, J., Huang, Y., Yan, M., 2009. Accumulation and resistance to copper of two biotypes of *Cynodon dactylon*. Bull. Environ. Contam. Toxicol. 82 (4), 454–459.

Weaver, R.W., Melton, J.R., Wang, D., Duble, R.L., 1984. Uptake of arsenic and mercury from soil by bermudagrass *Cynodon dactylon*. Environ. Pollut. A 33 (2), 133–142.

Wu, F.Y., Bi, Y.L., Leung, H.M., Ye, Z.H., Lin, X.G., Wong, M.H., 2010. Accumulation of As, Pb, Zn, Cd and Cu and arbuscular mycorrhizal status in populations of *Cynodon dactylon* grown on metal-contaminated soils. Appl. Soil Ecol. 44 (3), 213–218.

Xue, L., Liu, J., Shi, S., Wei, Y., Chang, E., Gao, M., Chen, L., Jiang, Z., 2014. Uptake of heavy metals by native herbaceous plants in an antimony mine (Hunan, China). Clean Soil Air Water 42 (1), 81–87.

Xu, J., Wang, Z., Cheng, J.J., 2011. Bermuda grass as feedstock for biofuel production: a review. Bioresourc. Technol. 102, 7613–7620.

Yu, J., Sun, L., Fan, N., Yang, Z., Huang, B., 2015. Physiological factors involved in positive effects of elevated carbon dioxide concentration on Bermuda grass tolerance to salinity stress. Environ. Exp. Bot. 115, 20–27.

Yuan, Y., Yu, S., Banuelos, G.S., He, Y., 2016. Accumulation of Cr, Cd, Pb, Cu, and Zn by plants in tanning sludge storage sites: opportunities for contamination bioindication and phytoremediation. Environ. Sci. Pollut. Res. 23 (22), 22477–22487.

Zhao, Y., Du, H., Wang, Z., Huang, B., 2011. Identification of proteins associated with water-deficit tolerance in C4 perennial grass species, *Cynodon dactylon, Cynodon transvaalensis* and *Cynodon dactylon*. Physiol. Plant 141 (1), 40–55.

Zhu, H., Yu, X., Xu, T., Wang, T., Du, L., Ren, G., Dong, K., 2015. Transcriptome profiling of cold acclimation in bermudagrass (*Cynodon dactylon*). Sci. Hor. 194, 230–236.

Moso bamboo (*Phyllostachys edulis* (Carrière) J.Houz.)–one of the most valuable bamboo species for phytoremediation

12

Purabi Saikia[a], Vimal Chandra Pandey[b,*]
[a]Department of Environmental Sciences, School of Natural Resource Management, Central University of Jharkhand, Ranchi, India; [b]Department of Environmental Science, Babasaheb Bhimrao Ambedkar University, Lucknow, Uttar Pradesh, India
*Corresponding author

1 Introduction

Bamboo is one of the most primitive, fastest growing, and diverse group of plants in the world belonging to the subfamily of *Bambusoideae* under the family *Poaceae*. Bamboo grows especially in East and Southeast Asian, African, Central and South American countries (Banik, 2000) in tropical and subtropical areas due to the wide tolerance to climatic and edaphic conditions. Some species also grow successfully in mild temperate regions in Europe and North America (Soderstrom and Ellis, 1988). Around 80% of the total bamboo species of the world are confined to China, India, and Myanmar (Sharma et al., 2016). Some 3 million years ago, bamboo vanished from Europe sometimes during the last ice age (Recht and Wetterwald, 1992). Various authors reported varied information regarding the number of species and genera of bamboos found worldwide, in China and India (Table 12.1). There is *ca.* 1500 species of bamboos under 87 genera are found worldwide (Ohrnberger, 1999; Li and Kobayashi, 2004). China has the world's richest bamboo resource with about 500 species under 39 genera (Di and Wang, 1996), which account for one-third of the world's total bamboo species. The bamboo plantation area is increasing with an average rate of 1%–2% in China (Zhou et al., 2011). About 25% of bamboos of the world are found in India particularly abundant in the Western Ghats, and Northeast India (Biswas, 1988; Rai and Chauhan, 1998) in the tropical, sub-tropical, and temperate regions where the mean annual rainfall and temperature ranges between 1200 and 4000 mm and 16°C–38°C, respectively (Tewari, 1992). Bamboos occupy 13% of the total forests area of the country (Varmah and Bahadur, 1980) and two-thirds of bamboos in the country are available in the Northeast India (Sharma et al., 2016). It can grow from the coastal plains to Himalaya upto an elevation of 3700 m (Mehra and Sharma, 1975). Bamboos are great colonizer and adaptable to a wide range of habitats and some bamboos have fast (culm) growth rates from *ca.* 7.5–100 cm per day (Buckingham et al., 2011). It characterized by profuse fibrous root/rhizome, hollow, cylindrical, and woody stem,

Phytoremediation Potential of Perennial Grasses. http://dx.doi.org/10.1016/B978-0-12-817732-7.00012-2

Table 12.1 Species distribution details of bamboos in worldwide, China and India as reported by various authors.

Region	Species distribution details of bamboos	References
Worldwide	ca. 1500 species under 87 genera of bamboo	Ohrnberger (1999); Li and Kobayashi (2004)
	ca. 1000 species of woody and perennial grasses in more than 100 genera	Lewington (1990)
	ca. 1250–1500 species of bamboo comprising ca. 75–107 genera distributed across ca. 31.5 million ha of land	Ohrnberger (1999); Scurlock et al. (2000)
	Over 1250 species under 75 genera	Soderstrom and Ellis (1988)
	ca. 2000 species are found worldwide belonging to over 70 genera and cover an area of 14 Mha worldwide	Dransfield (1992)
China	Over 500 species under 39 genera found in China which account for one-third of the world's total bamboo species.	Di and Wang (1996); Zhou et al. (2005)
	About 300 species of bamboo belonging to 44 genera are available in China	Xu et al. (2018)
	ca. 500–534 species that occupy ca. 4.84–5.71 million ha	Chen et al. (2009); Song et al. (2011)
India	ca. 128 species of bamboos distributed over ca. 5.48 Mha land area	Tewari, (1992)
	A total of ca. 148 species under 29 genera of bamboos are currently occurring in India both in wild and cultivation.	Sharma and Nirmala (2015)
	There are 124 indigenous and exotics species, under 23 genera found naturally and/or under cultivation	Naithani (1993)

petiolated leaf blades, and infrequent flowering (Soderstrom and Calderon, 1979). Sympodial and monopodial are the two main types of growth found in bamboos determined by root/rhizome structure. In general, sympodial bamboos are characteristics of tropical climate and monopodial bamboos are of temperate climate and are highly invasive and unmanageable (Uma Shaanker et al., 2004). Sympodial bamboo such as the genus *Bambusa* used to have pachymorph rhizomes that turn upward as soon as it formed and each rhizome develops into new culm close to the parent plant. Monopodial bamboo such as *Phyllostachys* develops from long, slender, leptomorph rhizomes that grow horizontally under the soil. Bamboos play an important role in local economies especially in the Asia-Pacific region and grown for industrial, agricultural, ornamental, and ecological purposes (Fu and Banik, 1996).

Phyllostachys edulis (Carrière) J.Houz. (Moso bamboo) is a monopodial, indigenous, widely distributed, highly productive, and economically important bamboo of China that covers about 70% of the country's total area of bamboo forests with an area of more than 3.37 Mha (Chen et al., 2009). It has accounted for 80% of bamboo forest regions over the world (Song et al., 2016, 2017). Its growth rate is markedly higher than any other tree species and known as the fastest growing plant in the world. It can grow up to 100 cm daily and 15–24 m high in a period of 40–60 days (Bai et al., 2016; Li et al., 2016; Sun et al., 2016). It reaches a mature state in less than 2 months with an average height of 15 m and mean shoot elongation and biomass accumulation rate are 17 cm day^{-1} and 96 g day^{-1}, respectively (Xu et al., 2011). It develops aerial culm from buds that develop at certain interval along the rhizome and its rhizome can grow more than 10 m in a year (Uma Shaanker et al., 2004). The recorded culm density ranged from 1350 to 4722 culm ha^{-1} in natural forests of China (Zhou and Jiang, 2004; Xu et al., 2018). It is one of the most valuable bamboos widely distributed in Asia, mainly throughout the tropical and subtropical zones (Song et al., 2013). In China, the establishment of *P. edulis* plantations is strongly recommended by the government because of its higher economic and ecological returns. Hence, it is an important forestry plant in southern China especially in Fujian province, where the bamboo industry generates 3.43 billion Yuan (about US$500 million) per year, and ranks among the five most important businesses (Zhang et al., 2001). In China, it is usually found in the elevations less than 800 m in mountain and hills (Zhang et al., 2015) but, can grow in an altitude between 10 and 1700 m above the sea level. The species originates from China and has been naturalized in some other neighboring countries, such as Korea and Vietnam. It was introduced in Japan to utilize the culm wood, and edible young sprouts during 1746 (Suzuki, 1978) and has escaped from the planted areas and invaded the broadleaved forests of Japan. *P. edulis* along with *Bambusa blumeana* introduced in Brazil for controlling soil erosion, preventing nutrient loss, and improving soil structure (Fu et al., 2000). It has some superior attributes in terms of adaptation to environmental stress, extensive competitive ability, fast growth rate, high biomass productivity, multi-purpose applications, such as for furniture, building materials, and decoration, ease in cultivation, short cutting time (4–5 years), and high ecological values (Wang et al., 2013) as compared to other bamboo species. *P. edulis* in China has threatened the associated rare/endangered species as it expands quickly (Song et al., 2011). In this chapter, effort has been made to provide critical information on the role of *Phyllostachys edulis*(Carrière) J.Houz. toward provisioning ecosystem services, and nature sustainability to enable planning for sustainable utilization and conservation of bamboo in general and *P. edulis*in particular.

2 Bamboo-provisioned ecosystem services

Bamboo is an ideal economic investment that has enormous potential for alleviating both environmental and social problems facing by the today's world. Bamboo provides numerous environmental services including utilization of bamboos in construction of rural households, and traditional crafts, large capacity for carbon storage, everyday use by billions of people in their day to day life (Nath and Das, 2008; Yen and Wang, 2013), generates more oxygen, protects against ultraviolet rays and is an important atmospheric and soil purifier. These are only a few among numerous reasons of maintaining and increasing carbon stocks of bamboos at forests, and rural agro-ecosystems. Bamboo forests can provide income to rural communities from dual sources (1) selective harvest and selling bamboo and bamboo-based products and, (2) under Clean Development Mechanism (CDM) and Reducing Emissions from Deforestation and Forest Degradation (REDD) program by Certified Emission Reductions (i.e., carbon credits) through various afforestation/reforestation mechanisms (Nath et al., 2015). Certified emission reductions can be traded in the national and international markets that have committed to reduce their carbon footprint. It has the potential of being effective in carbon sequestration, thus, helping in countering the emission of greenhouse gases, global warming, and climate change. Bamboo requires little fertilizer and pesticides for its management, and protects traditional houses from winds, and fulfills requirements of fuelwoods as an alternative to wood (Li and Kobayashi, 2004; Nath et al., 2009). Sustainable use of bamboo help in livelihood security of millions of rural communities through commercialization of bamboo based products by value addition, thus helping to decrease poverty and increase employment (Nath et al., 2009; Ly et al., 2012). Fifty-six different bamboo species are edible and young shoots used in preparation of traditional curries as well as pickles, while 12 species can be used as construction material for traditional houses and furniture, and 18 species to make paper and pulp (Li and Kobayashi, 2004) which consumes sizeable proportion of the total annual production. Bamboos are now being used for wall paneling, floor tiles, briquettes for fuel, rebar for reinforced concrete beams, etc. (Sharma et al., 2016). *P. edulis* is harvested throughout South East Asia for both construction purposes and young edible shoots (Xu et al., 2018). Bamboo and bamboo based products has resulted in an estimated global market value of US\$7 billion (Lobovikov et al., 2012) with ca. 1500 commercial applications (Scurlock et al., 2000). Bamboo is a timber substitute at forest ecosystem level, and important for rehabilitation of degraded land, and watershed protection (INBAR, 2015). Bamboo can bind up to 6 m^3 of soil and helps in preventing soil erosion (Zhou et al., 2005), conserving soil moisture, improves nutrients retention, and regulation of water flow in water bodies due to its extensive shallow fibrous, connected rhizome/root system and its habit of producing new culms, dense foliage, and accumulation of leafy mulch (Li and Kobayashi, 2004; Lobovikov et al., 2012). It grows well on steep hillsides, road embankments, gullies, or on the banks of ponds and streams. Therefore, there is a strong need to recognize the ecosystem services of bamboos for well-being of society, nature conservancy, and for considerations toward trading of carbon credits.

3   Major role of bamboo toward nature sustainability

3.1   Phytoremediation

Phytoremediation is an effective and ideal means for the recovery of various contaminated sites and a kind of environmentally sustainable and economically feasible remediation technique to remove heavy metals and other contaminants using plants (Tangahu et al., 2011; Ahmadpour et al., 2012). Metal hyper-tolerance is usually a heritable and constitutive species level trait (Macnair, 1993) and can be achieved through restricted uptake, detoxification through complexation or transformation, and intracellular compartmentalization (Clemens, 2001, 2006). Several plants can grow in heavy metal contaminated soils as they use avoidance strategy, limiting heavy metal uptake by chelating metal ions in rhizosphere, or binding in mycorrhizal fungi (Meier et al., 2012; Rajkumar et al., 2012). The high efficiency of phytoremediation depends on plant roots that accumulate bioavailable metals and transport them rapidly to the aboveground tissues. About 700 plants have been reported to be hyper-accumulators of different contaminants (Xi et al., 2010) with high efficiency of accumulation of heavy metals with better tolerance ability (Kamran et al., 2014; Salazar and Pignata, 2014). Different plant species have different strategies in selectively absorbing heavy metals to maintain their own ecological balance (Zaier et al., 2010). Metal hyper-accumulators are found in a large number of plant families and the most dominant is the *Brassicaceae* (Verbruggen et al., 2009; Krämer, 2010). They are generally slow growing plants with less biomass yields (Peuke and Rennenberg, 2005; Tandy et al., 2006; Epelde et al., 2008), requires more time to reduce metal concentrations in soil (McGrath and Zhao, 2003). Therefore, there is an urgent need to identify other fast growing plants with greater biomass production for phytoremediation of heavy metal contaminated sites. Bamboos are an appropriate option for phytoremediation due to its fast growth, high biomass production (Rajkumar et al., 2012), wide root system, quick harvest, and high tolerance to abiotic stresses with minimal maintenance (Yang et al., 2005). They also have the ability to re-establish at sites where natural vegetation has failed to grow. Although, heavy metal accumulation in *P. edulis* did not reach the standard of hyper-accumulators, it has been recognized as a potential phytoremediation plant due to its fast growth, huge biomass production upto $121.14\,t\,hm^{-2}$ (Wu et al., 2002), high tolerance to environmental stresses, and high endurance against heavy metal stress including Zn, Cd, Pb, Cr, Mn, Cu, etc. (Chen et al., 2015; Liu et al., 2015). The greater surface area of roots is known to contribute more to the absorption of heavy metals and nutrients (Li et al., 2014). *P. edulis* has well-developed, massive, and complex root pattern distributed in depth of 0–40 cm soil layer. Its roots have effectively accumulated bio available metals and rapidly transported them to rhizomes, stems, branches, and leaves. Normal growth of *P. edulis* was observed in old Pb/Zn mine areas of China (Zhang et al., 2006) may be due to selective absorption of essential nutrients to maintain appropriate nutrition of their photosynthetic organs (Zaier et al., 2010). High concentration of Pb ($>400\,mg\,kg^{-1}$) in soils can influence the cells metabolism (Ekmekçi et al., 2009) and inhibit growth of *P. edulis* but, low concentration ($<400\,mg\,kg^{-1}$) used to promote growth and exhibited positive stimulating effects (Zhong et al., 2017). On the other hand, under zinc stress ($100\,mg\,L^{-1}$), the seed germination rate in *P. edulis* did not differ significantly from that of the control, and the Zn concentrations in the root ranged from 2329 to 8642 mg kg^{-1} (Liu et al., 2014). The Cd accumulating ability is weaker (not good beyond 5 mg kg^{-1} Cd) in *P. edulis* as compare to hyper-accumulator plants like *Viola baoshanensis* and *Sedum alfredii* but, the much higher biomass, much developed root system make it suitable for phytoremediation of soils having low contamination of Cd (Li et al., 2014). Besides, it has strong ability to concentrate Cu and stronger absorbing capacity than those of hyper-accumulator plants (Chen et al., 2016; Bian et al., 2017). It also yields biomass for carbon dioxide fixation, thereby carrying out two ecological services, that is, in situ remediation and carbon sequestration (Yen and Lee, 2011).

3.2   Carbon sequestration

Natural forests play an important role in the global carbon cycle and in mitigating the warming effect through carbon sequestration (Keith et al., 2009). Bamboo has large capacity for biomass accumulation within a short period of time, and also has a high potential for carbon storage (Kleinhenz and Midmore, 2001). Biomass carbon stock and sequestration rate in woody bamboos are quite comparable with agroecosystem and forest ecosystems due to its rapid biomass accumulation and effective fixation of solar energy and carbon dioxide (Zhou et al., 2005). One quarter of the biomass in tropical regions and one-fifth in subtropical regions comes from bamboo (Anonymous, 1997). Bamboo forests are important carbon source and sink on the earth (Li et al., 2003) with a high carbon stock potential due to fast growth and regrowth (INBAR, 2010). Carbon accumulation rates basically depend on type of bamboo (whether monopodial or sympodial), prevailing environmental conditions, and management practices (irrigation, weeding, thinning, harvesting intensity). The carbon sequestration rates varied greatly in various types of bamboos worldwide and it is as high as 13–24 Mg C ha^{-1}year^{-1} (Nath et al., 2015). The carbon storage in bamboo forests in China (169–259 Mg C ha^{-1}) is much higher than in natural forests of China and global forests (i.e., 39 and 86 Mg C ha^{-1}, respectively) (Song et al., 2011). In case of sympodial bamboos, it is highest (24 Mg C ha^{-1}year^{-1}) in *Bambusa bambos* plantation in India (Shanmughavel and Francis, 1996) followed by (16 Mg C ha^{-1}year^{-1}) *B. oldhamii* plantation in Mexico (Castañeda-Mendoza et al., 2005). On the other hand, in monopodial bamboos, *P. bambusoides* plantation in Japan has the highest carbon sequestration rate (13 Mg C ha^{-1}year^{-1}) (Isagi et al., 1993). The annual carbon sequestration rate in *P. makinoi* is 8.13 Mg ha^{-1}year^{-1} (Yen et al., 2010). The fast growth of *P. edulis* forests in China sequestered 33% higher carbon (i.e., 5.10 Mg C ha^{-1}) than a tropical mountain rainforest in a single year (Zhou and Jiang, 2004; Kuehl et al., 2013). A positive net carbon production rate (8.5 tha^{-1}year^{-1}) is observed in *P. edulis* stand in Kyoto Prefecture, Japan which is comparable to natural forests in Japan with similar climates (Isagi et al., 1997). *P. edulis* is known as the most promising species for carbon sequestration with high biomass production and biomass accumulation and carbon storage mainly occurs in the initial growth stage. Its' carbon stock varies considerably (Chen et al., 2016; Yen, 2015) which could be attributed to exclusion of several influencing factors because of limited data availability. The annual carbon sequestration rate of *P. edulis* ranged from 6.0 to 7.6 Mg C ha^{-1} (Xu et al., 2018), and the carbon contents of leaves, branches, and stems were 45.44, 48.15, and 46.28%, respectively (Yen and Lee, 2011), suggesting *P. edulis* as a candidate species for carbon fixation. The total carbon stock (soil and biomass carbon) of *P. edulis* forests increased with latitude and ranged between 87.83 and 119.5 Mg C ha^{-1} (Xu et al., 2018).

3.3   Climate change mitigation

A major international policy goal in the context of global climate change is to find a low-cost method to sequester carbon (Montagnini and Nair, 2004). Rapid growth rates of bamboo favor the accumulation of organic carbon by photosynthesis in aboveground culms including the culm branches with their sheaths and leaves, and an underground network of roots and persistent rhizomes (Düking et al., 2011; Lobovikov et al., 2012). Bamboo is a perennial woody grass and is not adequately recognized as part of natural forests in major international policies such as the Kyoto Protocol and the Marrakech Accords (Lou et al., 2010). Under selective felling strategy, biomass and carbon stock in bamboo are also the permanent stock as harvesting of bamboos and subsequent loss of biomass and carbon are balanced by new culm produced in the clump every year (Nath and Das, 2011). Bamboo forests can sequester substantial quantities of carbon, thereby helping to mitigate the effects of climate change (Nath et al., 2015) due to its large areal distribution and their high growth rates. Plantation of bamboo and its management in homegarden have both ecological and economical benefits to the rural life and can become an effective choice for climate change mitigation strategy. There is a need for greater integration of bamboo into national and international policies to manage the effects of global climate change as it acts as an important carbon sink. Sustainable biomass removal can be included in future REDD+ to realize the potential of bamboo to combat deforestation (Hein and van der Meer, 2012). Therefore, reorganization of ecological implications of bamboo in sustainable forest management is an important consideration for REDD schemes.

3.4   Bamboo-based agroforestry system

Bamboo is being cultivated and used by humans since at least 6000 years back (Song et al., 2011), and it is now become integral part of routine life for billions of people worldwide (Lobovikov et al., 2007). In rural landscape of Asia Pacific region, farmers manages bamboo either in the traditional agroforestry system or in pure stands adjoining to their homegardens called "bamboo grove" (Nath and Das, 2011). Bamboos based agroforestry can play an important role in enhancing productivity, sustainability, and resource conservation. Bamboo is an important livelihoods option for rural communities (Lobovikov et al., 2012) and is selectively harvested annually. Selective felling does not affect the productivity of bamboo forest and therefore, its cultivation is a sustainable practice (Hoogendoorn and Benton, 2014). Bamboo plantation has inherent ecological consistency of local farming practices under traditional homegarden management system (Fig. 12.1A,B). Bamboo is one of the important components of the homegardens of Assam, Northeast India which provides a wide range of goods and services to the owners (Das and Das, 2005; Saikia et al., 2012). Bamboos have socioeconomic and ecological values and its management can provide benefits on a local, national, and global level through livelihood, economic, and environmental security for millions of rural people (Nath et al., 2009). Integrating bamboos with carbon trading promote the cultivation and management of woody bamboos in agroforestry and forest ecosystems and therefore, generating another income source for the rural communities. Cultivating bamboo as a cash crop alone may not be viable as the economic returns per unit area of bamboo are almost 8 times lesser than cash crops like cassava (Ly et al., 2012). Therefore, bamboo should be managed with an ecosystem perspective considering the socio-ecological interactions.

Figure 12.1 (A) Bamboo based agroforestry system (homegarden) and (B) Pure patch of Bamboo in homegardens of Assam, Northeast India.

4   Future research prospects

- Additional research is needed to determine bamboo biomass, vegetation, and soil carbon capture, and storage through incorporating improved methodological protocols to enable precise estimation of carbon storage, and sequestration rate of bamboo forests.

- More attention needs to be paid to determine the critical ecosystem services of bamboos in terms of carbon farming and subsequently carbon trading.
- Further research warrants emphasizing on hydrological and ecological functions of bamboo forest for erosion control, water conservation, etc.
- Long-term monitoring strategies are critical to investigate related toxicities and ecological impact when phytoremediation of heavy metals used to be done as it remains in the roots/shoots after accumulation.
- There is an urgent need for screening the high yielding species of bamboo for successful integration of bamboos in agroforestry systems, also soil-plant-water interaction in the bamboos based agroforestry needs to be studied to reduce competition with other traditional crops.

5 Conclusions

Bamboo is an important versatile forest and agroforestry based products used in industry, routine life and also for environment protection. Besides, it is a perfect solution for the environmental and social consequences of tropical deforestation as well as important alternatives for phytoremediation of contaminated sites due to higher biomass accumulation and fast growth. Considering its role in climate change adaptation, mitigation, its noteworthy contribution in socio-economic life of rural households, and numerous other environmental services, bamboos warrant serious consideration for carbon farming and carbon trading. Bamboo based agroforestry systems can help in augmenting the income of farmers besides conserving the resources efficiently. The soil-plant-water interactions in bamboo based agroforestry also need to be studied as bamboo used to be more competitive due to fast growth when grown in association with traditional crops in agroforestry.

References

Ahmadpour, P., Ahmadpour, F., Mahmud, T.M.M., Abdu, A., Soleimani, M., Tayefeh, F.H., 2012. Phytoremediation of heavy metals: a green technology. Afr. J. Biotechnol. 11, 14036–14043.

Anonymous, 1997. Healing degraded land. INBAR Magazine 5 (3), 40–45.

Bai, Q., Hou, D., Li, L., Cheng, Z., Ge, W., Liu, J., Li, X., Mu, S., Gao, J., 2016. Genome-wide identification and characterization of the Dof gene family in moso bamboo (*Phyllostachys heterocycla* var. *pubescens*). Genes Genom. 38, 733–745.

Banik, R.L., 2000. Silviculture and Field-Guide to Priority Bamboos of Bangladesh and South Asia. Government of the people's Republic of Bangladesh, Bangladesh Forest Research Institute, Chittagong, p. 82.

Bian, F., Zhong, Z., Zhang, X., Yang, C., 2017. Phytoremediation potential of moso bamboo (*Phyllostachys pubescens*) intercropped with Sedum plumbizincicola in metal-contaminated soil. Environ. Sci. Pollut. Res. 24 (35), 27244–27253, 10.1007/s11356-017-0326-2.

Biswas, S., 1988. Studies on bamboo distribution in North-eastern region of India. Indian For. 114 (9), 514–531.

Buckingham, K., Jepson, P., Wu, L., Ramanuja, R.I.V., Jiang, S., Liese, W., Lou, Y., Fu, M., 2011. The potential of bamboo is constrained by outmoded policy frames. Ambio 40, 544–548.

Castañeda-Mendoza, A., Vargas-Hernández, J.J., Gómez-Guerrero, A., Valdez-Hernández, J.I., Vaquera-Huerta, H., 2005. Acumulación de carbono en la biomasa aérea de una plantación de *Bambusa oldhamii*. Agrociencia 39, 107–116.

Chen, J., Shafi, M., Li, S., Wang, Y., Wu, J., Ye, Z., Peng, D., Yan, W., Liu, D., 2015. Copper induced oxidative stresses, antioxidant responses and phytoremediation potential of moso bamboo (*Phyllostachys pubescens*). Sci. Rep. 5, 135–154.

Chen, J., Shafi, M., Wang, Y., Wu, J., Ye, Z., Liu, C., Zhong, B., Guo, H., He, L., Liu, D., 2016. Organic acid compounds in root exudation of moso bam-boo (*Phyllostachys pubescens*) and its bioactivity as affected by heavy metals. Environ. Sci. Pollut. Res. 23, 20977–20984.

Chen, X., Zhang, X., Zhang, Y., Booth, T., He, X., 2009. Changes of carbon stocks in bamboo stands in China during 100 years. For. Ecol. Manag. 258, 1489–1496.

Clemens, S., 2001. Molecular mechanisms of plant metal homeostasis and tolerance. Planta 212, 475–486.

Clemens, S., 2006. Toxic metal accumulation, responses to exposure and mechanisms of tolerance in plants. Biochimie 88, 1707–1719.

Das, T., Das, A.K., 2005. Inventorying plant biodiversity in homegardens: a case study in Barak Valley, Assam, North East India. Curr. Sci. 89, 155–163.

Di, B.J., Wang, Z.P., 1996. Flora of China vol. 9, section 1, Science Press, Beijing (in Chinese).

Dransfield, S., 1992. The bamboos of Sabah. Sabah Forest Report, No. 141, Malaysia.

Düking, R., Gielis, J., Liese, W., 2011. Carbon flux and carbon stock in a bamboo stand and their relevance for mitigating climate change. Bamboo Sci. Cult. 24, 1–7.

Ekmekçi, Y., Tanyolaç, D., Ayhan, B., 2009. A crop tolerating oxidative stress induced by excess lead: maize. Acta Physiol. Plant. 31, 319–330.

Epelde, L., Hernández-Allica, J., Becerril, J.M., Blanco, F., Garbisu, C., 2008. Effects of chelates on plants and soil microbial community: comparison of EDTA and EDDS for lead phytoextraction. Sci. Total Environ. 401 (1–3), 21–28.

Fu, M., Banik, R.L., 1996. Bamboo production system and their management. In: Rao, I.V.R., Sastry, C.B., Widjaja, E. (Eds.), Proceedings on Bamboo, People and the Environment: Propagation and Management, Vol.1, Fifth International Bamboo Workshop and Fourth International Bamboo Congress, Bali, 19–22 June 1995, pp. 18–33.

Fu, M., Xiao, J., Lou, Y., 2000. Cultivation and Utilization on Bamboo. China Forestry Publishing House, Beijing.

Hein, L., van der Meer, J.P., 2012. REDD+ in the context of ecosystem management. Curr. Opin. Environ. Sustain. 4, 604–611.

Hoogendoorn, J.C., Benton, A., 2014. Bamboo and rattan production and the implications of globalization. In: Nikolakis, W., Innes, J. (Eds.), Forest and Globalization: Challenges and Opportunities for Sustainable Development, Vol. 711. Third Avenue, New York, p. 10017.

INBAR, 2015. The partnership for a better world-strategy to the year 2015. Beijing, China.

INBAR, 2010. Bamboo and climate change mitigation: a comparative analysis of Carbon sequestration, International Network for bamboo and Rattan (INBAR), Technical Report No. 32, Beijing, China, p. 47.

Isagi, Y., Kawahara, T., Kamo, K., 1993. Biomass and net production in a bamboo *Phyllostachys bambusoides* stand. Ecol. Res. 8, 123–133.

Isagi, Y., Kawahara, T., Kamo, K., Ito, H., 1997. Net production and carbon cycling in a bamboo *Phyllostachys pubescens* stand. Plant Ecol. 130, 41–52.

Kamran, M.A., Amna, Mufti, R., Mubariz, N., Syed, J.H., Bano, A., Javed, M.T., Munis, M.F., Tan, Z., Chaudhary, H.J., 2014. The potential of the flora from different regions of Pakistan in phytoremediation: a review. Environ. Sci. Pollut. Res. 21, 801–812.

Keith, H., Mackey, B.G., Lindenmayer, D.B., 2009. Re-evaluation of forest biomass carbon stocks and lessons from the world's most carbon-dense forests. PNAS 106, 11635–11640.

Kleinhenz, V., Midmore, D.J., 2001. Aspects of bamboo agronomy. Adv. Agron. 74, 99–153.

Krämer, U., 2010. Metal hyperaccumulation in plants. Annu. Rev. Plant Biol. 61 (1), 517–534.

Kuehl, Y., Li, Y., Henley, G., 2013. Impacts of selective harvest on the carbon sequestration potential in moso bamboo (*Phyllostachys pubescens*) plantations. For. Trees Livelihoods 22, 1–18.

Lewington, A., 1990. Plants for People. Oxford University Press, New York.

Li, Q., Song, X., Gu, H., Gao, F., 2016. Nitrogen deposition and management practices increase soil microbial biomass carbon but decrease diversity in moso bamboo plantations. Nat. Sci. Rep. 6, 28235.

Li, Q., Zhou, B.Z., Wang, X.M., Ge, X.G., Cao, Y.H., 2003. Drought effects on vegetation carbon storage in a moso bamboo forest in northern Zhejiang: results of a through fall exclusion experiment. Adv. Mater. Res. 864–867, 2715–2718.

Li, T.Q., Tao, Q., Shohag, M.J.I., Yang, X.E., Sparks, D.L., Liang, Y.C., 2014. Root cell wall polysaccharides are involved in cadmium hyperaccumulation in *Sedum alfredii*. Plant Soil 389, 387–399.

Li, Z.H., Kobayashi, M., 2004. Plantation future of bamboo in China. J. Forest. Res. 15, 233–242.

Li, Z.H., Kobayashi, M., 2004. Plantation future of bamboo in China. J. Forest. Res. 15, 233–242.

Liu, D., Chen, J., Mahmood, Q., Li, S., Wu, J., Ye, Z., Peng, D., Yan, W., Lu, K., 2014. Effect of Zn toxicity on root morphology, ultrastructure, and the ability to accumulate Zn in moso bamboo (*Phyllostachys pubescens*). Environ. Sci. Pollut. Res. 21, 13615–13624.

Liu, D., Li, S., Islam, E., Chen, J.R., Wu, J.S., Ye, Z.Q., Peng, D.L., Yan, W.B., Lu, K.P., 2015. Lead accumulation and tolerance of moso bamboo (*Phyllostachys pubescens*) seedlings: applications of phytoremediation. J. Zhejiang Univ. Sci. B 16, 123–130.

Lobovikov, M., Paudel, S., Piazza, M., Ren, H., Wu, J., 2007. World Bamboo Resources: A Thematic Study Prepared in the Framework of the Global Forest Resources Assessment 2005. FAO, Rome, Italy.

Lobovikov, M., Schoene, D., Lou, Y., 2012. Bamboo in climate change and rural livelihoods. Mitig. Adapt. Strateg. Glob. Change 17, 261–276.

Lou, Y., Li, Y., Buckingham, K., Henley, G., Zhou, G., 2010. Bamboo and Climate Change Mitigation: A Comparative Analysis of Carbon Sequestration. International Network for Bamboo and Rattan (INBAR), Beijing.

Ly, P., Pillot, D., Lamballe, P., de Neergaard, A., 2012. Evaluation of bamboo as an alternative cropping strategy in the northern central upland of Vietnam: above-ground carbon fixing capacity, accumulation of soil organic carbon, and socio-economic aspects. Agric. Ecosyst. Environ. 149, 80–90.

Macnair, M.R., 1993. The genetics of metal tolerance in vascular plants. New Phytol. 124 (4), 541–559.

McGrath, S.P., Zhao, F.J., 2003. Phytoextraction of metals and metalloids from contaminated soils. Curr. Opin. Biotech. 14, 277–282.

Mehra, P.N., Sharma, M.L., 1975. Cytological studies in some central and Eastern Himalayan grasses. V. The *Bambuseae*. Cytologia 40 (2), 463–467.

Meier, S., Borie, F., Bolan, N., Cornejo, P., 2012. Phytoremediation of metal-polluted soils by arbuscular *Mycorrhizal* fungi. Crit. Rev. Environ. Sci. Technol. 42, 741–775.

Montagnini, F., Nair, P.K.R., 2004. Carbon sequestration: an under exploited environmental benefit of agroforestry systems. Agroforest. Syst. 61, 281–295.

Naithani, H.B., 1993. Contributions to the Taxonomic Studies of Indian Bamboos. Ph.D. thesis. Vol. 1. H.N.B. Garhwal University, Srinagar, Garhwal.

Nath, A.J., Das, A.K., 2011. Carbon storage and sequestration in bamboo-based smallholder homegardens of Barak Valley, Assam. Curr. Sci. 100, 229–233.

Nath, A.J., Das, A.K., 2008. Bamboo resources in the homegardens of Assam: a case study from Barak Valley. J. Trop. Agric. 46, 46–49.

Nath, A.J., Das, G., Das, A.K., 2009. Above ground standing biomass and carbon storage in village bamboos in North East India. Biomass Bioenerg. 33, 1188–1196.

Nath, A.J., Lal, R., Das, A.K., 2015. Ethnopedology and soil properties in bamboo (*Bambusa* sp.) based agroforestry system in North East India. CATENA 135, 92–99.

Ohrnberger, D., 1999. The Bamboos of the World. Elsevier, Amsterdam, Netherlands.

Peuke, A., Rennenberg, H., 2005. Phytoremediation "viewpoint". EMBO J. 6, 497–501.

Rai, S.N., Chauhan, K.V.S., 1998. Distribution and growing stock of bamboos in India. Indian Forest. 124 (2), 89–97.

Rajkumar, M., Sandhya, S., Prasad, M.N., Freitas, H., 2012. Perspectives of plant-associated microbes in heavy metal phytoremediation. Biotechnol. Adv. 30, 1562–1574.

Recht, C., Wetterwald, M.F., 1992. Bamboos, first ed. Stuttgart, Germany.

Saikia, P., Choudhury, B.I., Khan, M.L., 2012. Floristic composition and plant utilization pattern in homegardens of Upper Assam, India. Trop. Ecol. 53 (1), 105–118.

Salazar, M.J., Pignata, M.L., 2014. Lead accumulation in plants grown in polluted soils. Screening of native species for phytoremediation. J. Geochem. Explor. 137, 29–36.

Scurlock, J.M., Dayton, D.C., Hames, B., 2000. Bamboo: An Overlooked Biomass Resource? Oak Ridge National Laboratory, Oak Ridge.

Shanmughavel, P., Francis, K., 1996. Above ground biomass production and nutrient distribution in growing bamboo (*Bambusa bambos* (L) Voss). Biomass Bioenerg. 10, 383–391.

Sharma, P., Sharma, K.P., Saikia, P., 2016. Diversity, uses and in vitro propagation of different bamboos of Sonitpur District, Assam. J. Ecosyst. Ecogr. 6 (2), 1–9.

Sharma, M.L., Nirmala, C., 2015. Bamboo Diversity of India: An Update. 10th World Bamboo Congress, Korea.

Soderstrom, T.R., Calderon, C.E., 1979. A commentary on the bamboos (*Poaceae: Bambusoideae*). Biotropica 11 (3), 161–172.

Soderstrom, T.R., Ellis, R.P., 1988. The woody bamboos (*Poaceae: Bambusoideae*) of Sri Lanka. In: A Morphological-Anatomical Study. Smithsonian Contributions of Botany, Vol. 72. Smithsonian Institution Press, Washington, D.C., pp. 30–36.

Song, Q., Ouyang, M., Yang, Q., Lu, H., Yang, G.Y., Chen, F.S., Shi, J.M., 2016. Degradation of litter quality and decline of soil nitrogen mineralization after moso bamboo (*Phyllostachys pubscens*) expansion to neighboring broadleaved forest in subtropical China. Plant Soil 404, 113–124.

Song, X., Li, Q., Gu, H., 2017. Effect of nitrogen deposition and management practices on fine root decomposition in Moso bamboo plantations. Plant Soil 410, 207–215.

Song, X., Zhou, G., Jiang, H., Yu, S., Fu, J., Li, W., Wang, W., Ma, Z., Peng, C., 2011. Carbon sequestration by Chinese bamboo forests and their ecological benefits: assessment of potential, problems, and future challenges. Environ. Rev. 19, 418–428.

Song, X.Z., Peng, C.H., Zhou, G.M., Jiang, H., Wang, W.F., Xiang, W.H., 2013. Climate warming-induced upward shift of Moso bamboo population on Tianmu Mountain, China. J. Mt. Sci. 10, 363–369.

Sun, H., Li, L., Lou, Y., Zhao, H., Gao, Z., 2016. Genome-wide identification and characterization of aquaporin gene family in moso bamboo (*Phyllostachys edulis*). Mol. Biol. Rep. 43, 437–450.

Suzuki, S., 1978. Index to Japanese *Bambusaceae*. Gakken, Tokyo.

Tandy, S., Schulin, R., Nowack, B., 2006. The influence of EDDS on the uptake of heavy metals in hydroponically grown sunflowers. Chemosphere 62 (9), 1454–1463.

Tangahu, B.V., Abdullah, S.R.S., Basri, H., Idris, M., Anuar, N., Mukhlisin, M., 2011. A review on heavy metals (As, Pb, and Hg) uptake by plants through phytoremediation. Int. J. Chem. Eng. 2011, 1–31.

Tewari, D.N., 1992. A Monograph on Bamboos. International Book Distributors, Dehra Dun, India, pp. 498.

Uma Shaanker, R., Ganeshaiah, K.N., Srinivasan, K., Ramanatha R.V., Hong, L.T., 2004. Bamboos and Rattans of the Western Ghats: Population biology, Socio-economics and Conservation strategies. ATREE, IPGRI and UAS, Bangalore.

Varmah, J.C., Bahadur, K.N., 1980. Country report and status of research on bamboos in India. Indian For. Rec.N. Ser. Bot. 6 (1), 1–28.

Verbruggen, N., Hermans, C., Schat, H., 2009. Molecular mechanisms of metal hyperaccumulation in plants. New Phytol. 181 (4), 759–776.

Wang, B., Wei, W.J., Liu, C.J., You, W.Z., Niu, X., Man, R.Z., 2013. Biomass and carbon stock in moso Bamboo forests in subtropical China: characteristics and implications. J. Trop. For. Sci. 25, 137–148.

Wu, J.S., Yu, Y.W., Zhu, Z.J., Xiao, X.F., 2002. Studies on the biomass of different forests in Huzhou city. Jiangsu For. Sci. Technol. 29 (4), 22–24, (in Chinese).

Xi, X.Y., Liu, M.Y., Huang, Y., et al., 2010. Response of flue-cured tobacco plants to different concentration of lead or cadmium. Fourth International Conference on Bioinformatics and Biomedical Engineering (ICBBE), IEEE, pp. 1–4.

Xu, M., Ji, H., Zhuang, S., 2018. Carbon stock of moso bamboo (*Phyllostachys pubescens*) forests along a latitude gradient in the subtropical region of China. PLoS ONE 13 (2), e0193024.

Xu, X., Du, H., Zhou, G., Ge, H., Shi, Y., Zhou, Y., Fan, W., Fan, W., 2011. Estimation of aboveground carbon stock of Moso bamboo (*Phyllostachys heterocycla* var. *pubescens*) forest with a Landsat Thematic Mapper image. Int. J. Remote Sens. 32, 1431–1448.

Yang, X., Feng, Y., He, Z., Stoffella, P.J., 2005. Molecular mechanisms of heavy metal hyperaccumulation and phytoremediation. J. Trace Elements Med. Biol. 18, 339–353.

Yen, T.M., 2015. Comparing aboveground structure and aboveground carbon storage of an age series of moso bamboo forests subjected to different management strategies. J. For. Res. 20, 1–8.

Yen, T.M., Ji, Y.T., Lee, J.S., 2010. Estimating biomass production and carbon stock for a fast-growing makino bamboo (*Phyllostachys makinoi*) plant based on the diameter distribution model. For. Ecol. Manag. 260, 339–344.

Yen, T.M., Lee, J.S., 2011. Comparing aboveground carbon sequestration between moso bamboo (*Phyllostachys heterocycla*) and Chinafir (*Cunninghamia lanceolata*) forests based on the allometric model. For. Ecol. Manag. 261, 995–1002.

Yen, T.M., Wang, C.T., 2013. Assessing carbon storage and carbon sequestration for natural forests, man-made forests, and bamboo forests in Taiwan. Int. J. Sustain. Dev. World 20, 455–460.

Zaier, H., Tahar, G., Lakhdar, A., Baioui, R., Ghabrichea, R., Mnasri, M., Sghair, S., Lutts, S., Abdellya, C., 2010. Comparative study of Pb-phytoextraction potential in *Sesuvium portulacastrum* and *Brassica juncea*: tolerance and accumulation. J. Hazard. Mater. 183, 609–615.

Zhang, H., Zhuang, S., Qian, H., Wang, F., Ji, H., 2015. Spatial variability of the topsoil organic carbon in the moso bamboo forests of Southern China in association with soil properties. PLOS ONE 10, e0119175.

Zhang, L., Ye, Z.Q., Li, T.Q., Yang, X.E., 2006. Studies on soil microbial activity in areas contaminated by tailings from Pb Zn mine. J. Soil Water Conserv. 20 (3), 136–140, (in Chinese).

Zhang, Y.X., Zhang, Z.Q., Zhang, X.J., Liu, Q.Y., Jin, J., 2001. Population dynamics of phytophagous and predatory mites (*Acari*: *Tetranychidae*, *Eriophyidae*, *Phytoseiidae*) on bamboo plants in Fujian, China. Exp. Appl. Acarol. 25, 383–391.

Zhong, B., Junren, C., Mohammad, S., Jia, G., Ying, W., Jiasen, W., Zhengqian, Y., Lizhi, H., Dan, L., 2017. Effect of lead (Pb) on antioxidation system and accumulation ability of moso bamboo (*Phyllostachys pubescens*). Ecotoxicol. Environ. Saf. 138, 71–77.

Zhou, B., Li, Z., Wang, X., Cao, Y., An, Y., Deng, Z., Geri, L., Wang, G., Gu, L., 2011. Impact of the 2008 ice storm on moso bamboo plantations in southeast China. J. Geophys. Res. 116, 1–10.

Zhou, B.Z., Fu, M.Y., Xie, J.Z., Yang, X.S., Li, Z.C., 2005. Ecological functions of bamboo forest: research and application. J. Forest. Res. 16, 143–147.

Zhou, G., Jiang, P., 2004. Density, storage and spatial distribution of carbon in *Phyllostachys pubescens* forest. Scient. Silvae Sin. 40, 20–24, (in Chinese with English abstract).

Zhou, G., Meng, C., Jiang, P., Xu, Q., 2011. Review of carbon fixation in bamboo forests in China. Bot. Rev. 77, 262–270.

The application of *Calamagrostis epigejos* (L.) Roth. in phytoremediation technologies

Dragana Ranđelović[a,b,*], Ksenija Jakovljević[c], Slobodan Jovanović[c]
[a]University of Belgrade, Faculty of Mining and Geology, Department for Mineralogy, Crystallography, Petrology and Geochemistry, Belgrade, Serbia; [b]Institute for Technology of Nuclear and other Mineral Raw Materials, Belgrade, Serbia; [c]University of Belgrade, Faculty of Biology, Institute of Botany and Botanical Garden, Belgrade, Serbia
*Corresponding author

1 Introduction

Different processes such as intensification of industrial activities, mining, smelting, agricultural activities through their effluences, fertilizers, pesticides, manures, lead to the release of a certain amount of heavy metals and metalloids in the environment, whose large part is kept in the soil. Some of these elements are toxic, some are mutagenic, teratogenic, cancerogenic, and all of them share the characteristic that they accumulate in the soil and represent a long-term source of pollutants entering the food chains (Ali et al., 2013). Soil pollution is a problem that has raised the attention of the scientific public over the last decades, primarily due to the indirect impact on human health.

The most effective, but at the same time for nature the least burdensome method of removing heavy metals and metalloids from soil is phytoremediation. Unlike physical and chemical removal of pollutants, phytoremediation cost less—only about 5% of the cost of alternative cleaning methods, according to Prasad (2003); does not lead to secondary pollution of the soil (as the case with chemical methods); it can improve both the physical (erosion protection) and the chemical properties of the soil (smaller quantity of heavy metals). Basically, phytoremediation refers to the use of plants in reducing, removing, immobilization, or degradation of heavy metals and metalloids (Peer et al., 2005). There are several different phytoremediation techniques: phytoextraction, phytostabilization, phytodegradation, rhizodegradation, phytofiltration, phytodesalinization, and phytovolatilization (Ali et al., 2013). The best known and most commonly used process is phytoextraction (phytoaccumulation or phytoabsorption), defined as uptake of contaminants and their transportation to the aerial parts of plants, which can be easily removed by harvesting (Kumar et al., 1995). The main requirement is the use of hyperaccumulator plant species (Panday and Chaney, 1983), but there are some additional criteria which plants have to meet: species may not be invasive or otherwise dangerous to the ecosystem which is introduced in; have to be fast-growing with high biomass and easy for mowing (Clemens et al., 2002). Besides phytoextraction, the

Phytoremediation Potential of Perennial Grasses. http://dx.doi.org/10.1016/B978-0-12-817732-7.00013-4

most commonly used phytoremediation technique, especially for the mine wastes, is phytostabilization. However, unlike the phytoextraction, phytostabilization refers to the adoption of heavy metals and metalloids from the soil, whereby the quantities adopted are retained in the root system and are not being transported to above-ground tissues. In this way, metals become less available to surrounding plants, and the possibility of their leaching is reduced. Before the implementation, plants have to fulfill certain criteria, like non-invasiveness, rapid growth, and metal tolerance (Mendez and Maier, 2007). One of the most important ways of pollutant removal is their degradation. Considering that heavy metals are nondegradable, only organic pollutants can be decomposed in this process. This degradation can take place within the plant itself or in rhizosphere, and therefore we distinguish two different processes, phytodegradation and rhizodegradation (Arthur et al., 2005) In phytodegradation, plants use their enzymes (oxygenase, dehalogenase, etc.) in order to transform pollutants in smaller particles that can be broken down or released, while rhizodegradation refers to decomposition of organic pollutants by microorganisms in the soil surrounding plant roots (Ali et al., 2013). Phytofiltration represents the use of aquatic plants in the removal of pollutants from solutions, most often different types of waste waters. Different plant parts can be used in this phytoremediation technique, but usually the pollutants are filtered through the roots (Olguín and Sánchez-Galván, 2012). When the soil is burdened with a significant amount of salt, which inhibits or prevents the survival of a large number of plants, its quality can be significantly improved in the process of phytosalinization. This remediation technique include the use of halophyte plants in removal of salt excess (Manousaki and Kalogerakis, 2010). The most controversial of all phytoremediation techniques is phytovolatization. To be specific, it represents uptake of pollutants by plants, their transformation to volatile form, and subsequent diffusion into the air. Thus, pollutants are only partially removed, while at the same time there is a decrease in air quality (Ali et al., 2013; Limmer and Burken, 2016).

In the recent years grasses (especially perennials) have been widely used in the phytoremediation process, having in mind that certain number of them fulfills implementation criteria (rapid growth, extensive root system, high biomass yield, metal tolerance, etc.) (Barbosa et al., 2015). For example, *Lolium perenne* has been used in refinery wastewater purification considering that can efficiently remove nutrient and pollutants from the water (Li et al., 2012); *Festuca arundinacea* is able to absorb significant quantities of heavy metals (especially Cd) from the soil, so it is also potential candidate for phytoremediation process (Żurek et al., 2014). Ability of *Phragmites australis* in removing heavy metals from aquatic environments is proven in many studies (Srivastava et al., 2014; Prica et al., 2019). *Miscanthus* species also became interesting from the phytoremediation point of view. Namely, they are able to survive in various habitat and soil types, often heavily degraded, at the same time displaying a series of benefits for the environment in which they grow [metal tolerance, high yields, and erosion protection (Bang et al., 2015; Barbosa et al., 2015)].

The first indication that species can be used in phytoremediation is its spontaneous appearance on the habitats polluted by heavy metals. One of these species is *Calamagrostis epigejos* (L.) Roth, which appears as a pioneer species on different heavily degraded habitats (mine waste sites, abandon land, etc.), where it not only survives successfully, but may also build very stable monodominant stands (Rebele and Lehmann, 2001). The constitutive tolerance of *C. epigejos* to heavy metal pollution was described by many authors (Rebele and Lehmann, 2001; Lehmann and Rebele, 2004; Kasowska et al., 2018). Similar was found in certain grass genera (*Festuca, Agrostis, Vetiveria*, etc.) as well the genera of other families (*Arabidopsis, Armeria, Plantago, Thlaspi*, etc.) (Smith and Bradshaw, 1979; Assunção et al., 2003; Remon et al., 2007; Olko et al., 2008; Rascio and Navari-Izzo, 2011; Suele et al., 2017). Bearing in mind its obvious plasticity, as well as pioneering nature, especially on highly contaminated habitats, *C. epigejos* can play significant role in phytoremediation process.

2 Morphology, propagation, and reproduction

C. epigejos is tall, robust, perennial caespitose grass. It is fast-growing and deep rooting plant as root reaches the depth of up to 2 m, with long and strong rhizomes (up to several meters long) and with culm up to 200 cm high and with 2–4 joints. Leaves are broadly linear, with flat or weakly convolute leaf-blades, green, upper leaf surface scabrid, 4–20 mm wide, and up to 105 cm long. Panicles up to 30 cm long and 3–6 cm wide, erect, dense, spikelets to 10 mm long, silvery-gray to brownish-purple color and with one floret each. Glumes lanceolate, up to 7.5 mm, subequal or lower slightly longer, scabrous on the keel. Lower glume is 1-nerved, and the upper on 1–3-nerved. Lemma lanceolate, nearly 1/2 shorter than the glumes, awn short, arise from the middle of lemma. Callus nearly twice longer then the lemma, with dense long hairs (Rozhevitz, 1934; Faille and Fardjah, 1977; Clarke, 1980; Fitter and Fitter, 1984; Janczyk-Weglarska, 1996; Rebele, 1996b; Rebele and Lehmann, 2001; Rutkowski, 2008) (Fig. 13.1).

The dominant type of reproduction is very intensive vegetative propagation. *C. epigejos* spreads vegetatively by forming ramets on the rhizomes, capable of an independent existence after founding. Ramets can be formed as intra- and extra-vaginal. Intravaginal formation involves creation of groups of closely attached ramets (clumps), while extravaginal ramets were formed on long rhizomes (Březina et al., 2006). New rhizomes, which grow and spread several meters in one direction, appear on the base of the ramets (Rebele and Lehmann, 2001). Clonal diversity depends on habitat conditions (Lehmann, 1997) and it is especially significant in areas heavily loaded by trace metals. Easy building up from rhizomes and rapid expansion were also observed in anthropogenic habitats (Kopecky, 1986). Although rare, sexual reproduction is not excluded. Seeds need open sites and adequate moisture over a long period of time in order to germinate (Faille and Fardjah, 1977). However, the dormancy is not obligatory (Rebele and Lehmann, 2001).

3 Ecology

Calamagrostis epigejos is generalist plant with wide ecological amplitude in terms of soil hummus and nutrient content, pH (1.9–8.5 in KCl) and soil moisture (from wet to dry). Consequently, it is able to grow in wide range of highly contrasting habitats,

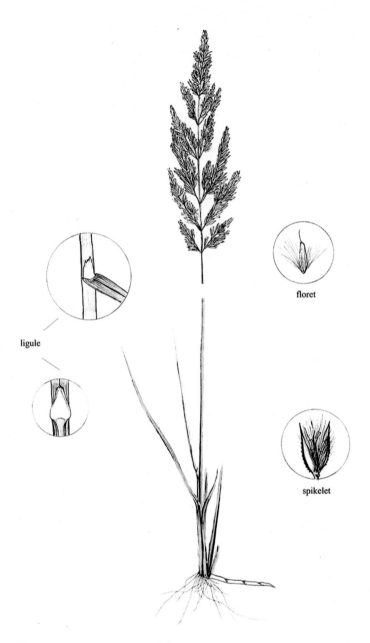

ligule

floret

spikelet

Figure 13.1 *Calamagrostis epigejos* (L.) Roth.

both natural and man-made (Rebele and Lehmann, 2001; Süß et al., 2004; Ranđelović et al., 2018). However, the most optimal growth and reproduction is achieved under moist, nutrient rich and full-light conditions (Brünn, 1999; Fiala et al., 2011; Stefanowicz et al., 2015).

Plant species that can grow in nutrient-poor habitats, like *C. epigejos*, generally have an ability to retain captured nutrients due to a longer life span of tissues (Schlapfer and Ryser, 1996). *C. epigejos* can provide nutrients from soils by an extensive system of below-ground organs and ability to use rhizomes for selective spreading of ramets into nutrient-rich patches (de Kroon and Hutchings, 1995; Lehmann and Rebele, 2002). Moreover, elevated atmospheric N deposition has been identified as a possible cause of increased abundance of *C. epigejos* over last decades (Kooijman et al., 1998). Studies of Rebele (1996b) and Brünn (1999) confirmed enhanced growth of *C. epigejos* in the case of good nutrient supply, mainly N and P. Inasmuch, *C. epigejos* can additionaly produce up to 2 m deep undergound biomass system (Conert, 1989), which enlarges it's water aquisition capability. Water use efficiency of *C. epigejos* increases with nitrogen fertilization, as stated by Süß et al. (2004). In mountain plant communities of Poland, the competitive strength of *C. epigejos* was enhanced by insolation and certain soil physicochemical properties (proportion of sand and silt, organic matter, pH, P, K), as found by Pruchniewicz et al. (2017).

However, Süß et al. (2004) showed that *C. epigejos* is unable to spread if the stress factors nutrient deficiency and drought are combined, while Pruchniewicz et al. (2017) noticed that development of *C. epigejos* patches in the forest communities may be inhibited by light deficiency and soil infertility. Moreover, researches of Fiala et al. (2011) stated that drought can substantially change the aboveground biomass production of *C. epigejos,* while Bakker et al. (2007) found that, at low nutrient availability, *C. epigejos* was not very sensitive to the treatments with different water levels.

Direct correlation between the nutrient amounts and biomass produced by *C. epigejos* had been already confirmed (Süß et al., 2004). In man-made habitats, which are generally less fertile, establishment of *C. epigejos* individuals was stimulated by sufficient influx of P and K, provided through anthropogenically caused eutrophication. Lack of water itself does not always have a pronounced negative impact, but in combination with nutrient deficiency, cover values of *C. epigejos* are significantly decreasing (Süß et al., 2004). Habitat conditions (above all nutrient status) to a significant extent determine total biomass and the ratio of bellow ground to above ground biomass of *C. epigejos*. The total biomass varies from 231 ± 32 g m^{-2} on the poor, up to 2760g m^{-2} on the soil rich in nutrients, with the higher ratio on the poor comparing to the nutrient rich soil [4.36 and 0.69, respectively, according to Rebele and Lehmann (2001)].

There is a broad range of natural habitats where *C. epigejos* successfully grows: river floodplains, sand dunes, fens, dry grasslands, forests (Rebele and Lehmann, 2001). It appears in a very wide range of altitudes, from the sea level (Baltic and North Sea coasts) up to 4100 m a.s.l. (above sea level) in the Karakorum (Rebele and Lehmann, 2001). In this sense, there are clear differences in altitude ranges in Europe and Asia. While in Europe *C. epigejos* occurs to the greatest extent in the vertical belt between lowland and submontane zone, reaching the highest altitude in the Swiss

Alps (Conert, 1989), it occurs pretty often at a much higher altitudes in Asia. It can be commonly found in the altitudes above 2000 m a.s.l., even more than 3000 m in Nepal and Tibet, up to the 4100 m. a.s.l. in Karakorum, according to Rebele and Lehmann (2001).

The basic ecological preferences of the species *C. epigejos* are summed up by ecological indices given by Ellenberg (Ellenberg et al., 1991). According to these indices, it prefers light places, but also grows in partly shady places. Regarding the moisture it shows broad amplitude, while according to indices slightly modified for British Isles (Hill et al., 1999) *C. epigejos* is indicator of damp places, mainly constantly moist or damp. Regarding the soil reaction, it is indicator of slightly acid to slightly basic conditions. However, *C. epigejos* was found on the soils with pH value 2.8 (Pruchniewicz et al., 2017). Likewise, according to indices, in terms of site fertility, it prefers rich places and it is absent from saline area, although numerous studies confirmed that *C. epigejos* successfully survives both in rich, as well as in very poor habitats, and even in habitats with increased salinity (Schlapfer and Ryser, 1996; Ranđelović, 2015; Pietrzykowski and Likus-Cieślik, 2018).

Regarding syntaxonomical position of communities in which *C. epigejos* is the dominant species, there is only one association (*Calamagrostietum epigeji* Juraszek 1928) described from inland sand dunes near Warsaw (Poland) on sites where pine forests had been clear cut (Rebele and Lehmann, 2001). Therefore this association was mainly classified in vegetation of forest clearings from the class *Epilobietea angustifolii* (Rebele and Lehmann, 2001; Kutyna et al., 2016; Sienkiewicz-Paderewska et al., 2017). However, it seems that syntaxonomic status and position of this association as well as other communities with significant participation of *C. epigejos* are not yet precisely defined. Namely, according to Rebele and Lehmann (2001), *C. epigejos* is a very abundant in other types of clear cut vegetation and communities of some other classes in Central Europe. Therefore plant sociologists do not accept *C. epigei* Juraszek 1928 as a valid association. It has been found that floristic composition and structure of the *C. epigeji* phytocoenoses varies, mainly regarding the soil conditions (particle size distribution, in particular) and the stage of succession taking place in the phytocoenosis (Kutyna et al., 2016). According to Chytrý and Tichý (2003), *C. epigejos* is diagnostic, constant, and dominant species in vegetation of *Epilobetea angustifolii* class in Czech Republic, but also the constant species in annual vegetation of disturbed sandy soils (Alliance *Salsolion ruthenicae*; Class *Chenopodietea*), and dominant species in perennial thermophilous ruderal vegetation on disturbed loamy soils (Alliance *Convolvulo-Agropyrion*; Class *Agropyretea repentis*). Additionally, Rebele and Lehmann (2001) reported that *C. epigejos* can be found in Central Europe as an accompanying species in associations of the vegetation classes such as *Ammophiletea*, *Scheuchzerio-Caricetea fuscae*, *Phragmitetea*, *Nardo-Callunetea*, *Sedo-Scleranthetea*, *Festuco-Brometea*, *Salicetea purpureae*, *Querco-Fagetea*, *Quercetea robori-petraeae*, *Vaccinio-Piceetea*, *Trifolio-Geranietea sanguinei*, *Molinio-Arrhenatheretea*, *Artemisietea*, and others. The same authors quoted that *C. epigejos* communities from natural habitats outside Central Europe have been reported from river flood-plains, fens, coastal dunes of Lake Baikal, subalpine grassland, and montane steppes in Siberia (Rebele and Lehmann, 2001). However, *C. epigejos* grows much often on different artificial habitats such as industrial heavily polluted sites around copper smelters or chemical plants, where it often forms monodominant stands (Rebele et al., 1993). Because the succession of vegetation may not proceed under these harsh environmental conditions, Rebele and Lehmann (2001) reasonably stand out opinion of Wolak (1980) which described such stands of *C. epigejos* as an "industrioclimax" vegetation.

4 Distribution and expansion

Calamagrostis epigejos is distributed in Europe, Africa, temperate Asia, tropical Asia, Australasia, and North America (Fig. 13.2) (Sharp and Simon, 2002). It is native to Euro-Asian region (Europe, North, Central and West Asia) and S&E Africa (as *C. epigejos* subsp. *capensis* (Stapf) Tzvelev) and belongs to Eurasian boreo-temperate geoelement (Meusel et al., 1965). According to Clarke (1980), it is widely distributed in Europe, but rarely in the south-west parts. Although native, it occur mostly synanthropic in Central and Western Europe, with increasing abundance in recent decades (Ten Harkel and van der Meulen, 1995; Rebele, 1996a; Süß et al., 2004). As

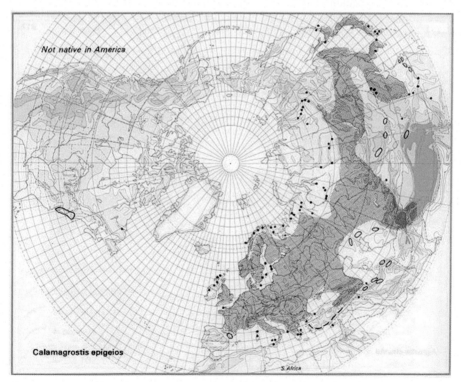

Figure 13.2 Distribution map of *Calamagrostis epigejos* (the map was kindly provided by Arne Anderberg from Swedish Museum of Natural History).

naturalized alien species, it can be found in North America where it has been introduced both accidentally and intentionally. First deliberate introduction was carried out in 1923 by the US Department of Agriculture, followed by its expansion all over the US territory and in Canada, giving the fact it was used for prevention of erosion and revegetation (Aiken et al., 1989; Rebele, 2000). According to Rebele and Lehmann (2001), *C. epigejos* can also be found in New Zealand and Tasmania as a synanthropic species.

Calamagrostis epigejos has been recognized for invading semi-natural grasslands in Europe as a competitive plant species, therefore taking part in declining of their species richness (Lepš, 2014; Pruchniewicz and Żołnierz, 2017). On open and nutrient-rich sites *C. epigejos* is frequently dominant and may form dense stands of low species diversity (Fig. 13.3). Expansion of *C. epigejos* leads to the competitive exclusion of characteristic species, therefore resulting with disappearance of certain native species (Sedlakova and Fiala, 2001). This species possess ability to spread and become dominate even in low-nutrient ability habitats (Kleijn et al., 2009) which discriminates her from majority of competitive meadows grasses. The study of Pruchniewicz et al. (2017) revealed a significant influence of the underground *C. epigejos* biomass expansion on the species diversity and composition of communities in the Sudetes Mountains, Poland. It has been found that expansion of *C. epigejos* leads to degradation of the vegetation from the *Arrhenatheretalia* order, *Molinio-Arrhenatheretea* class, primarily through reducing the diversity of species, but also the percentage of taxa representative of mesic meadows (Pruchniewicz and Żołnierz, 2017). Aggressive expansion of *C. epigejos* has been observed also in abandoned vineyards, agricultural

Figure 13.3 *Calamagrostis epigejos* forming dense stands on the abandoned agricultural land (Dubašnica, Serbia).

fields, alluvial meadows, dry grasslands, etc. (Haźi and Bartha, 2002; Holub, 2002; Luoto et al., 2003), causing numerous problems from the conservation points of view.

Being tall species, *C. epigejos* forms shadow to lower growing species. Additionally, by forming dense and thick layer of slowly degraded litter, germination of other species is prevented (Fiala et al., 2003; Holub et al., 2012; Těšitel et al., 2018). The most significant changes of habitat conditions caused by *C. epigejos* are reflected in preventing access to the air, causing dryness, extensive use of nutrients and consequently decreasing of their level for other species, which makes *C. epigejos* competitively strong (Wojcikowska-Kapusta et al., 2017). Holub (2002) also noticed that *C. epigejos* is probably able to attain higher biomass production and outcompete key plant species on natural floodplain meadows due to a higher nitrogen use efficiency. Growth strategy and the significant abiotic stress tolerance of *C. epigejos*, given by a higher water and nitrogen use efficiency, seems to be responsible for its competitive success in alluvial meadows of landscape affected by water control measures in Czech Republic, as found by Fiala et al. (2003). The fact that many high-natural values grasslands in Europe have been abandoned or have low-intensity land use further facilitate spreading and dominance of *C. epigejos* (Těšitel et al., 2017). Natural disturbances, as well as anthropogenic influence, may further enhance the spreading of *C. epigejos*. As a successful both primary and secondary colonizer (Ranđelović et al., 2014), this species was increasing its abundance after forest fires in *Pinus* stands of Lithuania (Marozas et al., 2007) and Poland (Loster et al., 2011), as well as after prescribed burnings of grasslands in Hungary (Végvári et al., 2011).

It is known that climate change will profoundly have an impact on plant community structure and composition, since the temperature and water availability are important factors that affect plant growth and interactions (Yang et al., 2011). Certain researches stated that changes in precipitation are likely to dominate plant responses in comparison to elevated temperature, but also that climate changes will influence the competitive interactions of dominant and subdominant species (Engel et al., 2009; Kardol et al., 2010). In this sense, distribution perspectives and competitive success of *C. epigejos* can be based on features such as growth strategy, nutrient use, and abiotic stress tolerance (Fiala et al., 2003), which will enable this species to spread even more in various habitats of Europe and North America (Aiken et al., 1989; Somodi et al., 2008). Additionally, in regard of climate change models forecast, increasing photosynthetic rates due to elevated concentration of CO_2 is especially pronounced for C3 plants, such as *C. epigejos*, at high temperatures and under water-limited conditions (Kirschbaum, 2004; Sage and Kubien, 2007).

5 Suppression and control

Certain researches aimed to find the most suitable method for controlling and suppressing the spread of *C. epigejos*. Investigations of Lehmann and Rebele (2002) on landfill site in Germany showed that the species can be controlled by regular mowing twice per year. Házi et al. (2011) stated that significant decrease of *C. epigejos* in mid-successional grasslands of Hungary appeared after 2 years of mowing.

Following these measures, species richness increased after 4 years, and their diversity after 8 years. Low intensity of mowing and grazing was found to be inefficient for suppression of *C. epigejos*, since the species is proven to be well adapted to non-frequent above-ground disturbance by mobilizing its below-ground resources (Fiala et al., 2003; Těšitel et al., 2017). Regular mowing has potential to speed up the succession process by opening a space for colonization of some valuable target species, as well as exhausting the nutrients stored in rhizomes of *C. epigejos* (Fiala et al., 2003; Házi et al., 2011). Intense management measures may also harm other sensitive species in present plant communities (Somodi et al., 2008). Prescribed burnings cannot be an alternative measure since burned areas become quickly invaded by species re-sprouting, therefore increasing the dominance of *C. epigejos* (Marozas et al., 2007). However, researches of Házi et al. (2011) revealed slow and spontaneous decrease of *C. epigejos* in the control plots, suggesting that *C. epigejos* can disappear spontaneously after 40–50 years of secondary grassland succession. Investigation of Pruchniewicz and Żołnierz (2019) on the effect of various restoration treatments in meadows degraded by *C. epigejos*, revealed that mowing and mulching with transferred hay could be useful measures from the point of maintaining species richness and diversity at the investigated habitats.

Other researches concentrated on the biological measures for control and suppression of *C. epigejos*, especially concerning its below-ground resource storage. Těšitel et al. (2017) found that hemiparasitic plant *Rhinanthus alectorolophus* to be an efficient tool for targeted biological control of *C. epigejos*, by interfering with the underground storage and clonal growth strategy of *C. epigejos*. Moreover, experiments showed that the effect of the hemiparasites on *C. epigejos* had increased success when combined with intensive mowing. Additional studies of Těšitel et al. (2018) also demonstrated the suitability of *Rhinanthus major* as an efficient tool to suppress *C. epigejos* and facilitate community restoration in species-rich dry grasslands of Čertoryje National Nature Reserve, Czech Republic. Additionally, *R. major* was found to have a broader and longer-term application potential than *R. alectorolophus* in ecological restoration and preservation of these communities. The application of the hemiparasitic species was found to be more effective than increased mowing intensity in terms of long-term suppressive effect on *C. epigejos*, as well as the preservation of the habitat's natural values. However, these kinds of promising applications must be confirmed by detailed experimental studies in order to minimize the risks of side-effects.

6 Phytoremediation

Calamagrostis epigejos often appears on contaminated or otherwise degraded sites where it forms more or less dense vegetation cover (Fig. 13.4). Moreover, *C. epigejos* is found to be a dominant species in central European post-mining lands at the initial stage of vegetation succession (Mudrák et al., 2010). As a competitive grass, it can arrest succession for a long time by forming a dense, compact sward on degraded lands. Kovář et al. (2004) conclude dominance of *C. epigejos* in the herbaceous plant layer in the area of the pyrite flotation basin in Chvaletice, in Czech Republic. In these

Figure 13.4 *Calamagrostis epigejos* on ruderal habitat in the peri-urban industrial zone (Belgrade, Serbia).

ecologically extreme conditions, in later phases of succession population of *C. epigejos* declines, so it is considered that on these kinds of highly degraded sites species plays a positive role, comparing to coal-mine sites where it captures the natural succession process (Prach and Pysek, 2001; Kovář et al., 2004). Thriving of *C. epigejos* on polluted sites is enhanced by different adaptation mechanisms, including phenotypic plasiticity and genotypic variability, low nutrient demands, antioxidative defence mechanism, arbuscular mycorrhizae, etc. (Dueck et al., 1986; Rebele and Lehmann, 2001; Lehmann and Rebele, 2004; Bert et al., 2012; Zieleźnik-Rusinowska et al., 2017).

Bearing in mind *C. epigejos* habitat preferences and its ecological plasticity, the same plasticity was also expected on physiological level. Several studies have investigated its behavior in conditions of increased concentrations of heavy metals in the habitat (Leyval et al., 1997; Gloser, 2002; Gajić et al., 2014), where significant changes in physiological characteristics were observed. Similarly to the other species, the photosynthetic activity in *C. epigejos* decreased with elevation of stress level (Gajić et al., 2014). Significant reduction of chlorophylls and carotenoids were also detected, which is not surprising considering their well-known sensitivity to heavy metals (Vajpayee et al., 2000). The presence of vesicular-arbuscular mycorrhizae on the roots of *C. epigejos* in areas polluted with Zn mitigates the negative effects of this metal on the plant, according to Dueck et al. (1986). Zieleźnik-Rusinowska et al. (2017) suggested that symbiosis with arbuscular fungi is important for the survival of plant in investigation of mycorrhizal colonization of *C. epigejos* on post-mining heaps and has role in oxidative stress mitigation. Research on saprophytic microorganisms inhabiting the populations of *C. epigejos* in various stages of succession at the pyrite waste landfill in Eastern Bohemia, Czech Republic (contaminated with high concentrations of Mn, Fe and Zn) revealed a connection between the succession stages and the diversity and quantity of the saprophytes in the root zone (Požárová et al., 2001).

Calamagrostis epigejos is proven to be among the few species successfully inhabiting severely degraded sites where it significantly contributes to the prevention of erosion by building a dense vegetation cover. In Canada, *C. epigejos* has been used for gold-mine tailings stabilizing through hand planting of rhizomes, cultivated for this purpose (Dore and McNeill, 1980). Moreover, *C. epigejos* was also found to be a species well adapted to the conditions of elevated soil salinity and sulfur concentration in S-contaminated post sulfur mine site Jeziórko in Southern Poland (Pietrzykowski and Likus-Cieślik, 2018).

The species is considered for use in phytoremediation of Pb, Zn, and Cu (Ansari et al., 2015). Study on the accumulation of heavy metals (Cu, Zn, Pb, and Cd) by *C. epigejos* in the industrial areas of Berlin performed by Rebele (1986) revealed that heavy metals were predominantly accumulated in roots. *C. epigejos* showed multiple tolerance to metal(oids) and low transfer factors for As, Mo, Cu, and Zn when colonizing fly ash deposit lagoons of thermoelectric plant in Obrenovac, Serbia. At the same time, the species had higher transfer factor for B and accumulated it in shoots in concentrations within the toxic range for plants, developing the adaptations in relation to the uptake of B from the ash (Mitrović et al., 2008). This species, however, is characterized by efficient regulatory mechanisms related to the process of adoption of toxic elements, as well as significant activation of mechanisms for antioxidant protection, which makes it a suitable species in the reclamation of fly ash lagoons (Gajić et al., 2014).

The potential use of *C. epigejos* for phytoremediation is not in the area of phytoextraction, considering the generally low concentrations of heavy metals in aerial parts, especially compared to some hyperaccumulator species. For example, while Cd concentrations in aboveground tissues of *Thlaspi caerulescens* and *Arabidopsis halleri* are up to 3000 and 500 mg kg^{-1}, respectively, in aerial parts of *C. epigejos*, they are conspicuously lower (Dahmani-Muller et al., 2001; Schwartz et al., 2003). The phytoextraction potential of *C. epigejos* for Cd within three growing seasons ranged from 0.11% to 0.25% of the total soil content, and the low transfer factor for Cd at higher soil concentrations indicated the exclusion mechanism, as stated by Lehmann and Rebele (2004).

However, *C. epigejos* showed certain potential in the field of phytostabilization, considering the fact it retains Zn, Cu, Cd, and Mn in the roots and reduces the pollution through their binding. This is particularly important for the available pool of stated elements. The role of *C. epigejos* as a pioneer species inhabiting mining areas after environmental accidents (Fig. 13.5) is found to be important since the species reduces the bioavailability and spreading of contaminants into environment and food chains in the critical phases of their leaching and pollution (Ranđelović et al., 2018). In case of copper mine wastes from Bor, Serbia (Fig. 13.6), *C. epigejos* was found as one of the most successful primary colonizers of psammophytic character, and therefore potential candidate for anthropogenically assisted natural recovery (Ranđelović

Figure 13.5 *Calamagrostis epigejos* begins to colonize the abandoned antimony flotation tailings after hazardous tailing dam spill (Krupanj, Serbia).

et al., 2014). However, these quantities are not so large to qualify *C. epigejos* as the first pick in selecting species for phytoremediation, so Madzhugina et al. (2008) suggest that for efficient field decontamination of Ni and other heavy metals, screening of cultivars or production of transgenic *C. epigejos* plants with elevated metal accumulation needs to be performed.

Tolerance and adaptation of *C. epigejos* to multi-contaminated sites is a good precondition for the colonization of various degraded habitats, where this species proves to be useful for reducing erosion and heavy metal leakage (Lehmann and Rebele, 2004). Moreover, the spontaneous growth of *C. epigejos* in neglected peri-urban and agricultural areas and near the roadsides (Fig. 13.7), contributes to the natural remediation of these sites (Ranđelović et al., 2018). Therefore it is also recommended as a potential beneficial component of grass mixtures for near-road lawns (Madzhugina et al., 2008).

Figure 13.6 *Calamagrostis epigejos* colonizing the copper mine wastes (Bor, Serbia).

Figure 13.7 *Calamagrostis epigejos* growing on the roadside (Zlot, Serbia).

7 Other uses

Perennial grasses are present in many habitats over the world, developing adaptations that enable them to survive and grow in most diverse environmental conditions. Besides, perennials often possess deep and extensive roots that access resource-poor conditions in soils, enabling them to enhance their competitive abilities. Perennial grasses are often characterized by high biomass production and low requirement for inputs, which makes them good candidates for practical application in sustainable technologies, such as biomass energy production. Moreover, simultaneous use of plant species for both phytoremediation and energy crops on contaminated and marginal soils presents a developing investigation field (Gomes, 2012; Pandey et al., 2016).

Biomass energy is an important part of renewable energy sources, having share of around 10% in global energy consumption (World energy, 2016). Intense researches aim to identify those plants that best fulfill the demands of bio-energy production, namely high biomass yields and appropriate biomass characteristics (Sanderson and Adler, 2008). Biomass production, water and nitrogen use, tolerance to biotic and abiotic stress and low soil fertility are critical factors in the selection of ideal bio-energy feedstocks (Bressan et al., 2011).

Compared with the biofuels based on annual grain crops, perennial biomass crops require fewer inputs, produce more energy, and reduce greenhouse gas emissions more than annual cropping systems (Adler et al., 2007). Due to their high yield potentials, low input demands and multiple ecological benefits, perennial rhizomatous grasses can significantly contribute to sustainable biomass production in Europe and the United States (Lewandowski et al., 2003). Furthermore, due to the recycling of nutrients by their rhizome systems, perennial grasses have a low demand for nutrient inputs (Christian et al., 1997). There are combined efforts to identify the most highly productive plant species that can be grown on the various types of marginal land and to optimize production practices (Jones et al., 2015). If appropriate crops are selected, even a restoration of degraded lands may be possible (McKendry, 2002).

Calamagrostis epigejos inhabits diverse industrial and polluted habitats. However, due to weak root to shoot transfer, aboveground tissue of *C. epigejos* can remain uncontaminated and safely used for other purposes. It's rapid growth coupled with fast biomass production candidates this species for potential use as energy crop. However, utilization of *C. epigejos* as energy plant is still under study.

Certain researches on the application of *C. epigejos* as potential source of renewable energy were conducted in Poland. *Calamagrostis epigejos* yields amounted to 4.03–4.47 t ha^{-1} of dry matter (Grzelak et al., 2016), which are somewhat lower values comparing to other perennial energy crops, such as *Miscanthus x giganteus* (with average dry matter yield of 17.7 t ha^{-1}, as found by Christian et al., 2008) or *Arundo donax* (dry matter yield of 14–35 t ha^{-1}, as found by Bacher et al., 2001). Research by Harkot et al. (2007) on the grasslands in south-eastern Poland determined combustion heat for *C. epigejos* at 18.4 MJ kg^{-1}, while the same parameter was determined as 17.19 MJ kg^{-1} d.m. by Patrzałek et al. (2011) and 18.6 MJ kg^{-1} d.m. by Grzelak et al. (2016). The calorific value of *C. epigejos* is 15.91 MJ kg^{-1} (Sierka and

Kopczyńska, 2014), which is comparable to the calorific value of energy crop *Miscanthus x giganteus* ranging from 15.31 to –16.55 MJ kg^{-1} d.m. in Serbia (Cvetković et al., 2016) and 16.55 to 18.13 MJ kg^{-1} d.m. in United Kingdom (Mos et al., 2013). Energy value of *C. epigejos* is found to be comparable to the coal (25 MJ kg^{-1}), as stated by Kozłowski and Swędrzyński (2010) and Patrzałek et al. (2012).

The proportion of cellulose and lignin content in biomass is one of the determining factors in identifying the suitability of plant species for subsequent processing as energy crop. The content of structural carbohydrates in *C. epigejos* found by Grzelak et al. (2016) was: 36.4%–40.3% of cellulose, 19.2%–23.6% of lignin, and 61.7%–66.5% of hemicellulose. Typical content of celluloses and lignin for grasses was found to be 25%–40% and 10%–30%, respectively (Sugathapala, 2013).

The ash content of biomass affects the handling and processing costs of energy conversion. High ash content is a non-desirable feature because it makes the automation of biomass combustion process more difficult. Ash content in different biomass fuels ranges from 0.1%–0.2% d.m. for wood up to 11.2% d.m. of annual grass *Triticum aestivum* (Sugathapala, 2013). A difference in ash content was also noticed between C3 and C4 perennial species (Samson and Mehdi, 1998). According to Radiotis et al. (1996), the content of ash in C4 perennials, such as *Panicum virgatum* and *Miscanthus sinensis* is 1.7% and 2% d.m., respectively, while in C3 perennials such as *Phallaris arundinacea* ash content reaches 6.3% d.m. Ash content in *C. epigejos* was found to range between 4.99% d.m. (Patrzałek et al., 2011) and 5.8% d.m. (Grzelak et al., 2016).

Chemical composition of biomass influences the technology of processing. Content of sodium, potassium, calcium, zinc, lead, and iron in the *C. epigejos* harvested from Pb-Zn post-flotation site and mine waste tailings is small compared to cultivated energy crops (Nowińska et al., 2012). Biomass moisture content also influences the management of energy crops, the heating value and the energy conversion process (Bartzanas et al., 2012; Sugathapala, 2013). In general, the moisture content should be less than 10%. The content of moisture in *C. epigejos* was found to be 3.63%–5.5% d.m. by investigations of Grzelak et al. (2016). Other relevant parameters influencing technology and effectiveness of combustion are contents of chlorine and sulfur. Both chlorine and sulfur induce corrosion of technological equipment used in combustion processes. Chlorine occurs in biomass, ranging to relatively low levels in wood (0.01% for typical biomass composition, according to Tilman et al., 2009) up to higher levels in herbaceous plants (over 1.0%, according to Parmar, 2017). Relatively low chlorine content of about 0.2% in d.m. has been found in biomass of *C. epigejos* (Harkot et al., 2007). Most biomass fuels have sulfur content below 0.2% (Parmar, 2017) while in case of energy crops the content of sulfur can reach up to 0.7% (Jagustyn et al., 2013). Content of sulfur in *C. epigejos* is 0.16%, as found by Harkot et al. (2007), which is considered to be a relatively low value.

Certain researches on converting *C. epigejos* biomass to liquid biofuels were also conducted. Rogalski et al. (2008) found that 318–459 L of ethanol can be produced from 1 ha of this species in Skoszewskie (Poland), emphasizing that *C. epigejos* can be used as potential source of biofuel. Researches of Kozłowski and Swędrzyński (2010) also showed that the chemical composition of *C. epigejos* makes this species a potential source of heating biomass, and to a lesser extent for the production of biogas. Although perennials produced lower ethanol yield potential then annual crops, they are adapted for growing in marginal or contaminated land, providing opportunities for economic return as well as phytostabilization or phytoremediation of such sites (Roozeboom et al., 2018). Moreover, investigations of Nowińska et al. (2012) on mine waste tailings concluded that the *C. epigejos* is potential energy crop, thanks to the species high calorific value and its chemical composition, mainly the low content of metals that are not interfering with the combustion process.

One of the limitations for the use of *C. epigejos* as an energy crop lies in its relatively low aboveground biomass. Researches on energy allocation in populations of *C. epigejos* in final growing period showed that the energy in aboveground mass was only 21.5% of the total energy standing crop, while below ground was 78.85% of the total energy standing crop (Guo et al., 2001). However, the issue of economic use of *C. epigejos* as an energy plant is still poorly recognized. In order for *C. epigejos* to become a potential source of renewable energy it is necessary to conduct further researches, such as the impact of regular mowing and biomass export, breeding of varieties adapted to the needs of biomass production and development of the crop management systems (Lewandowski et al., 2003; Sierka and Kopczyńska, 2014). Besides, due to the fact that certain studies on energy crop phytoremediation have already been conducted (Šyc et al., 2012; Werle et al., 2019), there is a potential of linking phytoremediation and energy crop usage of *C. epigejos,* which requires additional investigations in this direction.

References

Adler, P.R., Del Grosso, S.J., Parton, W.J., 2007. Life-cycle assessment of net greenhouse-gas flux for bioenergy cropping systems. Ecol. App. 17, 675–691.

Aiken, S.G., Dore, W.G., Lefkovitch, L.P., Armstrong, K.C., 1989. *Calamagrostis epigejos* (Poaceae) in North America, especially Ontario. Can. J. Bot. 67, 3205–3218.

Ali, H., Khan, E., Sajad, M.A., 2013. Phytoremediation of heavy metals—concepts and applications. Chemosphere 91 (7), 869–881.

Ansari, A., Gill, S., Gill, R., Lanza, G., Newman, L., 2015. Phytoremediation: Management of Environmental Contaminants. Springer, Switzerland.

Arthur, E.L., Rice, P.J., Rice, P.J., Anderson, T.A., Baladi, S.M., Henderson, K.L., Coats, J.R., 2005. Phytoremediation—an overview. Crit. Rev. Plant Sci. 24 (2), 109–122.

Assunção, A.G., Schat, H., Aarts, M.G., 2003. *Thlaspi caerulescens*, an attractive model species to study heavy metal hyperaccumulation in plants. New Phytol. 159 (2), 351–360.

Bacher, W., Sauerbeck, G., Mix-Wagner, G., El Bassam, N., 2001. Giant reed (*Arundo donax* L.) Network-Improvement, Productivity, Biomass Quality. Braunschweig (Germany): FAL- – Bundesforschungsanstalt fur Landwirtschaft, Final report, FAIR-CT96-2028, 72.

Bakker, C., Bodegom, P., Nelissen, H., Aerts, N., Ernst, W., 2007. Preference of wet dune species for waterlogged conditions can be explained by adaptations and specific recruitment requirements. Aquat. Bot. 86 (1), 37–45.

Bang, J., Kamala-Kannan, S., Lee, K.J., Cho, M., Kim, C.H., Kim, Y.J., Bae, J.H., Kim, K.H., Myung, H., Oh, B.T., 2015. Phytoremediation of heavy metals in contaminated water and soil using *Miscanthus* sp Goedae-Uksae 1. Int. J. Phytoremediation 17 (6), 515–520.

Barbosa, B., Boléo, S., Sidella, S., Costa, J., Duarte, M.P., Mendes, B., Cosentino, S.L., Fernando, A.L., 2015. Phytoremediation of heavy metal-contaminated soils using the perennial energy crops *Miscanthus* spp. and *Arundo donax* L. Bioenerg. Res. 8 (4), 1500–1511.

Bartzanas, T., Bochtys, D., Sorensen, C., Green, O., 2012. Moisture content evaluation of biomass using CDF approach. Sci. Agr. 69 (5), 287–292.

Bert, V., Lors, C., Ponge, J., Caron, L., Biaz, A., Dazy, M., Masfaraud, J., 2012. Metal immobilization and soil amendment efficiency at a contaminated sediment 4 landfill site: a field study focusing on plants, springtails, and bacteria. Environ. Pollut. 169, 1–11.

Bressan, R., Reddy, M., Chung, S., Yun, D., Hardin, L., Bohnert, H., 2011. Stress-adapted extremophiles provide energy without interference with food production. Food Security 3, 93–105.

Březina, S., Koubek, T., Münzbergová, Z., Herben, T., 2006. Ecological benefits of integration of *Calamagrostis epigejos* ramets under field conditions. Flora Morphol. Distribut. Funct. Ecol. Plants 201 (6), 461–467.

Brünn, S., 1999. Untersuchungen zum Mineralstoffhaushalt von *Calamagrostis epigejos* (L.) Roth in stickstoffbelasteten Kiefernwäldern. Berichte des Forschungszentrums Waldökosysteme Universität Göttingen, Reihe A, Bd. 160. University of Göttingen Göttingen, Germany.

Panday, V.C., Chaney, R.L., 1983. Plant uptake of inorganic waste constituents. In: Parr, J.F., Marsh, P.B., Kla, J.S. (Eds.), Land Treatment of Hazardous Wastes. Noyes Data, Park Ridge, NY, pp. 50–76.

Christian, D., Riche, A., Yates, N., 1997. Nutrient requirement and cycling in energy crops. In: Bassam, N., Behl, R., Prochnow, B. (Eds.), Sustainable Agriculture for Food, Energy and Industry. James & James (Ltd), London, pp. 799–804.

Christian, D.G., Riche, A.B., Yates, N.E., 2008. Growth, yield and mineral content of *Miscanthus* × *giganteus* grown as a biofuel for 14 successive harvests. Ind. Crops Prod. 28 (3), 320–327.

Chytrý, M., Tichý, L., 2003. Diagnostic, Constant and Dominant Species of Vegetational Classes and Alliances of the Czech Republic: A Statistical Revision, vol. 108, Masaryk University, Brno, Czech Republic.

Clarke, G.C.S., 1980. *Calamagrostis* Adanson. In: Tutin, T.G., Heywood, V.H., Burges, N.A., Moore, D.M., Valentine, D.H., Walters, S.M., Webb, D.A. (Eds.), Flora Europaea, vol. 5., Cambridge University Press, pp. 236–237.

Clemens, S., Palmgren, M.G., Krämer, U., 2002. A long way ahead: understanding and engineering plant metal accumulation. Trends Plant Sci. 7 (7), 309–315.

Conert, H.J., 1989. Calamagrostis villosa. In: Hegi, G. (Ed.), Illustrierte Flora von Mitteleuropa I (3). Paul Parey, Hamburg, pp. 365–367.

Cvetkovic´, O., Pivic´, R., Dinic´, Z., Maksimovic´, J., Trifunovic´, S., Dželetovic´, Ž., 2016. Hemijska ispitivanja miskantusa gajenog u Srbiji - Potencijalni obnovljiv izvor energije. Zaštita Materijala 57 (3), 412–417.

Dahmani-Muller, H., van Oort, F., Balabane, M., 2001. Metal extraction by *Arabidopsis halleri* on an unpolluted soil amended with various metal-bearing solids: a pot experiment. Environ. Pollut. 114, 77–84.

de Kroon, H., Hutchings, M., 1995. Morphological plasticity in clonal plants: the foraging concept reconsidered. J. Ecol. 83 (1), 143–152.

Dore, W.G., McNeill, J., 1980. Grasses of Ontario. Monograph 26. Agriculture Canada, Ottawa.

Dueck, Th.A., Visser, P., Ernst, W.H., Schat, H., 1986. Vesicular-arbuscular mycorrhizae decrease zinc-toxicity to grasses growing in zinc-polluted soil. Soil Biol. Biochem. 18 (3), 331–333.

Ellenberg, H., Düll, R.,Wirth,V., Werner, W., Pauliβen, D., 1991. Zeigerwerte von Pflan¸zen in Mitteleuropa (Scripta Geobotanica; 18), third Auflage. Verlag Erich Goltze KG, Göttingen.

Engel, E.C., Weltzin, J.F., Norby, R.J., 2009. Responses of an old-field plant community to interacting factors of elevated [CO_2], warming, and soil moisture. J. Plant Ecol. UK 2, 1–11.

Faille, A., Fardjah, M., 1977. Structure et evolution des peuplements de *Calamagrostis epigejos* (L.) Roth en foret de *Fontainebleau*. Oecol. Plant. 12, 323–341.

Fiala, K., Holub, P., Sedláková, I., Tuema, I., Záhora, J., Tesarová, M., 2003. Reasons and consequences of expansion of *Calamagrostis epigejos* in alluvial meadows of landscape affected by water control measures – a multidisciplinary research. Ekol. (Bratislava) Ecol. (Bratislava) 22, 242–252.

Fiala, K., Tůma, I., Holub, P., 2011. Effect of nitrogen addition and drought on above-ground biomass of expanding tall grasses *Calamagrostis epigejos* and *Arrhenatherum elatius*. Biologia 66 (2), 275–281.

Fitter, R., Fitter, A., 1984. Collins Guide to the Grasses, Sedges, Rushes and Ferns of Britain and Northern Europe. William Collins Sons & Co. Ltd, London, UK.

Gajić, G., Pavlović, P., Kostić, O., Jarić, S., Đurđević, L., Pavlović, D., Mitrović, M., 2014. Ecophysiological and biochemical traits of three herbaceous plants growing on the disposed coal combustion fly ash of different weathering stage. Arch. Biol. Sci. 65 (4), 1651–1667.

Gloser, V., 2002. Seasonal changes of nitrogen storage compounds in a rhizomatous grass *Calamagrostis epigeios*. Biol. Plant. 45 (4), 563–568.

Gomes, H., 2012. Phytoremediation for bioenergy: challenges and opportunities. Environ. Technol. Rev. 1 (1), 59–66.

Grzelak, M., Gaweł, E., Murawski, M., Waliszewska, B., Knioła, A., 2016. Warunki siedliskowe, plonowanie i moŻliwości energetyczne wykorzystania biomasy z dominacją trzcinnika piaskowego (*Calamagrostis epigejos*). Fragmen. Agron. 33 (3), 38–45, [in Polish].

Guo, J.X., Wang, R.D., Wang, W., 2001. Study on dynamics of calorific value, biomass and energy in *Calamagrostis epigejos* population in Songnen Grassland. Acta Bot. Sin. 43 (8), 852–856.

Harkot, W., Warda, M., Sawicki, J., Lipińska, H., Wyłupek, T., Czarnecki, Z., Kulik, M., 2007. MoŻliwości wykorzystania runi łąkowej do celów energetycznych. Łąkarstwo w Polsce 50, 58–67, [in Polish].

Haźi, J., Bartha, S., The role of *Calamagrostis epigejos* in the succession of abandoned vineyards in the Western Cserhát, Hungary. Third European Conference on Restoration Ecology, August 25–31, 2002, Budapest, pp. 126.

Házi, J., Bartha, S., Szentes, S., Wichmann, B., Penksza, K., 2011. Seminatural grassland management by mowing of *Calamagrostis epigejos* in Hungary. Plant Biosyst. Int. J. Deal. Aspects Plant Biol. 145 (3), 699–707.

Hill, M.O., Mountford, J.O., Roy, D.B., Bunce, R.G.H., 1999. Ellenberg's indicator values for British plants. ECOFACT Vol 2 Technical Annex. Inst. Terrestrial Ecol, Huntingdon, UK.

Holub, P., 2002. The expansion of *Calamagrostis epigejos* into alluvial meadows: comparison of aboveground biomass in relation to water regimes. Ecol. J. Ecol. Prob. Biosphere 21, 27–37.

Holub, P., Tůma, I., Záhora, J., Fiala, K., 2012. Different nutrient use strategies of expansive grasses *Calamagrostis epigejos* and *Arrhenatherum elatius*. Biologia 67 (4), 673–680.

Jagustyn, B., Patyna, I., Skawińska, A., 2013. Ocena właściwości fizykochemicznych Palm Kernel Shell jako biomasy agro stosowanej w energetyce. Chemik 67 (6), 552–559, [in Polish].

Janczyk-Weglarska, J., 1996. Strategier ozwoju osobniczego *Calamagrostis epigejos* (L.) Roth na tle warunkow ekologicznych poznanskiego przelomu Warty. Seria Biologia 56, Poznan. [in Polish].

Jones, M., Finnan, J., Hodk, T., 2015. Morphological and physiological traits for higher bio-mass production in perennial rhizomatous grasses grown on marginal land. Gcb Bioenerg. 7, 375–385.

Kardol, P., Campany, C.E., Souza, L., Norby, R., Weltzin, J., Classen, A., 2010. Climate change effects on plant biomass alter dominance patterns and community evenness in an experimental old-field ecosystem. Global Change Biol. 16 (10), 2676–2687.

Kasowska, D., Gediga, K., Spiak, Z., 2018. Heavy metal and nutrient uptake in plants colonizing post-flotation copper tailings. Environ. Sci. Pollut. Res. 25, 824–835.

Kirschbaum, M., 2004. Direct and indirect climate change effects on photosynthesis and transpiration. Plant Biol. 6, 242–253.

Kleijn, D., Kohler, F., Báldi, A., Batáry, P., Concepción, E.D., Clough, Y., Díaz, M., Gabriel, D., Holzschuh, A., Knop, E., Kovács, A., 2009. On the relationship between farmland biodiversity and land-use intensity in Europe. Proc. Royal Soc. London B Biol. Sci. 276 (1658), 903–909.

Kooijman, A.M., Dopheide, J.C.R., Sevink, J., Takken, I., Verstraten, J.M., 1998. Nutrient limitations and their implications on the effects of atmospheric deposition in coastal dunes; lime-poor and lime-rich sites in the Netherlands. J. Ecol. 86, 511–526.

Kopecky, K., 1986. Versuch einer Klassifizierung der ruderalen *Agropyron repens* und *Calamagrostis epigejos* Gesellschaften unter Anwendung der deduktiven Methode. Folia Geobotanica Phytotaxonomica 21, 225–242, [in German].

Kovář, P., Štěpánek, J., Kirschner, J., 2004. Clonal diversity of *Calamagrostis epigejos* (L.) Roth in relation to type of industrial substrate and successional stage. Natural Recovery of Human-Made Deposits in Landscape (Biotic Interactions and Ore/Ash-Slag Artificial Ecosystems), pp. 285–293.

Kozłowski, S., Swędrzyński, A., 2010. Możliwości wykorzystania trzcinnika piaskowego w kontekście jego biologicznych, chemicznych i fizycznych właściwości. Łąkarstwo w Polsce 65, 117–136, [in Polish].

Kumar, P.N., Dushenkov, V., Motto, H., Raskin, I., 1995. Phytoextraction: the use of plants to remove heavy metals from soils. Environ. Sci. Technol. 29 (5), 1232–1238.

Kutyna, I., Młynkowiak, E., Malinowska, K., 2016. Structure and floristic diversity of the community *Calamagrostietum epigeji* Juraszek 1928 within different biotopes. Folia Pomeranae Univ. Technol. Stet. 325 (37), 47–64, 1.

Lehmann, C., 1997. Clonal diversity of populations of *Calamagrosotis epigejos* in relation to environmental stress and habitat heterogeneity. Ecography 20, 483–490.

Lehmann, C., Rebele, F., 2002. Successful management of *Calamagrostis epigejos* (L.) Roth on a sandy landfill site. J. Appl. Bot. 76, 77–81.

Lehmann, C., Rebele, F., 2004. Assessing the potential for cadmium phytoremediation with *Calamagrostis epigejos*: a pot experiment. Int. J. Phytoremed. 6 (2), 169–183.

Lepš, J., 2014. Scale-and time-dependent effects of fertilization, mowing and dominant removal on a grassland community during a 15-year experiment. J. Appl. Ecol. 51 (4), 978–987.

Lewandowski, I., Scurlock, J., Lindvall, E., Christou, M., 2003. The development and current status of perennial rhizomatous grasses as energy crops in the US and Europe. Biomass Bioenerg. 25 (4), 335–361.

Leyval, C., Turnau, K., Haselwandter, K., 1997. Effect of heavy metal pollution on mycorrhizal colonization and function: physiological, ecological and applied aspects. Mycorrhiza 7 (3), 139–153.

Li, H., Hao, H., Yang, X., Xiang, L., Zhao, F., Jiang, H., He, Z., 2012. Purification of refinery wastewater by different perennial grasses growing in a floating bed. J. Plant Nutr. 35 (1), 93–110.

Limmer, M., Burken, J., 2016. Phytovolatilization of organic contaminants. Environ. Sci. Technol. 50 (13), 6632–6643.

Loster, S., Dzwonko, Z., Gawronski, S., 2011. Early post-fire vegetation regeneration in a Scots pine forest site in southern Poland. In: Zemanek, B. (Ed.), Geobotanist and Taxonomist. Institute of Botany, Jagiellonian University, Cracow, pp. 117–130.

Luoto, M., Rekolainen, S., Aakkula, J., Pykala, J., 2003. Loss of plant species richness and habitat connectivity in grasslands associated with agricultural change in Finland. Ambio 32, 447–452.

Madzhugina, Y., Kuznetsov, V., Shevyakova, N., 2008. Plants inhabiting polygons for megapolis waste as promising species for phytoremediation. Russ. J. Plant Physiol. 55 (3), 410–419.

Manousaki, E., Kalogerakis, N., 2010. Halophytes present new opportunities in phytoremediation of heavy metals and saline soils. Ind. Eng. Chem. Res. 50 (2), 656–660.

Marozas, V., Racinkas, J., Bartkevicius, E., 2007. Dynamics of ground vegetation after surface fires in hemiboreal *Pinus sylvestris* forests. For. Ecol. Manage. 250 (1), 47–55.

McKendry, P., 2002. Energy production from biomass (part 2): conversion technologies. Bioresour. Technol. 83 (1), 47–54.

Mendez, M.O., Maier, R.M., 2007. Phytostabilization of mine tailings in arid and semiarid environments—an emerging remediation technology. Environ. Health Perspect. 116 (3), 278–283.

Meusel, H., Jäger, E., Weinert, E., 1965. Vergleichende Chorologie der zentraleuropäischen Flora 1. Karten, Text. Gustav Fischer Verlag, Jena.

Mitrovic´, M., Pavlovic´, P., Lakušic´, D., Đurđevic´, L., Stevanovic´, B., Kostic´, O., Gajic´, G., 2008. The potential of *Festuca rubra* and *Calamagrostis epigejos* for the revegetation of fly ash deposits. Sci. Total Environ. 407 (1), 338–347.

Mos, M., Banks, S.W., Nowakowski, D.J., Robson, P.R.H., Bridgwater, A.V., Donnison, I.S., 2013. Impact of *Miscanthus* x *giganteus* senescence times on fast pyrolysis bio-oil quality. Bioresour. Technol. 129, 335–342.

Mudrák, O., Frouz, J., Velichová, V., 2010. Understory vegetation in reclaimed and unreclaimed post-mining forest stands. Ecol. Eng. 36, 783–790.

Nowińska, K., Kokowska-Pawłowska, M., Patrzałek, A. 2012. Metale w *Calamagrostis epigejos* i *Solidago* sp. ze zrekultywowanych nieuŻyt-ków poprzemysłowych. Infrastruktura i Ekologia Terenów Wiejskich. Nr 2012/ 03. [in Polish].

Olguín, E.J., Sánchez-Galván, G., 2012. Heavy metal removal in phytofiltration and phycoremediation: the need to differentiate between bioadsorption and bioaccumulation. New Biotechnol. 30 (1), 3–8.

Olko, A., Abratowska, A., Żyłkowska, J., Wierzbicka, M., Tukiendorf, A., 2008. *Armeria maritima* from a calamine heap—Initial studies on physiologic–metabolic adaptations to metal-enriched soil. Ecotoxicol. Environ. Saf. 69 (2), 209–218.

Pandey, V.C., Bajpai, O., Singh, N., 2016. Energy crops in sustainable phytoremediation. Renew. Sustain. Energ. Rev. 54, 58–73.

Parmar, K., 2017. Biomass- an overview on composition characteristics and properties. IRA Int. J. Appl. Sci. 7 (1), 42–51.

Patrzałek, A., Kokowska-Pawłowska, M., Nowińska, K., 2012. Wykorzystanie roślin dziko rosnących do celów energetycznych. Górnictwo i Geologia 7 (2), 177–185, [in Polish].

Patrzałek, A., Kozłowski, S., Swędrzyński, A., Trąba, Cz., 2011. Trzcinnik piaskowy jako potencjalna roślina energetyczna. Wydawnictwo Politechniki Śląskiej, Monografia, Gliwice. [in Polish].

Peer, W.A., Baxter, I.R., Richards, E.L., Freeman, J.L., Murphy, A.S., 2005. Phytoremediation and hyperaccumulator plants. In:. Tamas, M.J., Martinoia, E. (Eds.), Molecular Biology of Metal Homeostasis and Detoxification. Topics in Current Genetics, vol. 14, Springer, Berlin, Heidelberg, pp. 299–340.

Pietrzykowski, M., Likus-Cieślik, J., 2018. Comprehensive study of reclaimed soil, plant, and water chemistry relationships in highly s-contaminated post sulfur mine site Jeziórko (Southern Poland). Sustainability 10 (7), 2442.

Požárová, E., Herben, T., Grzndler, M., 2001. Soil saprotrophic microfungi associated with roots of *Calamagrostis epigeios* on an abandoned deposit of toxic waste from smelter factory processing pyrite raw materials. Microbial Ecol. 41 (2), 162–171.

Prach, K., Pysek, P., 2001. Using spontaneous succession for restoration of human-disturbed habitats: experience from Central Europe. Ecol. Eng. 17 (1), 55–62.

Prasad, M.N.V., 2003. Phytoremediation of metal-polluted ecosystems: hype for commercialization. Russ. J. Plant Physiol. 50 (5), 686–701.

Prica, M., Andrejić, G., Šinžar-Sekulić, J., Rakic´, T., Dželetović, Ž., 2019. Bioaccumulation of heavy metals in common reed (*Phragmites australis*) growing spontaneously on highly contaminated mine tailing ponds in Serbia and potential use of this species in phytoremediation. Bot. Serbica 43 (1), 85–95.

Pruchniewicz, D., Żołnierz, L., 2017. The influence of *Calamagrostis epigejos* expansion on the species composition and soil properties of mountain mesic meadows. Acta Soc. Bot. Pol. 86 (1), 3516.

Pruchniewicz, D., Żołnierz, L., Andonovski, V., 2017. Habitat factors influencing the competitive ability of *Calamagrostis epigejos* (L.) Roth in mountain plant communities. Turkish J. Bot. 41, 579–587.

Pruchniewicz, D., Żołnierz, L., 2019. The effect of different restoration treatments on the vegetation of the mesic meadow degraded by the expansion of *Calamagrostis epigejos*. Int. J. Agric. Biol. 22, 347–354.

Radiotis, T., Li, J., Goel, K., Eisner, R., 1996. Fiber characteristics, pulpability, and bleachability studies of switchgrass. Proceedings of the 1996 TAPPI Pulping conference, pp. 371–376.

Ranđelović, D., 2015. Geobotanička i biogeohemijska karakterizacija rudničke otkrivke u Boru i moguc´nost primene rezultata u remedijaciji. PhD Thesis, Univerisity of Belgrade. [in Serbian].

Ranđelović, D., Cvetković, V., Mihailović, N., Jovanović, S., 2014. Relation between edaphic factors and vegetation development on copper mine wastes: a case study from Bor (Serbia, SE Europe). Environ. Manage. 53 (4), 800–812.

Ranđelović, D., Jakovljević, K., Mihailović, N., Jovanović, S., 2018. Metal accumulation in populations of *Calamagrostis epigejos* (L.) Roth from diverse anthropogenically degraded sites (SE Europe, Serbia). Environ. Monitor. Assess. 190 (4), 183.

Rascio, N., Navari-Izzo, F., 2011. Heavy metal hyperaccumulating plants: how and why do they do it? And what makes them so interesting? Plant Sci. 180 (2), 169–181.

Rebele, F., 1986. Die Ruderalvegetation der Industriegebiete von Berlin (West) und deren Immissionsbelastung. Landschaftsentwicklung und Umweltforschung, 43, 1–224. Rebele, F., 1996a. *Calamagrostis epigejos* (L.) Roth auf anthropogenen Standorten – ein Überblick. Verhandlungen der Gesellschaft für Ökologie 26, 753–763.

Rebele, F., 1996b. Konkurrenz und Koexistenz bei ausdauernden Ruderalpflanzen. Verlag Dr. Kovac, Hamburg, Germany. [in German].

Rebele, F., 2000. Competition and coexistence of rhizomatous perennial plants along a nutrient gradient. Plant Ecol. 147 (1), 77–94.

Rebele, F., Lehmann, C., 2001. Biological flora of central Europe: *Calamagrostis epigejos* (L.) Roth. Flora 196 (5), 325–344.

Rebele, F., Surma, A., Kuznik, C., Bornkamm, R., Brej, T., 1993. Heavy metal contamination of spontaneous vegetation and soil around the copper smelter "Legnica". Acta Soc. Bot. Pol. 62 (1–2), 53–57.

Remon, E., Bouchardon, J.L., Faure, O., 2007. Multi-tolerance to heavy metals in *Plantago arenaria* Waldst. & Kit.: adaptative versus constitutive characters. Chemosphere 69 (1), 41–47.

Rogalski, M., Wieczorek, A., Szenejko, M., Kamińska, A., Miłek, E., 2008. Możliwości wykorzystania ekstensywnie użytkowanych łąk nadmorskich do celów energetycznych. Łąkarstwo w Polsce 11, 177–184, [in Polish].

Roozeboom, K., Wang, D., McGowan, A., Propheter, J., Staggenborg, S., Rice, C., 2018. Long term biomass and potential ethanol yields of annual and perennial biofuel crops. Agronom. J. 110, 1–10.

Rozhevitz, R.J., 1934. *Calamagrostis*. In: Komarov, V.L., (Ed.), Flora of the USSR 2. The Botanical Institute of the Academy of Sciences of the USSR, pp. 152–184.

Rutkowski, L., 2008. Klucz do oznaczania roślin naczyniowych Polski niżowej. Wydawnictwo Naukowe PWN, Warszawa [in Polish].

Sage, R.F., Kubien, D.S., 2007. The temperature response of C3 and C4 photosynthesis. Plant Cell Environ. 30, 1086–1106.

Samson, R., Mehdi, B., 1998. Strategies to reduce the ash content in perennial grasses. Proceedings of Resource Efficient Agricultural Production-(REAP), Ste. Anne de Bellevue, Québec, Canada, pp. 3–8.

Sanderson, M., Adler, P., 2008. Perennial forages as second generation bioenergy crops. Int. J. Mol. Sci. 9 (5), 768–788.

Schlapfer, B., Ryser, P., 1996. Leaf and root turnover of three ecologically contrasting grass species in relation to their performance along a productivity gradient. Oikos 75 (3), 398–406.

Schwartz, C., Echevarria, G., Morel, J.L., 2003. Phytoextraction of cadmium with *Thlaspi caerulescens*. Plant Soil 249, 27–35.

Sedlakova, I., Fiala, K., 2001. Ecological problems of degradation of alluvial meadows due to expanding *Calamagrostis epigejos*. Ekologia Bratislava 20, 226–233.

Sharp, D., Simon, B.K., AusGrass: grasses of Australia, CSIRO Publishing, Melbourne, Australia.

Sienkiewicz-Paderewska, D., Narewska, S., Narewski, D., Olszewski, T., Paderewski, J., 2017. Ecological and syntaxonomical spectra of grass species of the grassland communities occurring in the lower section of the Bug River Valley. Grassland Sci. Poland 20, 149–162.

Sierka, E., Kopczyńska, S., 2014. Participation of *Calamagrostis epigejos* (L.) Roth in plant communities of the Bytomka river valley in terms of its biomass use in power industry. Environ. Socioecon. Studies 2, 38–44.

Smith, R.A., Bradshaw, A.D., 1979. The use of metal tolerant plant populations for the reclamation of metalliferous wastes. J. Appl. Ecol. 16, 595–612.

Somodi, I., Virágh, K., Podani, J., 2008. The effect of the expansion of the clonal grass *Calamagrostis epigejos* on the species turnover of a semi-arid grassland. Appl. Veg. Sci. 11 (2), 187–192.

Srivastava, J., Kalra, S.J., Naraian, R., 2014. Environmental perspectives of *Phragmites australis* (Cav.) Trin. Ex. Steudel. Appl. Water Sci. 4 (3), 193–202.

Stefanowicz, A.M., Kapusta, P., Błońska, A., Kompała-Bąba, A., Woźniak, G., 2015. Effects of *Calamagrostis epigejos*, *Chamaenerion palustre* and *Tussilago farfara* on nutrient availability and microbial activity in the surface layer of spoil heaps after hard coal mining. Ecol. Eng. 83, 328–337.

Suele, A.L., Hasan, S., Kusin, F.M., Yusuff, F.M., Ibrahin, Z.Z., 2017. Phytoremediation potential of *Vetiver* grass (*Vetiveria zizanioides*) for treatment of metal-contaminated water. Water Air Soil Pollut. 228 (4), 158.

Sugathapala, A., 2013. Technologies for converting waste agricultural biomass to energy, United Nations Environmental Programme Division of Technology, Industry and Economics International Environmental Technology Centre, Osaka.

Süβ, K., Storm, C., Zehm, A., Schwabe, A., 2004. Succession in inland sand ecosystems: which factors determine the occurrence of the tall grass species *Calamagrostis epigejos* (L.) Roth and *Stipa capillata* L. Plant Biol. 6 (4), 465–476.

Šyc, M., Pohořelý, M., Kameníková, P., Habart, J., Svoboda, K., Punčochář, M., 2012. Willow trees from heavy metals phytoextraction as energy crops. Biomass Bioenerg. 37, 106–113.

Ten Harkel, M.J., van der Meulen, F., 1995. Impact of grazing and atmospheric nitrogen deposition on the vegetation of dry coastal dune grasslands. J. Veg. Sci. 6, 445–452.

Těšitel, J., Mládek, J., Fajmon, K., Blažek, P., Mudrák, O., 2018. Reversing expansion of *Calamagrostis epigejos* in a grassland biodiversity hotspot: Hemiparasitic *Rhinanthus major* does a better job than increased mowing intensity. Appl. Veg. Sci. 21 (1), 104–112.

Těšitel, J., Mládek, J., Horník, J., Těšitelová, T., Adamec, V., Tichý, L., 2017. Suppressing competitive dominants and community restoration with native parasitic plants using the hemiparasitic *Rhinanthus alectorolophus* and the dominant grass *Calamagrostis epigejos*. J. Appl. Ecol. 54, 1487–1495.

Tilman, D.A., Duong, D., Miller, B., 2009. Chlorine in solid fuels fired in pulverized fuel boilers - sources, forms, reactions, and consequences: a literature review. Energ. Fuels 23 (7), 3379–3391.

Vajpayee, P., Rai, U.N., Choudhary, S.K., Tripathi, R.D., Singh, S.N., 2000. Management of fly ash landfills with *Cassia surattensis* Burm: a case study. Bull. Environ. Contamin. Toxicol. 65, 675–682.

Végvári, Zs., Ilonczai, Z., Boldogh, S., 2011. A tüzek hatása. In: Viszló, L. (Ed.), A természetkímélő gyepgazdálkodás: Hagyományőrző szemlélet, modern eszközök. Pro Vértes Természetvédelmi Közalapítvány, Csákvár, pp. 189–209, [in Hungarian].

Werle, S., Tran, K., Magdziarz, A., Sobek, S., Pogrzeba, M., Løvås, T., 2019. Energy crops for sustainable phytoremediation–fuel characterization. Energ. Proc. 158, 867–872.

Wojcikowska-Kapusta, A., Urban, D., Baran, S.T., Bik-Małodzińska, M.A., Żukowska, G.R., Pawłowski, A.R., Czechowska-Kosacka, A., 2017. Evaluation of the influence of composts made of sewage sludge, ash from power plant, and sawdust on floristic composition of plant communities in the plot experiment. Environ. Protect. Eng. 43 (2), 129–141.

Wolak, J., 1980. Reaction of ecosystems to large concentrations of air-pollution. Berichte des Internationalen Symposiums der Internationalen Vereinigung für Vegetationskunde ("Epharmonie", Rinteln 1979). Vaduz, pp. 301–308.

World energy council, World energy resources, 2016. Available from: https://www.worldenergy.org/wpcontent/uploads/2017/03/WEResources_Bioenergy_2016.pdf.

Yang, H., Wu, M., Liu, W., Zhang, Z., Zhang, N., Wan, S., 2011. Community structure and composition in response to climate change in a temperate steppe. Global Change Biol. 17, 452–465.

Zieleźnik-Rusinowska, P., Gieroń, Z., Kornacka, P., Szopiński, M., Sitk, K., Pasierbiński, A., Agnieszka Błońska, A., Rajtor, M., Magurno, F., Pawlikowska, S., Śmietana, M., Woźniak, G., Małkowski, E., 2017. Mycorrhizal colonization of *Daucus carota* and *Calamagrostis epigejos* growing on post-mining heap and its role in oxidative stress. In: Sierka, E., Nadgórska-Socha, A., (Ed.), Aktualne Problemy Ochrony Środowiska. Ocena Stanu, Zagrożenia Zasobów i Stosowane Technologie. Uniwersytet Śląski, Katowice, pp. 106, [in Polish].

Żurek, G., Rybka, K., Pogrzeba, M., Krzyżak, J., Prokopiuk, K., 2014. Chlorophyll a fluorescence in evaluation of the effect of heavy metal soil contamination on perennial grasses. Plos One 9 (3), pe91475.

Potential of Napier grass (*Pennisetum purpureum* Schumach.) for phytoremediation and biofuel production

14

Vimal Chandra Pandey[a,*], Divya Patel[b], Shivakshi Jasrotia[c], D.P. Singh[a]
[a]Department of Environmental Science, Babasaheb Bhimrao Ambedkar University, Lucknow, Uttar Pradesh, India; [b]Department of Biotechnology, Sant Gadge Baba Amravati University, Amravati, Maharashtra, India; [c]Department of Clinical Research, Delhi Institute of Pharmaceutical Sciences and Research, Government of N.C.T. of Delhi, India
[*]Corresponding author

1 Introduction

Plants and related remediation with biofuel production have been a quite a keen and focused research highlight of achieving a solution to both the economic and environmental aspects solving waste/heavy metal pollution issues by utilizing non-renewable fossil fuels, plant species etc. (Nimmanterdwong et al., 2017). This is further in relation to reduce the pollution level and load/dependence on fossil fuel, practices are being encouraged to search and provide effective strategies and models. Keeping the sustainable quotient in mind, one must urge to identify biofuels drawn from feedstock as such which could produce (1) low greenhouse gas GHG, (2) less impact on food security. Also, there has been much focused research on using plant species for phytoremediation purposes with eco-friendly solutions.

This section will throw light on the attempts of using the technique of phytoremediation and further using the rich biomass further for biofuels. Phytoremediation is the in situ application of plants and their associated micro flora for environmental clean-up (Antoniadis et al., 2017). Phytoremediation techniques use any one of the six mechanisms such as phytoaccumulation/phytoextraction, phytotransformation, phytostabilization, phytovolatilization, phytostimulation, and rhizofiltration (Rahman and Hasegawa, 2011). Among these, in situ-type phytoextraction mechanism is mostly preferred for heavy metals as in vivo and in vitro techniques are more expensive (Susarla et al., 2002). The following few paragraphs would deal in this sector of plant species used for phytoremediation and further biofuel production with special focus on Napier Grass (*Pennisetum purpureum* Schumach.). Grasses like *P. purpureum* have been studied for their effective use for biofuel production and phytoremediation. Carvalho et al. (2016) and Mohapatra et al. (2017) study brought forth the potential of *P. purpureum*.

Napier Grass (*P. purpureum* also commonly known as Elephant Grass) is a typical grass found in Africa (Kenya). It has also been spotted in Thailand as a widely used feedstock (Sawasdee and Pisutpaisal, 2014). Napier grass is a promising energy crop

Phytoremediation Potential of Perennial Grasses. http://dx.doi.org/10.1016/B978-0-12-817732-7.00014-6

in the tropical region (Phitsuwan et al., 2016) specifically a C4 tropical species that has been popularly used for forage/fodder by farmer since it has high dry matter productivity, sustainability over several years in low-altitudinal sites. It is a fast growing perennial grass which is deeply rooted and can attain a height of up to 4 m. It can propagate via underground stems to give a ground cover. This grass has a special feature of getting little damage from serious pests (Ishii et al., 2015). Recently, this grass has gained attention due to its potential as a bioethanol feedstock and for phytoremediation. Napier grass contains 30.9% total carbohydrates, 27% protein, lipid 14.8%, total ash 18.2%, fiber 9.1% (dry weight) (Sawasdee and Pisutpaisal, 2014). Other than grasses, aquatic macrophyte and microphyte have also been studies for phytoremediation of water bodies contaminated with high arsenic concentration (Jasrotia et al., 2017).

Declining petroleum resources and higher demand for energy due to increased industrialization and motorization have led to attempts to identify alternative energy sources (Phitsuwan et al., 2016). Tilman et al. (2009) pointed out the usage of biomass as a feedstock in a bio refinery to convert the biological materials into fuels and chemicals is still in a nascent state. He further added, industries can either directly use food crops as feedstock or replace existing arable land for food crops with energy crops which would cause higher food prices and triggers the farmers to clearing more forest to grow more food crops. Biggest disadvantage in this aspect is the mass land usage for being able to use the biomass from energy crops as feedstock's for a bio refinery. Two alternates can further be considered: (1) focus on biomass residues, such as straw, husk, and other agricultural co-products or wastes are the type of promising feedstock for advanced biofuels, (2) use of perennial warm-season grasses such as Napier grass (*P. purpureum*), *Miscanthus* (Morandi et al., 2016), Indian grass, and switchgrass (Felix and Tilley, 2009). In the present chapter, the various aspect of Napier grass such as origin and geographical distribution, ecology, taxonomy, and morphological description, propagation, important features of Napier grass, multiple uses, phytoremediation, bioenergy production, conclusion, and future prospects will be described in detail to understand the possibility of Napier grass in phytoremediation with bioenergy production.

2 Origin and geographical distribution

Napier grass (*P. purpureum* Schumach.) is a fast-growing C4 perennial grass native to Sub-Saharan Africa that is widely grown across the tropical and subtropical regions of the world such as Asia, Australia, the Middle East, and the Pacific islands. Napier grass is a monocotyledonous flowering plant belonging to the family *Poaceae* (the family of grasses) and the genus *Pennisetum* (Sandhu et al., 2015; Singh et al., 2013). *Pennisetum* is a highly diverse genus consisting of a heterogeneous group of approximately 140 species (Brunken, 1977; Donadío et al., 2009; Dos Reis et al., 2014) with different basic chromosome numbers of 5, 7, 8, or 9, a range of ploidy levels from diploid to octoploid, both sexual and apomictic reproductive behaviors and life cycles of an annual, biennial, or perennial nature (Martel et al., 1997). Napier grass is a stem–forming tall grass which attracted recent interest as a bio-energy crop due to its vigorous biomass production (Waramit and Chaugool, 2014). Fig. 14.1 represents the

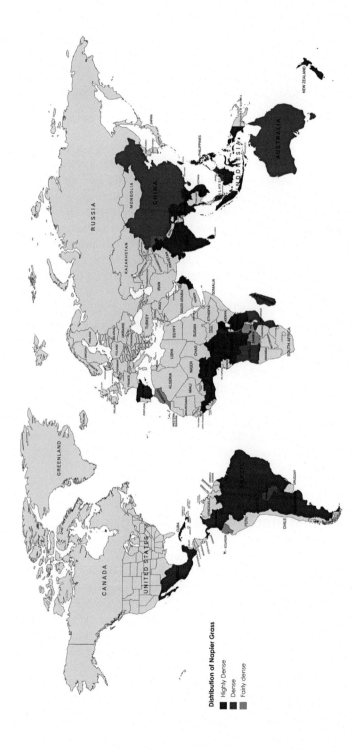

Figure 14.1 Distribution of Napier grass (*Pennisetum purpureum* Schumach.) across the world.
*Source:*Modified from Negawo et al., 2017.

distribution of Napier grass worldwide in three categories such highly dense, dense, and fairly dense.

3 Ecology

P. purpureum is a significant forage grass with perennial nature, and is closely linked to pearl millet (*P. americanum*). It is a rhizomatous large grass and occurs in agricultural fields, pastures, waterways, floodplains, wetlands, swamps, riverbanks, roadsides, forest edges, and degraded lands. It is well-adapted to drought conditions and also reported for its colonization in arid lowlands for instance habitats on Galápagos Islands (McMullen, 1999). It has been noticed to be adapted to grow across a wide range of soil conditions and agro-ecologies, from sea level to 2100 m, and it can withstand minor dry spells, although it grows best in areas where the annual rainfall is between 750 and 2500 mm (Singh et al., 2013). Given its wide agro-ecological adaptation, Napier grass has been naturalized in areas of Central and South America, tropical parts of Asia, Australia, the Middle East, and the Pacific islands (Cook et al., 2005; Kandel et al., 2015).

Association with other plants—Though Napier grass is mostly grown in pure stands, it can be cultivated in association with legumes plants such as tropical Kudzu (*Pueraria phaseoloides*), butterfly pea (*Centrosema pubescens*), soybean (*Neonotonia wightii*), and leucaena (*Leucaena leucocephala*). Such associations provide higher nutritional value than Napier grass alone and can produce higher dry matter yields, suppress weeds, and increase soil fertility. Napier grass is irregularly intercropped with banana and cassava plants in the home gardens (Mannetje, 1992). Three legumes, that is, *Desmodium intortum*, *Macrotyloma axillare*, *Neonotonia wightii* are associated with Napier grass in central Kenya. On comparison basis, it was recoded that *D. intortum* offered the finest option but *N. wightii* presented the lowest performance (Mwangi et al., 2004).

4 Taxonomy and morphological description

P. purpureum, also known as Napier grass, elephant grass, is a very large, robust, rhizomatous, tufted perennial grass (1–7 m tall). It has bamboo-like clumps, generally 2–4 m but sometimes up to 7.5 m in height, with rounded, rough, and perennial nature, branched toward top, about 3 cm diameter near the base, internodes more or less bluish, young nodes with white hairs, later becoming smooth and glabrous (hairless). It forms dense and thick clumps or colonies from the base, up to 1 meter across. Napier grass has a vigorous and deep root system that develops from the nodes of the creeping stolons. Leaves are flat, linear to tapering, hairy at the base, often bluish-green color, up to 40–120 cm long and 1–5 cm wide, pilose near the base. The leaf margins are generally rough and finely toothed. The leaf has a noticeable midrib, whitish above, strongly keeled below. The inflorescence of Napier grass is a stiff and dense terminal bristly spike, yellow-brown to purple-tinged in color, up to 20 cm long and 2 cm wide. Spikelets are 4–6 mm long, lonely or in groups of 2–6 on hairy axis, surrounded by 2 cm long sparsely plumose bristles and fall with the spikelets at maturity, outermost glume minute or absent. Its seeds are very small.

There are two major categories of Napier grass (also called elephant grass, possibly owing to its large size) cultivars based on their morphology, the normal or tall (up to 4–7 m) varieties (e.g., "Australiano," "Bana," and "French Cameroon") and the dwarf or semi-dwarf (<2 m) varieties (e.g., "Mott") (Rengsirikul et al., 2013). The normal varieties have been reported to produce up to twice as much yield as the dwarf ones. However, dwarf varieties also have a number of positive attributes, including enhanced overwintering capacity in the border areas between subtropics and temperate zones, better nutritive value, and ease of management and harvesting. Therefore, different cultivars of Napier grass can be adopted by farmers depending on their situation and ultimate use of the crop.

5 Propagation

Napier grass is widely grown in tropical and subtropical regions of the world, for use predominantly as animal fodder. Napier can be more commonly distributed by vegetative cuttings and tillers, since the grass cannot produce many seeds and those produced are normally very small, light, of poor quality and the spikelets are prone to shattering. Thus, the seeds are considered inappropriate for propagation as they produce weak seedlings and, as Napier grass is an open pollinated crop, the seedlings are also highly heterozygous. Consequently, propagation by stem cuttings is currently the dominant practice for the distribution of Napier grass. Napier grass is easily propagated by planting cuttings directly in the field (Lounglawan et al., 2014). As reviewed in Sandhu et al. (2015), a number of genebanks (e.g., the International Center for Tropical Agriculture (CIAT); the Commonwealth Scientific and Industrial Research Organisation (CSIRO); the International Livestock Research Institute (ILRI); and the National Bureau of Plant Genetic Resources (NBPGR)) are involved in conserving a substantial amount of tropical and sub-tropical forage genetic resources. Over 300 accessions of Napier grass are currently being maintained in different genebanks. A broad array of Napier grass accessions are currently being maintained by the ILRI forage genebank in the field at Debre Zeit and Zwai, Ethiopia with considerable diversity in growth and form. However, germplasm available from genebanks has so far been largely underutilized. Plant breeding and selection in Napier grass has primarily been aimed at improving different agronomic traits such as disease resistance, yield, nutritional quality, growth habit (dwarfing), palatability, and abiotic stress tolerance (Harris et al., 2010). Napier grass is cross-compatible with the closely related species pearl millet (*Pennisetum glaucum*) ($2n = 2x = 14$, genome AA); the resultant hybrids are triploid and sterile and can only be propagated by vegetative means which, although labor intensive, ensure a true-to-type variety. A number of agronomically important traits, nutritional quality and palatability, for example, have been introgressed into the genome of Napier grass from pearl millet through conventional plant breeding and

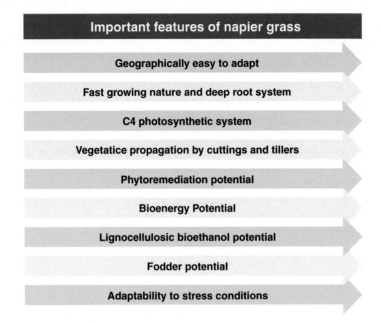

Figure 14.2 Important features of Napier grass (*Pennisetum purpureum* Schumach).

hybrids have become a crucial part of the forage crop value chain in Africa, Asia, and South America.

6 Important features of Napier grass

The Napier grass (*Pennisetum purpureum* Schumach.) has adaptability to various stress conditions due to having some important features such as C4 photosynthetic system, vegetative propagation, drought tolerance, adaptableness for weed control in animal burial site (Ishii et al., 2016), etc. Besides these features, Napier grass has some important features such as fast growing nature, high biomass productivity, phytoremediation potential, bioenergy potential, lignocellulosic bioethanol potential, and fodder potential. Fig. 14.2 depicts the important features of Napier grass.

7 Multiple uses

Napier grass is one of the most important fodder crops, particularly in Eastern and Central African smallholder farming communities. It is mainly used to feed livestock in cut and carry feeding systems. It is a multipurpose forage crop that can be grazed directly or made into silage or hay and there have also been reports of using Napier grass as fish food, for example, for feeding grass carp and tilapia in Nepal and Bangladesh.

Young shoots of Napier grass were used as a cooked vegetable. These varied uses provide an indication of the diversity of roles that Napier grass could contribute to the reduction of poverty and nutritional insecurity. In addition to its value as a forage crop, Napier grass can also be used to make fences, as a windbreak, to demarcate boundaries among neighboring farmers, and the dried material can be used as a fuel source. In crop land management systems, it is used as a mulch to control weed infestation and soil erosion and as a trap plant in the push–pull strategy, a pest management practice which uses repellent intercrop "push" plants and attractant trap "pull" plants for insect pest control in Africa, particularly for the maize stem borer. Plants are also used to scavenge pollutants, such as heavy metals, and Napier grass has been used in phytoremediation strategies, for example, for the clean up of cadmium-affected soil, reducing the concentration of cadmium to a depth of 15 cm in soil. With growing global interest in reducing fossil fuel consumption and concerns about the impacts of climate change, the search for alternative biofuel sources has led to the promotion of large biomass plants as second- or next-generation biofuel crops. Napier grass, with its perennial nature and fast growing characteristics, has been reported to produce a dry matter yield of up to 78 tons ha^{-1}annum^{-1}(35–41 tons ha^{-1} average). Rengsirikul et al. (2013) estimated a maximum ethanol production of 350–460 L ton^{-1} dry matter from Napier grass varieties grown in Thailand, and an estimated ethanol yield of 329 L ton^{-1} dry matter. Lima et al. (2014) demonstrated that this potential was 6% and 15% higher than for the tropical forages *Brachiaria brizantha* and *Panicum maximum*, respectively, around 15% higher than Eucalyptus bark and 17% higher than for sugarcane. Consequently, the potential exists for the use of Napier grass for phytoremediation purposes, after which the large harvest could go into processing plants for biofuel production (Table 14.1).

8 Phytoremediation

Soil is one of the important natural resources for essential elements and biodiversity. Due to industrialization, many countries, including the United States, Germany, United Kingdom, China, and India are greatly affected by urban soil contamination (Meuser, 2010; Patel et al., 2018). As a result one of the main constituent, that is, heavy metals of soil exceeded its limit and has gained a serious attention at the global perspective. Heavy metal can be very toxic even in low concentration and are not easily degradable. It can easily bio-accumulate and cause food chain contamination and harmful to humans and other living organisms. In soil heavy metals often exist in small amounts and in plants some of the trace metals play an essential role in promoting biological growth. Several soil remediation technologies have been used over the last few decades and phytoremediation considered to be one of the most cost effective and eco-friendly solution for soil metal contamination (Pandey et al., 2015; Purakayastha and Chhonkar, 2010; Pandey and Bajpai, 2019).

The choice of plant species plays a very crucial role in phytoremediation strategies. There should be some special characteristics, such as high accumulation capacity of

Table 14.1 List of multiple use of Napier grass "Elephant grass" (*Pennisetum purpureum* Schumach.) toward ecosystem services.

S. no.	Role of Napier grass toward environmental sustainability	Country	Main use	References
1	Swine wastewater treatment with Napier grass using vertical subsurface flow constructed wetland	Thailand	Swine wastewater treatment	Klomjek (2016)
2	Using elephant grass as a substrate for cellulase and xylanase production	Brazil	Production of cellulases and xylanases	Menegol et al. (2017)
3	The endophytic bacteria obtained from Napier grass improve plant growth and salt tolerance of hybrid *Pennisetum*	China	Isolation of bacteria for plant growth	Li et al. (2016)
4	Phytoremediation Potential of Napier grass for cadmium pollutant	Japan	Phytoremediation of cadmium	Ishii et al. (2015)
5	Phytoremidiation potential of Napier grass for heavy metals cultivated on tannery sludge	Bangladesh	Phytoremediation of tannery sludge	Juel et al. (2018)
6	Study of "Napier grass" delignification for production of cellulosic derivatives	Brazil	Production of cellulosic derivatives	de Araújo Morandim-Giannetti et al. (2013)
7	Use of *calliandra*–Napier grass contour hedges to control erosion	Kenya	Soil erosion	Angimaa et al. (2002)
8	Feasibility of biogas production from perennial Napier Grass	Thailand	Biogas production	Sawasdee and Pisutpaisal (2014)
9	Bioethanol production from the phytoremediated biomass of Napier grass with heavy metals	Taiwan	Bioethanol production from phytoremediated biomass	Ko et al. (2017)
10	Impacts of cultivation methods and planting season on biomass yield of Napier grass under rainfed conditions	Thailand	Biomass production	Haegele and Arjharn (2017)

11	Copper tolerance of the biomass crops Elephant grass, Vetiver grass and the upland reed in soil culture	China	Copper tolerant biomass grass for remediation	Liu et al. (2009)
12	Elephant grass ecotypes for bioenergy production via direct combustion of biomass	Brazil	Bioenergy production	de Carvalho Rocha et al. (2017)
13	Lignocellulosic butanol production from Napier grass	Taiwan	Butanol production	He et al. (2017)
14	Potential of fermentable sugar production from Napier grass residue as a substrate to produce Bioethanol	Thailand	Bioethanol production from Napier grass residue	Pensri et al. (2016)
15	Bio-oil production of Napier grass through rotating cylindrical reactor	Thailand	Bio-oil production	Singbua et al. (2017)
16	Yield and quality of elephant grass biomass produced in the Cerrados region for bioenergy	Brazil	Biomass production	Flores et al. (2012)
17	Effects of inter-cutting interval on biomass yield, growth components, and chemical composition of Napier grass cultivars as bioenergy crops	Thailand	Biomass yield	Rengsirikul et al. (2011)
18	Biomass yield, chemical composition, and potential ethanol yields of 8 cultivars of Napier grass	Thailand	Ethanol yield	Rengsirikul et al. (2013)
19	Phytoextraction of heavy metal by Napier grass from tailing waste	China	Phytoextraction	Ma et al. (2016)
20	Environmental performance assessment of Napier grass for bioenergy production	Thailand	Bioenergy production	Nimmanterdwong et al. (2017)
21	Bioenergy potentials of Elephant grass	Nigeria	Bioenergy	Ohimain et al. (2014)
22	Adaptability of Napier grass for weed control in site of animals buried after foot-and-mouth disease infection	Japan	Weed control	Ishii et al. (2016)
23	Potential of Napier grass with cadmium-resistant bacterial inoculation on cadmium phytoremediation and its possibility to use as biomass fuel	Thailand	Phytoremediation and biofuel	Wiangkham and Prapagdee (2018)

contaminant, fast growing, high biomass production, and of course survival in variable climate conditions etc. Compared to trees and shrubs, herbaceous plants, especially grasses have characteristics of rapid growth, large biomass, strong resistance, and effective stabilization to soils. This usually results in excellent restoration effects in degraded and mined lands, particularly in the tropics and subtropics with high temperature and precipitation (Loch, 2000; Ye et al., 2000). Tropical grasses emerged as a preferable choice for phytoremediation over other shrubs and trees because of high survival rate (Ali et al., 2013; Sinha et al., 2013; Ng et al., 2016). Grasses that use the C4 photosynthetic pathway are naturally efficient at cellulosic biomass production. Perennials particularly can benefit the ecosystem by sequestering more carbon, reducing the need for tillage, producing less soil erosion, requiring less displacement of food production, and requiring less consumption of fertilizers and pesticides (Sheehan et al., 2004; Moore et al., 2008). Several perennial herbaceous grasses have been studied in recent years as potential dedicated cellulosic biomass crops, including Elephant grass or Napier grass (*Pennisetum purpureum* Schumach). In such studies, it was shown that they are able to survive under unfavorable conditions, such as heavy metal stress, and are able to consistently produce high biomass (Du et al., 2006). However, few studies have been conducted to examine their growth and biomass as potential dedicated energy plants under copper (Cu) stress. Studies evaluated their potentials in areas polluted by heavy metals, such as farmlands near to Cu mining and smelting operations, or mine tailing areas, it is necessary to understand their abilities to resist or tolerate Cu stress, and the relationships between their biomass and calorific value under these circumstance (Liu et al., 2009). Napier grasses as a metal accumulator plays an important role. In case of aquatic environment, one of the most a detectable heavy metal is Pb. Its contamination in aquatic and coastal environment such as sediment, sea water, and brackish water has been acknowledged as the major problem worldwide that affects health of living being (Harguinteguy et al., 2015). However, the limitation of the phytoremediation is long time required for the clean up contaminated site and toxic plant left over after the remediation approach. Therefore, utilization and effective remediation of contaminated land synergistic bonding by using energy crop for bioenergy production has been considered as sustainable approach of phytoremediation (Pandey et al., 2016). Napier grass was suggested as a good candidate for Pb phytoremediation in a study examined under the brackish water that showed removal in high phytotoxicity. We can say that on the other hand, Napier grass is high tolerated to high level of Pb concentration (Hongsawat et al., 2018). Elephant grass has been used for fodder since it has high dry matter productivity, sustainability over several years in low-altitudinal sites of Kyushu, Japan and little damage from serious pests. Cadmium (Cd), a heavy metal, poses risks to human health through the consumption of contaminated agricultural products. Since much arable land in Japan has been polluted by Cd, mainly from mining activity, Napier grass cv. Wruk Wona showed higher potential for removing Cd contamination.Cd uptake was higher in those Napier cut twice than in those cut once. Thus, soil Cd concentration was reduced by 4.6% following two harvests in a single growing season, suggesting that Napier grass is a potential Cd phytoextractor (Ishii et al., 2015).

Sewage sludge promotes improvements in the physical-chemical and biological characteristics of the soil and increments in the yield of crops. Due to these properties, it is considered as a waste rich in organic matter and contains essential elements to plant nutrition. Regardless of all benefits, this waste may have in its composition high concentrations of organic contaminants, such as chlorobenzenes (CBs). CBs are present in the composition of cleaning material, solvents, pharmaceutical products, graphic materials, pesticides, deodorants, dyes, pesticides, degreasers, lubricants, and air purifiers so are generally found at high concentrations in sewage sludge (Kamarei et al., 2010). Low biodegradability of CBs are matter of great environmental concern (Schroll et al., 2004) and the harmful effects of CBs are prolonged duration, due to their high stability in the soils (Koe and Shen, 1997). *P. purpureum* cultivation in sewage sludge for approximately 5 months promotes reductions in the concentrations of 1, 3, 5-CB in the layers with greater concentration of roots, compared with uncultivated sewage sludge, though it does not influence the concentration of 1,4–CB. In general, the concentrations of 1, 4–CB and 1,3,5-CB, in sewage sludge, decrease with the increase in the *P. purpureum* cultivation period; thus, cultivation for at least 150 days is advisable to maintain the levels of the contaminants within the safest limits for the agricultural use of the waste (Alvarenga et al., 2017). Tannery sludge contains high concentration of heavy metals like Cr, As, Pb, Ni, Cu, Zn, and Cd due to the use of basic chromium salt, different syntans, dyes, pigments, retanning agents during the leather production (Juel et al., 2017). In recent years, some methods are available for reuse or recycling of tannery sludge but most of the tannery sludge is disposed of as landfilling without any treatment. This disposal can alter the physicochemical properties and fertility of the soil. When crops are grown on such kind of contaminated soil, the heavy metal can be accumulated in crops grown on such contaminated soil especially in its vegetative parts with incorporation into the food chain and leading to bio-amplification (Pergent and Martini, 1999). Napier grass has phytoremediation potential due to its long, deep root system and resistance to an extensive variety of unfavorable climatic and edaphic conditions. Very few researchers have investigated Napier grass in phytoremediation of heavy metals cultivated in contaminated sites. Napier grass was good uptaker of Cr, and Zn. (Juel et al., 2018). Napier grass has also played significant role in rehabilitation as well as phytoremediation of oil shale mined land. Oil shale waste dump makes a harsh habitat for vegetation development, but herbaceous plants could grow on it, especially after it was amended by fertilizer or fishpond sludge. Application of fertilizer resulted significant biomass of grass for remediation of Pb and Cd removal from contaminated soil. Therefore it would be possible and promising to rehabilitate the oil shale mined land into grassland for husbandry and wildlife. Fertilizer was a good amendment for oil shale waste, and it could enhance plant cover and biomass, and abate ability of plants to take up heavy metals, thereby reducing the risk of heavy metals' toxicity to animals and man through food chains. However, fertilizer would bring a negative impact on survival rate of plants if it were applied prior to planting plants. Fertilization did not reduce the amount of heavy metals accumulated in plants; the main reason that concentration of heavy metals in plants was decreased by fertilizer application was because an increase of biomass diluted the concentrations of metals in plants (Xia, 2004).

Most of the heavy metals get stored in the plant's stem due to its high biomass, although the highest concentration of the metal occurred in the fibrous roots. Studies have been made on hybrid giant Napier (HGN) grass for its phytoextraction potential for removing heavy metals from contaminated tailings. Soil Study indicated that heavy metal was phytoextracted by fibrous roots, then transported from roots to shoots and stored to all parts of the plant including fibrous roots, tap roots, stem, and leaves. Cultivation of HGN in heavy metal contaminated tailings reduces heavy metal content and increases soil nutrients. Finally, although more heavy metal was absorbed by the HGN grown in the tailings, the mass ratio among the four parts of the HGN in the control soil and the tailings is very close. This indicates that the plant's metabolic function plays a critical role in determining the metal distribution in the plant (Ma et al., 2016). Livestock farming, especially on swine farms, causes a large amount of wastewater which contains a high concentration of organic substances, solids, and nutrients. Without appropriate wastewater treatment methods, the effluent can contaminate water resources. In order to control water quality of the wastewater discharge from the swine farms, Thailand has developed effluent standards. Currently, only the large-sized swine farms are able to comply with the standards. Many small-sized farms fail to do so due to their budget limitations and lack of ability to operate the complicated systems. A simple operation and low cost wastewater treatment system can be an important alternative to improve water quality for these small-sized swine farms. In case of wastewater treatment, (Goorahoo et al., 2006) reported that Napier grass was used to reduce excess nutrients from diary effluent in order to reuse wastewater for irrigation. The grass showed a decent potential to absorb significant amounts of the excess nutrients in the wastewater. In India, a vertical subsurface flow constructed wetlands (VSF CW) planted with Napier grass was used to treat grey water in which the effluent reached the USEPA standard for water reuse (Pillai and Vijayan, 1997). It can be concluded from the study done by (Klomjek, 2016) that the Giant and Dwarf Napier grasses can be used in the VSF CW to treat swine wastewater. The VSF CW exhibited a prominent pollutant removal performance, especially for the BOD and TKN. Phytoremediation technology proved itself a eco-friendly and cost effective way to clean-up the contaminant sites but with a very well-known limitation that is long time requirement for the clean-up site and toxic plant left over after the remediation approach. Consequently, the sustainable approach of phytoremediation using energy crops for bioenergy production is being adopted in remediation methodologies. In view of this since recent years, cellulosic biomass such as Napier grass has been identified as good candidates for bioenergy production due to their high growth rate and high biomass production.

9   Bioenergy production

Bio-refineries have become promising alternative strategies for industrial development. To achieve the energy demand and climate change mitigation goals, the idea of producing energy from biological materials has been promoted. The dependency on non-renewable energy resources can be reduced by developing biomass utilization systems and which may provide effective strategies for both the economic and environmental aspects of utilizing fossil fuels. The ideal raw material for biofuels must produce lower greenhouse gases than conventional fossil (Nimmanterdwong et al., 2017). Biomass residues, such as straw, husk, and other agricultural co-products or wastes are the promising feedstock for advanced biofuels. Perennial warm season grasses like Napier grass (*P. purpureum*) give an alternative option to produce bioenergy because these crops could produce good yields even under severe conditions (Campbell et al., 2008). In Nigeria, wild strains of elephant grass (Napier) occur as invasive weed especially in disturbed freshwater swamps of Bayelsa State. In these places, weed control is done mostly manually and sometimes by herbicides. This practice takes much more money, labor, and herbicides to control invading grasses such as elephant grass. Instead of burning these grasses and waste its potential energy to the open air, the loss of useful energy can be transformed into bioenergy. A study was undertaken by Ohimain et al. (2014) to assess the productivity and bioenergy potentials of the grass. Utilization of Napier grass in such way definitely helps country like Nigeria to overcome poor socioeconomic condition, electricity issues, cooking, and transportation fuel. Along with all the benefits this research could also be beneficial for smallholder farmers, who instead of spending money to control the invading elephant grass, can earn money for the conversion of Napier grass to fuel ethanol and bagasse for power generation. Thailand government has promoted Napier grass as bioenergy crop, which has been widely used to feed local cattle and this is a potential lignocellulosic bioenergy feedstock (Nimmanterdwong et al., 2015). Several perennial grasses have been studied for their use as energetic feedstock due to high lignocellulosic production, making it an alternative bioenergy source. Study to investigate the aptitude of elephant grass for bioenergy production via direct combustion has also been made by Rocha et al. (2017).

Napier grass meets the requirement of lignocellulosic bioethanol production, because it has low lignin-content and a relatively high herbage mass per year and per area. Therefore, pre-treatment, saccharification, and fermentation processes for ethanol production from Napier grass have been extensively studied. The simultaneous saccharification and co-fermentation (SSCF) was most advantageous process since the SSCF can prevent contamination risks of other microorganism and can construct simple processing procedure (Yasuda et al., 2014). Bioethanol is one of the most significant renewable fuels. The major sources of bioethanol production are food crops such as corn, sugarcane, rice, wheat, and sugar beet. However, utilization of food crops to produce bioethanol could affect the food sources and disrupt the food to population ratio. Utilization of lignocellulosic materials such as wheat straw, grass, and crop residues to produce bioethanol has been developed for second-generation fuel, since those resources are abundant, cheap, and renewable. Napier Pakchong 1 grass (NPG) residue is a lignocellulosic waste obtained from the process of biogas production that can be used as an alternative material for bioethanol production. Study on the potential of fermentable sugar production from NPG residue was made by Pensri et al. (2016). In phytoremediation potential of Napier grass we have discussed that using plants to absorb and accumulate heavy metals from polluted soil, followed by the recycling of explants containing heavy metals, can help achieve the goal of returning contaminated soil to low heavy metal content soil. After phytoremediation, the influence of bioethanol production from Napier grass is the important topic of investigation. In study conducted by Han Ko et al. (2017), the fermentation efficiency of the heavy-metal containing biomass was higher than the control biomass. It can be confirmed that the heavy metals had a positive effect on bacteria fermentation. However, the fermentation efficiency was lower for biomass with severe heavy metal pollution. Thus, the utilization of Napier grass phytoremediation for bioethanol production has a positive effect on the sustainability of environmental resources. Lignocellulosic butanol production from Napier grass using semi-simultaneous saccharification fermentation. Compositional analysis has also been done. By comparing acid- and alkali-pretreatment, it was found that the alkali-pre-treatment process is favorable for Napier grass. Semi-simultaneous saccharification fermentation (sSSF) was carried out, where enzymatic hydrolysis and ABE fermentation were operated in the same batch. Finally, the efficiency of butanol production from Napier grass was calculated at 31% (He et al., 2017). Bio-oil production is another aspect in terms of bioenergy production from Napier grass by using different reactors and methodologies have been studied in recent years. Bio-oil has a high potential of becoming an alternative source of energy in future. Bio-oil is generated from the pyrolysis process of biomass. Pyrolysis is the process where thermal de-composition of organic compounds in the absence oxygen which produces bio-oil, char, and non-condensable gas (NGC). Study conducted by (Singbua et al., 2017) resulted that the maximum yield of bio-oil was obtained at a reactor temperature of 520°C. From the economic and cold efficiency analysis, it was found that the production cost of bio oil production and parameters quality can be achieved cost effectively. At present, many researchers are focusing on the types of reactors in order to develop of their systems because reactors have the greatest influence on bio-oil yields. It has been indicated in the literature that having a good heat transfer between the biomass and the heat source in reactor can increase the yield of bio-oil (Butler et al., 2011; Treedet and Suntivarakorn, 2012; Lelaphatikul, 2012). Biogas production can also be applicable where the feasibility of biogas production from Napier grass is one of the new approaches. Napier grass, a tropical plant, can grow up in drought and dry conditions. Napier grass contains 30.9% total carbohydrates, 27% protein, lipid 14.8%, total ash 18.2%, fiber 9.1% (dry weight). Its organic compositions are an ideal feedstock for biogas production. Economic analysis of biogas obtained from the experiment with liquid petroleum gas (LPG) with benefit/cost ratio (B/C ratio) greater than 1 suggested that the Napier grass is considered as a potential energy crop (Sawasdee and Pisutpaisal, 2014).

The potential of Napier grass as a bioenergy producing candidate has much more opportunities. In upcoming researches, one can explore this grass in such a way that it will contribute to achieve high level of bioenergy production after clean-up of contaminated sites as well as to achieve the socioeconomic aspects economically.

10 Conclusion and future prospects

Napier grass (*Pennisetum purpureum*) is a well recognized as bioenergy and fodder feed stock because of its high growth rate and adaptability nature. The grass has some significant characteristics like fast growing nature, high biomass productivity, deep root system, tolerance to a wide range of stress conditions. The utilization of Napier grass for phytoremediation and bioethanol production has a positive effect on the sustainability of environmental resources (Ko et al., 2017). Such research should be encouraged toward environment benefits and making earth a better habitat.

The literature suggests that, there is still a research gap in spite of the valuable characters of the grass, for example, studies regarding the phytomanagement of a wide range of degraded lands through using the cultivation of grass are required. No studies have been done concerning the heavy metals remediation potential of this grass in the presence of suitable microbial associations at different climatic regions. Furthermore, molecular studies will be helpful for revealing the mechanisms toward degrading or complexing the pollutants inside the grass in order to make the pollutants harmless and also makes the grass pollutant tolerant. Further, harvesting of the grass biomass for multiple uses namely, bioenergy production, pulp-paper production, is the following field of investigation which should be focused on along with management of the grass biomass after pollutants accumulation.

Acknowledgment

Financial assistance given to Dr. V.C. Pandey under the Scientist's Pool Scheme (Pool No. 13 (8931-A)/2017) by the Council of Scientific and Industrial Research, Government of India is gratefully acknowledged. The authors declare no conflicts of interest.

References

Ali, H., Khan, E., Sajad, M.A., 2013. Phytoremediation of heavy metals: concepts and applications. Chemosphere 91, 869–881.

Alvarenga, A.C., Sampaio, R.A., Pinho, G.P., Cardoso, P.H.S., Sousa, I. de P., Mário, H.C., Barbosa, M.H.C., 2017. Phytoremediation of chlorobenzenes in sewage sludge cultivated with *Pennisetum purpureum* at different times. Revista Brasileira de Engenharia Agrícola e Ambiental 21 (8), 573–578, DOI: 10.1590/1807-1929/agriambi.v21n8p573-578.

Angimaa, S.D., Stott, D.E., O'Neill, M.K., Ong, C.K., Weesies, G.A., 2002. Use of calliandra–Napier grass contour hedges to control erosion in central Kenya. Agriculture. Ecosys. Environ. 91, 15–23.

Antoniadis, V., Shaheen, S.M., Boersch, J., Frohne, T., Du Laing, G., Rinklebe, J., 2017. Bioavailability and risk assessment of potentially toxic elements in garden edible vegetables and soils around a highly contaminated former mining area in Germany. J. Environ. Manag. 186, 192–200.

Brunken, J.N., 1977. A systematic study of *Pennisetum* sect. *Pennisetum* (*Graminae*). Am. J. Bot. 64 (2), 161–176.

Butler, E., Devlin, G., Meier, D., McDonnell, K.A., 2011. Review of recent laboratory research and commercial developments in fast pyrolysis and upgrading. Renew. Sust. Energ. Rev. 15, 4171–4186.

Campbell, J.E., Lobell, D.B., Genova, R.C., Field, C.B., 2008. The global potential of bioenergy on abandoned agriculture lands. Environ. Sci. Technol. 42, 5791–5794.

Carvalho, A.R., Fragoso, R., Gominho, J., Saraiva, A., Costa, R., Duarte, E., 2016. Water energy nexus: anaerobic co-digestion with elephant grass hydrolyzate. J. Environ. Manag. 181, 48–53.

Cook, B.G., Pengelly, B.C., Brown, S.D., Donnelly, J.L., Eagles, D.A., Franco, M.A., Hanson, J., Mullen, B.F., Partridge, I.J., Peters, M., Schultze-Kraft, R., 2005. Tropical Forages: an interactive selection tool. CD-ROM. Brisbane, Australia: CSIRO, Queensland, Australia: Department of Primary Industries and Fisheries, Cali, Colombia: CIAT and Nairobi, Kenya: ILRI.

Tilman, D., Socolow, R., Foley, J.A., Hill, J., Larson, E., Lynd, L., Pacala, S., Reilly, J., Searchinger, T., Somerville C., Williams, R., 2009. Energy. Beneficial biofuels—the food, energy, and environment trilemma, Science, 325(5938), 270–271.

de Araújo Morandim-Giannetti, A., Albuquerque, T.S., de Carvalho, R.K.C., Araújo, R.M.S., Magnabosco, R., 2013. Study of "napier grass" delignification for production of cellulosic derivatives. Carbohydr. Polymers 92, 849–855.

de Carvalho Rocha, J.R., do, A.S., Machado, J.C., Carneiro, P.C.S., Carneiro, J., da, C., Resende, M.D.V., Pereira, A.V., de Souza Carneiro, J.E., 2017. Elephant grass ecotypes for bioenergy production via direct combustion of biomass. Ind. Crops Prod. 95, 27–32.

Donadío, S., Giussani, L.M., Kellogg, E.A., Zuolaga, F.O., Morrone, O., 2009. A preliminary molecular phylogeny of *Pennisetum* and *Cenchrus* (*Poaceae-Paniceae*) based on the trnL-F, rpl16 chloroplast markers. Taxon 58, 392–404.

Dos Reis, G.B., Mesquita, A.T., Torres, G.A., Andrade-Vieira, L.F., Vander Pereira, A., Davide, L.C., 2014. Genomic homology between *Pennisetum purpureum* and *Pennisetum glaucum* (Poaceae). Comp. Cytogenet. 8, 199.

Du, Ry., Nei, Cr., Lin, Cx., Liu, Y., 2006. Effects of soil cadmium on the growth of two potential energy plants (in Chinese with abstract in English). Ecol. Environ. 15, 735–738.

Felix, E., Tilley, D.R., 2009. Integrated energy, environmental and financial analysis of ethanol production from cellulosic switchgrass. Energy 34, 410–436.

Flores, R.A., Urquiaga, S., Alves, B.J.R., Collier, L.S., Boddey, R.M., 2012. Yield and quality of elephant grass biomass produced in the Cerrados region for bioenergy. Eng. Agríc. 32 (5), 831–839, set./out. 2012.

Goorahoo, D., Cassel, F.S., Adhikari, D., Morton, R., 2006. Update On Elephant Grass Research And Its Potential As A Forage Crop. In: California Alfalfa and Forage Symposium; 2005 Dec 12–14; Visalia, Ca.

Haegele, M.T., Arjharn, W., 2017. The effects of cultivation methods and planting season on biomass yield of Napier grass (*Pennisetum purpureum* Schumach.) under rainfed conditions in the northeast region of Thailand. Field Crops Res. 214, 359–364.

Harguinteguy, C.A., Pignata, M.L., Fernández-Cirelli, A., 2015. Nickel, lead and zinc accumulation and performance in relation to their use in phytoremediation of macrophytes *Myriophyllum aquaticum* and *Egeria densa*. Ecological Engineering 82, 512–513.

Harris, K., Anderson, W., Malik, R., 2010. Genetic relationships among Napier grass (*Pennisetum purpureum Schum.*) nursery accessions using AFLP markers. Plant Genetic Res. 8, 63–70.

He, C.-R., Kuo, Y.-Y., Si-Yu Li, S.-Y., 2017. Lignocellulosic butanol production from Napier grass using semi-simultaneous saccharification fermentation. Bioresour. Technol. 23, 101–108.

Hongsawat, P., Suttiarporn, P., Wutsanthia, K., Kongsiri, G., 2018. Optimization of Lead Removal via Napier Grass in Synthetic Brackish Water using Response Surface Model. IOP Conf. Series: Earth and Environmental Science 120 (2018) 012020 doi:10.1088/1755-1315/120/1/012020.

Ishii, Y., Hamano, K., Kang, D.J., Idota, S., Nishiwaki, A., 2015. Cadmium phytoremediation potential of napiergrass cultivated in Kyushu, Japan. Appl. Environ. Soil Sci. 2015, 6. Available from: http://dx.doi.org/10.1155/2015/756270.

Ishii, Y., Iki,Y., Inoue, K., Nagata, S., Idota, S., Yokota, M., Nishiwaki, A., 2016. Adaptability of Napiergrass (*Pennisetum purpureum Schumach.*) for weed control in site of animals buried after foot-and-mouth disease infection. Scientifica 2016, 8. Available from: http://dx.doi.org/10.1155/2016/6532160.

Jasrotia, S., Kansal, A., Mehra, A., 2017. Performance of aquatic plant species for phytoremediation of arsenic-contaminated water. Appl. Water Sci. 7, 889–896, DOI 10.1007/s13201-015-0300-4.

Juel, M.A.I., Mizan, A., Ahmed, T., 2017. Sustainable use of tannery sludge in brick manufacturing in Bangladesh. Waste Manage. 60, 259–269, http://Doi.Org/Http://Dx.Doi.Org/10.1016/J. Wasman.2016.12.041.

Juel, M.A.I., Dey, T.K., Akash, M.I.S., Das, K.K., 2018. Heavy metals phytoremidiation potential of napier grass (Pennisetum purpureum) cultivated on tannery sludge. Proceedings of the Fourth International Conference on Civil Engineering for Sustainable Development (ICCESD 2018), 9–11 February 2018, KUET, Khulna, Bangladesh.

Kamarei, F., Ebrahimzadeh, H., Yamini, Y., 2010. Optimization of temperature-controlled ionic liquid dispersive liquid phase micro extraction combined with high performance liquid chromatography for analysis of chlorobenzenes in water samples. Talanta 83, 36–41, https://doi.org/10.1016/J. Talanta.2010.08.035.

Kandel, R., Singh, H.P., Singh, B.P., Harris-Shultz, K.R., Anderson, W.F., 2015. Assessment of genetic diversity in Napier Grass (*Pennisetum purpureum Schum.*) using microsatellite, single-nucleotide polymorphism and insertion-deletion markers from Pearl Millet (*Pennisetum glaucum* (L.) R. Br.). Plant Mol. Biol. Rep. 34, 265–272.

Klomjek, P., 2016. Swine wastewater treatment using vertical subsurface flow constructed wetland planted with Napier grass. Sustain. Environ. Res. 26, 217–223.

Ko, C.-H., Yu, F.-C., Chang, F.-C., Yang, B.-Y., Chen, W.-H., Hwang, W.-S., Tu, T.-C., 2017. Bioethanol production from recovered napier grass with heavy metals. J. Environ. Manage. 203, 1005–1010, doi:10.1016/j.jenvman.2017.04.049.

Koe, L.C.C., Shen, W., 1997. High resolution GC-MS analysis of Vocs in wastewater and sludge. Environ. Monitor. Assess. 44, 549–561, https://Doi.Org/10.1023/A:1005779529320.

Lelaphatikul, W., 2012. Effect of the height of bed on exhaust gas in a rice husk fired cyclone combuster. KKU English J. 39, 69–76.

Lima, M.A., Gomez, L.D., Steele-King, C.G., Simister, R., Bernardinelli, O.D., Carvalho, M.A., Rezende, C.A., Labate, C.A., McQueen-Mason, S.J., Polikarpov, I., 2014. Evaluating the composition and processing potential of novel sources of Brazilian biomass for sustainable biorenewables production. Biotechnol. Biofuels 7, 10, doi: 10.1186/1754-6834-7-10.

Liu, X., Shen, Y., Lou, L., Ding, C., Cai, Q., 2009. Copper tolerance of the biomass crops Elephant grass (*Pennisetum purpureum* Schumach), Vetiver grass (*Vetiveria zizanioides*) and the upland reed (*Phragmites australis*) in soil culture. Biotechnol. Adv. 27, 633–640.

Loch, R.J., 2000. Effects of vegetation cover on runoff and erosion under simulated rain and overland flow on a rehabilitated site on the meandu minejj, Tarong, Queensland. Aust. J. Soil Res. 38, 299–312.

Lounglawan, P., Lounglawan, W., Suksombat, W., 2014. Effect of cutting interval and cutting height on yield and chemical composition of King Napier grass (*Pennisetum purpureum* × *Pennisetum americanum*). APCBEE Procedia 8, 27–31.

Ma, C., Ming, H., Lin, C., Naidu, R., Bolan, N., 2016. Phytoextraction of heavy metal from tailing waste using Napier grass. Catena 136, 74–83, http://dx.doi.org/10.1016/j.catena.2015.08.001.

Mannetje, L. (1992). Pennisetum Purpureum Schumach. Record from Proseabase. Mannetje, L.'t and Jones, R. M. (Editors). PROSEA (Plant Resources of South-East Asia) Foundation, Bogor, Indonesia. pp. 122–124.

Martel, E., De Nay, D., Siljak-Yakoviev, S., Brown, S., Sarr, A., 1997. Genome size variation and basic chromosome number in Pearl millet and fourteen related *Pennisetum species*. J. Hered. 88, 139–143.

McMullen, C.K., 1999. Flowering Plants of the Galápagos. Comstock Publisher Assoc, Ithaca, New York, USA, 370 pp.

Menegol, D., Scholl, A.L., Dillon, A.J.P., Camassola, M., 2017. Use of elephant grass (*Pennisetum purpureum*) as substrate for cellulase and Xylanase production in solid-state Cultivation by *Penicillium echinulatum*. Braz. J. Chem. Eng. 34 (3), 691–700.

Meuser, H., 2010. Causes of soil contamination in the urban environment. In: Contaminated Urban Soils. Environmental Pollution, 18(1), Springer, Dordrecht, pp. 29–94. https://doi.org/10.1007/978-90-481-9328-8_3.

Moore, K.J., Fales, S.L., Heaton, E.A., 2008. Biorenewable energy: new opportunities for grassland agriculture. In: Multifunctional Grasslands in a Changing World, Igc/Irc Conference Hohhot, China, pp. 1023–1030.

Mohapatra, S., Dandapat, S.J., Thatoi, H., 2017. Physicochemical characterization, modelling and optimization of ultrasono-assisted acid pretreatment of two *Pennisetum* sp. using Taguchi and artificial neural networking for enhanced delignification. J. Environ. Manag. 187, 537–549.

Morandi, F., Perrin, A., Østergård, H., 2016. Miscanthus as energy crop: Environmental assessment of a miscanthus biomass production case study in France. J. Clean. Prod. 137, 313–321, DOI: 10.1016/j.jclepro.2016.07.042.

Mwangi, D.M., Cadish, G., Thorpe, W., Giller, K.E., 2004. Harvesting management options for legumes intercropped in napier grass in the central highlands of Kenya. Trop. Grassl. 38, 234–244.

Negawo, A.T., Teshome, A., Kumar, A., Hanson, J., Jones, C.S., 2017. Opportunities for Napier Grass (*Pennisetum purpureum*) improvement using molecular genetics. Agronomy 7 (28), doi:10.3390/agronomy7020028.

Ng, C.C., Law, S.H., Amru, N.B., Motior, M.R., Mhd Radzi, B.A., 2016. Phyto-assessment of soil heavy metal accumulation in tropical grasses. J. Anim. Plant Sci. 5, 686–696.

Nimmanterdwong, P., Chalermsinsuwan, B., Østergård, H., Piumsomboon, P., 2017. Environmental performance assessment of Napier grass for bioenergy production. J. Clean. Prod. 165, 645–655.

Nimmanterdwong, P., Chalermsinsuwan, B., Piumsomboon, P., 2015. Emergy evaluation of biofuels production in Thailand from different feedstocks. Ecol. Eng. 74, 423–437.

Ohimain, E.I., Kendabie, P., Nwachukwu, R.E.S., 2014. Bioenergy potentials of Elephant grass, *Pennisetum purpureum Schumach*. Ann. Res. Rev. Biol. 4 (13), 2215–2227.

Pandey, V.C., Bajpai, O., 2019. Phytoremediation: From Theory Toward Practice. In: Pandey, V.C., Bauddh, K. (Eds.), Phytomanagement of Polluted Sites. Elsevier. Amsterdam, Netherlands, pp. 1–49.

Pandey, V.C., Pandey, D.N., Singh, N., 2015. Sustainable phytoremediation based on naturally colonizing and economically valuable plants. J. Clean. Prod. 86, 37–39.

Pandey, V.C., Bajpai, O., Singh, N., 2016. Energy crops in sustainable phytoremediation. Renew. Sustain. Energy Rev. 54, 58–73.

Patel, D., Thakare, P.V., Jasrotia, S., 2018. Multivariate Analysis of Heavy Metals in Topsoil: An Impact of Thermal Power Plants of Maharashtra (India). In: Siddiqui, N.A., Tauseef, S.M., Bansal, Kamal (Eds.), Advances in Health and Environment Safety. Springer, Singapore, pp. 151–168, DOI: 10.1007/978-981-10-7122-5_17.

Pensri, B., Aggarangsi, P., Chaiyaso, T., Chandet, N., 2016. Potential of fermentable sugar production from Napier cv. Pakchong 1 grass residue as a substrate to produce bioethanol. Energ. Procedia 89, 428–436.

Pergent, G., Martini, C.P., 1999. Mercury levels and fuxes in *Posidonia Oceanica* meadows. Environ. Pollut. 106, 33–37.

Phitsuwan, P., Sakka, K., Ratanakhanokchai, K., 2016. Structural changes and enzymatic response of Napier grass (*Pennisetum purpureum*) stem induced by alkaline pretreatment. Bioresour Technol. 218, 247–256, DOI: 10.1016/j.biortech.2016.06.089.

Pillai, J.S., Vijayan, N., 1997. Wastewater treatment: an ecological sanitation approach in a constructed wetland. Int. J. Innovat. Res. Sci. Eng.Technol. 2 (10), 5193–5204.

Purakayastha, T.J., Chhonkar, P.K., 2010. Phytoremediation of Heavy Metal Contaminated Soils. In: Sherameti, I., Varma, A. (Eds.), Soil Heavy Metals, Soil Biology, Vol. 19. Springer, Berlin, Germany, pp. 389–429. Available from: https://doi.org/10.1007/978-3-642-02436-8_18.

Rahman, M., Hasegawa, H., 2011. Aquatic arsenic: phytoremediation using floating macrophytes. Chemosphere 83, 633–646.

Rengsirikul, K., Ishii, Y., Kangvansaichol, K., Sripichitt, P., Punsuvon, V., Vaithanomsat, P., Nakamanee, G., Tudsri, S., 2013. Biomass yield, chemical composition and potential ethanol yields of 8 cultivars of Napier grass (*Pennisetum purpureum* Schumach.) harvested 3-monthly in Central Thailand. J. Sustain. Bioenerg. Syst. 3, 107–112, http://dx.doi.org/10.4236/jsbs.2013.32015.

Rengsirikul, K., Ishii, Y., Kangvansaichol, K., Pripanapong, P., Sripichitt, P., Punsuvon, V., Vaithanomsat, P., Nakamanee, G., Tudsri, S., 2011. Effects of inter-cutting interval on biomass yield, growth components and chemical composition of napiergrass (*Pennisetum purpureum* Schumach) cultivars as bioenergy crops in Thailand. Japanese Society of Grassland Science. Grassland Sci. 57, 135–141, doi: 10.1111/j.1744-697X.2011.00220.x.

Rocha, J.R., do, A.S., de, C., Machado, J.C., Carneiro, P.C.S., Carneiro, J., da, C., Resende, M.D.V., Pereira, A., Vander, Carneiro, J.E., de, S., 2017. Elephant grass ecotypes for bioenergy production via direct combustion of biomass. Ind. Crops Prod. 95, 27–32.

Sandhu, J.S., Kumar, D., Yadav, V.K., Singh, T., Sah, R.P., Radhakrishna, A., 2015. Recent trends in breeding of tropical grass and forage species. In: Proceedings of the 23rd International Grassland Congress, New Delhi, India, 20-24 November, Vijay, D., Srivastava, M.K., Gupta, C.K., Malaviya, D.R., Roy, M.M., Mahanta, S.K., Singh, J.B., Maity, A., Ghosh, P.K., Eds.; Range Management Society of India: Jhansi, India, pp. 337–348. in press.,

Sawasdee, V., Pisutpaisal, N., 2014. Feasibility of biogas production from napier grass. Energy Procedia 61, 1229–1233, doi: 10.1016/j.egypro.2014.11.1064.

Schroll, R., Brahushi, F., Dörfler, U., Kühn, S., Fekete, J., Munch, J.C., 2004. Bio mineralisation of 1, 2, 4-trichlorobenzene in soils by an adapted microbial population. Environ. Pollut. 127, 395–401, https://Doi.Org/10.1016/J. Envpol.2003.08.012.

Sheehan, J., Andy, A., Keith, P., Kendrick, K., John, B., Mariew, et al., 2004. Energy and environmental aspects of using corn stover for fuel ethanol. J. Ind. Ecol. 7, 117–146.

Singbua, P., Treeedet, W., Duangthong, P., Seithtanabutara, V., Suntivarakorn, R., 2017. Bio-oil production of Napier grass by using rotating cylindrical reactor. Energy Procedia 138, 641–645.

Singh, B.P., Singh, H.P., Obeng, E., 2013. Elephant Grass in Biofuel Crops, Production, Physiology, Genetics. In: Singh, B.P. (Ed.), CAB International: Fort Valley State University, Fort Valley, GA, USA, pp. 271–291.

Sinha, S., Mishra, R.K., Sinam, G., Mallick, S., Gupta, A.K., 2013. Comparative evaluation of metal phytoremediation potential of trees, grasses, and flowering plants from tannery wastewater contaminated soil in relation with physicochemical properties. Soil Sediment. Contam. 22 (8), 958–983.

Susarla, S., Medina, V., Mc Cutcheon, S.C., 2002. Phytoremediation: an ecological solution to organic chemical contamination. Ecol. Eng. 18, 647–658.

Treedet, W., Suntivarakorn, R., 2012. A comparison of energy conversion between pyrolysis and gasification process for bio-fuel production from sugar cane trash. J. bio. Mater. Bio. 6, 622–626.

Tilman, D., Socolow, R., Foley, J.A., Hill, J., Larson, E., Lynd, L., Pacala, S., Reilly, J., Searchinger, T., Somerville, C., Williams, R., 2009. Beneficial biofuels—the food, energy, and environment trilemma. Science 325, 370–371.

Waramit, N., Chaugool, J., 2014. Napier grass: a novel energy crop development and the current status in Thailand. J. ISSAAS Int. Soc. Southeast Asian Agric. Sci. 20, 139–150.

Wiangkham, N., Prapagdee, B., 2018. Potential of Napier grass with cadmium-resistant bacterial inoculation on cadmium phytoremediation and its possibility to use as biomass fuel. Chemosphere 201, 511–518.

Xia, H.P., 2004. Ecological rehabilitation and phytoremediation with four grasses in oil shale mined land. Chemosphere 54, 345–353.

Yasuda, M., Ishii, Y., Ohta, K., 2014. Napier grass (*Pennisetum purpureum* Schumach) as raw material for bioethanol production: pretreatment, saccharification, and fermentation. Biotechnol. Bioprocess Eng. 19, 943–950, 10.1007/s12257-014-0465-y.

Ye, Z.H., Wong, J.W.C., Wong, M.H., 2000. Vegetation response to lime and manure compost amendments on acid lead/zinc mine tailings: a greenhouse study. Rest. Ecol. 8, 289–295.

Role of microbes in grass-based phytoremediation

Madhumita Roy[a], Vimal Chandra Pandey[b,*]
[a]Department of Microbiology, Bose Institute, Kolkata, India; [b]Department of Environmental Science, Babasaheb Bhimrao Ambedkar University, Lucknow, Uttar Pradesh, India
*Corresponding author

1 Introduction

One of the greatest burning issues that the planet is currently facing is environmental pollution, which is causing grave and irreparable damage to our mother Earth (Pandey et al., 2015a). Pollutants can be naturally occurring substances or energies produced in excess of natural levels, or man-made xenobiotic compounds. A pollutant can be a toxic chemical (HM, radionuclides, organophosphorus compounds, gases, polycyclic aromatic hydrocarbon, polychlorinated biphenyls) or geochemical substance (dust, sediment), biological organism or product, or physical substance (heat, radiation, noise) that is discharged by choice or unwittingly by man into the atmosphere with actual or potential adverse, harmful, unpleasant, or inconvenient effects. An ever-increasing disposal of industrial, domestic and agricultural wastes, urban effluents and consumer goods, excessive farming have resulted in the wide-scale contamination of soil and water with organic compounds and HMs and causing detrimental effects on ecosystems and human health (Pandey and Singh, 2019). Conventional soil remediation methods are expensive, labor-intensive, and sometimes environmentally destructive in nature. In the last decade, the urge to find alternative treatment technology has highlighted the solar power driven extraordinary capacity of plants and their associated microorganisms for environmental cleanup. Plants deploy a wide range of physiological, biochemical, and molecular mechanisms to counter the deleterious effects of environmental pollutants by exploiting the natural uptake mechanisms of plant root systems, translocation, bioaccumulation, or detoxifying abilities to clean up the surrounding environments. The evapotranspirational activity of green plants works as a natural pump-and-treatment system. This plant mediated remediation or phytoremediation technologies are based on the joint collaborative action of plants and rhizospheric microbes and hold great promise for the remediation of contaminated land and water (Pandey et al., 2015a; Pandey and Bajpai, 2019; Praveen et al., 2019). The specific advantages, limitations and economics of phytoremediation have been extensively reviewed (Burken et al., 2000; Salt et al., 1998; Pandey and Bajpai, 2019; Pandey and Souza-Alonso, 2019). Success of phytoremediation depends on the interaction of plants with the surrounding soil medium, the contaminants and the microbes present surrounding the roots (Praveen et al., 2019). The root soil interface known as the rhizosphere is different from the root-free bulk soil and serves as the most dynamic microenvironment where roots, soil,

Phytoremediation Potential of Perennial Grasses. http://dx.doi.org/10.1016/B978-0-12-817732-7.00015-8

and microbial communities cross talk with each other (Praveen et al., 2019). Even in natural environments the role of plant growth promoting bacteria (PGPR), phosphorous solubilizing bacteria, arbuscular mychorrhizal fungi in the rhizosphere soil are very important for maintaining biogeochemical cycling of nutrients, soil fertility, and plant health. The mutuality between plants and their associated rhizospheric microbial communities is complicated. Uptake efficiency of xenobiotics by plants can be augmented by these soil-borne rhizospheric bacteria. In general, for successful rhizoremediation, a plant species must have a dense and highly branched root system and they provide a key point for assessment of the phytoremediation potential of a particular plant species (Praveen et al., 2019; Pandey and Bajpai, 2019). Among different types of grass species, perennial grass species has been found to be most suitable candidates for phytoremediation of most organic pollutants and HMs. Perennial grasses have been widely used for centuries as fodder crops, and now they are again gaining attention as bioenergy crops. Switchgrass (*Panicum virgatum*), miscanthus (*Miscanthus* spp.), reed canary grass (*Phalaris arundinacea*), vetiver grass (*Vetiveria zizanioides* L.*)*, elephant grass (*Pennisetum purpureum Schumach*), and giant reed (*Arundo donax*) are some very important perennial grass species in terms of their phytoremediation abilities of different organic and inorganic pollutants. Phytoremediation by these grass species are safe as they are non-food crops and their bioenergy production would contribute to the economic sustainability of phytomanagement of polluted sites (Pandey et al., 2016; Pandey and Souza-Alonso, 2019). There are wide-ranging dumpsites across the nations that can be managed by using perennial grasses via the restoration and bioenergy production (Pandey, 2017). Different phytoremediation methods and bioenergy production from the biomass is shown in Fig. 15.1.

2   Perennial grasses: suitable agents for phytomanagement

Grasses are the largest and most economically important and ecologically special plants belonging to the family Poaceae (formerly called the Gramineae). Poaceae or Gramineae is a massive and ubiquitous family of monocotyledonous flowering plants that represents the domesticated cereal crops, such as maize, wheat, rice, barley, and millet as well as forage, building materials (bamboo, thatch, straw) and biofuel grasses and the grasses of natural grassland and cultivated lawns. In terms of the world food service for humans and domestic livestock, no other plant family is as important as the grasses. According to their uses, grasses are classified into three main divisions (grains like rice, wheat, corn; forage plants and lawn grasses) and four minor divisions (ornamentals; soil-binders; sugar producing grasses like sugarcane; and textile grasses like reed and bamboo) and some unclassified. Weeds are another important category of grasses that cause heavy economic loss. Examples include crabgrass or *Digitaria* spp. (a familiar weed in lawns), Barnyard grass or *Echinochloa crusgalli* (grows in fields of cultivated rice), ylang-ylang or *Imperata cylindrica* (a weed of various types of cultivated lands in tropical Asia).

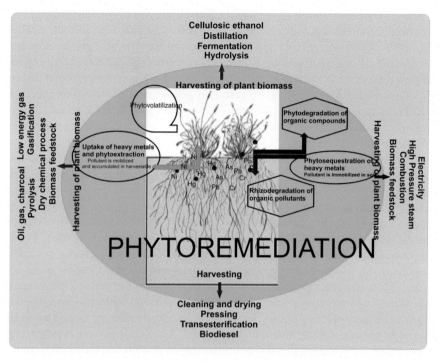

Figure 15.1 Biofuel production from biomass of perennial grasses.

Botanists so far have recognized over nine thousand species of grasses. Grasslands where the grasses dominate such as savannah and prairie are estimated to constitute 40.5% of the land area of the Earth, excluding Greenland and Antarctica. Grasses play an important part of wetlands, forests, and tundra vegetation. Grasses range from tiny herbs (> 5 cm) to tall giant bamboos (> 40 m). Among the grasses bamboos are the only woody plants. In the tall-grass prairies, some grasses can attain height as long as 6.5 ft. Examples of the tall grasses are the wild rye (*Elymus virginicus*), big blue-stem (*Andropogon gerardi*), indian grass (*Sorghastrum nutans*), dropseed (*Sporobolus asper*), panic grass (*Panicum virgatum*), and needle grass (*Stipa spartea*). Somewhat drier sites support mixed-grass prairies containing shorter species, for example, grama grass (*Bouteloua gracilis*), little blue-stem (*Andropogon scoparius*), green needlegrass (*Stipa viridula*), and wheat grass (*Agropyron smithii*). The driest habitats of the world support short-grass prairies like grama grasses (*Boutelouadactyloides* and *B. gracilis*), dropseed (*Sporobolus cryptandrus*), etc. Some particular species of grasses grow more frequently in marshes, including the reed (*Phragmites communis*) which can attain a height greater than 13 ft. (North America's tallest grass). The reed is widely occurring species, found across all the continents. Some seaside sandy habitats are populated by beach grass (*Ammophila breviligulata*), sand-reed (*Calamovilfa longifolia*), and beach rye (*Elymus mollis*) grass. Salt-marshes have cord grasses, such as *Spartina alterniflora* and *S.patens*. Kentucky blue-grass is one of the most common species in lawns, and it also occurs widely in disturbed habitats.

Grasses are either annual or perennials. Annual grasses (most agricultural crops) grow from seed to seed in one year, dying at the end of the year. Cereals such as wheat, oats, barley, spelt, corn and rye and some lawn grasses are all annual grasses. Perennial grasses die back to the soil surface at the end of one growing season (generally winter) and then regenerate the next season by shoots developing from underground rhizome or root systems. Pasture grasses are perennial grasses used as nutritious fodder for agricultural animals. Examples of some pasture grass include cock's-foot (*Dactylis glomerata*), meadow fox-tail (*Alopecurus pratensis*), rye-grasses (*Lolium perenne* and *L. multiflorum*) and timothy (*Phleum pratense*). Some essential oil producing aromatic grasses are Citronella grass (*Cymbopogon nardus*), lemon-grass (*C. citratus*), ginger-grass (*C. martinii*), and sweet grass (*Hierochloe odorata*). They are cultivated for industries producing insect repellent, perfume, and medicines. The annuals or perennials again are classified into C3 and C4 types. The C3 grasses (involves a 3-carbon molecule in the first product of carbon fixation) are referred to as "cool-season" grasses, while the C4 (involves a 4-carbon molecule that then enters the C3 cycle) are considered "warm-season" grasses. The annual cool-season grasses (or annual C3 grasses) are oat, wheat, rye, and annual bluegrass (annual meadow grass, *Poa annua*). Most important perennial cool-season/C4 grasses are orchard grass (cocksfoot, *Dactylis glomerata*), fescue (*Festuca* spp.), Kentucky bluegrass and perennial ryegrass (*Lolium perenne*). Maize, Sudan grass, and Pearl millet are among warm season annual grasses, and Big bluestem, Indiangrass, Bermuda grass and switchgrass are the perennial warm-season grasses. In comparison to annuals and biennials, perennial grasses remain dormant over long periods of time and then again continue growth and reproduction. Perennials are either evergreen perennials (retain their foliage all year round) or deciduous perennials (grow and bloom during the warm part of the year and foliage dies in the winter). In many parts of the world, seasonality is expressed as wet and dry periods rather than warm and cold periods, and deciduous perennials lose their leaves in the dry season.

Perennial plants have deep, extensive, fibrous root systems which can hold soil together to prevent erosion, capture dissolved N_2 before it can contaminate ground and surface water, and out-compete weeds (reducing the need for herbicides). Herbaceous plants have an average maximum rooting depth of 2.4 ± 0.1 m, but prairie grasses can reach depths of 4–6 m (Jackson et al., 1996). However, if contaminants occur at more depths than roots of grasses can reach, then remediation with grasses may be limited. However, grasses typically provide more uniform coverage of the soil surface, diminishing surface runoff and erosion (Collins, 2007). Naturally growing native perennial grass species show a great promise in rehabilitation of many degraded landscapes. For example, Mishra et al. (2017) has shown important role of *Saccharum bengalense Retz.* along with other native grass species in the restoration of red mud deposit (Mishra et al., 2017). *Saccharum munja, Saccharum spontaneum,* and *Ricinus communis* L. showed good prospect in restoration of fly ash lagoons (Pandey et al., 2012; Pandey, 2013; Pandey et al., 2015b). Aromatic grasses have been suggested for ecological and socio-economic sustainability of phytomanagement of fly ash dumps, red mud dumps, coal mine spoils, etc. as well as contaminated sites (Verma et al., 2014; Pandey and Singh, 2015; Pandey et al., 2019).

Most important physiological or morphological feature of perennial grasses is highly branched fibrous root system without a well-defined central taproot. They distribute themselves horizontally through the production of underground stems known as rhizomes, or prostrate stems aboveground known as stolons which are the points from where new grass shoots emerge. Root system of grasses has made them superior agents for phytoremediation. The roots of grasses have the highest rooting surface area (per m^3 of soil) of any type of plant and may enter the soil up to a depth of 3 m (Aprill and Sims, 1990). They also show an inherent genetic diversity, which give them a competitive advantage in becoming established under hostile soil condition (Aprill and Sims, 1990). One advantage of using not so deep rooted grasses than deep rooted plants is that the deep rooted plants occasionally create macro pores which may facilitate the transport of some contaminants to groundwater (Roulier et al., 2008). In a study by Knechtenhofer et al. (2003), it was found that soil preferential flow path below 20 cm was associated with roots surrounded by relatively wide root channels. Kretzschmar et al. (1999) showed that lead (Pb) may be transported as aqueous ions or bound by colloidal particles by these large macropores. Metal accumulating grasses or metal sequestrating grasses have several benefits over metal hyperaccumulating plants. Over 400 species of plants are known as natural metal hyperaccumulators that represents about 0.2% of all angiosperms. Sadly, most of these plants are characterized by slow growth and restricted biomass production. Due of these limitations such plants can't be used to take away HMs from soil. For instance, lead (Pb) phytoremediation technology can only be practically possible if one can employ high biomass plants, which are capable of accumulating more than 1% Pb in shoots and produce more than 20 t of biomass ha^{-1} year $^{-1}$(Szczygłowska et al., 2011). Because of the long remediation times required for phytoremediation, (years to decades), Meers et al. (2005) suggested that phytoextraction could only be considered feasible when the biomass produced during the phytoremediation process could be economically valorized. Most of the grasses in contrast to angiosperms are high biomass–producing crops. Various authors (Ginneken et al., 2007; Paquin et al., 2004; Balsamo et al., 2014) have shown the importance of different biofuel producing metal accumulating perennial grasses and rape seed varieties for phytoremediation of contaminated sites.

Based on biomass producing ability and ability to phytoremediate sites polluted with multiple HMs and other aromatics, vetiver grass (Danh et al., 2009), *Miscanthus* and (*Arundo donax*) are especially suitable. Various greenhouse and field studies showed that Vetiver grass (*Vetiveria zizanioides*) can generate high biomass and withstand extreme weather conditions, such as prolonged drought, flood, sinking, and temperature fluctuations. It can grow in highly acidic and alkaline soil and tolerate high concentration of salinity and HMs (As, Cd, Cr, Cu, Hg, Ni, Pb, Se, and Zn) (Danh et al., 2009). Vetiver can accumulate HMs, particularly lead (shoot 0.4% and root 1%) and zinc (shoot and root 1%). The majority of HMs is stored in roots thus suitable for phytostabilization, and for phytoextraction with addition of chelating agents. Vetiver can also absorb and promote biodegradation of organic wastes (2,4,6-trinitroluene, phenol, ethidium bromide, benzo[a]pyrene, atrazine). Although Vetiver is not very good as HM accumulator, its tolerance to a wide range of extreme environmental condition make vetiver a choice plant for phytoremediation of HMs and organic wastes. *Miscanthus* spp. and *Arundo donax* L. both are lignocellulosic and strong perennial nonfood crops. They give high biomass yields; greater vegetation cover that offers to prevent wind, water, and biological erosion; and greater performance under different soil and ecological conditions including multi metal contamination (Barbosa et al., 2015). These perennial energy crops display high water and N_2 efficiencies; low susceptible to pests and diseases; low need for pesticides and fertilizers; and deep, dense, and extensive root systems that can control the leaching of contaminants through soil and all these contribute to better phytomanagement. Details of bioenergy production from bioenergy crop can be found in the review of Lewandowski et al. (2003).

So, in outline, perennial crops (grasses and herbs) show better characteristics for the phytoremediation process since plants display (1) rapid growth, (2) high biomass yields, (3) deep and extensive root systems, (4) known agronomic techniques, (5) tolerance to contamination (6) greater rooting surface area and root biomass, and (7) tolerance to drought, acidic soils, and cold temperatures. Their genetic potential to tolerate, extract, and/or stabilize HMs coupled with their high biomass production for bioenergy, fiber, and other economic value added products has drawn attention to managers to apply them in various commercial phytomanagement program. Moreover, cultivation of perennial crops in contaminated soils avoids the land use conflict with food crops. Finally, together with the reduction and mitigation of the risk posed by the HMs, new jobs and markets for their products might be created in the region contributing economic development of the community/locality.

3 Phytoremediation strategies

There are several direct and indirect contaminant attenuation mechanisms, which assist plants in the elimination of toxicants/management of contaminated sites. Attenuation mechanisms of pollutants involved in phytoremediation are complex and not limited only to the direct metabolism of the pollutant by the plant. Some indirect attenuation mechanisms are also involved in phytoremediation, for example involvement of contaminant metabolism by plant associated microbes, and plant-induced changes in the polluted environment. In terrestrial plant species, transport of pollutants to the plant is dominated by the uptake of water by roots, and distribution within the plant relies on xylem or phloem transport. Various terms, reflecting each specific attenuation mechanism, have been extensively used to better describe specific applications of phytoremediation. These include phytoextraction, phytosequestration, phytodegradation, phytotransformation, phytovolatilization, and rhizodegradation (Burken et al., 2000; Brooks et al., 1977).

Phytoextraction—Phytoextraction is the term used for accumulation of certain pollutants during the natural uptake mechanism of a plant. The process normally occurs with HMs, radionuclides, and certain organic compounds that are resistant to plant metabolism. Brooks et al. (1977) coined the term hyperaccumulator for plants that accumulate >1000 mg kg^{-1} Ni on a dry matter basis. Such hyperaccumulation is only possible when plants grow vigorously and produce over 3 t dry matter/hectare able to accumulate large quantities of the pollutant(s) in the harvestable plant tissue (>1000 mg kg^{-1}). Several naturally occurring metal hyperaccumulators are present that can accumulate 10–500 times higher concentrations of metals than crop plants. After a certain time period, the plants are harvested and disposed of or processed by incineration or, in the case of organic pollutants, composted for recycling. After incineration of plants, careful disposal of the hazardous biomass in a sealed container provides effective site decontamination of bioavailable contaminants including PAHs and nitro aromatics (Huang et al., 1997; Pradhan et al., 1998). Unfortunately, most hyperaccumulator species are slow growing and have limited biomass production. As total metal extraction is the product of biomass and tissue concentration, the speed of overall metal removal from large sites is long time taking and accordingly limited. Field experiments by Lombi et al. (2000) and McGrath et al. (2000) highlight this problem, showing that metal removal efficiency is in general not high enough to remediate contaminated soils. Subsequently, research focused more on high biomass plants, such as tobacco (*Nicotiana tabacum*) (Fässler et al., 2010; Kayser et al., 2000), maize (*Zea mays*) (Fässler et al., 2010; Keller et al., 2003), Indian mustard (*Brassica juncea*) (Keller et al., 2003), and sunflower (*Helianthus annuus*) (Fässler et al., 2010), etc. They are fast growing, deep rooted, easy to propagate, able to uptake large concentration of metal, and relatively high biomass producing plant. However, it was again realized though that regardless of the plants used, to effectively decrease phytoremediation time, the rate of contaminant accumulation needs to be increased more. One approach to achieve this was by increasing the bioavailability of contaminants for plant uptake thorough artificial soil acidification or solubilization of contaminants by means of chelating agents. Several synthetic chelators like aminopolycarboxylic acid (APCA), diethylene triamino pentaacetic acid (DTPA), ethylene diamine tetraacetic acid (EDTA), trans-1,2-cyclohexylene dinitrilo tetraacetic acid (CDTA), N'-bis (2-hydroxyphenyl) acetic acid (EDDHA), ethylenediamine-N, and others displayed potential to significantly increase TE uptake by plants (Evangelou et al., 2007). However, with more research directed toward chelate-assisted phytoextraction, it was found that soil microbes were getting affected due to toxic effect of the chelating agents. It caused ecological risks as the mobilized contaminants could leach into groundwater or surface water (Evangelou et al., 2007; Meers et al., 2005). Then biodegradable chelating agents for example nitrilotriacetic acid (NTA) or ethylene diamine disuccinate (EDDS) were introduced. However, the degradation rates of biodegradable chelating agents such as NTA and EDDS were still too low to significantly reduce the risk of leaching (Evangelou et al., 2007; Meers et al., 2005). The risk of leaching was caused by the fact that in order to achieve plant shoot concentration of >1000 mg kg^{-1}, chelants were applied (1) in a single large dose, to break down the endodermis in order to increase the uptake via the limited apoplastic pathway, and (2) to large excess, as most chelants are nonspecific; hence, soil components such as Ca and Fe compete with targeted TE, thus reducing the efficiency of the applied chelants.

Phytosequestration or phytoimmobilization or phytocontainment—The numerous drawbacks in the development of phytoremediation has led to a change in focus from phytoextraction to phytostabilization. The aim of phytostabilization is to prevent the mobilization or leaching of pollutants by wind and water to other places or underground water reservoir and surface runoff. Ultimate aim is to minimize the transfer of contaminants into the food chain or drinking water supply by using plants with minimal uptake of contaminants. In contrast to phytoextraction where high accumulation of contaminants was desirable, in phytostabilization plants are selected based on their abilities to exclude contaminants from their aerial parts. The desired character for phytoextraction is that plants through their extensive root system and microbial communities should phytoimmolize the contaminants. Root exudates form an important role to bind contaminants and absorb them to the roots. Ideally, they should not translocate the contaminants to shoot or should do it only minimum. In recent years, the perception of contaminated soils has changed. For decades, such soils were regarded only as a source of hazard but now contaminated lands are viewed as an extensive underutilized resource, which should be used in a sustainable way to grow plants for a large variety of profitable purposes (Roy et al., 2018). The idea of phytomanagement emerged from this concept of containment instead of uptake, translocation and accumulation. Phytomanagement describes the engineering of soil, plant, microbial systems to restore lands with maximizing economic and/or ecological benefits while minimizing risks. The main component of phytomanagement is that it should either cost less than other remediation technologies or be a profitable operation, by producing valuable phyto-products (Robinson et al., 2009). Thus, the goal of phytomanagement is to produce economic returns on a polluted land without causing harmful effects on human health and nature (Pandey and Souza-Alonso, 2019).

Phytodegradation/phytotransformation—It refers to plant-mediated degradation of contaminants. Phytodegradation has been studied extensively to understand the fate of herbicides in crop plants. In phytodegradation/phytotransformation, contaminants are taken up from soil/water, metabolized in plant tissues and broken up to less toxic or non-toxic compounds within the plant by several metabolic processes via the action of enzymes produced by the plant and endophytes (Burken et al. 2000). As the overall metabolic processes or detoxification steps involved in phytodegradation has similarity with human metabolism/detoxification of xenobiotic chemicals; a "green liver" conceptual model is often used to describe phytodegradation (Sandermann, 1994). The uptake of hydrophobic organic chemicals is very efficient while extremely hydrophobic or hydrophilic compounds are not very good agents for phytoremediation. Such contaminants neither are absorbed strongly to the surface of the roots nor transported actively through plant membranes (Burken et al. 2000).

Phytovolatilization—Phytovolatilization is one type of phytotransformation in which volatile chemicals or their metabolicintermediates are released into the environment through plant transpiration.The intrinsic ability of a plant to volatilize a contaminant that has been taken up through its roots can be exploited as a natural air-stripping pump system. Phytovolatilization is most suitable to contaminants like BTEX, TCE, vinyl chloride, and carbon tetrachloride (Zhang et al., 2001).

Rhizodegradation—In rhizoremediation, the contaminant is transformed by microbes in the rhizosphere (i.e., the microbe-rich zone in intimate contact with the vascular root system of the plant). Rhizosphere is the zone of soil under the direct influence of plant roots, and extends a few millimeters from the root surface. The rhizosphere microbial community is comprised of microorganisms (bacteria, fungi, mycorrhiza) with different types of metabolism and adaptive responses to variation in environmental conditions. In the rhizosphere, plant roots along with the plant growth promoting microbes manipulate existing soil redox conditions, organic content, moisture, and other soil properties. Rhizodegradation is important for total petroleum hydrocarbons bioremediation (Tangahu et al., 2011). The fate of PAHs and other organic contaminants in the environment is dependent on both abiotic and biotic processes, including chemical oxidation, bioaccumulation, and microbial transformation. Microbe-mediated transformation or complete degradation is seemed as most influential and significant cause of PAH removal (Cerniglia, 1997). In all the earlier-mentioned phytoremediation methods, microbes present in the rhizosphere and the endophytic environment plays a significant role. Plant microbiome consists of the rhizobacteria (resides in plant rhizosphere and phyllosphere) and endophytes (resides in the endosphere or internal plant tissue). Most of the rhizobacteria are PGPR (plant growth promoting rhizobacteria) that have three characteristics: capability to colonize the root; capability for survival, proliferation and competition in microhabitats associated with the root surface; and ability to promote plant growth (Gamalero et al., 2004). One group of PGPR bacteria are important for nutrient cycling. They fix atmospheric nitrogen for meeting nitrogen requirement of plants, synthesize siderophores for iron supply to plant cells, solubilize phosphate for plant uptake, synthesize phytohormones such as auxins, cytokinins and gibberelins, solubilize minerals such as phosphorous, making them more readily available for plant growth and synthesizing the enzyme ACC-deaminase, which can lower ethylene levels. The second group is involved in the biocontrol of plant pathogens by means of antibiotic production, depletion of Fe from the rhizosphere, induce systemic resistance and produce enzymes for inhibiting growth of phytopathogenic fungi (Glick, 2010). Importance of PGPR bacteria on phytoremediation can be found in many reviews (Santoyo et al., 2016). A number of reports states that some PGPR species HM resistant and possess bioremediation potential (Mukherjee et al., 2017; Roy and Roy, 2018). Such naturally occurring rhizobacteria could assist phytoremediation both indirectly, by increasing the overall fertility of the contaminated soil and enhancing plant growth through nutrient uptake and control of pathogenity, and, also directly, catabolizing certain organics and/or intermediate partly oxidized biodegradation products. Recent evidence also emphasizes that phytoremediation success strongly depends on the plant microbiome activities. Numerous works have been done showing how bacterization of seeds improved phytoremediation ability of many plants. For example, in one study the plant growth promoting bacteria, such as *Vicia faba, Enterobacter cloacae, Rhizobium leguminosarum,* and *Pseudomonas* showed an increased remediation of copper toxicity from the soil (Fatnassi et al., 2015). Various dichromate-resistant strains, arsenic resistant strains have been found from HM enriched fly ash sites or tanneries waste that can offer cost-effective option for the cleanup operations of dichromates from different industrial effluents (Roychowdhury et al., 2016, 2018; Hemambika et al., 2013). Several growth promoting bacteria including *Bacillus endophyticus* NBRFT4, *Paenibacillus macerans* NBRFT5, and *Bacillus pumilus* NBRFT9 enhanced the uptake of metals (through mobilization) from the soil and promoted plant growth during phytoremediation of HMs with *S. munja.* (Tiwari et al., 2013). Visioli et al. (2015) and Berg et al. (2012)

have reviewed the importance of omics study and advanced microscopy to understand the interaction between metal hyperaccumulators and the bacterial rhizobiome. Environmental scanning electron microscopy is a recent powerful approach for the in situ analysis of colonization of plants and the survival of microbial inoculums under real conditions and it bypasses the need to use biological specimens sample preparation (Stabentheiner et al., 2010; Visioli et al., 2014).

4 Importance of microbial role in grass–based phytoremediation

A wealth of experimental evidences has been gathered on the role of plant-microbe interactions for remediation of various xenobiotics (see review by Khan et al., 2013; Gupta et al., 2000; Weyens et al., 2009). Sorkhoh et al. (2011) has shown how plant endobiome participates actively in bioremediation and phytovolatilization of volatile oil hydrocarbons. American grass (*Cynodon* sp.) and broad bean (*Vicia faba*) enrich themselves with these particular oil eating bacteria in their phylloshere when they are grown in oil contaminated soil instead of unpolluted soil. These bacteria help the plants to overcome the toxicity of these hydrocarbons by degrading the phytoaccumulated hydrocarbons in their leaves and thus detoxify the plant body from harmful effects of the hydrocarbons.

In another recent work, Eskandary et al. (2017) inoculated seeds of one grass species *Festuca arundinacea* (cool-season perennial C3 species of bunchgrass) with multiple PAH degrading *B. licheniformis* and *B. mojavensis* and observed plants from bacterized seeds accumulated more PAH than uninoculated plants. Moreover, they injected the bacteria into rhizosphere of the potted plants. PAH concentration significantly decreased in the rhizosphere soil of the treated grass. In such plant–bacteria symbiosis, plant supplied the bacteria with special carbon sources that stimulated them to degrade the PAHs in the soil. In response, bacteria supported the plant to conquer contaminated-induced stress responses, and this improved plant growth and development. In addition, plants were benefitted from the bacteria with hydrocarbon-degradation ability, leading to enhanced hydrocarbon mineralization and lowering of both phytotoxicity and evapotranspiration of volatile hydrocarbons and improved the efficiency of phytoremediation. Su and Zhu (2008) also reported the contributions of plant uptake and plant-boosted rhizosphere associated microbial biodegradation of PAHs in rice seedling (*O. sativa*).

In a critical review, Mandal et al. (2018) evaluated all pros and cons of polyaromatic hydrocarbon accumulation in leaves and has raised the issue that there is a possibility of reentry of PAH from leaves of agricultural/medicinal crops grown in PAH polluted sites into biological system. So here comes the suitability of using bioenergy non-food grasses for cleanup of PAH contaminated sites. Same is applicable to heavy metal cleanup also. Grobelak and Napora (2015) have shown the suitability of using grass species to phytoimmbolize metal contaminations. They have shown that additives like a combined application of triple superphosphate or potassium phosphate fertilizers and sewage sludge can increase phytostabilization efficiency. These additives also caused an increase in biomass of grass species (tall fescue) used. Many authors have recommended this grass species for phytostabilization of heavy metals since they accumulate the bioavailable metals only in the roots and very little is transported to above ground parts (László, 2005). Other authors have also recommended use of environmental friendly biosolids during phytostabilization of heavy metals (Kacprzak et al., 2014; Basta and Sloan, 1999).

5 Phytoremediation of different types of pollutants through perennial grass species

Petroleum hydrocarbons are the most widespread class of organic contaminants and have mutagenic and carcinogenic properties. Total petroleum hydrocarbons (TPHs) represents volatile mono aromatic compounds such as benzene, toluene, ethyl-benzene, and xylene (BTEX), polycyclic aromatic hydrocarbons (PAHs), and aliphatic compounds. One source of these pollutants is underground leakingstorage tanks (USTs). Over 500,000 instances of fuel leaks from USTs were reported in the United Sates between 1984 and 2011 (EPAUST, 2011).Phytoremediation of aromatic pollutants is a multidisciplinary treatment technique with the central thrust on plant physiology. Grass has been shown to be more effective inthe degradation of PAHs than many other herbaceous plants (Olson et al., 2007). Two of the most common perennial grasses used for rhizoremediation of aromatic pollutants are tall fescue (*Festuca arundinacea Schreb.*) and perennial ryegrass (*Lolium perenne L.*). They offer deep and extensive root systems, robust growth after establishment, and tolerance to drought, acidic soils, and cold temperatures (Aprill and Sims, 1990; Muratova et al., 2008). Perennial ryegrass, in particular, has been shown to be more resilient in maintaining root biomass than some crop and legume species in oil-sludge-contaminated soil (Muratova et al., 2008). Many researchers have suspected a relationship between plant biomass or root surface area and contaminant degradation, but there have been limited observed correlations (Parrish et al., 2004). While grasses have more fibrous root systems, they typically do not root as deeply as trees. Herbaceous plants have an average maximum rooting depth of 2.4 ± 0.1 m, except for prairie grasses that can reach depths of 4–6 m (Jackson et al., 1996). If contaminants occur at depths greater than roots, then remediation with grasses may be limited. However, grasses typically provide more uniform coverage of the soil surface, diminishing surface runoff and erosion (Collins, 2007). Phytoremediation of aromatic pollutants aims to degrade them into less toxic or non-toxic harmless compounds and limit their further movement by sequestration and accumulation. Organic compounds can be translocated to other plant tissues and subsequently volatilized; they may undergo partial or complete degradation or they may be transformed to less phytotoxic compounds and bound in plant tissues. The process of phytoremediation begins with contaminant transport to the plant. After that they may undergo some degree of transformation in plant cells before being sequestered in vacuoles or bound to insoluble cellular structures such as lignin. In constructed wetlands, contaminants can enter the macrophytes through the roots or can partition from the water column directly into plant tissues. Uptake in terrestrial plants has been studied for many plant contaminant combinations, and quantitative models to predict uptake rates have been documented (Trapp, 1995). These models are based on flow in the transpiration stream and organic partition relationships for various xenobiotic compounds. The "green liver" model is often used as a comparison to mammalian metabolism of xenobiotics with plants (Sandermann, 1994). Metabolism of foreign compounds in plant systems is generally considered to be a "detoxification" process that is similar to the metabolism of xenobiotic compounds in humans (Burken et al., 2000). The detoxification of pollutants is carried out in three stages: transformation, conjugation, and sequestration. The initial reactions of reduction, oxidation, and hydrolysis give rise to more simple compounds that are amenable to subsequent conjugation reactions. After conjugation of the xenobiotic compound with an organic molecule of the plant, toxicity is reduced to the plant by various processes of sequestrations (Trapp, 1995).The oxidation reactions are the principle transformation reactions in the metabolism of pesticides and certain nitro aromatic compounds (Komoßa et al., 1995). The storage of certain conjugates in plant organelles like vacuoles is a common mechanism, whereupon the compound is no longer capable of interfering with cell function. The xenobiotic conjugates can also be combined into biopolymers such as lignin where they are considered as nonextractable or bound residues, whereas in the plants of aquatic system they can be excreted for storage outside the plant. The uptake of hydrophobic organic chemicals is very efficient while extremely hydrophobic or hydrophilic compounds are not very good candidates for phytoremediation. Such contaminants cannot be easily translocated within the plant, as they are either bound strongly to the surface of the roots or are not sorbed by roots and are actively transported through plant membranes (Burken et al., 2000).

Enzymes are released into the rhizosphere either by the plant or microbes to decompose humus, organic compounds and convert many biounavailable compounds to more bioavailable forms, etc. (Khaziev, 1980). Plant enzymes that play important rolesin xenobiotic degradation are mono- and dioxygenases, dehydrogenases, dehalogenases, carboxylesterases, hydrolases, peroxidases, nitroreductases, nitrilases, and phosphatases (Mansuy, 1998; Pilon-Smits, 2005). The action of plantenzymes to degrade xenobiotics is essentiallyaccidental, as their natural function is other. For example, cytochrome P450 monooxygenases is a family of enzymes that catalyze most oxidation steps in the secondary metabolism, a complex set of reactions that allows plants to communicate with their bioenvironment, (e.g., attract mutualists like pollinators, certain soil microorganisms, and deter herbivores andpathogens). The reactions catalyzed by plant P450s extend from simple hydroxylation or epoxidation steps, to more complex phenol coupling, ring formation, and modification or decarboxylation of appropriate substrates (Mansuy, 1998).

5.1   Polycyclic aromatic hydrocarbons

Polycyclic aromatic hydrocarbons (PAHs) pollution has become an increasingly environmental hazardous issue with industrial development, increasing petroleum energy consumption. Waste water, waste gas, waste residue from petroleum industries pollute all chambers of the environment. Many PAHs are regarded as priority pollutants because of their health hazard potential, environmental persistence, bio-accumulative characteristics, and toxicity (Singer et al., 2004). High hydrophobicity and low bio-degradation properties lead to their accumulation in the environment. PAHs can remain adsorbed onto soil particles due to poor water solubility and a high octanol-water partition coefficient. Soil becomes the primary carrier of these compounds and is able to remain bioavailable for long periods of time. The stimulation of elimination of PAHs from soil could be brought about through various mechanisms of plant–soil interactions, including: (1) enhance in soil microbial activity, (2) increase in microbial association with the root and toxic compounds, and (3) changes in the physicochemical characteristics of the polluted soil. Many reviews are present showing the role of plants and microbes in the degradation of PAH (Alkorta and Garbisu, 2001; Singh and Jain, 2003). PAH degradation was noticed in several plant cell tissue cultures (Kolb and Harms, 2000). In fields and lab studies, rhizodegradation has been found to be primary responsible for decreasing PAH level within 6–12 months (Lu et al., 2010; Wei and Pan, 2010). Among plants, grasses with a high root surface are especially efficient for bringing down PAH level as their fibrous root system provides a great surface area for microbes and rhizodegradation. It appears that the host plant can selectively pick up microbes with degradative genes out of a huge pool of candidates in the bulk soil (Sipila et al., 2008), but a detailed understanding on this is yet to be done. Bell et al. (2013) revealed a high diversity of hydrocarbonoclastic bacteria in the rhizosphere of Salix growing on hydrocarbon contaminated soil in Canada (Bell et al., 2013). Phytoremediation research has also focused plant-endophytes to improve contaminant biodegradation (Ijaz et al., 2015). It has been observed that longer the contact time between the contaminant and endophytic microorganisms better is the degradation of the contaminant thereby reducing the risk of phytotoxicity and otherwise evapotranspiration of volatile organic contaminants like TCE-dissipation (Barac et al., 2004). In one study hydrocarbon degradation was found more effective when endophytes were inoculated rather than rhizospheric strains in Italian rye grass (Andria et al., 2009). The endophytes showed a higher level of root-colonization, stable interaction with plant, gene-expression, and maintenance. Recent high number of publications reflects the full ongoing extent of these extraordinary plant–endophyte relationships (Ijaz et al., 2015). Although most studies of PAH phytoremediation are based on potted plants or soils freshly spiked with PAHs under greenhouse or laboratory conditions, few studies are present on the removal and plant uptake of PAHs under field conditions. In one field study, Ryegrass (*Lolium perenne L.*) was chosen in an agricultural field near a steelworks of Nanjing, east China where concentration of 12 PAHs (3–6-ring) was about 1400 μg kg^{-1} soil (oven dry basis), including benzo(a) pyrene at 112 μg kg^{-1}. Here two treatment set ups: unplanted control (CK) and soil planted with perennial ryegrass (PR) was performed in a fully randomized layout of field plots with four replicates of each treatment. The whole plants were harvested 8 months after sowing and seeds were sown again as before. The second harvest was made after 1 year. Oven dried plant samples were taken and divided into shoots and roots that were ground for the analysis of PAHs. Enhanced soil dehydrogenase and soil peroxidase activities were noticed. At the same time, higher microbial presence was also noticed. Enzymes play key role in PAH biodegradation during phytoremediation. Dehalogenase, nitroreductase, peroxidase, and laccase are the most important enzymes in this category. Table 15.1 shows the different

Table 15.1 Polycyclic aromatic hydrocarbon degradation by various grass species in field based studies.

Pollutant type	Plant species involved	Extent of remediation	Mechanism of action	References
PAH contaminated soil	*Echinacea purpurea, Festuca arundinacea Schred, FirePhoenix.* and *Medicago sativa* L.		Enzymatic reactions of polyphenol oxidase (except for Fire Phoenix), dehydrogenase (except for Fire Phoenix), and urease (except *Medicago sativa* L.) were prominent during PAH degradation period. So enzymatic activities were responsible for PAH degradation	Liu et al. (2015)
Perennial ryegrass (*Lolium perenne* L.)		Two seasons of grass cultivation the mean concentration of 12 PAHs in soil decreased by 23.4% compared with the initial soil. The 3-, 4-, 5-, and 6-ring PAHs were dissipated by 30.9%, 25.5%, 21.2%, and 16.3% from the initial soil, respectively	Enhanced microbial activity and presence of different enzymes were noticed in the rhizosphere	Fu et al. (2012)
Clemenson, South Carolina.	Annual ryegrass (*Lolium multiflorum*)	The pyrene level was found to decrease below its detection limit after 301 days		Lalande et al. (2003)
Lead	*Alternanthera philoxeroides, Sanvitalia procumbens,* and *Portulaca grandiflora*	Lead concentration decreased 62.61-23.18 mg kg⁻¹ from initial 75 mg kg⁻¹. *A. philoxeroides* was the greatest accumulator of lead	Dense roots with long stolons and large surface areas of *A. philoxeroides* caused more lead accumulation than taprooted *P. granaiflora*	Cho-Ruk et al. (2006)
Pyrene	Alfalfa (*Medicago sativa* L.).			Fan et al. (2008)
Diesel contaminated subsurface soil	Perennial ryegrass (*L. perenne*)			Kechavarzi et al. (2007)

Location	Plants	Results	Mechanism	Reference
Petroleum hydrocarbon contaminated soil of Japan	Italian ryegrass (*Lolium multiflorum*), Sorghum (*Sorghum vulgae*), Maize (*Zea mays*), alfalfa (*Medicago sativa*), Bermuda grass (*Cynodon dactylon*), Southern crabgrass (*Digitaria ciliaris*), Red clover (*Trifolium pretense*), rice (*Oryza sativa*), Spinach (*Spinacia oleracea*), Kudzu (*Pueraria lobata*), Beggar (*Bidens frondosa*)	Total petroleum concentration decreased in all. Beggar ticks showed highest remediation followed by alfalfa, Bermuda grass, and kudzu. TPH was reduced from 9800 mg kg^{-1} to 4410, 4083, 4330, and 3730 mg kg^{-1}.	Enhanced microbial activity and rhizodegradation	Kaimi et al. (2007)
PAH contaminated soil of manufactured gas plant (MGP) in Newark, NJ	alfalfa (*Medicago sativa*), switch grass (*Panicum virgatum*), and little bluestem grass (*Schizachyrium scoparium*)	At initial t-PAH concentration of ~200 mg kg^{-1} in soil treated with switch grass, alfalfa, and little blue stem, respective reductions in t-PAHs were 57%, 56%, and 47%. Similarly, respective reductions in c-PAHs were 30%, 28%, and 28%. In an unplanted control, t-PAH reduction was only 26%, and a reduction in the c-PAHs was not observed	Rhizodegradation	Pradhan et al. (1998)
Shengli Oil Field in Dongying City, Shandong Province, China	(*Echinacea purpurea*), 3(*Festuca arundinacea* Schred), (*Fire Phoenix*), and (*Medicago sativa* L.) showed degradation rate greater than 90% after 150 d, reaching 92.91%, 97.59%, 99.4%, and 98.11%, respectively	(*Echinacea purpurea*), (*Festuca arundinacea* Schred), (*Fire Phoenix*) and (*Medicago sativa* L.) exhibited good potential to remediate PAH-contaminated soils	Enzymatic catalysis particularly oxidoreductases and hydrolases	Liu et al. (2015)
Sandy loamy soil spiked with PAHs	Medicago sativa, Lolium perenne, and Festuca arundinacea	Phenanthrene, fluoranthene, benzopyrene	Abiotic processes and rhizodegradation	Afegbua and Batty (2018)

pot scale and field scale application of perennial grasses for phytoremediation of polycyclic aromatic hydrocarbons.

5.2   Explosives/nitroaromatics, herbicides, and polychlorinated biphenyls

The main three classes of explosives are nitroaromatics, nitramines, and nitrate esters. Soil contamination with explosives is largely due to manufacturing, storage, testing, and inappropriate waste disposal of explosive chemicals. The primary explosives at hazardous waste sites are TNT (2,4,6-trinitrotoluene), Royal Demolition eXplosive-RDX (hexahydro-1,3,5-trinitro-1,3,5-triazine) and High Melting eXplosive-HMX (octahydro-1,3,5,7-tetranitro-1,3,5,7-tetrazine). TNT is a nitroaromatic constituent of many explosives. TNT is stable, environmentally long lasting but is readily acted upon by alkalis to form unstable compounds that are very sensitive to heat and impact. Health effects due to exposure to TNT include anemia, abnormal liver function, skin irritation, and cataracts. RDX is a nitramine used widely as an explosive and as a constituent in plastic explosives. RDX causes seizures at high dose and nervous system disorder at low dose. HMX is a nitramine that burst at high temperatures. It is used in nuclear devices, plastic explosives, and rocket fuels. Insufficient studies on the effects of HMX to the health of humans and animals have been performed.

Plants, as mostly sessile organisms, have evolved complex detoxification systems to deal with a diverse range of toxic chemicals. The plasticity of this system also enables plants to detoxify relatively recently produced, synthetic pollutants such as explosives. There have been several recent reviews on the phytoremediation of explosives and the metabolism techniques of them in plants (Rylott and Bruce, 2008; Hannink et al., 2002). Groom et al. (2002) performed phytoremediation field trials in explosive HMX contaminated soil from an anti-tank firing range in Alberta, Canada. Here, wheat and ryegrass demonstrated rapid growth in the presence of HMX and accumulated significant quantities of this explosive, identifying these plants as candidates for HMX phytoremediation (Groom et al., 2002).The United States Army is endeavoring to clean TNT and RDX from polluted wetlands and from groundwater polluted with residues of the explosives TNT, RDX, HMX, and dinitrotoluene using a range of plants. The sampling and screening of many plant species growing on explosive-polluted soils has yielded significant information contributing to understanding of TNT uptake by plants in the natural environment. Even some grass species like reed canary grass (*Phalaris arundianasea*), bromegrass (*Bromus inermis*) tested positive for, RDX and other explosives in the Joliet Army Ammunition Plant, Illinois and Iowa Army Ammunition Plant (IAAP), Iowa. Soils contaminated with pollutants such as TNT, TCE, and PCP have been successfully remediated using parrot-feather (*Myriophyllum aquaticum*) (Macek et al., 2000). Best et al. (1997) and Larson et al. (1999a, b) successfully identified plants with the ability to remove TNT and RDX from aquatic systems; however, the fate of transport and the metabolic pathways were not determined. Plants with the greatest potential for use in wetland systems were the submerged plants coon tail and American pondweed and the emergent plants common arrowhead, reed canary grass and fox sedge. Sung et al. (2002) conducted a 2-year field study to remediate soil contaminated with TNT and 2,2,5,5'-tetrabromobiphenyl using Johnson grass and Canadian wildrye and decreased their level below their detection limits after treatment.

Polychlorinated biphenyl or PCBs that have chlorine atoms covalently attached to the main carbon backbone and consists of one to several aromatic rings were once widely used in industries due of their extreme chemical stability and insulating properties. But now they have been banned in many industrialized nations. But significant amounts are still released into the environment from old electrical equipment and other sources. By contrast to other persistent organic pollutants very little is known about the metabolism of these compounds in plants (Pilon-Smits, 2005).

Plants were also found suitable to degrade various herbicides. A mixed stand of big bluestem, switch grass, and yellow indiangrass developed a rhizosphere with microflora that detoxified pesticide residues. Atrazine is a widely used herbicide used to prevent broadleaf weeds in crops such as maize and sugarcane and on turf, such as golf courses and residential lawns. Specific atrazine degrading bacteria have been found in the rhizosphere of crops. The free enzyme atrazine chlorohydrolase enhanced the rate of biotransformation of atrazine in soil in the bacterium. Metolachlor degradation was accelerated significantly by the prairie grass/rhizosphere effect (Henderson et al., 2006). Carbamazepine was taken and processed by *Phragmites australis* (Sauvêtre and Schröder, 2015).

5.3 Heavy metals

Heavy metal pollution coming from natural and anthropogenic sources is a global environmental pollution due to its deleterious effect on human health, biodiversity, and ecosystem (Raskin and Ensley, 2000). Among the HMs (essential) copper (Cu), cobalt (Co), chromium (Cr), iron (Fe), manganese (Mn), molybdenum (Mo), nickel (Ni), selenium (Se), and zinc (Zn) are necessary for living organisms and they participate in different enzymatic and redox reactions (Zenk, 1996). However, lead (Pb), cadmium (Cd), mercury (Hg), arsenic (As) (metalloid), etc., (nonessential) have no known metabolic functions. These four HMs (Pb, Cd, Hg, and As) are regarded as the most dangerous HMs by the Agency for Toxic Substances and Disease Registry, based on the frequency of occurrence, toxicity, and the potential for human exposure. HMs exist in colloidal, ionic, particulate, and dissolved phase. Metals have high affinity for humic acids, organic matters of clays, and oxides coated with organic matters. HMs are present as soluble, partially soluble, and insoluble forms. The soluble forms are generally ions or unionised organometallic chelates or complexes and their solubility is controlled by pH, oxidation state of the mineral components, redox potential of the system, amounts of the metals, cation exchange capacity, and organic carbon content. Neutral to basic conditions causes strong adsorbtion of cationic metals to the clay fractions and can be absorbed by various hydrous oxides of Fe, Al, or Mn. Elevated salt concentration creates increased competition between cations and metals for binding sites (Tangahu et al., 2011).

The idea of using green plants to remediate HMs from polluted soils came from the discovery of certain hyperaccumulators (Tangahu et al., 2011). Hyperaccumulators are plants that are capable of accumulating more than 100 times more potentially phytotoxic elements than non-accumulators species (Raskin and Ensley, 2000). Plant species that accumulate greater than 100 mg kg^{-1} dry weight Cd, or greater than 1000 mg kg^{-1} dry weight Cu, Ni, and Pb or greater than 10,000 mg kg^{-1} Mn and Zn in their shoots when they grow on metal rich soils are called hyperaccumulators (Roy et al., 2015). Currently, more than 420 species belonging to 45 plant families are recorded as HM hyperaccumulators and this number is likely to increase in future (Ent et al., 2013). The Brassicaceae family contains a large number of hyperaccumulating species against large number of metals. Number of Ni hyperaccumulator plants (more than 300) is the maximum among all.

One drawback of hyperaccumulating plant species is that they are generally slow growers with a low biomass production. This is the drawback of their phytoextraction applications, since the metal removal capacity is determined by the harvestable plant biomass multiplied by the concentration of HMs contained within this biomass. Another drawback is the fact that extreme concentrations in the soil may be a prerequisite for vigorous plant growth and the ability to compete with other fast growing, non-accumulating plant species. Moreover, agronomic practices for most of these species are not well established (e.g. pest and weed control). Also, the metal-selectiveness (only 1 or 2 metals) of hyperaccumulators reduces the overall suitability of hyperaccumulators for phytoextraction purposes. There is also risk of environmental contamination from disposed plant matter. In contrast many accumulator species are available (e.g., *Brassica*) which have all the traits to use for practical phytoremediation. Other metal tolerant grasses are available that can phytoimmbolize HMs at the soil surface. Moreover, their high biomass producing ability makes them ideal candidate for bioenergy generation.

There are various mechanisms by which plants alleviate the toxicity of HMs. In general, metal tolerance of hyperaccumulation includes sequestration of metals in specific intracellular compartments where metals can do the least harm, changing the speciation status to less toxic forms by redox mechanisms, precipitation, or binding of metals to ligands, as well as export of metals out of the cells and plants (Leitenmaier and Küpper, 2013). One important mechanism is triggering the production of oligopeptide ligands known as phytochelatins (PCs) and metallothioneins (MTs) (Cobbett, 2000). PCs are functionally equivalent to MTs. Metallothioneins (MTs), are small gene encoded, Cys-rich polypeptides. These peptides bind and formstable complex with the HM and thus neutralize the toxicity of the metal ions. Phytochelatins (PCs) are synthesised with glutathione as building blocks resulting in a peptide with structure Gly-(y-Glu-Cys-)n; {where, $n = 2$–11}. Appearance of phytochelating ligands has been reported in hundreds of plant species exposed to HMs (Rauser, 1999). Chelators have been isolated from various metallophytic plants. Chelating agents like EDTA were applied to Pb contaminated soils to increases bioavailable lead and greater accumulation of Pb was really observed in the plants (Huang et al., 1997). The addition of chelates to Pb contaminated soil increased lead concentration in shoots of corn *(Zea mays)* and pea *(Pisun sativum)* from less than 500 mg kg^{-1} to more than 10,000 mg kg^{-1}. Similar enhanced phytoextraction of uranium was found using citric acid (Jagetiya and Sharma, 2013). Order of effectiveness for different chelates used for Pb was EDTA >Hydroxyethylethylene-diaminetriacetic acid (HEDTA) > Dieth ylenetriaminepentaaceticacid (DTPA) > Ethylenediamine di(o-hyroxyphenylacetic acid) EDDHA (Huang et al., 1997).

Plant associated microbes play a crucial role in acclimatizing the plants to metal-liferous environments and enhancing their metal tolerance abilities. Several reviews have pointed out importance of PGPR microbes in increasing HM stress tolerances of plants (Mishra et al., 2017). Table 15.2 shows the different pilot scale and lab scale applications of phytoremediation of heavy metals by perennial grasses.

6   Pros and cons of phytoremediation with perennial grasses

The general limitations of phytoremediation have been discussed by many authors (Macek et al., 2000; Cunningham and Ow, 1996). But the main risk of phytoremedia-tion is contamination from the litter and postharvest biomass management or biomass disposal. Trace element contaminated litter might reenter the environment through dispersal via wind or water (Perronnet et al., 2000). Thus, prevention measures should be taken to control litter dispersion, especially in situations where wind and water can be expected. Grasses offer advantage over tress as litter dispersion is minimal and the dense grass cover over forms a shield over the entire area minimizing pollut-ants dispersal. The root system forms an interlocking net that holds the soil together. The extraction of transpiration water causes the formation of pores that are easily drained and facilitates soil aeration. This also reduces surface runoff and thus soil ero-sion. During dry periods vegetation protects the soil surface from drying. Soil water consumption for transpiration also reduces contaminant leaching (Pilon-Smits, 2005). Root exudates also promote the formation of soil organic matter and thus increase the sorption capacity of the soil, increase colloidal nature of soil decreasing soil erosion.

Another risk of phytoremediation is associated with bioaccumulation of contami-nants and particularly HMs like selenium or arsenic into vegetable or other food crops. Seleniferous soils are distributed widely in the Great Plains of the USA, Canada, South America, China, and Russia (White et al., 2007). While most soils contain only 0.01–2.0 mg Se kg^{-1}, the Se concentration of seleniferous soils can reach up to 1200 mg Se kg^{-1}. In these cases instead of using hyperaccumulators grasses like *Brassica* sp. or barley (*Hordeum vulgare*), the aim should be not cleansing but only control by the production of biofortified products (Banuelos and Mayland, 2000). In the seleniferous soil of Central Valley, Banuelos and Mayland (2000) produced Se-enriched canola (*Brassica napus*) and utilized it as Se-biofortified forage to feed Se-deficient animals. Similarly, in India, plants such as barley (*Hordeum vul-gare*), bajra (*Pennisetum typhoides*), cowpea (*Vigna sinensis*), guar (*Cyamopsis tetragonoloba*), rapeseed (*Brassica napus*), raya (*Brassica juncea*), spearmint (*Mentha viridis*), sugarcane (*Saccharum officinarum*), sunflower (*Helianthus ann-uus*), and wheat (*Triticum aestivum*) were grown on various seleniferous soils to produce Se-enriched food, fodder, or fertilizer (Banuelos and Dhillon, 2011). In

Table 15.2 Case studies of heavy metal phytoremediation using various hyperaccumulator and accumulator grass species.

Plant used	Contaminant to remove	Efficiency of phytoextraction	References
Soybean (*Glycine max*)	As, Cd, Cr, Pb, Zn	Bioaccumulation of the HMs in shoot was Cd 0.57, Cr 0.02,Pb 0.01–0.03, Zn 0.15–0.66	Murakami and Ae (2009); Zhuang et al. (2013)
Corn (*Zea mays*)	Cd, Cr, Pb, Zn	Bioaccumulation of the HMs in grain was Cd 0.57, Cr 0.02, Pb 0.01–0.13, Zn 0.54–4.95	Salazar et al. (2012); Zhuang et al. (2013)
	As, Cd, Cr, Pb, Zn	HMs accumulated both in shoot. Bioaccumulation factor was As 0.03, Cd 0.1–1.88, Pb 0.05–1.13, Zn 0.2–3.7	Kayser et al. (2000); Chen et al. (2004); Chiu et al. (2005)
Sorghum (*Sorghum bicolor* (L.) (Moench),	As, Cd, Cr, Pb, Zn	HMs accumulated both in shoot and fruit. Bioaccumulation factor was As (0.02–0.03), Cd (0.05–0.1), Pb 0.01–0.02 Zn 0.09–0.15 (Shoot)	Murillo et al. (1999); Chen et al. (2004); Marchiol et al. (2007)
Wheat (*Triticum aestivum* L.)	As, Cd, Cr, Pb, Zn	HMs accumulated in shoot. Bioaccumulation factor was As 0.04–0.11, Cr 0.5–1.3, Pb 0.02, Zn 0.2	Chen et al. (2004); Bermudez et al. (2011); Jamali et al. (2009)
	Cu, Cd, Cr, Zn, Fe, Ni,Mn, and Pb	Maximum accumulation of Fe followed by Mn and Zn in root >shoot >leaves >seeds.	Chandra et al. (2009)
Wheat (*Triticum vulgare*, sort Umanka—terrestrial)	Ag, As, B, Ba, Be, Bi, Ca, Cd, Co, Cr, Cu, Fe, K, Li, Mg, Mn, Mo, Na, Ni, P, Pb, Rb, S, Sb, Se, Sr, Th, Tl, U, V, and Zn	Concentrations of Ag, Cd, Cu, Pb, Sb, and Zn in the initial contaminated soil were 3–6 times higher than those in the initial clean soil. In particular, contents of Cu, Mo, Ni, Pb, Sb, and Zn in roots of the wheat grown in the contaminated soil were higher than those in the roots of the plants grown in the clean soil. Moreover, all the elements except Pb transferred more easily from roots to leaves.	Shtangeeva et al. (2004)
Pusa Bold—terrestrial	Arsenic (As) as of sodium (meta-) arsenite (50 µM, 150 µM, and 300 µM)	Increased tolerance in P. Bold may be due to the defensive role of antioxidant enzymes, induction of MAPK, and upregulation of PCS transcript which is responsible for the production of metal-binding peptides	Hammer et al. (2003)

Plant	Metals	Findings	References
Giant reed (*Arundo donax*)	As 0.012 Cr Pb 0.007–0.0005 Zn	HMs accumulated in shoot. Bioaccumulation factors were As 0.012, Pb 0.007–0.0005, Zn (0.04–0.008)	Boularbah et al. (2006); Guo and Miao (2010)
Smilo grass (*Piptatherum miliaceum*)	As, Cd, Cr, Pb, Zn	HMs accumulated in shoot. Bioaccumulation factors were As 0.09, Cd 0.07, Pb 0.003, Zn 0.02	Marchiol et al. (2013)
Elephant grass (*Pennisetum purpureum* Schumach)	As, Cd, Cr, Pb, Zn	HMs accumulated in shoot. Bioaccumulation factor was 0.5	Amonoo-Neizer et al. (1996)
Vetiver grass (*Vetiveria zizanioides* L.)	As, Cd, Cr, Pb, Zn	HMs accumulated in shoot. Bioaccumulation factor was As 0.04, Cd 1.25, Pb 0.004–0.07, Pb 0.004–0.07	Lai and Chen (2004); Chiu et al. (2005); Rotkittikhun et al. (2007)
[a]Indian mustard (*Brassica juncea*) terrestrial	Mercury as $HgCl_2$ (0,0.05, 0.5, 1, 2.5, 5,10 mg/L)	Roots-concentrated Hg 100–270 times above initial solution concentrations. Mercury was more toxic to plants at 5 and 10 mg/L. The plants translocated little Hg to the shoots, which accounted for just 0.7%–2% of the total Hg in the plants. Most Hg volatilization occurred from the roots. Volatilized Hg was predominantly in the Hg(0) vapor form. Phytofiltration was the primary mechanism.	Moreno et al. (2008)
[a]*Brassica juncea* (Indian mustard), *Brassica rapa* (field mustard), and *Brassica napus* (rape) terrestrial	Cd, Cr, Cu, Ni, Pb, and Zn	*Brassica rapa* exhibited the highest affinity for accumulating Cd and Pb from the soil, either with/without additional use of mobilizing soil amendments. Two *Brassica* species (*Brassica napus* and *Raphanus sativus*) were moderately tolerant when grown on a multi-metal contaminated soil. The distribution of HMs in the organs of crops decreased in the following order: leaves >stems >roots >fruit shell >seeds.	Ginneken et al. (2007)
[a]*Brassica campestris* L. terrestrial	Cu, Cd, Cr, Zn, Fe, Ni, Mn, and Pb	Maximum accumulation of Fe followed by Mn and Zn in root >shoot >leaves >seeds.	Chandra et al. (2009)

(Continued)

Table 15.2 Case studies of heavy metal phytoremediation using various hyperaccumulator and accumulator grass species. (*Cont.*)

Plant used	Contaminant to remove	Efficiency of phytoextraction	References
[a]Indian mustard (*Brassica juncea var. megarrhiza*)—terrestrial	Pb as Pb(NO$_3$)^2	*Brassica juncea* is one plant which accumulates high levels of Pb and other HMs. The results indicate that lead nitrate obviously inhibits the root, hypocotyls, and shoot growth of *Brassica juncea* at the concentration of 10^{-3} M Pb $^{2+}$*Brassica juncea* has the ability to accumulate Pb primarily in its roots, transport, and concentrate it in its hypocotyls and shoots in much lesser concentrations.	Liu et al. (2000)
[a]Rapeseed (*Brassica napus*)	As, Cd, Cr, Pb, Zn	HMs accumulated in shoot. Bioaccumulation factor was Pb 0.03, Zn 0.16	Solhi et al. (2005)
Rice (*Oryza sativa* L.)—terrestrial	Cd, Cr, Pb, As, and Hg	Rice grain contained significant lower amounts of five metals than straw and root in all sampling sites. Rice root accumulated Cd, As, and Hg from the paddy soil. The rice plant transported As very weakly, whereas Hg was transported most easily into the straw and grain among studied HMs.	Liu et al. (2007)
Grasses (mixture of selected species), sorghum (*Sorghum bicolor* L.) and sudan grass (*Sorghum sudanense*)	As, Cd, Mo, Pb, Zn (soil), and As, Cd, Pb,Ni, Zn (water)	The integrated phytoremediation treatments not produced high biomass, and decreased the HM content in the plants.	Murányi and Ködöböcz (2008)
Aromatic grasses of tannery effluent contaminated sites	Cr, Ni, Pb	Suwarna variety was best among seven aromatic grasses tested. In Suwarna variety reduction potential of chromium, lead, and cadmium were 70.07%, 85.29%, and 77.36%.	Pandey et al. (2019)

[a] Perennial herbs.

China, clover (*Trifolium repens*) and alfalfa (*Medicago sativa*) were used to produce Se-biofortified fodder (Yuan et al., 2012). However, consumption of Se-enriched food or fodder is needed under medical supervision as overdose of Se would be toxic. So, Se concentrations of food and feed products have to be determined and controlled and Se-enriched biomass should be used carefully. Some plants can hyperaccumulate Se to concentrations above the safety threshold for human or animal consumption. In search for alternative uses, Dhillon et al. (2007) incorporated Se-rich plant materials (up to 20 t ha^{-1}) into a non-seleniferous agricultural soil and produced wheat (*Triticum aestivum* L.) grains and straw with increased but safe levels of Se supplement in the diets of animals and humans living in Se-deficient areas. Using Se-rich plant material as an organic Se fertilizer for growing other crops could thus be a safe alternative for utilizing plant material that otherwise is too dangerous as direct Se source in animals and humans nutrition.

Hyperaccumulation of HMs into agricultural crops entails potential ecological risks. Tobacco (*Nicotiana tabacum*) accumulates Cd to relatively high levels (1000 mg kg^{-1}) compared to other species (Kayser et al., 2000; Keller et al., 2003; Wenger et al., 2002). The food crop *Glycine max* reaches higher Zn shoot concentrations than other species (Murakami and Ae, 2009). *Brassica napus* was reported to accumulate more Pb than *Triticum* spp., *Zea mays* L., and sorghum (Tangahu et al., 2011). Concerning potential risks deriving from biomass produced on polluted soils, it can be decided from the available literature that *N. tabacum* and *Saccharum* spp. would be unsuitable for Cd-polluted soils, *G. max* for Zn-polluted soils, and *Triticum* spp. for As-polluted soil (Vamerali et al. 2010; Tangahu et al., 2011).

The success of phytomanagement depends on the selection of the appropriate plants, suitable cultivation methods, post-harvest management of the biomass for economic benefit. Before performing phytomanagement a number of parameters must be optimized under pot and greenhouse condition. Results of many pot and field studies in which plants have been grown on polluted soils should be integrated into model-based decision support systems (DSS), such as Phyto-DSS (Robinson et al., 2003), REC-Phyto-DSS (Onwubuya et al., 2009), and Phyto-3 (Bardos et al., 2011). All the mentioned DSS used either a multi-criteria analysis or life cycle analysis or both. Phyto-DSS was a generic tool that could predict the efficiency of metal phytoextraction and evaluate its economic feasibility. REC-Phyto-DSS was a European DSS that specifically focused on "gentle" site remediation techniques and in particular on phytoextraction and phytostabilization. It was implemented in the Dutch REC (Risk reduction, Environmental merits, and Cost) framework. Phyto-3 was designed for US conditions. It provided guidance for regulators and practitioners, evaluating options of remedial phytomanagement and the production of biomass. While the existing DSSs provided a good basis for the assessment of contaminated sites, none of them have a sufficient focus on the economic revenue (Cano-Reséndiz et al., 2011). Thus, to encourage efficiency and increase the self-sustainability and keeping possible risks derived from phytomanagement sites at a minimum, development of a universal DSS for phytomanagement is need of the hour.

Another crucial question is the fate of HMs in biodiesel. In their study on the accumulation and distribution of HMs in oil crops, Angelova et al. (2004) determined the content of the HMs (Cd, Cu, and Pb) in plant organs and in the oil of rape seed (*Brassica napus* L.) grown in a polluted area. The distribution of HMs showed the following order: leaves > stems > roots > fruit shell > seeds. Although the concentration effect of HMs was the lowest in the seeds (which contain the pure plant oil), the quantities of HMs in the rape seed oil were higher than the accepted maximum permissible concentrations for human consumption. So, the exhaust fumes from biodiesel produced from such rape seed plants would have hazardous effect. Further scientific research to investigate this issue is essential. It is crucial that the remediation effect of the plant will not be tackled by higher heavy-metal emissions of vehicles, running on biodiesel obtained from phytoremediation plants.

One of the key parameters to the acceptance of phytoremediation pertains to the measurement of its performance, ultimate utilization of by-products and its overall economic viability. To date, commercial phytoremediation has been constrained by the expectation that site remediation should be achieved in a time comparable to other clean-up technologies. Except few field studies, most of the phytoremediation experiments have been performed in the lab scale, where plants are fed HM diets. While the lab based results are promising, but in real soil, the situation is different where many metals are tied up in insoluble forms, and they are less accessible. Despite these and other successes, phytoremediation is largely confined in lab scale studies. A recent literature review found that the number of academic papers mentioning phytoremediation each year has been climbing up steadily since 1990s, but the number of phytoremediation-related patents or actual application of it to remediate renowned polluted sites are missing or falling far beyond (Beans, 2017; Gerhardt et al., 2017). After understanding phytoremediation potential, commercialization of the technology started in the early 1990s. The Departments of Energy and Defense offered funding for popularizing this technology. Academic research spurred several phytoremediation-based businesses, including one called Phytotech, Inc. But investors soon lost interest and government funding waned. In 1999, Phytotech sold its phytoremediation technology to Edenspace. Other early companies folded, including Verdant Technologies, an offshoot of research at the University of Washington, and Phytoworks, an offshoot of University of Georgia research due to the time-intensive nature of phytoremediation efforts.

Genetic engineering can generate plants that can remove and break down toxins quickly. But lengthy environmental impact assessments would make it difficult to get a genetically engineered plant to market and it may cost $100–150 million and 10 years to commercialize a transgenic agricultural crop (Beans, 2017).

7 Conclusions

Phytoremediation is a solar power driven low cost viable option to decontaminate polluted soil and water, when the biomass produced during the phytoremediation process could be economically valorized in the form of bioenergy. The use of low metal accumulating bioenergy crops might be suitable for this purpose. Although there are real risks associated with phytoremediation but traditional methods of remediation including, excavation and landfilling, soil incineration, soil washing, and vitrification have more real risks, and are potentially more dangerous than phytoremediation. In most cases phytoremediation risks are small compared to the risks of doing nothing or the financial and engineering risks of "dig and haul." Confinement strategies such as onsite processing, harvesting before seed set, and volunteer management, may reduce the likelihood of HM laden pollen and seed dispersal thus reducing potential risks. Because the remediation industry is compliance driven, phytoremediation technologies must demonstrate their effectiveness at meeting State and Federal regulations. So for proper commercialization, changes in regulatory status and/or continuing technical improvements will be necessary.

References

Afegbua, S.L., Batty, L.C., 2018. Effect of single and mixed polycyclic aromatic hydrocarbon contamination on plant biomass yield and PAH dissipation during phytoremediation. Environ. Sci. Pollut. R. 25, 18596–18603.

Alkorta, I., Garbisu, C., 2001. Phytoremediation of organic contaminants in soils. Bioresour. Technol. 79, 273–276.

Amonoo-Neizer, E.H., Nyamah, D., Bakiamoh, S.B., 1996. Mercury and arsenic pollution in soil and biological samples around the mining town of Obuasi. Ghana. Water. Air. Soil. Pollut. 91, 363–373.

Andria, V., Reichenauer, T.G., Sessitsch, A., 2009. Expression of alkane monooxygenase (alkB) genes by plant-associated bacteria in the rhizosphere and endosphere of Italian ryegrass (*Lolium multiflorum* L.) grown in diesel contaminated soil. Environ. Pollut. 157, 3347–3350.

Angelova, V., Ivanova, R., Ivanov, K., 2004. Heavy metal accumulation and distribution in oil crops. Commun. Soil. Sci. Plant. Anal. 35, 2551–2566.

Aprill, W., Sims, R.C., 1990. Evaluation of the use of prairie grasses for stimulating polycyclic aromatic hydrocarbon treatment in soil. Chemosphere 20 (1–2), 253–265.

Balsamo, R.A., Kelly, W.J., Satrio, J.A., Ruiz-Felix, M.N., et al., 2014. Utilization of grasses for potential biofuel production and phytoremediation of heavy metal contaminated soils. Int. J. Phytoremediation 17, 448–455.

Banuelos, G.S., Mayland, H.F., 2000. Absorption and distribution of selenium in animals consuming canola grown for selenium phytoremediation. Ecotoxicol. Environ. Saf. 46, 322–328.

Banuelos, G.S., Dhillon, K.S., 2011. Developing a sustainable phytomanagement strategy for excessive selenium in western United States and India. Int. J. Phytoremediation 13, 208–228.

Barac, T., Taghavi, S., Borremans, B., Provoost, A., Oeyen, L., Colpaert, J.V., Vangronsveld, J., van der Lelie, D., 2004. Engineered endophytic bacteria improve phytoremediation of water-soluble, volatile, organic pollutants. Nat. Biotechnol. 22, 583–588.

Barbosa, B., Boléo, S., Sidella, S., Costa, J., 2015. Phytoremediation of heavy metal-contaminated soils using the perennial energy crops *Miscanthus* spp. and *Arundo donax* L. Bioenerg. Res. 8, 1500–1511.

Bardos, R.P., Bone, B., Andersson-Skold, Y., Suer, P., Track, T., Wagelmans, M., 2011. Crop-based systems for sustainable risk-based land management for economically marginal damaged land. Remediation 21, 11–33.

Basta, N.T., Sloan, J.J., 1999. Bioavailability of metals in strongly acidic soils treated with exceptional quality biosolids. J. Environ. Qual. 28, 633–638.

Beans, C., 2017. Core concept: phytoremediation advances in the lab but lags in the field. PNAS. 114, 7475–7477.

Bell, T.H., Yergeau, E., Maynard, C., Juck, D., Whyte, L.G., Greer, C.W., 2013. Predictable bacterial composition and hydrocarbon degradation in Arctic soils following diesel and nutrient disturbance. ISME J. 7, 1200–1210.

Berg, B., Ekbohm, G., Soderstrom, B., Staaf, H., 1991. Reduction of decomposition rates in scotspine needle litter due to heavy-metal pollution. Water. Air. Soil. Pollut. 59, 165–177.

Berg, J., Brandt, K.K., Al-Soud, W.A., Holm, P.E., Hansen, L.H., Sørensen, S.J., et al., 2012. Selection for Cu-tolerant bacterial communities with altered composition, but unaltered richness, via long term Cu exposure. Appl. Environ. Microbiol. 78, 7438–7446.

Bermudez, G.M.A., Jasan, R., Plá, R., Pignata, M.L., 2011. Heavy metal and trace element concentrations in wheat grains: assessment of potential non-carcinogenic health hazard through their consumption. J. Hazard. Mater. 193, 264–271.

Best, E.H.P., Zappi, M.E., Fredrickson, H.L., Sprecher, S.L., Larson, S.L., Ochman, M., 1997. Screening of aquatic and wetland plant species for the phytoremediation of explosives-contaminated groundwater from the Iowa Army Ammunition Plant. Ann. NY Acad. Sci. 829, 179–194.

Boularbah, A., Schwartz, C., Bitton, G., Aboudrar, W., Ouhammou, A., Morel, J.L., 2006. Heavy metal contamination from mining sites in South Morocco: 2. Assessment of metal accumulation and toxicity in plants. Chemosphere. 63, 811–817.

Brooks, R.R, Lee, J., Reeves, R.D., Jaffre, T., 1977. Detection of nickeliferous rocks by analysis of herbarium specimens of indicator plants. J. Geochem. Explor. 7, 49–57.

Burken, J.G., Shanks, J.V., Thompson, P.L., 2000. Phytoremediation and plant metabolism of explosives and nitroaromatic compounds. In: Spain, J.C., Hughes, J.B., Knackmuss, H.J. (Eds.), Biodegradation of Nitroaromatic Compounds and Explosives. Lewis, Washington, D.C., pp. 239–275.

Cano-Reséndiz, O., de la Rosa, G., Cruz-Jiménez, G., Gardea-Torresdey, J.L., Robinson, B.H., 2011. Evaluating the role of vegetation on the transport of contaminants associated with a mine tailing using the Phyto-DSS. J. Hazard. Mater. 189, 472–478.

Cerniglia, C.E., 1997. Fungal metabolism of polycyclic aromatic hydrocarbons: past, present and future applications in bioremediation. J. Ind. Microbiol. Biotechnol. 19, 324–333.

Chandra, R., Bharagava, R.N., Yadav, S., Mohan, D., 2009. Accumulation and distribution of toxic metals in wheat (*Triticum aestivum* L.) and Indian mustard (*Brassica campestris* L.) irrigated with distillery and tannery effluents. J. Hazard. Mater. 162 (2–3), 1514–1521.

Chen, Y.H., Li, X.D., Shen, Z.G., 2004. Leaching and uptake of heavy metals by ten different species of plants during an EDTA-assisted phytoextraction process. Chemosphere. 57, 187–196.

Chiu, K.K., Ye, Z.H., Wong, M.H., 2005. Enhanced uptake of As, Zn, and Cu by *Vetiveria zizanioides* and *Zea mays* using chelating agents. Chemosphere. 60, 1365–1375.

Cho-Ruk, K., Kurukote, J., Supprung, P., Vetayasuporn, S., 2006. Perennial plants in the phytoremediation of lead-contaminated soils. Biotechnology 5, 1–4.

Cobbett, C.S., 2000. Phytochelatins and their role in heavy metal detoxification. Plant. Physiol. 123, 825–832.

Collins, C.D., 2007. Implementing phytoremediation of petroleum hydrocarbons. Methods Biotechnol. 23, 99.

Cunningham, S.D., Ow, D.W., 1996. Promises and prospects of phytoremediation. Plant. Physiol. 110, 715–719.

Daane, L.L., Harjono, I., Zylstra, G.J., Maggblom, M.M., 2001. Isolation and characterization of polycyclic aromatic hydrocarbon-degrading bacteria associated with the rhizobia of salt marsh plants. Appl. Environ. Microbiol. 67, 2683–2691.

Danh, L.T., Truong, P., Mammucari, R., Tran, T., Foster, N., 2009. Vetiver grass, *Vetiveria zizanioides*: a choice plant for phytoremediation of heavy metals and organic wastes. Int. J. Phytoremediation 11 (8), 664–691.

Dhillon, S.K., Hundal, B.K., Dhillon, K.S., 2007. Bioavailability of selenium to forage crops in a sandy loam soil amended with Se-rich plant materials. Chemosphere 66, 1734–1743.

Ent, A., Baker, A.M., Reeves, R., Pollard, A.J., Schat, H., 2013. Hyperaccumulators of metal and metalloid trace elements: facts and fiction. Plant. Soil. 362, 319–334.

Environmental Protection Agency Underground Storage Tanks (EPA UST). Semiannual Report Of UST Performance Measures, End of Fiscal Year 2011 [Internet] [cited 2011 12/23]. Available from: http://epa.gov/oust/cat/ca 11 34.pdf.

EPA UST (Environmental Protection Agency Underground Storage Tanks), 2011. Semiannual Report of UST Performance Measures, End of Fiscal Year. [Internet] [cited 2011 12/23]. Available from: http://epa.gov/oust/cat/ca_11_34.pdf.

Eskandary, S., Tahmourespour, A., Hoodaji, M., Abdollahi, A., 2017. The synergistic use of plant and isolated bacteria to clean up polycyclic aromatic hydrocarbons from contaminated soil. J. Environ. Health. Sci. Eng. 15, 12.

Evangelou, M.W.H., Bauer, U., Ebel, M., Schaeffer, A., 2007. The influence of EDDS and EDTA on the uptake of heavy metals of Cd and Cu from soil with tobacco Nicotiana tabacum. Chemosphere 68, 345–353.

Fan, S.X., Li, P.J., Gong, Z.Q., Ren, W.X., He, N., 2008. Promotion of pyrene degradation in rhizosphere of alfalfa (*Medicago sativa* L.). Chemosphere 71, 1593–1598.

Fässler, E., Robinson, B.H., Stauffer, W., Gupta, S.K., Papritz, A., Schulin, R., 2010. Phytomanagement of metal-contaminated agricultural land using sunfl ower, maize and tobacco. Agric. Ecosyst. Environ. 136, 49–58.

Fatnassi, I.C., Chiboub, M., Saadani, O., Jebara, M., Jebara, S.H., 2015. Phytostabilization of moderate copper contaminated soils using co-inoculation of *Vicia faba* with plant growth promoting bacteria. J. Basic. Microbiol. 55 (3), 303–311.

Fu, D., Teng, Y., Shen, Y., et al., 2012. Dissipation of polycyclic aromatic hydrocarbons and microbial activity in a field soil planted with perennial ryegrass. Front. Environ. Sci. Eng. 6, 330.

Gamalero, E., Lingua, G., Capri, F.G., et al., 2004. Colonisation pattern of primary tomato roots by Pseudomonas fluorescens A6RI characterised by dilution plating, flow cytometry, fluorescence, confocal and scanning electron mycroscopy. FEMS Microbiol. Ecol. 48, 79–87.

Gerhardt, K.E., Gerwing, P.D., Greenberg, B.M., 2017. Opinion: Taking phytoremediation from proven technology to accepted practice. Plant. Sci. 256, 170–185.

Ginneken, L.V., Meers, E., Guisson, R., Ruttens, A., et al., 2007. Phytoremediationfor heavy metal-contaminated soils combined with bioenergy production. J. Environ. Eng. Landsc. 15, 227–236.

Glick, B.R., 2010. Using soil bacteria to facilitate phytoremediation. Biotechnol. Adv. 28, 367–374.

Grobelak, A., Napora, A., 2015. The chemophytostabilisation process of heavy metal polluted soil. PLoS ONE 10 (6), e0129538.doi:10.1371/journal.pone.0129538.

Groom, C.A., Halasz, A., Paquet, L., Morris, N., Olivier, L., Dubois, C., Hawari, J., 2002. Accumulation of HMX (octahydro-1,3,5,7-tetranitro-1,3,5,7-tetrazocine) in indigenous and agriculturalplants grown in HMX-contaminated anti-tank firing range soil. Environ. Sci. Technol. 36, 112–118.

Guo, Z.H., Miao, X.F., 2010. Growth changes, tissues anatomical characteristics of giant reed (*Arundo donax* L.) in soil contaminated with arsenic, cadmium, lead. J. Cent. South. Univ. Technol. 17, 770–777.

Gupta, S.K., Herren, T., Wenger, K., Krebs, R., Hari, T., 2000. In situ gentle remediation measures for heavy metalpolluted soils. In: Terry, N., Bañuelos, G., (Eds.), Phytoremediation of Contaminated Soil and Water, Lewis Publishers, Boca Raton, FL, USA, pp. 303–322.

Hemambika, B., Balasubramanian, V., Kannan, V.R., James, R.A., 2013. Screening of chromium-resistant bacteria for plant growth-promoting activities. Soil. Sediment. Contam. 22, 717–736.

Hammer, D., Kayser, A., Keller, C., 2003. Phytoextraction of Cd and Zn with *Salix viminalis* in field trials. Soil Use Manage. 19 (3), 187–192.

Henderson, K.L., Belden, J.B., Zhao, S., Coats, J.R., 2006. Phytoremediation of pesticide wastes in soil. Z Naturforsch C. 61 (3–4), 213–221.

Hannink, N., et al., 2002. Phytoremediation of explosives. Crit. Rev. Plant Sci. 21, 511–538.

Huang, J.W., Chen, J., Berti, W.R., Cunningham, S.D., 1997. Phytoremediation of lead contaminated soils-role of synthetic chelates in lead phytoextraction. Environ. Sci. Technol. 31, 800–806.

Ijaz, A., Imran, A., Anwar ul Haq, M., Khan, Q., Afzal, M., 2015. Phytoremediation: recent advances in plant-endophytic synergistic interactions. Plant. Soil, 1–17.

Jackson, R., Canadell, J., Ehleringer, J., Mooney, H.A., Sala, O., Schulze, E., 1996. A global analysis of root distributions for terrestrial biomes. Oecologia 108 (3), 389–411.

Jagetiya, B., Sharma, A., 2013. Optimization of chelators to enhance uranium uptake from tailings for phytoremediation. Chemosphere 91 (5), 692–696.

Jamali, M.K., Kazi, T.G., Arain, M.B., Afridi, H.I., Jalbani, N., Kandhro, G.A., Shah, A.Q., Baig, J.A., 2009. Heavy metal accumulation in different varieties of wheat (*Triticum aestivum* L.) grown in soil amended with domestic sewage sludge. J. Hazard. Mater. 164, 1386–1391.

Kacprzak, M., Grobelak, A., Grosser, A., Prasad, M.N.V., 2014. Efficacay of biosolids in assisted phytostabilization of metalliferous acidic sandy soils with five grass species. Int. J. Phytoremed. 16, 593–608.

Kaimi, E., Mukaidani, T., Tamaki, M., 2007. Screening of twelve plant species for phytoremediation of petroleum hydrocarbon-contaminated soil. Plant Prod. Sci. 10 (2), 211–218.

Kayser, A., Wenger, K., Keller, A., Attinger, W., Felix, H.R., Gupta, S.K., Schulin, R., 2000. Enhancement of phytoextraction of Zn, Cd, and Cu from calcareous soil: the use of NTA and sulfur amendments. Environ. Sci. Technol. 34, 1778–1783.

Kechavarzi, C., Pettersson, K., Leeds-Harrison, P., Ritchie, L., Ledin, S., 2007. Root establishment of perennial ryegrass (L. perenne) in diesel contaminated subsurface soil layers. Environ. Pollut. 145, 68–74.

Keller, C., Hammer, D., Kayser, A., Richner, W., Brodbeck, M., Sennhauser, M., 2003. Root development and heavy metal phytoextraction efficiency: comparison of different plant species in the field. Plant. Soil 249, 67–81.

Khan, S., Afzal, M., Iqbal, S., Khan, Q.M., 2013. Plant-bacteria partnerships for the remediation of hydrocarbon contaminated soils. Chemosphere 90, 1317–1332.

Khaziev, F.K.H., 1980. Soil Enzyme Activities. Science Press, Beijing, pp. 21–24.

Kolb, M., Harms, H., 2000. Metabolism of fluoranthene in different plant cell cultures and intact plants. Environ. Toxicol. Chem. 19 (5), 1304–1310.

Komoßa, D., Langerbartels, C., Sanderman, H., 1995. Metabolic processes for organic chemicals in plants. In: Trapp, S., McFarland, J.C. (Eds.), Plant Contamination-Modeling and Simulation of Organic Chemical Process. CRC Press, Boca Raton, FL, pp. 69–103.

Kretzschmar, R., Borkovec, M., Grolimund, D., Elimelech, M., 1999. Mobile subsurface colloids and their role in contaminant transport. Adv. Agron. 66, 121–193.

Lai, H.Y., Chen, Z.S., 2004. Effects of EDTA on solubility of cadmium, zinc, and lead and their uptake by rainbow pink and vetiver grass. Chemosphere 55, 421–430.

Lalande, T.L., Skipper, H.D., Wolf, D.C., Reynolds, C.M., Freedman, D.L., Pinkerton, B.W., Hartel, P.G., Grimes, L.W., 2003. Phytoremediation of pyrene in a Cecil soil under field conditions. Int. J. Phytoremed. 5, 1–12.

Larson, S.L., Jones, R.P., Escalon, B.L., Parker, D., 1999a. Classification of explosives transformation products in plant tissues. Environ. Toxicol. Chem. 18, 1270–1276.

Larson, S.L., Weiss, C.A., Escalon, B.L., Parker, D., 1999b. Increased extraction efficiency of acetonitrile/water mixtures for explosives determination in plant tissues. Chemosphere 38, 2153–2162.

László, S., 2005. Stabilization of metals in acidic mine spoil with amendments and red fescue (Festuca rubra L.) growth. Environ. Geochem. Health 27, 289–300.

Leitenmaier, B., Küpper, H., 2013. Compartmentation and complexation of metals in hyperaccumulator plants. Front. Plant Sci. Available from: https://doi.org/10.3389/fpls.2013.00374.

Lewandowski, I., Scurlock, J.M.O., Lindvall, E., Christou, M., 2003. The development and current status of perennial rhizomatous grasses as energy crops in the US and Europe. Biomass. Bioenerg. 25, 335–361.

Li, Y., Liang, F., Zhu, Y.F., Wang, F.P., 2013. Phytoremediation of a PCB contaminated soil by alfalfa and tall fescue single and mixed plants cultivation. J. Soil. Sediment. 13, 925–931.

Liu, D., Jiang, W., Liu, C., Xin, C., Hou, W., 2000. Uptake and accumulation of lead by roots, hypocotyls and shoots of Indian mustard [Brassica juncea (L.)]. Bioresour. Technol. 71, 273–277.

Liu, R., Dai, Y., Sun, L., 2015. Effect of rhizosphere enzymes on phytoremediation in PAH contaminated soil using five plant species. PLoS ONE 10 (3), e0120369. doi:10.1371/journal. pone.0120369.

Liu, W.X., Shen, L.F., Liu, J.W., Wang, Y.W., Li, S.R., 2007. Uptake of toxic heavy metals by rice (Oryza sativa L.) cultivated in the agricultural soil near Zhengzhou City, People's Republic of China. Bull. Environ. Contamin. Toxicol. 79, 209–213.

Livingstone, D.R., 1998. The fate of organic xenobiotics in aquatic ecosystems: quantitative and qualitative differences in biotransformation by invertebrates and fish. Comp. Biochem. Physiol. A: Mol Integ. Physiol. 120, 43–49.

Lombi, E., Zhao, F.J., Dunham, S.J., McGrath, S.P., 2000. Cadmium accumulation in populations of Thlaspi caerulescens and Thlaspi goesingense. New. Phytol. 145, 11–20.

Lu, M., Zhang, Z.Z., Sun, S.S., Wei, X.F., Wang, Q.F., Su, Y.M., 2010. The use of goosegrass (Eleusine indica) to remediate soil contaminated with petroleum. Water. Air. Soil. Pollut. 209 (1–4), 181–189.

Macek, T., Mackova, M., Kas, J., 2000. Exploitation of plants for the removal of organics in environmental remediation. Biotechnol. Adv. 18, 23–34.

Mandal, V., Chouhan, K.B.S., Tandey, R., Sen, K.K., Kala, H.K., Mehta, R., 2018. Critical analysis and mapping of research trends and impact assessment of polyaromatic hydrocarbon accumulation in leaves: let history tell the future. Environ. Sci. Pollut. Res. Int. 25 (23), 22464–22474.

Mansuy, D., 1998. The great diversity of reactions catalyzed by cytochromes P450. Comp. Biochem. Physiol. C. Pharmacol. Toxicol. Endocrinol. 121 (1–3), 5–14.

Marchiol, L., Fellet, G., Perosa, D., Zerbi, G., 2007. Removal of trace metals by Sorghum bicolor and Helianthus annuus in a site polluted by industrial wastes: a field experience. Plant. Physiol. Biochem. 45, 379–387.

Marchiol, L., Fellet, G., Boscutti, F., Montella, C., Mozzi, R., Guarino, C., 2013. Gentle remediation at the former "Pertusola Sud" zinc smelter: evaluation of native species for phytoremediation purposes. Ecol. Eng. 53, 343–353.

McGrath, S.P., Dunham, S.J., Correll, R.L., 2000. Potential for phytoextractionof zinc and cadmium from soils using hyperaccumulator plants. In: Terry, N., Bañuelos, G. (Eds.), Phytoremediation of Contaminated Soil and Water. Boca Raton, FL, Lewis, pp. 109–128.

Meers, E., Ruttens, A., Hopgood, M., Lesage, E., Tack, F.M.G., 2005. Potential of *Brassica rapa*, *Cannabis sativa*, *Helianthus annuus* and *Zea mays* for phytoextraction of heavy metals from calcareous dredged sediment derived soils. Chemosphere 61, 561–572.

Mishra, J., Singh, R., Arora, N.K., 2017. Alleviation of heavy metal stress in plants and remediation of soil by rhizosphere microorganisms. Front. Microbiol. 8, 1706.

Mishra, T., Pandey, V.C., Singh, P., Singh, N.B., Singh, N., 2017. Assessment of phytoremediation potential of native grass speciesgrowing on red mud deposits. J. Geochem. Explor. 182, 206–209.

Moreno, F.N., Anderson, C.W., Stewart, R.B., Robinson, B.H., 2008. Phytofiltration of mercury-contaminated water: volatilization and plant-accumulation aspects. Environ. Exp. Bot. 62 (1), 78–85.

Mukherjee, P., Roychowdhury, R., Roy, M., 2017. Phytoremediation potential of rhizobacterial isolates from Kans grass (*Saccharum spontaneum*) of fly ash ponds. Clean Technol. Environ. Policy 19 (5), 1373–1385.

Murakami, M., Ae, N., 2009. Potential for phytoextraction of copper, lead, and zinc by rice (*Oryza sativa* L.), soybean (*Glycine max* [L.] Merr.), and maize (*Zea mays* L.). J. Hazard. Mater. 162, 1185–1192.

Muranyi,A., Ködöböcz, L., 2008. Heavy metal uptake byplants in different phytoremediation treatments. In: Proceedings of the Seventh Alps-Adria Scientific Workshop, Stara Lesna, Slovaki. Available from: http://www.mokkka.hu/publications/0387.117 MOKKA AM LK.pdf.

Muratova, A.Y., Dmitrieva, T., Panchenko, L., Turkovskaya, O., 2008. Phytoremediation of oil-sludge contaminated soil. Int. J. Phytoremed. 10 (6), 486–502.

Murillo, J.M., Maranon, T., Cabrera, F., Lopez, R., 1999. Accumulation of heavy metals in sunflower and sorghum plants affected by the Guadiamar spill. Sci. Total. Environ. 242, 281–292.

Olson, P.E., Castro, A., Joern, M., DuTeau, N.M., Pilon-Smits, E.A.H., Reardon, K.F., 2007. Comparison of plant families in a greenhouse phytoremediation study on an aged polycyclic aromatic hydrocarbon contaminated soil. J. Environ. Qual. 36 (5), 1461–1469.

Onwubuya, K., Cundy, A., Puschenreiter, M., Kumpiene, J., Bone, B., Greaves, J., et al., 2009. Developing decision support tools for the selection of "gentle" remediation approaches. Sci. Total. Environ. 407, 6132–6142.

Pandey, J., Verma, R.K., Singh, S., 2019. Screening of most potential candidate among different lemongrass varieties for phytoremediation of tannery sludge contaminated sites. Int. J. Phytoremed. 21 (6), 600–609, https://doi.org/10.1080/15226514.2018.1540538.

Pandey, V.C., 2013. Suitability of *Ricinus communis* L. cultivation for phytoremediation of fly ash disposal sites. Ecol. Eng. 57, 336–341.

Pandey, V.C., Pandey, D.N., Singh, N., 2015a. Sustainable phytoremediation based on naturally colonizing and economically valuable plants. J. Clean. Prod. 86, 37–39.

Pandey, V.C., Singh, K., Singh, R.P., Singh, B., 2012. Naturally growing *Saccharum munja* on the fly ash lagoons: a potential ecological engineer for the revegetation and stabilization. Ecol. Eng. 40, 95–99.

Pandey, V.C., Singh, N., 2015. Aromatic plants versus arsenic hazards in soils. J. Geochem. Explor. 157, 77–80.

Pandey, V.C., Rai, A., Korstad, J., 2019. Aromatic crops in phytoremediation: from contaminated to waste dumpsites. In: Pandey, V.C., Bauddh, K. (Eds.), Phytomanagement of Polluted Sites. Elsevier, Amsterdam, Netherlands, pp. 255–275.

Pandey, V.C., 2017. Managing waste dumpsites through energy plantations. In: Bauddh, K., Singh, B., Korstad, J. (Eds.), Phytoremediation Potential of Bioenergy Plants. Springer, Singapore, pp. 371–386.

Pandey, V.C., Bajpai, O., Pandey, D.N., Singh, N., 2015b. *Saccharum spontaneum*: an underutilized tall grass for revegetation and restoration programs. Genet. Resour. Crop Evol. 62 (3), 443–450.

Pandey, V.C., Bajpal, O., Singh, N., 2016. Energy crops in sustainable phytoremediation. Renew. Sust. Energ. Rev. 54, 58–73.

Pandey, V.C., Bajpai, O., 2019. Phytoremediation: From Theory Toward Practice. In: Pandey, V.C., Bauddh, K. (Eds.), Phytomanagement of Polluted Sites. Elsevier, Amsterdam, Netherlands, pp. 1–49.

Pandey, V.C., Singh, V., 2019. Exploring the potential and opportunities of current tools for removal of hazardous materials from environments. In: Pandey, V.C., Bauddh, K. (Eds.), Phytomanagement of Polluted Sites. Elsevier, Amsterdam, Netherlands, pp. 501–516.

Pandey, V.C., Souza-Alonso, P., 2019. Market opportunities in sustainable phytoremediation. In: Pandey, V.C., Bauddh, K. (Eds.), Phytomanagement of Polluted Sites. Elsevier, Amsterdam, Netherlands, pp. 51–82.

Paquin, D.G., Campbell, S., Li, Q.X., 2004. Phytoremediation in subtropical Hawaii—A review of over 100 plant species. Remediation 14, 127–139.

Parrish, Z.D., Banks, M.K., Schwab, A.P., 2004. Effectiveness of phytoremediation as a secondary treatment for polycyclic aromatic hydrocarbons (PAHs) in composted soil. Int. J. Phytoremed. 6 (2), 119–137.

Perronnet, K., Schwartz, C., Gerard, E., Morel, J.L., 2000. Availability of cadmium and zinc accumulated in the leaves of *Thlaspi caerulescens* incorporated into soil. Plant. Soil 227, 257–263.

Pilon-Smits, E., 2005. Phytoremediation. Annu. Rev. Plant. Biol. 56, 15–39.

Pradhan, S.P., Conrad, J.R., Paterek, J.R., Srivastava, V.J., 1998. Potential of phytoremediation for treatment of PAHs in soil at MGP sites. J. Soil Contamin. 7 (4), 467–480.

Praveen, A., Pandey, V.C., Marwa, N., Singh, D.P., 2019. Rhizoremediation of polluted sites: harnessing plant–microbe interactions. In: Pandey, V.C., Bauddh, K. (Eds.), Phytomanagement of Polluted Sites. Elsevier, Amsterdam, Netherlands, pp. 389–407.

Rascio, N., Navari-Izzo, F., 2011. Heavy metal hyperaccumulating plants: how and why do they do it? And what makes them so interesting? Plant. Sci. 180, 169–181.

Raskin, I., Ensley, B.D., 2000. Phytoremediation of toxic metals: using plants to clean up the environment. Wiley, New York.

Rauser, W.E., 1999. Structure and function of metal chelators produced by plants: the case for organic acids, amino acids, phytin, and metallothioneins. Cell. Biochem. Biophys. 31, 19–48.

Robinson, B., Fernandez, J.E., Madejon, P., Maranon, T., Murillo, J.M., Green, S., Clothier, B., 2003. Phytoextraction: an assessment of biogeochemical and economic viability. Plant. Soil 249, 117–125.

Robinson, B.H., Banuelos, G., Conesa, H.M., Evangelou, M.W.H., Schulin, R., 2009. The phytomanagement of trace elements in soil. Crit. Rev. Plant. Sci. 28, 240–266.

Rotkittikhun, P., Chaiyarat, R., Kruatrachue, M., Pokethitiyook, P., Baker, A.J.M., 2007. Growth and lead accumulation by the grasses *Vetiveria zizanioides* and *Thysanolaena maxima* in lead-contaminated soil amended with pig manure and fertilizer: a glasshouse study. Chemosphere 66, 45–53.

Roulier, J.L., Tusseau-Vuillemin, M.H., Coquery, M., Geffard, O., Garric, J., 2008. Measurement of dynamic mobilization of trace metals in sediments using DGT and comparison with bioaccumulation in *Chironomus riparius*: first results of an experimental study. Chemosphere 70 (5), 925–932.

Roy, M., Dutta, S., Mukherjee, P., Giri, A.K., 2015. Integrated phytobial remediation for sustainable management of arsenic in soil and water. Environ. Int. 75, 180–198.

Roy, M., Roychowdhury, R., Mukherjee, P., 2018. Remediation of fly ash dumpsites through bioenergy crop plantation and generation: a review. Pedosphere 28 (4), 561–580.

Roy, S., Roy, M., 2018. Characterization of plant growth promoting feature of a neutromesophilic, facultatively chemolithoautotrophic, sulphur oxidizing bacterium *Delftia* sp. strain SR4 isolated from coal mine spoil. Int. J. Phytoremed. 21 (6), 531–540, DOI:10.1080/15226514.2018.1537238.

Roychowdhury, R., Mukherjee, P., Roy, M., 2016. Identification of chromium resistant bacteria from dry fly ash sample of mejia MTPS thermal power plant, West Bengal. India. Bull. Environ. Contam. Toxicol. 96 (2), 210–216.

Roychowdhury, R., Roy, M., Rakshit, A., Sarkar, S., Mukherjee, P., 2018. Arsenic bioremediation by indigenous heavy metal resistant bacteria of fly ash pond. Bull. Environ. Contam. Toxicol. 101 (4), 527–535.

Rylott, E.L., Bruce, N.C., 2008. Plants disarm soil: engineering plants for the phytoremediation of explosives. Trends. Biotechnol. 27 (2), 73–81.

Salazar, M.J., Rodriguez, J.H., Nieto, G.L., Pignata, M.L., 2012. Effects of heavy metal concentrations (Cd, Zn and Pb) in agricultural soils near different emission sources on quality, accumulation and food safety in soybean [*Glycine max* (L.) Merrill]. J. Hazard. Mater. 233–234, 244–253.

Salt, D.E., Smith, R.D., Raskin, I., 1998. Phytoremediation. Annu. Rev. Plant. Physiol. Plant. Mol. Biol. 9, 643–668.

Sandermann, H., 1994. Higher plant metabolism of xenobiotics: the green liver concept. Pharmacogenetics 4, 225–241.

Santoyo, G., Moreno-Hagelsieb, G., Orozco-Mosqueda, M.D.C., Glick, B.R., 2016. Plant growth-promoting bacterial endophytes. Microbiol. Res. 183, 92–99.

Sauvêtre, A., Schröder, P., 2015. Uptake of carbamazepine by rhizomes and endophytic bacteria of *Phragmites australis*. Front. Plant Sci. 6, 83.

Shtangeeva, I., Laiho, J.V.P., Kahelin, H., Gobran, G.R., 2004. Phytoremediation of metal-contaminated soils. Symposia Papers Presented Before the Division of Environmental Chemistry, American Chemical Society, Anaheim, Calif., USA. Available from: http://ersdprojects.science.doe.gov/workshop pdfs/california 2004/p050.pdf.

Singer, A.C., Thompson, I.P., Bailey, M.J., 2004. The tritrophic trinity: a source of pollutant degradaing enzymes and its implications for phytoremediation. Curr. Opin. Microbiol. 72, 39–244.

Singh, O.V., Jain, R.K., 2003. Phytoremediation of toxic aromatic pollutants from soil. Appl. Microbiol. Biotechnol. 63, 128–135.

Sipila, T.P., Keskinen, A.K., Akerman, M.L., Fortelius, C., Haahtela, K., Yrjala, K., 2008. High aromatic ring-cleavage diversity in birch rhizosphere: PAH treatment-specific changes of I.E.3 group extradiol dioxygenases and 16S rRNA bacterial communities in soil. ISME J. 2, 968–981.

Solhi, M., Shareatmadari, H., Hajabbasi, M., 2005. Lead and zinc extraction potential of two common crop plants, *Helianthus annuus* and *Brassica napus*. Water Air Soil Pollut. 167, 59–71.

Sorkhoh, D.M., Al-Mailem, N., Ali, H., Al-Awadhi, S., Salamah, M., Eliyas, S.S., Radwan, 2011. Bioremediation of volatile oil hydrocarbons by epiphytic bacteria associated with American grass (*Cynodon* sp.) and broad bean (*Vicia faba*) leaves N. A. Int. Biodeterior. Biodegrad. 65, 797–802.

Stabentheiner, E., Zankel, A., Pölt, P., 2010. Environmental scanning electron microscopy (ESEM) a versatile tool in studying plants. Protoplasma. 246, 89–99.

Staicu, L.C., van Hullebusch, E.D., Lens, P.N.L., Pilon-Smits, E.A.H., Oturan, M.A., 2015. Electro coagulation of colloidal biogenic selenium. Environ. Sci. Pollut. Res. Int. 22, 3127–3137.

Su, Y.H., Zhu, Y.G., 2008. Uptake of selected PAHs from contaminated soils by rice seedling (*Oryza sativa*) and influence of rhizosphere on PAH distribution. Environ. Pollut. 155, 359–365.

Sung, K., Munster, C.L., Rhykerd, R., Drew, M.C., Yavuz, C.M., 2002. The use of box lysimeters with freshly contaminated soils to study the phytoremediation of recalcitrant organic contaminants. Environ. Sci. Technol. 6, 2249–2255.

Szczygłowska, M., Piekarska, A., Konieczka, P., Namieśnik, J., 2011. Use of brassica plants in the phytoremediation and biofumigation processes. Int. J. Mol. Sci. 12 (11), 7760–7771.

Tangahu, B.V., Abullah, S.R.S., Basri, H., Idris, M., Anuar, N., Mukhlish, M., 2011. A review on heavy metals (As, Pb, and Hg) uptake by plants through phytoremediation. Int. J. Chem. Eng. 2011, 1–31.

Thijs, S., Sillen, W., Rineau, F., Weyens, N., Vangronsveld, J., 2016. Towards an enhanced understanding of plant–microbiome interactions to improve phytoremediation: engineering the metaorganism. Front. Microbiol. 7, 341, https://doi.org/10.3389/fmicb.2016.00341.

Tiwari, S., Singh, S.N., Garg, S.K., 2013. Microbially enhanced phytoextraction of heavy-metal fly-ash amended soil. Commun. Soil Sci. Plant Anal. 44 (21), 3161–3176.

Trapp, S., 1995. Model for uptake of xenobiotics into plants. In: Trapp, S., McFarlane, J.C. (Eds.), Plant Contamination: Modeling and Simulation of Organic Chemical Process. CRC Press, Boca Raton, FL, pp. 107–152.

Vamerali, T., Bandiera, M., Mosca, G., 2010. Field crops for phytoremediation of metal-contaminated land. A review. Environ. Chem. Lett. 8, 1–17.

Visioli, G., D'Egidio, S., Sanangelantoni, A.M., 2015. The bacterial rhizobiome of hyperaccumulators: future perspectives based on omics analysis and advanced microscopy. Front. Plant Sci. 5, 752.

Visioli, G., D'Egidio, S., Vamerali, T., Mattarozzi, M., Sanangelantoni, A.M., 2014. Culturable endophytic bacteria enhance Ni translocation in the hyperaccumulator *Noccaea caerulescens*. Chemosphere 117, 538–544.

Verma, S.K., Singh, K., Gupta, A.K., Pandey, V.C., Trivedi, P., Verma, R.K., Patra, D.D., 2014. Aromatic grasses for phytomanagement of coal fly ash hazards. Ecol. Eng. 73, 425–428.

Wei, S.Q., Pan, S.W., 2010. Phytoremediation for soils contaminated by phenanthrene and pyrene with multiple plant species. J. Soils. Sediments. 10 (5), 886–894.

Wenger, K., Gupta, S.K., Furrer, G., Schulin, R., 2002. Zinc extraction potential of two common crop plants, *Nicotiana tabacum* and *Zea mays*. Plant. Soil. 242, 217–225.

Weyens, N., van der Lelie, D., Taghavi, S., Newman, L., Vangronsveld, J., 2009. Exploiting plant-microbe partnerships to improve biomass production and remediation. Trends Biotechnol. 27, 591–598.

White, P., Broadley, M.R., Bowen, H.C., Johnson, S.E., 2007. Selenium and its relationship with sulfur. In: Hawkesford, M.J., De Kok, L.J. (Eds.), Sulfur in Plants. Springer, Dordrecht, pp. 225–252.

Yuan, L., Yin, X., Zhu, Y., Li, F., Huang, Y., Liu, Y., Lin, Z., 2012. Selenium in plants and soils, and selenosis in Enshi, China: implications for selenium biofortification. In: Yin, X., Yuan, L. (Eds.), Phytoremediation and Biofortification. Springer, Dordrecht, pp. 7–31.

Zenk, M.H., 1996. Heavy metal detoxification in higher plants: a review. Gene 179, 21–30.

Zhang, C., Daprato, R.C., Nishino, S.F., Spain, J.C., Hughes, J.B., 2001. Remediation of dinitrotoluene contaminated soils from former ammunition plants: soil washing efficiency and effective process monitoring in bioslurry reactors. J. Hazard. Mater. 87, 139–154.

Zhuang, P., Li, Z.-A., Zou, B., Xia, H.-P., Wang, G., 2013. Heavy metal contamination in soil and soybean near the Dabaoshan Mine, South China. Pedosphere 23, 298–304.

Case studies of perennial grasses—phytoremediation (holistic approach)

Vimal Chandra Pandey[a,*], Divya Patel[b], Deblina Maiti[c], D.P. Singh[a]
[a]Department of Environmental Science, Babasaheb Bhimrao Ambedkar University, Lucknow, Uttar Pradesh, India; [b]Department of Biotechnology, Sant Gadge Baba Amravati University, Amravati, Maharashtra, India; [c]Central Institute of Mining and Fuel Research, Dhanbad, Jharkhand, India
[*]Corresponding author

1 Introduction

Urbanization, industrialization, and risky agricultural practices have led to the emergence of highly toxic organic (trichloroethylene, methyl tertiary-butyl ether, and trinitrotoluene) and inorganic pollutants (metals, metalloids and radionuclides like Cd, Pb, Cr, Ni, Se, Zn, Cu, As, Co, Hg, U, Ra, Se, Cs, and Sr) which has deteriorated the soil health in many sites across the world and has had a tremendous negative influence on the ecosystems (Cerne et al., 2011; Pandey et al., 2011). Most importantly, these pollutants are non-biodegradable and eventually tend to bio-accumulate in living organisms from the contaminated soil or water and gradually incur very hazardous metabolic disorders in them (FAO and ITPS, 2015). The most important pollutants are trace elements, mineral oils, and polycyclic aromatic hydrocarbons, which emerge from common polluting sites like overburden dumps, fly ash deposits, and red mud deposits (Mench et al., 2010). As also mentioned in previous chapters, phytoremediation is 10-fold cheaper than conventional technologies for remediation of earlier-mentioned types contaminated soils and wastes, therefore in this chapter we have outlined few most effective grasses which can be beneficially used for remediating/ecorestoring such sites. Phytoremediation by these grasses can be effectively done on vast acres of contaminated lands. To be more effective, mixed plantation of the later mentioned grasses can yield simultaneous ecological and socioeconomic benefits through the various phyto-products (Pandey et al., 2015a, 2016). Besides creating greenery they can restore landscaping values, biodiversity, and increase substrate fertility. The financial burden to maintain the site after plantation of these grasses can be substantially lesser compared to other horticultural sites because minimum after care is required for their future propagation as they can survive in harsh environment (Koelmel et al., 2015). It would be interesting to read the importance of 13 perennial grasses through some case studies done in the past.

Phytoremediation Potential of Perennial Grasses. http://dx.doi.org/10.1016/B978-0-12-817732-7.00016-X

2 Potential case studies of perennial grasses in phytoremediation

2.1 Bermuda grass case

Bermuda grass is one of the most commonly found creeper grass scientifically known as *Cynodon dactylon*. It has shown immense potential for bioaccumulation of heavy metals from waste materials, for example, fly ash (Maiti et al., 2016). The latter researchers had analyzed the bioavailable metal concentrations in fly ash and topsoil and the corresponding total metal concentration in grass tissues from fly ash as well as topsoil. Their results demonstrated similar pattern of bioaccumulation of Zn, Mn, Ni, and Cu in the grasses growing on ash or soil while different trends were observed for Pb, Co, and Cd. However, fly ash also led to more translocation of Cd, Mn, Zn, and Cu in the aerial parts of the grass which indeed is a risk factor for metal transfer through food chain.

In another study, Basumatary and Bordoloi (2016) evaluated the efficiency of the grass for remediation of petroleum-contaminated soils. This study was conducted in Upper Assam, India wherein implementation of two levels of fertilizer were examined on the plants grown in net house. The plants were maintained for 180 days to allow reduction of petroleum from soil. Though the contaminants led to decrease in plant biomass, root growth, and height by 20%–30% yet lower level fertilizer application led to higher biomass production. This study revealed that the grass can tolerate a petroleum concentration of up to 7.5% on dry weight basis and serve as an alternative for remediation of hydrocarbons from soil.

2.2 Vetiver grass case

Vetiver grass has extraordinary properties, which make it an outstanding candidate for phytoremediation of wastelands, most importantly landfills. It has tolerance to biotic as well as abiotic stresses such as pests, drought, flood, metals, salts, hydrocarbons, etc. In many places like Republic Gulf Pines landfill, it has been used as a part of phytoremediation project, wherein acres of lands are planted with the grass to remediate million gallons of leachate every year. This green technology could save $8 million in contrast to traditional procedures and was later on awarded from American Academy of Environmental Engineers. Consequently, Vetiver based projects were started in Mexico and other places throughout the world for handling domestic/industrial leachates. The results from Mexico demonstrated that the grass could effectively stabilize the steep slopes, utilized fresh leachates as a resource and prevented leachate outbreaks. Thus, use of Vetiver is an innovative technology, which will change the path of handling of landfill leachates by the solid waste industries (Granley and Truong, 2012).

In another study, Vetiver was used along with Lemon grass for the reclamation of Cu tailings and to assess the consequences of manures on the reclamation project. A pot scale study was carried out to remediate metals like Cu, Ni, Mn, and Zn from the tailings of Rakha mines, Jharkhand, India. These tailings were indelibly piled in tailings ponds near the mines and caused massive environmental degradation. The results from the study revealed that there was a significant increase in nutritional status of the tailings after implementation of the project. The DTPA extractable Cu and Ni were found to reduce in the amended tailings, while Mn and Zn was elevated significantly. The grasses on the other hand accumulated lower amounts of Cu and Ni unlike the larger amounts of Mn and Zn. Plant biomass increase was directly proportional to the application rates of manure mixtures and it was concluded that the usage of chicken manure at 5% (w/w) can lead to optimum growth of Vetiver in such tailings (Das and Maiti, 2009).

2.3 Lemon grass case

Lemon grass is another metal-tolerant aromatic grass whose metal tolerant index is >100%. In a study done by Gautam et al. (2017), they evaluated the effects of red mud on Lemon grass growth, which is otherwise not used for vegetation establishment due to its harsh nature. Different concentrations of red mud (0%, 5%, 10%, and 15% w/w) were used in soil along with biowastes [cow dung and sewage sludge] and it was observed that the total plant biomass increased under all treatments, maximally at 5% red mud with 91% sewage sludge and 51 % cow dung compared to control (no redmud and bio wastes). The grass also acted as a potential phytostabilizer of Fe, Mn, and Cu in roots and translocated Al, Zn, Cd, Pb, Cr, As, and Ni into the shoots.

In another study (Kumar and Maiti, 2015), Lemon grass was checked for phytoremediation of a metal contaminated mine waste by applying divergent parts of chicken manure, farmyard manure, and soil. The study site was based on the abandoned chromite-asbestos mines of Singhbhum, Jharkhand, India. The existing dumps were creating massive environmental pollution, through the mine wastes, which had very high concentration of Cr and Ni in contrast to the soil. The results from the study showed that manure application significantly improved the waste characteristics, plant growth, and reduced the availability of Cr or Ni. Particularly, maximum biomass production was found in combination of 90% mine waste + 2.5% chicken manure and farmyard manure each + 5% garden soil. This combination also resulted in lower metal build-up in the grass tissues.

2.4 Phragmites grass case

Studies regarding phytoremediation of both metal and organic pollutants simultaneously are less in literature. The grass *Phragmites australis* can be an efficient plant for the purpose. It is a common wetland plant, which in some studies has shown remediation of both metal and organic pollutants simultaneously, for example, cadmium pentachlorophenol (Hechmi et al., 2015). This was shown in a greenhouse study in which Cd in 0, 10, and 20 mg/kg with or without pentachlorophenol in 0, 50, and 250 mg kg^{-1} was added to soil. The treatments were overgrown with this grass and maintained for 75 days. The results showed that the plant biomass was significantly influenced by interaction of both the pollutants, for example, Cd had a stronger effect than pentachlorophenol, but the coexistence of pentachlorophenol at low level with Cd, had reduced Cd toxicity as well as accumulation in the plants, and in turn improved plant growth.

Another case study revealed the accumulation characteristics of copper in wild *Phragmites australis*, growing in Liao River estuary wetland. The experiment was done on an experimental test pool in Shenyang Agricultural University while the materials like soil, irrigating water, and plants were obtained from Liao River estuary wetland. The results revealed Cu concentration of 16–49 mg/kg and 0.8–89 mg/kg in soil and the grass, respectively. The enrichment coefficients of Cu in tissues of the grass changed with its growth and were greater than 1 in the whole plant. Thus the study concluded that the grass can be used for Cu phytoremediation in contaminated areas as it is a Cu hyperaccumulator (Su et al., 2018).

2.5   Calamagrostis epigejos *grass case*

Calamagrostis epigejos is recognized as a natural bioindicator for environmental pollution, especially heavy metals in soils. An experiment was carried out to examine the accumulation potentiality of the grass from five experimental sites vulnerable to anthropogenic activities. Significant differences were observed between uptake, accumulation and translocation of metals and the results corresponded to the level of metal pollution as well as the soil physicochemical characteristics. The grass also has the potential for metal phytostabilization because it takes up a considerable portion of available metals from soil and accumulates it in its roots (Randelovic et al., 2018).

In another study, the grass was studied with another grass *Festuca rubra*. *C. epigejos* was found to naturally colonize the experimental sites (two weathered fly ash deposit lagoons). The lagoons were weathered for 5 and 13 years and had excess of As, Mo, and Cu. *C. epigejos* could show a constant photosynthetic efficiency under the harsh conditions of fly ash, unlike *F. rubra* displayed differences in their density along with damage symptoms like necrosis, chlorosis, and wilting. The later symptom was due to build-up of boron in lethal concentrations in the tissues, apart from the Cu and Mn deficiency. On the oldest fly ash dump, the coverage area of *F. rubra* was 9.5% compared to the coverage area of *C. epigejos* (87.5%). The influential expansion and survival of *C. epigejos* is due to multiple tolerances to various stressed conditions at waste deposits and its dominating nature. Therefore, characteristics of these kinds of naturally colonized grass species can be implemented for next-generation biological restoration programs of waste deposits. Their extensive root growth can binding the ash and gradually creates conditions for colonization with other species (Mitrovic et al., 2008).

2.6   Bamboo grass *case*

Bamboo is a grass with high biomass production potential and stress tolerant nature, which is a perfect model to be used for phytoremediation purposes. Different parts of the plants including roots, shoots, rhizomes, leaves, and fibers can assist in environmental clean-up from wastewater, air, and soil. Simultaneously this also helps in carbon sequestration. It has over 500 species under 48 genera, which efficiently grows in tropical as well as sub-tropical regions. In China, a planned study was done in which bamboo was grown to cover a major part of forests. The study investigated the phytoremediation mechanism by the plants and then introduced it as a successful plant for phytoremediation projects (Emamverdian et al., 2018).

In other study bamboo was used to reduce the risk of chromium exposure from lands containing large volumes of Cr-contaminated waste, which were burnt and disposed since decades. Cr in the contaminated site exceeded the recommended limit of 100 mg/kg and was significantly higher than the control soils. The results showed 100% survival rate of all the species of the plant compared to other plants. The bamboo species demonstrated high phytoextraction potential, thus advocating its application in phytoremediation (Were et al., 2017).

2.7 Napier grass case

Napier grass or *Pennisetum purpureum* is a C4 tropical species, having lofty dry matter productivity, resistance to biotic stresses and sustainability over several years. In recent times, the grass has accomplished great attention due to its prospective nature as a bio-ethanol feedstock. Additionally it has high metal phytoextraction potential due to its well-developed root systems and biomass-producing tendency. This was shown by a study done by Ma et al. (2016). They used hybrid giant Napier grass to examine the removal of metals from contaminated tailings. The plants were maintained for 2 year, which could remove 12%–26% of metals (Zn, Cu, Mn, Pb, Cr, Cd, As) from the tailings compared to control soil (16%–74%). Results showed that most of the metals were accumulated in the plant's shoot due to its larger biomass, yet the highest concentration of the metals was found in its roots. The plants harvested from the polluted tailings were having half amount of biomass compared to the control due to metal phytotoxicity of the former. However, the amount of biomass produced in contaminated condition was significant enough to be used for energy production in large-scale projects.

In another study, Napier grass cultivar Wruk Wona was grown as a crop on cadmium-contaminated soils under two harvesting schedules. The results showed that the yearly dry matter productivity was not hampered significantly by two harvesting schedules, but the concentration of Cd and its uptake was higher after second harvesting; higher content of Cd was observed in the herbage from the second cutting. This resulted in decrease in the soil Cd level by 4.6% (Ishii et al., 2015).

2.8 Miscanthus grass case

Miscanthus is also a C4 grass, native to Asia and is relatively cold tolerant compared to other C4 species. It gives high productivity from low input requirements, which make it an excellent candidate for bioenergy feedstock production. It can be widely grown on heavy metal contaminated soils as a biomass crops to create an alternative commercial opportunity as the option for growing of food crops on such sites is ruled out. To test the plant's phytoremediation potential, a study was done with seed based *Miscanthus* hybrids and commercial clones on agricultural fields, contaminated with

Cd, Pb, and Zn. Randomized plots were designed with a density of 2 plants per square meter. After the growing season plants were harvested 2 times in autumn (October) as well as winter (March) to ascertain changes in heavy metal uptake. The results showed their ability to persist on heavy metal polluted soil, but indicated their sensitivity to frost. However, the study recommended high probability that the tested genotypes could perform better in winter under more suitable conditions (Krzyzak et al., 2017).

In another study, the efficiency of *Miscanthus* to grow on extremely marginal soils was tested. The study site was a land degraded by lignite mining. Field experiments were conducted in three blocks located on mining waste dumps. The site was monitored for the first 3 years after *Miscanthus* cultivation, for the changes in plant growth and yield due to fertilization. The types of treatment used in addition to the control were sewage sludge or sewage sludge + mineral fertilizer or sewage sludge + twice the dose of mineral fertilizer. The results indicated significant differences and increase in plant biomass production every year. The biomass production in control was 0.96 kg/plant while that under sewage sludge treatment was 1.5 kg/plant. The plant growth also improved the organic carbon levels and decreased the overall pH (Jezowski et al., 2017).

2.9 Switchgrass case

Switchgrass or *Panicum virgatum* is a well-known, high biomass yielding perennial grass often found occurring on marginal agricultural areas with minimal inputs. Its bioremediation potential for lead and cadmium was analyzed in the presence of related microbes like arbuscular mycorrhizal fungi and *Azospirillum* in pot trials. Results indicated that the microbes helped in improving the root length, root, and shoot biomass of the plant. In addition, the bioconcentration and translocation factors for metals were 0.23–0.25 and 16–17, respectively, which was also 45% higher than control. Lower values of BCF at highest concentration of metals revealed the capability of the grass for phytostabilization of metals (Arora et al., 2016).

In another study, phytoremediation potential of the grass for cadmium, chromium, and zinc was analyzed through considering factors like plant metal content and biomass yield. Using these data linear and exponential decay models was prepared in a study, which is more appropriate for understanding the relationship between these two factors. Through these models, the maximum amount of Cd extracted by the grass approached to 40 μg pot^{-1}, while that for chromium, and zinc was 56 μg/pot and 358 μg/pot, respectively. It was also indicated to follow the optimal time of harvest to obtain maximum metal removal. Hence, it is feasible to implement switchgrass for metal phytoextraction from contaminated sites (Chen et al., 2012).

2.10 Reed canary grass case

Phalaris arundinacea or reed canary grass can also efficiently accumulate various trace metals in its tissues, and for this reason it is often used for biomonitoring as well as phytoremediation purposes. Thus, in a study, Polechonska and Klink (2014) analyzed Zn, Mn, Fe, Pb, Cu, Ni, Cd, Co, and Cr concentration in different tissues of the grass, along with the water and bottom sediments where the grass was found to grow along the Bystrzyca River; the grass mainly grew in parts of the river which was highly influenced by anthropogenic activities. The metal concentrations in plant samples were more in root followed by leaf and stem. This limited translocation of the metals from roots to shoots makes it important for phytostabilization of metal contaminated bottom sediments. Highest levels of Zn and Pb were found in grasses obtained from agricultural fields and cities while the lowest metal concentration was observed in plants obtained from dams, reservoirs, and forests.

In another study, *P. arundinacea*, along with other plants (*Salix viminalis, Zea mays*) was used to see their effectiveness for nickel phytoremediation from contaminated soil. Thus, a 2 year plot experiment was conducted with above plants on Ni contaminated Haplic Luvisol soil which had 0, 60, 100, or 240 mg/kg of Ni. The plants were found to phytostabilize the metal; for example, *Z. mays* accumulated high levels of Ni in its roots, but showed little tolerance. Thus *Z. mays* can be grown on moderate contaminated soil so that the plant yield is not affected and the biomass can be safely used as forage. On the other hand, *S. viminalis* and *P. arundinacea* accumulated very less Ni in their roots and demonstrated high tolerance to Ni. Thus the latter plants can be grown on highly contaminated soil so the contamination can be stopped from spreading (Korzeniowska and Stanislawska-Glubiak, 2018).

2.11   Festuca rubra *grass case*

Festuca rubra L. (Red fescue) has high adaptive capacity to various kinds of harsh environments, thus it is a suitable species for ecorestoration and large-scale reclamation purpose. This was shown by a long-term research done by Gajic et al. (2016) to see the adaptive response of *F. rubra* and its remediation potential on fly ash dumps of a power plant in Serbia. They carried out a field research on passive fly ash lagoons of 3–11 years old compared to a control site, that is, a Botanical Garden in Belgrade. The fly ash was having alkaline pH, low nitrogen, Mn, Zn, and high concentration of As, B. However, the grass growth marked notable improvement in the earlier-mentioned characteristics over time, which was reflected in the higher values of hygroscopic water, clay, silt, adsorbed bases, and reduced concentrations of total and available As, Cu, B, Mn, and Zn. The plant is a metal excluder and retains the metals in its extensive roots compared to the leaves and in turn stabilizes loose material like fly ash. The grass also demonstrated large amount of MDA, and lesser values of chlorophyll as well as total carotenoids, which indicated oxidative stress in the cells due to fly ash toxicity. This also proves active adaptive mechanisms of the plant to sustain the harsh environment as reported by the increased levels of anthocyanins, ascorbic acid, phenolics, and total radical scavenging activity in the cells.

In another experiment, the suitability of the grass was tested among five plant species under the temperate climate of Canada. Laboratory scale pot experiments were carried out for the purpose and the other four plants were *Lolium perenne, Helianthus annuus, Poa pratensis,* and *Brassica napus*. The soils used in the pots were treated with different metals like Cu, Mn, Pb, and Zn. The plants were harvested after 90 and

120 days after sowing for detailed metal analysis. The results showed that *Lolium* was better for phytostabilization of Pb and Cu, *Poa* for Zn and *Festuca* for Mn. *Festuca* also showed maximum accumulation for Cu, Poa for Mn and Helianthus for Zn and Pb. Metal removal was highest after 120 days than compared to the harvested plants after 90 days while their concentrations were more in roots than shoots (Padmavathiamma and Li, 2009).

2.12   Saccharum *species grass case*

Saccharum spontaneum L. is a perennial tall grass, which invades abandoned lands in tropical countries and is multifunctional in nature. In recent years, it has attracted serious attention for ecological restoration of waste lands mainly fly ash dumps, due to its propagation properties which makes it a dominant plant in the habitat. Pandey et al. (2015) conducted a grass vegetation study on a naturally colonized fly ash dump in Uttar Pradesh, India and reported similar results. They also reported improvement in physicochemical properties of fly ash due to grass growth, which suggested its ecological suitability for such areas.

Apuan et al. (2018) from their study reported that *S. spontaneum* to be tolerant to high concentrations of chromium and can be considered as a characteristic metallophyte for phytostabilization of the metal. They concluded the earlier-mentioned results after analysis on 10 plant species, which included *S. spontaneum*. These plants were found growing on a ferrochromium stockpile aged 5 years, in Manticao, Misamis Oriental. Their results reported a Cr concentration of 811.0 mg kg^{-1} in the grass roots which outperformed other plant species in phytostabilization of the waste.

2.13   Sewan grass case

Sewan grass (*Lasiurus scindicus*) is a perennial grass that can live up to 2 decades. It is a bushy, multi-branched desert grass with ascending to erect wiry stems, height up to 1–1.6 m. It is the most common grass in north India, and is popular in dry areas of North Africa, Sudan and Sahelian regions, East Africa, and Asia. It grows mainly in the Rajasthan desert area. Being rich in protein it is used as a fodder. Optimal growth conditions are found on sandy soils or light sandy soils having pH of 8.5 as well as on yearly rainfall below 250 mm. It is highly tolerable of drought. Sharma and Pandey (2017) conducted a Sewan grass based phytoremediation of Pb contaminated soil and water over a 105 day pot trial period. It was noticed that high accumulation level of lead in root after 45 and 65 days. The results showed that Sewan grass has phytoremediation potential and could be used in restoring Pb polluted soil as phytostabilizer.

Tarafdar and Rao (1997) observed that Sewan grass has potential to grow naturally on gypsum mined spoil. The root infection (percentage) by AM fungi was higher in Sewan grass grown on mined spoils than normal soil. Grass grown on mined spoil had significantly higher K, Ca, Mg, and Fe concentrations and lower concentrations of N, P, Zn, and Cu. The results revealed the possibility of using AM fungi in Sewan grass for rehabilitating waste dumpsites like gypsum mined spoils.

3   Conclusion and future prospects

The researchers across the world have made several phytomanagement of polluted sites with perennial grasses over the years of practices and have gained useful insights from the evaluation of their phytoremediation programs. C4 grasses are more efficient than C3 grasses. The C4 grasses fix CO_2 more effectively, and have some benefits over C3 grasses: higher yield, higher water use efficiency, more resistant to periodical shortage of water in soil, and are not susceptible to lodging. Grasses are used to produce second-generation bioethanol and biodiesel from lignocellulosis. Besides, the perennial grasses have been well identified for phytoremediation of polluted sites with multiple benefits from ecologically to socio-economically. In Indian context, *Saccharum munja, S. spontaneum*, and *Vetiveria zizanioides* are underutilized and neglected grass species. These grasses have a lot of potential toward phytoremediation with economic returns in term of phytoproducts, biomass, carbon sequestration, biodiversity conservation, climate change mitigation, and ecosystem restoration. We suggest for starting wide research in phytoremediation programs for suitable utilization of the potential of these grasses in the sustainability of environment and human well-being.

Acknowledgment

Financial assistance given to Dr. V.C. Pandey under the Scientist's Pool Scheme (Pool No. 13 (8931-A)/2017) by the Council of Scientific and Industrial Research, Government of India is gratefully acknowledged. The authors declare no conflicts of interest.

References

Apuan, D., Fruto, J., Perez, T., Claveria, R.J., Tan, M., Apuan, M.J., Perez, E., 2018. Phytostabilization potential of *Saccharum spontaneum* in chromium contaminated soil. JBES 13 (4), 116–123.

Arora, K., Sharma, S., Monti, A., 2016. Bio-remediation of Pb and Cd polluted soils by switchgrass: a case study in India. Int. J. Phytoremediation 18 (7), 704–709, DOI: 10.1080/15226514.2015.1131232.

Basumatary, B. and Bordoloi, S., 2016. Phytoremediation of Crude Oil-Contaminated Soil Using Cynodon dactylon (L.) Pers. In: Phytoremediation (pp. 41-51). Springer, Cham.

Cerne, M., Smodis, B., Strok, M., 2011. Uptake of radionuclides by a common reed (*Phragmitesaustralis* (Cav.) Trin. ex Steud.) grown in the vicinity of the former uranium mine at Zirovskivrh. Nucl. Eng. Des. 24, 1282–1286.

Chen, B.C., Lai, H.-Y., Juang, K.W., 2012. Model evaluation of plant metal content and biomass yield for the phytoextraction of heavy metals by switchgrass. Ecotoxicol. Environ. Saf. 80, 393–400.

Das, M., Maiti, S.K., 2009. Growth of *Cymbopogon citratus* and *Vetiveria zizanioides* on Cu mine tailings amended with chicken manure and manure-soil mixtures: a pot scale study. Int. J. Phytoremediation 11 (8), 651–663.

Emamverdian, A., Ding, Y., Mokhberdoran, F., Xie, Y., 2018. Antioxidant response of bamboo (*Indocalamus latifolius*) as affected by heavy metal stress. J. Elem. 23 (1), 341–352.

FAO and ITPS, 2015.Status of the World's Soil Resources (SWSR)—Main Report. Food and Agriculture Organization of the United Nations and Intergovernmental Technical Panel on Soils, Rome, Italy (ISBN 978-92-5-109004-6).

Gajic, G., Djurdjevic, L., Kostic, O., Jaric, S., Mitrovic, M., Stevanovic, B., Pavlovic, P., 2016. Assessment of the phytoremediation potential and an adaptive response of *Festucarubra* L. sown on fly ash deposits: native grass has a pivotal role in ecorestoration management. Ecol. Eng. 93, 250–261.

Gautam, M., Pandey, D., Agrawal, M., 2017. Phytoremediation of metals using lemongrass (*Cymbopogon citratus* (DC) Stapf.) grown under different levels of red mud in soil amended with biowastes. Int. J. Phytoremediation 19 (6), 555–562.

Granley, B.A., Truong, P.N., 2012. A changing industry: on-site phytoremediation of landfill leachate using trees and grasses–case studies. In: Global waste management symposium 2012, September 30–October 3, 2012, Arizona Grand Resort, Phoenix, USA.

Hechmi, N., Ben Aissa, N., Abdenaceur, H., Jedidi, N., 2015. Uptake and bioaccumulation of pentachlorophenol by emergent wetland plant *Phragmites australis* (common reed) in cadmium co-contaminated soil. Int. J. Phytoremediation 17 (2), 109–116.

Ishii, Y., Hamano, K., Kang, D.J., Idota, S., Nishiwaki, A., 2015. Cadmium phytoremediation potential of Napier grass cultivated in Kyushu, Japan. Appl. Environ. Soil Sci. 2015 (1), 1–6.

Jezowski, S., Mos, M., Buckby, S., Cerazy-Waliszewska, J., Owczarzak, W., Mocek, A., Kaczmarek, Z., McCalmont, J.P., 2017. Establishment, growth, and yield potential of the Perennial grass *Miscanthus × giganteus* on degraded coal mine soils. Front. Plant Sci. 726, 1–8, doi: 10.3389/fpls.2017.00726.

Koelmel, J., Prasad, M.N.V., Pershell, K., 2015. Bibliometric analysis of phytotechnologies for remediation: global scenario of research and applications. Int. J. Phytorem. 17 (2), 145–153.

Korzeniowska, J., Stanislawska-Glubiak, E., 2018. Phytoremediation potential of *Phalaris arundinacea*, *Salix viminalis*, and *Zea mays* for nickel-contaminated soils. Int. J. Environ. Sci. Technol. 16 (4), 1999–2008, https://doi.org/10.1007/s13762-018-1823-7.

Krzyzak, J., Pogrzeba, M., Rusinowski, S., Clifton-Brown, J., Mccalmont, J.P., Kiesel, A., Mangold, A., Mos, M., 2017. Heavy metal uptake by novel *Miscanthus* Seed-based hybrids cultivated in heavy metal contaminated soil. Civil Environ. Eng. Rep. 26 (3), 121–132, DOI: 10.1515/ceer-2017-0040.

Kumar, A., Maiti, S.K., 2015. Effect of organic manures on the growth of *Cymbopogon citratus* and *Chrysopogon zizanioides* for the phytoremediation of chromite-asbestos mine waste: a pot scale experiment. Int. J. Phytoremediation 17 (5), 437–447.

Ma, C., Ming, H., Lin, C., Naidu, R., Bolan, N., 2016. Phytoextraction of heavy metal from tailing waste using Napier grass. Catena 136, 74–83.

Maiti, S.K., Kumar, A., Ahirwal, J., Das, R., 2016. Comparative study on bioaccumulation and translocation of metals in Bermuda grass (*Cynodon Dactylon*) naturally growing on fly ash lagoon and topsoil. Appl. Ecol. Environ. Res. 14 (1), 1–12.

Mench, M., Lepp, N., Bert, V., Schwitzguebel, J.P., Gawronski, S.W., Schroder, P., 2010. Successes and limitations of phytotechnologies at field scale: outcomes, assessment and outlook from COST Action 859. J. Soils Sediments 10, 1039–1070.

Mitrovic, M., Pavlovic, P., Lakusic, D., Djurdjevic, L., Stevanovic, B., Kostic, O., Gajic, G., 2008. The potential of *Festuca rubra* and *Calamagrostis epigejos* for the revegetation of fly ash deposits. Sci. Total Environ. 407 (1), 338–347.

Padmavathiamma, P.K., Li, L.Y., 2009. Phytoremediation of metal-contaminated soil in temperate humid regions of British Columbia, Canada. Int. J. Phytoremediation 11 (6), 575–590.

Pandey, V.C., Bajpai, O., Pandey, D.N., Singh, N., 2015. *Saccharum spontaneum*: an underutilized tall grass for revegetation and restoration programs. Genet. Resour. Crop Evol. 62, 443–450, 10.1007/s10722-014-0208-0.

Pandey, V.C., Bajpai, O., Singh, N., 2016. Energy crops in sustainable phytoremediation. Renew. Sustain. Energy Rev. 54, 58–73.

Pandey, V.C., Pandey, D.N., Singh, N., 2015a. Sustainable phytoremediation based on naturally colonizing and economically valuable plants. J. Clean. Prod. 86, 37–39.

Pandey, V.C., Singh, J.S., Singh, R.P., Singh, N., Yunus, M., 2011. Arsenic hazards in coal fly ash and its fate in Indian scenario. Resour. Conserv. Recycl. 55, 819–835.

Polechonska, L., Klink, A., 2014. Trace metal bioindication and phytoremediation potentialities of PhalarisarundinaceaL. (reed canary grass). J. Geochem. Explor. 146, 27–33.

Randelovic, D., Jakovljevic, K., Mihailovic, N., Jovanovic, S., 2018. Metal accumulation in populations of *Calamagrostis epigejos* (L.) Roth from diverse anthropogenically degraded sites (SE Europe, Serbia). Environ. Monitor. Assess. 190 (4), 183.

Su, F., Wang, T., Zhang, H., Song, Z., Feng, X., Zhang, K., 2018. The distribution and enrichment characteristics of copper in soil and Phragmites australis of Liao River estuary wetland. Environ. Monitor. Assess. 190 (6), p. 365.

Were, F.H., Wafula, G.A., Wairungu, S., 2017. Phytoremediation using bamboo to reduce the risk of chromium exposure from a contaminated Tannery site in Kenya. J. Health Pollut. 7 (16), 12–25, doi: 10.5696/2156-9614-7.16.12.

Tarafdar, J.C., Rao, A.V., 1997. Mycorrhizal colonization and nutrient concentration of naturally grown plants on gypsum mine spoils in India. Agric. Ecosyst. Environ. 61, 13–18.

Sharma, P., Pandey, S., 2017. Phytomanagement of heavy metal lead by Fodder grass *Lasiurus scindicus* in polluted soil and water of Dravyawati river. Am. J. Environ. 13 (2), 167–171, DOI: 10.3844/ajessp.2017.167.171.

Index

Note: Page numbers followed by "f" indicate figures and "t" indicate tables.

A

Abiotic stress tolerance, Bermuda grass, 230
Accumulator plants, 1. *See also* Metal hyperaccumulator(s)
Acrylamide, removal, vetiver grass in, 37
Adaptive agricultural practices, vetiver grass, 47
Adsorption, 1
Aeluropus lagopoides, 98
Aerva persica, 68
Agency for Toxic Substances and Disease Registry, 319
Agrochemicals, phytoremediation of, vetiver grass in, 37
Agroforestry system, bamboo-based, 251, 252f
Agropyron, 119
Agropyron repens, 120, 122
Agropyron smithii (Western wheat grass), 5t, 305
Agrostis, 119
Agrostis alba (European grass), 5t
Agrostis capillaries, 122
Agrostis castellana (colonial bent grass), 5t
Agrostis curtisii, 122
Agrostis durieui, 122
Agrostis palustris, 120
Agrostis stolonifera, 122
Agrostis tenuis (colonial bent grass), 5t
Agrotechnical technology, 130, 131f
    F. rubra L. on fly ash deposits (TENT, Serbia), 130, 131f
    hand sowing, 130
    hydroseeding, 130
    machine sowing, 130
Alfalfa, intercropped with tall fescue, 11
Aliphatic compounds, 313
Alopecurus, 119

Alopecurus pratensis (meadow fox-tail), 122, 306
Amaranthus ascendens, 98
American Academy of Environmental Engineers, 338
Aminopolycarboxylic acid (APCA), 308
Ammophila breviligulata (beach grass), 305
Andropogon, 119
Andropogon gerardii (big bluestem), 9, 19, 120, 181, 305
Andropogon scoparius (little blue-stem), 305
Animal bedding, *Miscanthus* use in, 88
Animal shades and huts, perennial grasses for, 14
Annual grass(es), 306. *See also* Biennial grass(es); Perennial grass(es)
Anthocyanins, in *F. rubra* L. leaves grown on fly ash deposits, 140t, 142
Anthoxathum, 119
Antibiotics, phytoremediation, vetiver grass in, 37
Antioxidant(s), *F. rubra* L. grown on fly ash deposits, 139, 140t, 142
APCA. *See* Aminopolycarboxylic acid (APCA)
Arbuscular mychorrhizal fungi (AMF), 67, 69f, 185, 303
Archanara geminipuncta, 108
Arctagrostis latifolia, 18
Arctophila fulva, 18
Areca catechu L. (Arecanut), 47
Arecanut (*Areca catechu* L.), 47
Aristida setifolia, 122
Aromatic grasses, 2, 4, 306. *See also* Perennial grass(es); *specific names*
    Cymbopogon flexuosus. See Cymbopogon flexuosus (lemon grass)
    essential oils from, 17
    metal accumulation, 195
    for phytoremediation, 11, 165

Aromatic pollutants, phytoremediation of, 313
Arrenatherum, 119
Arrhenatherum elatius, 122
Arsenate (AsV), 133
Arsenic (As), 115, 319
　in *F. rubra* L. grown on fly ash deposits,
　　133, 134t
　toxicity, 133
Arsenite (AsIII), 133
Arthrocnemum macrostachyum, 98
Arundo donax (giant reed), 6t, 9, 16, 19, 21,
　　303, 307
　biomass yields, 307
　in pulp and paper manufacturing, 17
Asbestos, 115
Ascorbic acid (AsA), in *F. rubra* L. leaves
　　grown on fly ash deposits, 140t, 142
Aster squamatus, 98
Atrazine, 37, 185, 319
Atriplex canescens, 98
Avena, 119
Avena sativa, 181
Avena sterilis, 122
Axonopus grass, 9
Ayurveda, 239
Azolla caroliniana, 165
Azolla filiculoides, 98
Azospirillum, 74
　in rhizosphere of sewan grass, 74, 75f
Azotobacter, 74
　in rhizosphere of sewan grass, 74, 75f

B

Bacillus endophyticus NBRFT4, 310
Bacillus pumilus NBRFT9, 310
Bamboo. *See Phyllostachys pubescens* (Moso
　　bamboo)
"Bamboo grove", 251
Bamboo-provisioned ecosystem services, 247
Bambusa bambos, 250
Bambusa blumeana, 246
Bambusa oldhamii, 250
Bambusa sp, 4
Bambusoideae, 245
Bassia indica, 98
Bauxite residue (red mud), phytoremediation
　　of, 202
BCF. *See* Bioconcentration factor (BCF)
Beach grass *(Ammophila breviligulata)*, 305

Beach rye *(Elymus mollis)*, 305
Beckmania eruciformis, 122
Benzene, 115
Benzene, toluene, ethyl-benzene, and xylene
　　(BTEX), 313
Bermuda grass *(Cynodon dactylon). See
　　Cynodon dactylon* (Bermuda grass)
Biennial grass(es), 306. *See also* Annual
　　grass(es); Perennial grass(es)
Big bluestem *(Andropogon gerardii). See
　　Andropogon gerardii* (big bluestem)
Bioconcentration factor (BCF), 117, 132
Biodiesel, HMs in, 325
Biodiversity conservation, perennial grasses,
　　12
Bioenergy production
　Calamagrostis epigejos, 273
　Napier grass, 294
　perennial grasses as source, 12, 16
　Saccharum spp., 220
　switchgrass, 187
Bioethanol, production
　Napier grass, 295
　Saccharum spp., 216f, 218
Bio-fuel(s), 17
　from Bermuda grass, 237
　from biomass of perennial grasses, 302f,
　　303
　herbaceous plants for, 179
　Miscanthus as, 83
Biofuels Feedstock Development Program, 187
Biogas production
　Napier grass, 295
　Phragmites species, 108
Bioindicator, reed canary grass as, 169
Biomass, from grasses, 15, 307
　Bermuda grass, 237
　bio-fuel production from, 302f, 303
　Calamagrostis epigejos, 267, 273
　giant reed, 307
　Miscanthus, 83, 84, 307
　Napier grass, 294
　Phragmites species, 108, 109f
　productivity, sewan grass, 70, 72f
　reed canary grass, 170, 171
　Saccharum spp., 220
　sustainability aspects, 19
　switchgrass, 187
　vetiver grass, 51, 52

Biomass cropping, 3
Biomass yield, reed canary grass, 170
Bio-oil production, Napier grass, 295
Biorecultivation, 129
    of *F. rubra* L. on fly ash deposits (TENT,
        Serbia), 129
        grass-legume mixture, fertilization, and
            agrotechniques, 130t
Bioremediation, 31. *See also*
        Phytoremediation; Remediation
        technology(ies)
Blue gamma grass *(Bouteloua gracillis). See*
        Bouteloua gracillis (blue gamma grass)
Boron (B)
    in *F. rubra* L. grown on fly ash deposits,
        134t, 136
    role of, 136
    toxic effects, 136
Botanical description. *See* Morphology
Bouteloua curtipendula, 120
Bouteloua gracillis (blue gamma grass), 5t,
        120, 305
Brachiaria, 119
Brachiaria mutica (paragrass), 5t
Brassica campestris, 322t
Brassicaceae, 248
Brassica juncea (Indian mustard), 308, 322t
Brassica napus, 322t, 343
Brassica rapa, 322t
Briza maxima, 122
Broad bean *(Vicia faba)*, 310, 312
Bromosopsis inermis, 122
Bromus, 119
Bromus catharticus, 122
Bromus hordeaceus, 122
BTEX. *See* Benzene, toluene, ethyl-benzene,
        and xylene (BTEX)
Buchloe dactyloides (buffalo grass), 5t, 120
Budding, vetiver propagation, 34
Buffalo grass *(Buchloe dactyloides)*, 5t, 120
Building materials, *Miscanthus* use in, 88

C

Cadmium (Cd), 115, 319
    removal, by switchgrass, 184
Calamagrostis, 119
Calamagrostis canadensis, 18
Calamagrostis epigejos (Eurasian grass), 4,
        6t, 9, 18, 120f, 122

biomass energy, 273
biomass production, 267, 273
case study, 340
communities, 264
distribution and expansion, 265f, 266f
ecological preferences, 264
ecology, 261
Festuca rubra and, 340
limitations, 263, 275
morphology, 261, 262f
other uses, 273
overview, 259
phytoremediation, 11, 268, 269f, 271f,
        272f
propagation, 261
as renewable energy source, 273
reproduction, 261
restoration of degraded environments, 10
suppression and control, 267
tolerance and adaptation, 271
Calamovilfa longifolia (sand-reed), 305
Canadian wild rye *(Elymus canadensis)*, 5t
Carbamazepine, 319
Carbon dioxide (CO_2), 12
Carbon neutral plant, *Miscanthus as*, 88. *See
        also Miscanthus*
Carbon sequestration
    bamboo in, 250
    Bermuda grass in, 238
    C. flexuosus in, 203
    Miscanthus, 88
    by perennial grasses, 13
    Phragmites species in, 107
    reed canary grass in, 171
    Saccharum spp. in, 219
    switchgrass in, 188
    vetiver grass in, 46
Carex albula, 9
Carex aquatalis, 18
Carex chungii, 9
Carex muskingumensis, 9
Carotenoids, total, in *F. rubra* L. leaves
        grown on fly ash deposits, 140t, 141
Case study(ies), perennial grasses
    bamboo grass, 340
    Bermuda grass, 338
    Calamagrostis epigejos grass, 340
    Festuca rubra L. grass, 343
    future prospects, 345

Case study(ies), perennial grasses (*cont.*)
  heavy metal phytoremediation, 322t
  lemon grass, 339
  Miscanthus grass, 341
  Napier grass, 341
  overview, 337
  phragmites grass, 339
  reed canary grass, 342
  Saccharum spp. grass, 344
  Sewan grass, 344
  switchgrass grass, 342
  vetiver grass, 338
CDTA. *See* Trans-1,2-cyclohexylene dinitrilo
      tetraacetic acid (CDTA)
CEED. *See* Crop, Expansion, Encapsulation,
      and Delivery system (CEED)
Cenchrus ciliaris, 16, 63, 122
Centrosema pubescens (butterfly pea), 286
Ceratophyllum demersum, 98
Cereals, 306
Certified emission reductions, 247
Cesium (^{137}Cs), 45
C3 grasses, 19, 306. *See also specific names*
C4 grasses, 19, 63, 65, 81, 83, 84, 182, 188,
      212, 289, 306. *See also specific
      names*
Chasmanthium latifolium, 9
Chelating agents, 308, 320. *See also*
      Phytochelatins (PCs)
  biodegradable, 308
Chenopodium sp., 98
Chernobyl disaster, 45
Chloris, 119
Chloris gayana, 120
Chloris virgate Sw., 63
Chlorobenzenes (CBs), 293
Chlorophylls, content in *F. rubra* L. leaves
      grown on fly ash deposits, 139, 140, 141
Chlorosis, 143
Chromite-asbestos mine, phytoremediation
      of, lemon grass, 199
Chromium (Cr), 115, 319
  removal, by switchgrass, 185
Chromium (VI) toxicity, lemon grass and, 198
Chrysopogon zizanioides L. Roberty. *See*
      Vetiveria zizanioides L. Nash
Citronella grass *(Cymbopogon nardus). See*
      Cymbopogon nardus (Citronella
      grass)

Clean Development Mechanism (CDM), 247
Cleanup technology, 116
Climate change, 80
Climate change mitigation
  bamboo in, 250
  perennial grasses and, 12
C4 NADP-ME type anatomy, leaves, 65
CO_2. *See* Carbon dioxide (CO_2)
Cobalt (Co), 319
Cochin grass, 197
Cocksfoot grass *(Dactylis glomerata). See*
      Dactylis glomerata (cocksfoot grass)
Coelachyrum brevifolium Hochst. & Nees., 63
Collar propagation. *See also* Propagation
  Miscanthus, 82f, 83
Colonial bent grass
  Agrostis castellana, 5t
  Agrostis tenuis, 5t
Common reed, 21
  *Phragmites australis. See Phragmites
      australis* (common reed)
  *Phragmites communis. See Phragmites
      communis* (common reed)
Commonwealth Scientific and Industrial
      Research Organisation (CSIRO), 287
Complexation, 1
Conjugation, 313
Conservation reserve program (CRP)
      lands, 188
Contaminants. *See also* Pollutants; *specific
      names*
  examples, 115
  removal
    methods, 79
    Miscanthus, 85
    vetiver grass for, 31
Contaminated sites
  cleanup solutions, management, and
      assessment, 116
  Miscanthus in phytoremediation, 86t
  perennial grasses on, 119f
  phytoremediation, *F. rubra* L. in, 132
    fly ash deposits, 133, 134t
    metal(loid)s, 132
  technical reclamation of, 128
Contamination. *See also* Pollution
  causes, 79, 115
  cleanup solutions, management, and
      assessment, 116

heavy metal, 79. *See also* Heavy metal(s)
land. *See* Land contamination
Conventional soil remediation methods,
    limitations, 303
Convolvulus arvensis, 98
Cool-season grasses. *See* C3 grasses
Copper (Cu), 115, 319
    deficiency, 136
    in *F. rubra* L. grown on fly ash deposits,
        134t, 136
    removal, by switchgrass, 185
    role of, 136
Copper mine tailings, phytostabilization of,
    lemon grass, 200
Cortaderia selloana, 9
Cosmetics, vetiver grass in, 50
Crafts, grasses as raw material for, 14
Crop, Expansion, Encapsulation, and
        Delivery system (CEED), *Miscanthus*
        propagation, 82
Crude oil, remediation, grass species for, 120
Cultural programs, perennial grasses in, 15
Cuttings and culms method, vetiver
        propagation, 34
Trans-1,2-cyclohexylene dinitrilo tetraacetic
        acid (CDTA), 308
Cymbopogon citratus, 122, 196, 197, 201,
        306
    morphology, 198
    for phytoremediation, 11
    restoration of degraded environments, 10
Cymbopogon flexuosus (lemon grass), 6t, 16
    attributes and uses, 196t
    botanical description, 198
    chromium (VI) toxicity and, 198
    cultivation, 196
    ecology, 196
    ecorestoration of dumping site rich in
        sulfur content, 201
    future prospects, 205
    growing on degraded land, 197f
    implementation strategies, 205
    lemongrass-legume cover establishment on
        degraded land for stabilization, 202
    medicinal use, 204
    multiple uses, 199
        carbon sequestration, 203
        essential oil production, 202
        phytoremediation, 199f

origin and distribution, 197f
other commercial uses, 204
overview, 195
phytoremediation, 198, 199
    of bauxite residue (red mud), 202
    of chromite-asbestos mine, 199
phytostabilization of copper mine
        tailings, 200
propagation, 198
reclamation and re-vegetation of fly ash
        disposal sites, 201
socio-economic development, 204
Cymbopogon genus, 2, 196
Cymbopogon martinii (ginger-grass), 306
Cymbopogon nardus (Citronella grass), 197,
        306
in pulp and paper manufacturing, 17
Cynodon aethiopicus, 122
Cynodon dactylon (Bermuda grass), 4–6, 9,
        10, 13, 16, 98, 120
    abiotic stress tolerance of, 230
    in ayurvedic medicines, 239
    case study, 338
    common names, 228
    drought stress in, 230
    ecology, 228
    ethanolic extracts, 239
    flooding-induced responses, 230
    geographical distribution and occurrence,
        228, 229f
    grass-legume cover, 234
    metal accumulation potential of, 234
    morphology, 229
    multiple uses, 232, 233t
        biomass and biofuel, 237
        carbon sequestration, 238
        green capping, 237
        in Hindu rituals, 238, 239f
        in medicine, 239
        phytoremediation of toxic elements,
            234–236
        soil conservation, 237
        wasteland's restoration, 232
    nitric oxide application, 230
    origin, 228
    overview, 227
    propagation, 229, 230f
    salinity stress, 231
    temperature range, 231

Cynodon sp., 119, 312
Cyperus esculentus, 10, 13
Cytochrome P450 monooxygenases, 314

D

Dactylis, 119
Dactylis glomerata (cocksfoot grass), 19,
    120, 122, 306
Dactyloctenium aegyptium, 10
Decision support systems (DSS), 325
Dendrocalamus sp., 4
Derelict lands, green cover development,
    Phragmites, 107
Desert plantsm, rhizosphere microorganisms
    of, 64
Desmodium intortum, 286
Detoxification, of pollutants, 313
    conjugation, 313
    sequestration, 313
    transformation, 313
Dianthus gratianopolitanus, 9
Dichanthium annulatum, 122
Diesel, 115
Diethylene triamino pentaacetic acid
    (DTPA), 308
Digitaria, 119
Digitaria ciliaris, 120
Digitaria spp. (crabgrass), 304
Dinitrotoluene, 318
1,1-diphenyl-2-picrylhydrazyl (DPPH), 142
D1 protein, 141
Dropseed *(Sporobolus asper)*, 305
Drought stress, Bermuda grass, 230
DTPA. *See* Diethylene triamino pentaacetic
    acid (DTPA)

E

East Indian lemongrass, 197
Echinochloa crusgalli (Barnyard grass), 304
Ecological aspects, perennial grasses, 10
    biodiversity conservation, 12
    carbon sequestration, 13
    climate change mitigation, 12
    habitat corridors, 14
    phytoremediation, 11
    restoration, 10
    soil erosion control, 13
    wild life shelter, 13

Ecology. *See also* Morphology
    Bermuda grass, 228
    C. flexuosus, 196
    Calamagrostis epigejos, 261
    Napier grass, 286
    red fescue, 125, 126f
        habitats and plant communities, 126
    reed canary grass, 167
    Saccharum spp., 212
    sewan grass, 64
    switchgrass, 180, 181f
    vetiver grass, 35
Economic aspects, perennial grasses,
    15. *See also* Phytocommerce/
        phytoeconomics
    bioenergy feedstock, 16
    essential oil, 17
    industrialization, 17
    job creation and poverty alleviation, 17
    low input and minimum maintenance, 15
    medicinal use, 16
    pulp and paper manufacturing, 17
Economic return. *See also* Phytocommerce/
        phytoeconomics
    from vetiver grass, 48, 50f
        energy, 51
        essential oil, 48
        medicines, 50
        other uses, 51
        perfumery and cosmetics, 50
Ecophysiological adaptation, 149
Ecorestoration, 117. *See also* Restoration
    of degraded land, nitrogen role in, 202
    described, 118
    of dumping site rich in sulfur content,
        lemon grass in, 201
    F. rubra L. (techniques), 128
        agrotechnical technology, 130, 131f
        biorecultivation on fly ash deposits
            (TENT, Serbia), 129
        seed production, 129
    of fly ash dumps, *Saccharum* spp. in, 220
    perennial grasses on contaminated
        sites, 119f
    Phragmites species, 106
    of waste land, *Saccharum* spp. role in,
        218, 219f
Ecosystem services, bamboo-provisioned,
    247

EDDHA. *See* N'-bis (2-hydroxyphenyl)
acetic acid (EDDHA)
EDDS. *See* Ethylene diamine disuccinate
(EDDS)
EDS program (energy-dispersive
spectrometry), 145
EDTA. *See* Ethylene diamine tetraacetic acid
(EDTA)
Elephant grass. *See Pennisetum purpureum*
(Napier grass)
Elymus, 119
Elymus canadensis (Canadian wild rye), 5t
Elymus mollis (beach rye), 305
Elymus virginicus (Wild rye), 305
Elytrigia, 119
Endophytes, 310
PAHs degradation, 314
Energy. *See also* Bioenergy
demand for, 80
renewable source of, 80
vetiver grass as source of, 51
Energy-dispersive spectrometry (EDS), 145
English ryegrass *(Lolium perenne). See
Lolium perenne* (English ryegrass)
Enterobacter cloacae, 310
Enzymes, 314
Epilobietea angustifolli, 264
Epilobium angustifolium, 18
Equisetum arvense, 18
Eragrostis curvula (weeping lovegrass), 19
Eragrostis nutans, 10
Eremochloa, 119
Erianthus ravennae, 9
Eriophorum augustifolium, 18
Erosion control, red fescue for, 127, 128
Essential oil(s), 2, 17
aromatic grasses (examples), 306
from *C. flexuosus*, 202
form perennial grasses, 17
from vetiver grass, 48
Esthetic uses, of grasses, 15
Ethidium bromide (EtBr), remediation,
vetiver grass in, 37
Ethylene diamine disuccinate (EDDS), 308
Ethylenediamine-N, 308
Ethylene diamine tetraacetic acid (EDTA),
308, 320
Eulaliapsis binata (Sabai), 15, 122
in pulp and paper manufacturing, 17

EUNIS classification, 126, 127
Eurasian grass *(Calamagrostis epigejos). See
Calamagrostis epigejos* (Eurasian
grass)
EU Reed Canary Grass Project (1995-99),
172
European Environment Agency, 126
European grass *(Agrostis alba)*, 5t
European OPTIMA project, 220
Evapo-transpirational activity, of green
plants, 303
Excluder grass plants, 1
Excluders, 117
Explosives/nitroaromatics, 318

F
FA. *See* Fly ash (FA)
Facilitation, 10
Feed, *Miscanthus* use for, 88
Festuca, 119
Festuca arundinacea (tall fescue), 5t, 6t, 9,
120, 260, 312, 313
intercropped with alfalfa, 11
Festuca elata, 122
Festuca hervierri, 9
Festuca lemanii, 9
Festuca longifolia, 122
Festuca pratensis (Meadow fescue), 5t, 122
Festuca pseudotricophylla, 122
Festuca rubra L. (red fescue), 4, 5t, 14, 120,
122
Calamagrostis epigejos and, 340
case study, 343
as cover crop in orchards, 127
ecology, 125, 126f
ecorestoration techniques, 128
agrotechnical technology, 130, 131f
biorecultivation on fly ash deposits
(TENT, Serbia), 129
seed production, 129
for erosion control, 127, 128
EUNIS classification, 126, 127
geographical distribution, 123
for grazing, 127, 128
habitats and plant communities, 126
Europe, 126
Serbia, 127
morphology, 123f, 125
multiple uses and management, 127

Festuca rubra L. (red fescue) (*cont.*)
  physiological and morphological response,
      139
    leaf morphology, 143, 144f
    photosynthesis, pigments, and
        antioxidants, 139, 140t
    SEM analysis of leaf surface structure,
        145–147
  in phytoremediation of contaminated
      sites, 132
    fly ash deposits, 133, 134t
    metal(loid)s, 132
  Red List Category and conservation, 124
  reproduction, 125
  in soil erosion control, 13
  subspecies, 124
  taxonomy, 123f
  as turfgrass, 128
Fimbristylis bisumbellata, 10
Fimbristylis umbellata, 13
The Finish Agro-Fiber Project (1993-95),
    172
Flooding-induced responses, Bermuda
    grass, 230
Flower(s), vetiver grass, 33
Fly ash (FA). *See also* Heavy metal(s) (HM);
    Metal(loid)s; Organic pollutants
  characteristics, 220
  deposits and coalmine land in Serbia,
      121f
    biorecultivation of *F. rubra* L. on
        (TENT, Serbia), 129
    grass-legume mixture, fertilization, and
        agrotechniques, 130t
  phytoremediation
    Calamagrostis epigejos, 270
    F. rubra L. in, 133, 134t
    grasses for, 122
    vetiver grass in, 45
  reclamation and re-vegetation of disposal
      sites, 201
  toxic substances in, 220
  uses, 45, 220
Fly ash dumps
  ecorestoration
    Bermuda grass, 234, 235f
    Saccharum spp. in, 220
  reclamation and re-vegetation of disposal
      sites, 201

Fodder, grasses as, 14
  Saccharum spp. for, 216f, 217
  sewan grass, 63, 67, 70
Food crops, 4
Food Safety and Standards Authority of
    India, 11
France Indian verbena, 197
Fuels, 115. *See also* Bio-fuel(s)

G
Gasoline, 115
Genetic diversity, sewan grass, 73
Genetic engineering, 326
Geographical distribution. *See also*
    Propagation
  C. flexuosus, 197f
  and expansion, *Calamagrostis epigejos*,
      265f, 266f
  moso bamboo, 245, 246
  Napier grass, 283f, 284
  and occurrence, Bermuda grass, 228, 229f
  Phragmites species, 97, 99, 100f
  red fescue, 123
  reed canary grass, 166
  Saccharum spp., 215
  sewan grass, 64
  switchgrass, 179, 182
  vetiver grass, 35, 36f
GHGs. *See* Greenhouse gases (GHGs)
Giant reed (*Arundo donax*). *See Arundo
    donax* (giant reed)
Ginger-grass (*Cymbopogon martinii*), 306
Global warming, 12
Glycine max (soybean), 322t
Grading, 128
Gramineae. *See* Poaceae family
Grass communities
  red fescue, 126
  Sewan grass, 68, 71f
Grass(es). *See also* Perennial grass(es);
    specific names
  accumulators, 1
  annual, 306
  aromatic, 2, 4
  biennials, 306
  biomass from, 15
  C3. *See* C3 grasses
  C4. *See* C4 grasses
  characteristics, 2, 9

classification, 1, 304
esthetic uses, 15
excluders, 1
as fodder, 14
indigenous and site-specific, 4
as initial colonizers, 2
metal uptake by roots from soil, 1, 2f
monoculture, 4
native, 3
natural colonization, 3
non-aromatic, 4
ornamental, in park, 9
pasture, 306
perennials, 306
for phytoremediation, 1. *See also*
    Phytoremediation
plantation, 3
as raw material for crafts, 14
regeneration potential, 3
rhizospheric metal bioavailability, 1, 2f
role of, 305
root structure, 2
species, 305
Grasslands, 305
Grass-legume mixture, on fly ash deposits
    (TENT, Serbia), 130t
Grazing
    red fescue for, 127, 128
    sewan grass, 66
Great Indian Desert, 64
Green capping, Bermuda grass in, 237
Green cover development of derelict lands,
    Phragmites, 107
Greenhouse gases (GHGs), 12
Greenhouse trial (Australia), 45
"Green liver" model, 313
Green needlegrass *(Stipa viridula)*, 305
Growth promoting bacteria, 310
Guinea grass *(Panicum maximum)*, 5t, 287

H

Habitat corridors, perennial grasses as, 14
Halocnemum strobilaceum, 98
Hand sowing, 130
Harvesting, *Miscanthus*, 83
Heavy metal(s) (HM), 319, 322t. *See also*
    Fly ash (FA); Metal(loid)s; Organic
    pollutants
    biodiesel in, 325

case studies, 322t
concentration, *L. sindicus*, 67, 70f
contamination, 79
hyperaccumulation, 325
hyperaccumulators. *See* Metal
    hyperaccumulator(s)
most dangerous, 319
necessary, 319
perennial grass species, 303
phytoremediation, 307
    bamboo, 248
    Bermuda grass, 234–236
    Calamagrostis epigejos, 270
    Miscanthus, 85
    oligopeptide ligands production and,
        320
    Phragmites species, 105
    vetiver grass, 36, 307
Helianthus annuus (sunflower), 308, 343
Herbaceous Energy Crops Program
    (HECP), 179
Herbaceous plants, for biofuel, 179
Herbicide(s), 115, 318
    reduction in use, M*iscanthus* and, 88
HGN grass. *See* Hybrid giant Napier (HGN)
    grass
Hierochloe alpine, 18
Hierochloe ordorata (sweet grass), 18, 306
High Melting eXplosive-HMX (octahydro-
    1,3,5,7-tetranitro-1,3,5,7-tetrazine),
    318
Hindu rituals, Bermuda grass in, 238, 239f
Hiroshima attack, 45
HMX. *See* High Melting eXplosive-HMX
    (octahydro-1,3,5,7-tetranitro-1,3,5,7-
    tetrazine)
Holcus lanatus, 122
Hordeum jubatum, 18
Hordeum vulgare, 4
Huts and animal shades, perennial grasses
    for, 14
Hybrid giant Napier (HGN) grass, 294
Hydrogel, 130
Hydrophytic properties, vetiver grass, 52
Hydroseeding, 130
Hyperaccumulator(s), metal, 248, 259, 307,
    308, 319. *See also* Phytoextraction;
    specific names
    drawback of, 320

I

IAA. *See* Indole acetic acid (IAA)
Imperata cylindrica (ylang-ylang), 98, 304
Indian grass *(Sorghastrum nutans)*, 181, 305
Indian mustard *(Brassica juncea)*, 308, 322t
Indigofera cordifolia, 68
Indole acetic acid (IAA), 34, 220
Industrialization. *See also* Land
       contamination
   impact of, 31, 115
   perennial grasses and, 17
Industrial revolution, 31
Inflorescence, *Phragmites australis*, 99, 101f
Inorganic pollutants, 117. *See also*
       Metal(loid)s
Insecticides, 115
Intercropping, vetiver grass, 47
International Center for Tropical Agriculture
       (CIAT), 287
International Livestock Research Institute
       (ILRI), 287
Inula crithmoides, 98
Iowa Army Ammunition Plant (IAAP),
       Iowa, 318
Ipomoea carnea, 165
Iron (Fe), 319
Irrigation, 128
ISSR markers, 73
Italian ryegrass *(Lolium multiflorum). See*
       Lolium multiflorum (Italian ryegrass)

J

Jatropha curcas, 83
Job creation, perennial grasses production
       and, 17
Johnson grass *(Sorghum halepense)*, 5t
Joliet Army Ammunition Plant, Illinois, 318
Juncus acutus, 98

K

Kans grass *(Saccharum spontaneum). See*
       Saccharum spontaneum (Kans grass)
Kentucky bluegrass *(Poa pratensis). See Poa*
       pratensis (Kentucky bluegrass)
"KHUS". *See Vetiveria zizanioides* L. Nash
Khusinol, 48
"King grass". *See Lasiurus sindicus* Henrard
       (Sewan grass)

L

Land contamination, 115. *See also* Pollution
   causes, 115, 116f
   cleanup solutions, management, and
       assessment, 116
   effects of, 115, 116f
   sources, 115, 116f
Lasiurus sindicus-Cenchrus biflorus, 68
Lasiurus sindicus-Cymbopogon
       jawarancusa-Eleusine compressa, 68
Lasiurus sindicus-Cymbopogon
       schoenanthus, 68
Lasiurus sindicus-Dactyloctenium sindicum-
       Aristida adscensionis, 68
Lasiurus sindicus-Eleusine compressa, 68
Lasiurus sindicus Henrard (Sewan grass), 4,
       6t, 13, 14, 21
   biomass productivity, 70, 72f
   case study, 344
   as drought-tolerant plant, 66, 67
   ecology, 64
   features, 63, 66
   as fodder, 63, 67, 70
   forage yields, 71, 72f
   future prospects, 75
   genetic diversity and conservation, 73
   geographical distribution, 64
   grass communities associated with, 68, 71f
   growth period, 64
   heavy metals concentration, 67, 70f
   morphological description, 65
   multiple uses, 67
   nutrients concentration, 67, 69f
   origin, 64
   overview, 63
   in Pb phytostabilization, 67, 68f
   phenology, 64
   in phytoremediation, 67, 68f
   propagation, 66
   rhizospheric microbiology, 64, 74, 75f
   rhizospheric moisture and grazing, 66
   root infection by AM fungi in, 67, 69f
   senescence period, 64
Lasiurus sindicus-Panicum turgidum, 68
Leaching, risk of, 308
Lead (Pb), 115, 307, 319
   phytoextraction, by switchgrass, 185
   phytostabilization, 307
   Sewan grass in, 67, 68f

Leaf(ves)
  C4 NADP-ME type of anatomy, 65
  F. rubra L. grown on fly ash deposits
    morphology, 143, 144f
    SEM analysis of surface structure,
      145–147
  sewan grass, 65
  vetiver grass, 33f
Leersia oryzoides (rice cutgrass), 5t
Lemna gibba, 98
Lemon grass
  in carbon sequestration, 203
  case study, 339
  chromium (VI) toxicity and, 198
  *Cymbopogon flexuosus. See Cymbopogon
    flexuosus* (lemon grass)
  Cymbopogon sp., 2
  ecorestoration of over burden dumping site
    rich in sulfur content, 201
  essential oils from, 17, 202
  implementation strategies, 205
  medicinal use, 16, 204
  origin and distribution, 197f
  other commercial uses, 204
  phytoremediation, 198, 199
    of bauxite residue (red mud), 202
    of chromite-asbestos mine, 199
  phytostabilization of copper mine
    tailings, 200
  propagation, 198
  reclamation and re-vegetation of fly ash
    disposal sites, 201
  socio-economic development, 204
Lemongrass-legume cover, on degraded land
  for stabilization, 202
Leucaena leucocephala (leucaena), 220,
  286
LHCPII, 141
Liao River estuary wetland, case study, 339
Limbarda crithmoides, 98
Limonium pruinosum, 98
Liquid petroleum gas (LPG), 295
Little blue-stem *(Andropogon scoparius)*,
  305
Lolium, 119
Lolium multiflorum (Italian ryegrass), 9,
  120, 306
Lolium perenne (English ryegrass), 5t, 6t, 14,
  120, 122, 260, 306, 313, 314, 343

Long life span, vetiver grass, 52
Low input and minimum maintenance, in
  phytoremediation, 15
Luzula confusa, 18
Lygeum, 119

M

Machine sowing, 130
Macrotyloma axillare, 286
Madagascar lemongrass, 197
Maize *(Zea mays)*, 4, 308, 320, 322t
Malabar grass, 197
Malondialdehyde (MDA), in *F. rubra* L.
  leaves grown on fly ash deposits,
  140t, 141
Malva parviflora, 98
Manganese (Mn), 319
  in *F. rubra* L. grown on fly ash deposits,
    134t, 137
  role of, 137
MDA. *See* Malondialdehyde (MDA)
Meadow fescue *(Festuca pratensis). See
  Festuca pratensis* (Meadow fescue)
Meadow fox-tail *(Alopecurus pratensis)*,
  122, 306
Medicinal use(s)
  Bermuda grass in, 239
  C. flexuosus, 204
  perennial grasses, 16
  Saccharum spp., 216f, 217
  vetiver grass, 50
Melica ciliate, 122
Mentha longifolia, 98
Mercury (Hg), 115, 319
Metal accumulation techniques, 195
Metal-binding compound protein
  (MPT), 138
Metal-chelate complexes, 132
Metal hyperaccumulator(s), 248, 307, 308,
  319. *See also* Phytoextraction
  drawback of, 320
Metal hyper-tolerance, 248
Metal(loid)s. *See also* Fly ash (FA); Heavy
  metal(s) (HM); Organic pollutants
  examples, 115, 116f
  phytoremediation, 117
    Bermuda grass, 234–236
    bioconcentration factor (BCF), 117, 132
    F. rubra L. in, 132

Metal(loid)s. *See also* Fly ash (FA); Heavy
    metal(s) (HM); Organic pollutants
    (*cont.*)
  Miscanthus, 85
  plant tolerance and, 132
  translocation factor (TF), 117, 132
  transport, 132
Metallothioneins (MTs), 320
Metal-tolerant plants, perennial grasses as, 9.
    See also specific names
Metal uptake, by grass roots from soil, 1, 2f
Methane production, *Phragmites* species, 108
Microbes, role in grass-based
    phytoremediation, 303
  endophytes, 310
  "green liver" model, 313
  growth promoting bacteria, 310
  importance of, 312
  perennial grasses as suitable agents,
    304, 305f
  PGPR. *See* Plant growth promoting
    rhizobacteria (PGPR)
  pollutants types and, 313
    explosives/nitroaromatics, 318
    heavy metals, 319
    herbicides, 318
    polychlorinated biphenyls, 318
    polycyclic aromatic hydrocarbons
      (PAHs), 314, 316t
    pros and cons, 321
    rhizobacteria, 310
    strategies, 308
Micropropagation, 34. *See also* Propagation
  Miscanthus, 82
Mining waste, in Serbia, 120f
Miscanthus, 4, 6t, 16, 19, 21, 119, 260,
    303, 307
  AGB-1, 87
  in animal bedding, 88
  as biofuel crop, 83
  biology, 81
  biomass yields, 307
  in building and packing, 88
  as carbon neutral plant, 88
  carbon sequestration, 13, 88
  case study, 341
  environmental consideration, 88
  for feed, 88
  harvesting, 83

herbicides use reduction, 88
inflorescence, 81
multiple uses, 85f, 88, 89f
overview, 79
phytoremediation, 79, 84, 85f
  contaminated sites, 86t
  metal(loid)s, 85
  organic pollutants, 87
power generation, 83
propagation, 81
  CEED, 82
  collar propagation, 82f, 83
  micropropagation, 82
  nodal stem cutting, 82f
  rhizome, 82
  seed propagation, 82
remediation of contaminated sites with,
    85, 86t
restoration of degraded environments, 10
with SWOT analysis, merits and demerits,
    89, 90f
taxonomy, 81
for wild life shelter, 13
Miscanthus floridulos, 81
Miscanthus sacchariflorus, 80–82
Miscanthus sinensis, 9, 80–82, 85, 87
Miscanthus x giganteus, 13, 19, 80–82,
    85, 120f
Mo-enzyme (Moco), 138
Molecular markers technique, 73
Molinia caerulea, 9
Molybdenum (Mo), 319
  in *F. rubra* L. grown on fly ash deposits,
    134t, 138
  role of, 138
Monoculture, 4
Morphology. *See also* Ecology
  Bermuda grass, 229
  Calamagrostis epigejos, 261, 262f
  Cymbopogon citratus, 198
  Cymbopogon flexuosus, 198
  F. rubra L. grown on fly ash deposits, 139
    leaf, 143, 144f
    SEM analysis of leaf surface structure,
      145–147
  Napier grass, 286
  Phragmites species, 99–102
  red fescue, 123f, 125
  reed canary grass, 167

Saccharum spp., 213, 214f
sewan grass, 65
   leaves, 65
switchgrass, 182
vetiver grass, 33f, 34t
   flowers, 33
   leaves, 33f
   roots, 33, 34t
   stems, 33
Moso bamboo *(Phyllostachys pubescens)*.
   See Phyllostachys pubescens (Moso
   bamboo)
MTP8, 138
MTs. *See* Metallothioneins (MTs)
Mulching, 128

N

Nagasaki attack, 45
Napier grass *(Pennisetum purpureum)*. *See*
   Pennisetum purpureum (Napier grass)
Napier Pakchong 1 grass (NPG) residue, 295
Naproxen, remediation, vetiver grass in, 45
Nassella tenuissima, 9
National Bureau of Plant Genetic Resources
   (NBPGR), 287
National Development Programs, 205
Native species, perennial grasses, 18. *See*
   also specific names
Natural colonization, 3
Natural Resources Stewardship Circle
   (NRSC), 51
Nature sustainability. *See also specific*
   species names
   bamboo' role in, 248
      agroforestry system, 251, 252f
      carbon sequestration, 250
      climate change mitigation, 250
      phytoremediation, 248
   biomass from grasses, 19
N'-bis (2-hydroxyphenyl) acetic acid
   (EDDHA), 308, 320
Needle grass
   Stipa capillata, 5t
   Stipa spartea, 305
Neonotonia wightii (soybean), 286
Newell, L.C., 179
New Energy Farms, the, 82
Nickel (Ni), 115, 319
Nicotiana tabacum (tobacco), 308

Nitramines, 318
Nitrate esters, 318
Nitric oxide, application, Bermuda grass, 230
Nitrilotriacetic acid (NTA), 308
Nitroaromatics, 308, 318
Nitrogen, role in eco-restoration of degraded
   land, 202
Nodal stem cutting, *Miscanthus* propagation,
   82f
Non-aromatic grasses, 4, 306
NTA. *See* Nitrilotriacetic acid (NTA)
Nuclear power plants, 45
Nuclear wastes, impacts, 45
Nurse plants, 10
Nutrient loads, removal, vetiver grass
   in, 46
Nutrients, concentration, *L. sindicus*, 67, 69f

O

OAA. *See* Oxaloacetic acid (OAA)
Ochthochloa compressa, 68
Ocimum tenuiflorum (Tulsi), 238
OEC (oxygen evolving complex), 139
Oil contaminated soil, remediation, vetiver
   grass in, 45
Oligopeptide ligands, production, 320
Organic pollutants, phytoremediation.
   See also Fly ash (FA); Heavy metal(s)
   (HM); Metal(loid)s
   Miscanthus, 87
   perennial grass species, 303
   Phragmites species, 105
   vetiver grass in, 37, 307
Organochlorines, 115
Organophosphates, 115
Ornamental grasses, in park, 9. *See also*
   specific names
Oryza sativa (rice), 4, 322t
OsnRAMP5 protein, 137
Oxaloacetic acid (OAA), 35
Oxygen evolving complex (OEC), 139

P

Packing materials, *Miscanthus* use in,
   88. *See also* Pulp and paper
   manufacturing
Paenibacillus macerans NBRFT5, 310
PAHs. *See* Polycyclic aromatic hydrocarbons
   (PAHs)

PAHs/PCB pollutants, rhizomediation, grass species for, 120
Palmasora, 46
Panicoidaea, 196
Panicum, 119
Panicum genus, 182
Panicum maximum (Guinea grass), 5t, 287
Panicum turgidum, 68
Panicum virgatum L. (Switchgrass), 5t, 6t, 9, 13, 14, 16, 21, 120, 303, 305
    bioenergy production, 187
    as candidate energy crop, 182
    carbon sequestration, 188
    case study, 342
    ecology, 180, 181f
    ecotypes, 182
    features, 180f
    future perspectives, 189
    geographic distribution, 179, 182
    history, 179
    limiting factors, 184
    morphological description, 182
    multiple uses, 183
    origin, 182
    overview, 179
    physiological adaptation, 188
    phytoremediation, 184
    plant propagation and regeneration, 183
    in pulp and paper manufacturing, 17
Paper manufacturing. *See* Pulp and paper manufacturing
Paragrass *(Brachiaria mutica)*, 5t
Paraquat, 115
Paspalum, 119
Paspalum notatum, 120
Pasture grasses. *See* Non-aromatic grasses
PBBs. *See* Polybrominated biphenyls (PBBs)
PCBs. *See* Polychlorinated biphenyls (PCBs)
PCs. *See* Phytochelatins (PCs)
Pea *(Pisum sativum)*, 320
Pennisetum divisum (Gmel.) Henrard, 63
Pennisetum genus, 119, 284
Pennisetum glaucum, 287
Pennisetum pedicellatum, 16
Pennisetum purpureum (Napier grass), 4, 6t, 14, 16, 21, 303, 322t
    association with other plants, 286
    bioenergy production, 294
    case study, 341

categories, 287
ecology, 286
features, 288f
future prospects, 296
geographical distribution, 283f, 284
hybrid giant Napier (HGN) grass, 294
morphology, 286
multiple uses, 288, 290t
origin, 284
overview, 283
phytoremediation, 289
propagation, 287
in pulp and paper manufacturing, 17
taxonomy, 286
Perennial grass(es), 306. *See also* Annual grass(es); Biennial grass(es); Grass(es); *specific names*
    as bio-energy feedstock source, 12
    biofuel production from biomass of, 302f, 303
    case studies
        bamboo grass, 340
        Bermuda grass, 338
        Calamagrostis epigejos grass, 340
        Festuca rubra L. grass, 343
        future prospects, 345
        lemon grass, 339
        Miscanthus grass, 341
        Napier grass, 341
        overview, 337
        phragmites grass, 339
        reed canary grass, 342
        Saccharum spp. grass, 344
        Sewan grass, 344
        switchgrass grass, 342
        vetiver grass, 338
    on contaminated sites, 119f
    development of, 227
    ecological aspects, 10
        biodiversity conservation, 12
        carbon sequestration, 13
        climate change mitigation, 12
        habitat corridors, 14
        phytoremediation, 11
        restoration, 10
        soil erosion control, 13
        wild life shelter, 13
    economic aspects, 15
        bioenergy feedstock, 16

essential oil, 17
industrialization, 17
job creation and poverty alleviation, 17
low input and minimum maintenance,
    15
medicinal use, 16
pulp and paper manufacturing, 17
ecosystem services by, 10, 11f
features, 119
as genetic resources for phytoremediation,
    4
  benefits, 4
  categories, 4
  examples, 5t, 6t
  ornamental grasses in park, 9
  as phytoremediator, 9
  specific species, 6t
  uses, 6t
importance of, 10
as metal-tolerant plants, 9
mixed crops of, 4
perennial growth, 19
in phytoremediation, 11, 303. *See also*
    specific names
  coupling with perennial native
      grasses, 18
  factors, 18
  future prospects, 21
  improvements, 20
  limitations, 321
  parameters, 326
  policy framework, 21
  practices of, 21
  reasons behind, 18
root systems, 306, 307
societal aspects, 14
  cultural programs, 15
  as fodder, 14
  huts and animal shades, 14
  as raw material for crafts, 14
  rope manufacturing, 15
  as suitable agents for phytomanagement,
      304
sustainable biomass source, 19
Perfumery, vetiver grass in, 50
Persistent organic pollutants (POPs), 115
Petroleum hydrocarbons, 313
PGPB. *See* Plant growth promoting bacteria
    (PGPB)

PGPR. *See* Plant growth promoting
    rhizobacteria (PGPR)
Phalaris arundinacea L. (reed canary grass),
    4, 6t, 13, 16, 19, 21, 303
  biomass yield, 170
  botanical description, 167
  case study, 342
  ecology, 167
  features, 168
  geographical distribution, 166
  medicinal use, 16
  multiple uses, 171
    biomass production, 171
    carbon sequestration, 171
    pulp and paper industry, 172
  origin, 166
  overview, 165
  in phytoremediation, 168
    as bioindicator, 169
    biomass yield, 170
    photosynthesis, rate of, 170
    as phytoextractor, 169
    as phytostabilizer, 169
    propagation, establishment, and biomass
        production, 170
    tolerance to wide-ranging stress
        conditions, 170
  propagation, 168
  in soil erosion control, 13
  tolerance to wide-ranging stress
      conditions, 170
Phalaris genus, 167
Pharmaceutical products, 115
Phenol, removal, vetiver grass in, 37
Phenolics content, in *F. rubra* L. leaves
    grown on fly ash deposits, 140t, 142
Phleum, 119
Phleum pretense, 120, 122, 306
Phosphorous solubilizing bacteria, 303
Photosynthesis
  F. rubra L., 139, 140t
  reed canary grass, 170
Phragmites australis (common reed), 4, 6t,
    14, 15, 97, 99, 120, 260
  biomass production, 108
  carbon sequestration, 107
  case study, 339
  inflorescence, 99, 101f
  medicinal use, 16

Phragmites australis (common reed) (*cont.*)
  in phytoremediation
    heavy metals, 105
    organic pollutants, 105
    polluted water detoxification, 106
    rhizome and stolon, 99, 102f
    seed dispersal, 102
    sub species, 98
  for thatching and building rafts, 14
Phragmites communis (common reed), 19,
    120f, 305
  in phytoremediation, organic
    pollutants, 105
Phragmites frutescens H. Scholz, 97
Phragmites japonicus Steud., 97
Phragmites karka, 97
  in phytoremediation, 11
    organic pollutants, 105
Phragmites mauritianus Kunth, 97
Phragmites species, 119
  adaptive features, 103f
  botanical description, 99–102
  case study, 339
  colonization, 102
  ecology, 98
  ecosystem services by, 104t
  future perspectives, 110
  general aspects, 98
  geographical distribution, 97, 99, 100f
  habitat and propagation, 101
  inflorescence, 99, 101f
  multiple uses and management, 103, 104t
    biomass production, 108, 109f
    carbon sequestration, 107
    ecological restoration, 106
    green cover development of derelict
      lands, 107
    heavy metals remediation, 105
    organic pollutants remediation, 105
    other uses, 109
    phytoremediation, 103
    polluted water detoxification, 106
    soil formation, 107
  origin, 99
  overview, 97
  pollutants remediation, 11
  rhizome and stolon, 99, 102f
  species associated with growth of, 98
  taxonomy, 98

Phyllostachys edulis, 13
Phyllostachys makinoi, 250
Phyllostachys pubescens (Moso bamboo), 4,
    6t, 21, 305
  bamboo-provisioned ecosystem services,
    247
  case study, 340
  geographical distribution, 245, 246
  multiple uses, 247
  overview, 245
  role in nature sustainability, 248
    agroforestry system, 251, 252f
    carbon sequestration, 250
    climate change mitigation, 250
    phytoremediation, 248
    root/rhizome structure, 245
    socioeconomic and ecological values, 251
    sympodial bamboo, 245
    as wood replacement, 17
Physiology. *See also* Ecology; Morphology
  adaptations, switchgrass, 188
  F. rubra L. grown on fly ash deposits, 139
    photosynthesis, pigments, and
      antioxidants, 139, 140t
  vetiver grass, 35
Phyto-3, 325
Phytochelatins (PCs), 320. *See also*
    Chelating agents
Phytocommerce/phytoeconomics
  vetiver grass, 48, 50f
    energy, 51
    essential oil, 48
    medicines, 50
    other uses, 51
    perfumery and cosmetics, 50
Phytocontainment, 309
Phytodegradation/phytotransformation, 117,
    118, 165, 195, 259, 283, 308, 310.
    See also Phytoremediation
  grass species for, 120
Phytodesalinization, 259
Phyto-DSS, 325
Phytoextraction, 3, 117, 165, 195, 259, 283,
    308. *See also* Phytoremediation
  Calamagrostis epigejos in, 270
  chelate-assisted, 308
  defined, 259
  perennial grass species for, 120
  reed canary grass in, 169

Phytoextractor, reed canary grass as, 169
Phytofiltration, 259
Phytoimmobilization, 309
Phytomanagement, 309
    fly ash deposits, vetiver grass in, 45
    goal of, 309
    perennial grasses as suitable agents
        for, 304
    vetiver grass in, 31
Phytoremediating agents, 3
Phytoremediation, 117. *See also* Remediation
        technology(ies)
    advantages, 31, 165, 195
    aromatic grasses, 2
    bamboo' role in, 248
    of bauxite residue (red mud), 202
    C. flexuosus, 198, 199f
    Calamagrostis epigejos, 268, 269f, 271f,
        272f
    of chromite-asbestos mine, 199
    contaminated sites, *F. rubra* L. in, 132
        fly ash deposits, 133, 134t
        metal(loid)s, 132
    coupling, with perennial native grasses, 18
    defined, 1
    described, 1
    enhanced, 20
    factors, 303
    grasses root structure and, 2
    lemon grass, 198, 199
    limitations, 321
    Napier grass, 289
    need for, 4
    parameters, 326
    perennial grasses as genetic resources
        for, 4
        benefits, 4
        categories, 4
        examples, 5t, 6t
        ornamental grasses in park, 9
        as phytoremediator, 9
        specific species, 6t
        uses, 6t
    perennial grasses in, 11. *See also* Case
        study(ies); *specific names*
        factors, 18
        future prospects, 21
        improvements, 20
        policy framework, 21
        practices of, 21
        reasons behind, 18
        as suitable agents for, 304
    perennial grasses on contaminated sites,
        119f
    PGPR bacteria in, 310. *See also* Microbes,
        role in grass-based phytoremediation
    Phragmites species, 103
    phytodegradation/phytotransformation.
        See Phytodegradation/
        phytotransformation
    phytoextraction. *See* Phytoextraction
    phytostabilization. *See* Phytostabilization
    phytovolatilization. *See* Phytovolatilization
    plants used for, 1, 289
    of pollutants. *See* Pollutants,
        phytoremediation of
    preparations before, 4
    reed canary grass in, 168
        as bioindicator, 169
        biomass yield, 170
        photosynthesis, rate of, 170
        as phytoextractor, 169
        as phytostabilizer, 169
        propagation, establishment, and biomass
            production, 170
        tolerance to wide-ranging stress
            conditions, 170
    rhizodegradation/phytostimulation. *See*
        Rhizodegradation/phytostimulation
    rhizofiltration, 165
    sewan grass in, 67, 68f
    strategies used for, 165, 308
    switchgrass, 184
    technologies, 117, 259, 283, 308. *See also*
        specific techniques
    of toxic elements, Bermuda grass in,
        234–236
Phytoremediator, perennial grasses as, 9
Phytosequestration, 308, 309
Phytostabilization, 117, 165, 195, 259, 283,
        309. *See also* Phytoremediation
    of copper mine tailings, 200
    perennial grass species for, 120
    reed canary grass in, 169
Phytostabilizer, reed canary grass as, 169
Phytostimulation. *See* Rhizodegradation/
        phytostimulation
Phytotech, Inc., 326

Phytotransformation. *See* Phytodegradation/
   phytotransformation
Phytovolatilization, 165, 195, 259, 283,
   308, 310
Pigments, *F. rubra* L. grown on fly ash
   deposits, 139, 140t
Piptatherum, 119
Piptatherum miliaceum (smilo grass), 322t
Pisun sativum (Pea), 320
Plant growth promoting bacteria
   (PGPB), 118
Plant growth promoting rhizobacteria
   (PGPR), 303. *See also specific names*
   characteristics, 310
   in nutrient cycling, 310
   in phytoremediation, 310
Plant-microbe interactions, role of, 312. *See
   also* Microbes; role in grass-based
   phytoremediation
Plant tolerance, metal(loid)s
   phytoremediation and, 132
Pluchea dioscoridis, 98
Poa, 119
Poa arctica, 18
Poaceae family, 1, 32, 35, 64, 67, 81, 97,
   119, 182, 196, 212, 245. *See also*
   Perennial grass(es)
   overview, 304
Poa compressa, 122
Poa lanata, 18
Poa pratensis (Kentucky bluegrass), 5t, 120,
   122, 305, 343
Policy framework, perennial grasses in
   phytoremediation, 21
Pollutants, phytoremediation of, 313. *See
   also* Contaminants; *specific pollutants*
   aromatic, 313
   defined, 303
   detoxification, 313
   explosives/nitroaromatics, 318
   heavy metals, 319
   herbicides, 318
   polychlorinated biphenyls, 318
   polycyclic aromatic hydrocarbons (PAHs),
      314, 316t
   through perennial grass species, 313
   vetiver grass in, 36, 38t
      agrochemicals and antibiotics, 37
      heavy metals, 36

   organic pollutants, 37
   radioactive substances and oil
      contaminated soil, 45
Polluted water detoxification, *Phragmites*
   species in, 106
Pollution. *See also* Contamination; Land
   contamination
   causes, 79, 115
   cleanup solutions, management, and
      assessment, 116
   effects of, 31
   sources, 115, 116f
Poltava State Agrarian Academy (PSAA),
   Ukraine, 180, 181f
Polybrominated biphenyls (PBBs), 115
Polychlorinated biphenyls (PCBs), 115, 186,
   318, 319
Polychlorinated dibenzofurans (PCDFs),
   115, 120
Polycyclic aromatic hydrocarbons (PAHs),
   115, 308, 312–314, 316t
   degradation, 314, 316t
   endophytes in, 314
   stimulation of elimination, mechanisms,
      314
Polypogon monspliensis, 98
POPs. *See* Persistent organic pollutants
   (POPs)
Poverty alleviation, perennial grasses
   production and, 17
Power generation, *Miscanthus*, 83
Prairie cord grass *(Spartina pectinata).
   See Spartina pectinata* (prairie
   cord grass)
Precipitation, 1
Propagation. *See also* Geographical
   distribution
   Bermuda grass, 229, 230f
   C. flexuosus, 198
   Calamagrostis epigejos, 261
   methods, 34, 81
   Miscanthus, 81
      CEED, 82
      collar propagation, 82f, 83
      micropropagation, 82
      nodal stem cutting, 82f
      rhizome, 82
      seed propagation, 82
   Napier grass, 287

Phragmites species, 101
  reed canary grass, 168, 170
  Saccharum spp., 215
  seed, 168
  sewan grass, 66
  switchgrass, 183
  vegetative, 168, 170
  vetiver grass, 33, 34
Pseudomonas, 310
Pseudomonas koreansis AGB-1, 87
Pueraria phaseoloides (Kudzu), 286
Pulp and paper manufacturing, 17
  Arundo donax, 17
  Cymbopogon nardus, 17
  Eulaliapsis binata, 17
  Napier grass, 17
  perennial grasses, 17
  reed canary grass, 172
  Saccharum spp. in, 216f, 217
  Switchgrass, 17

R

Radioactive substances, remediation, vetiver
    grass in, 45
Rajasthan Forest Department, 68
RAPD-ISSR analysis, 73
RAPD markers, 73
Ratooning, vetiver propagation, 34
Raw material for crafts, grasses as, 14
RDX. *See* Royal Demolition eXplosive-RDX
    (hexahydro-1,3,5-trinitro-1,3,5-
    triazine)
Reactive oxygen species (ROS), 133
Reclamation, of fly ash disposal sites,
    lemongrass, 201
REC-Phyto-DSS, 325
Red fescue (*Festuca rubra* L.). *See Festuca
    rubra* L. (red fescue)
Red List Category and conservation, *F. rubra*
    L., 124
Red mud (bauxite residue), phytoremediation
    of, 202
Redox reactions, 1
Reducing Emissions from Deforestation and
    Forest Degradation (REDD) program,
    247, 250
Reed canary grass (*Phalaris arundinacea*
    L.). *See Phalaris arundinacea* L.
    (reed canary grass)

Remediation technology(ies). *See also*
    Phytoremediation
  cost-benefit analyses, 116
  eco-efficiency, 116
  ecorestoration, 117
  phytodegradation/phytotransformation,
    117, 118, 165
  phytoextraction, 117, 165
  phytoremediation, 117
  phytostabilization, 117, 165
  rhizodegradation/phytostimulation, 117,
    118
  selection, key factors, 116
Reproduction. *See also* Ecology;
    Morphology
  Calamagrostis epigejos, 261
  red fescue, 125
  vetiver grass, 33, 34
Republic Gulf Pines landfill, 338
Restoration. *See also* Ecorestoration
  ecological, Phragmites species, 106
  by perennial grasses, 10
  species selection for, 227
  wasteland, Bermuda grass, 232
Re-vegetation, of fly ash disposal sites, 201
Rhinanthus alectorolophus, 268
Rhinanthus major, 268
Rhizobacteria, 310
Rhizobium leguminosarum, 310
Rhizodegradation/phytostimulation, 117,
    118, 259, 283, 308, 310. *See also*
    Phytoremediation
Rhizofiltration, 165, 195, 259
Rhizomediation
  grasses used for, 313
  of PAHs/PCB pollutants, grass species for,
    120
Rhizomes, 307. *See also* Root(s)/root system
  Miscanthus propagation through, 82
  Phragmites australis, 99, 102f
Rhizoremediation, 303
Rhizosphere, 310
  enzymes released into, 314
  metal bioavailability, 1, 2f
  microbiology, sewan grass, 64, 74, 75f
  role of, 303
Rhizospheric moisture, sewan grass, 66
Rice cutgrass (*Leersia oryzoides*), 5t
Ricinus communis, 165, 306

Root(s)/root system
  features, 306
  grasses, structure, 2
  metal uptake by, 1, 2f
  perennial grass(es), 306, 307
  rhizospheric metal bioavailability, 1, 2f
  vetiver grass, 33, 34t, 52
Rope manufacturing, perennial grasses
    for, 15
Royal Demolition eXplosive-RDX
    (hexahydro-1,3,5-trinitro-1,3,5-
    triazine), 318
Rumex dentatus, 98

S

Sabai *(Eulaliapsis binata). See Eulaliapsis
    binata* (Sabai)
Saccharum arundinaceum, 9
Saccharum bengalense, 10, 306
Saccharum munja, 10, 122, 165, 212, 213, 306
  in carbon sequestration, 219
  ecology, 212
  as fodder ingredient, 216f, 217
  medical applications, 216f, 217
  morphology, 213, 214f
  propagation, 215
  in pulp and paper manufacturing, 216f, 217
  stress tolerant genes, 218
  in wastelands ecorestoration, 218, 219
Saccharum officinarum, 212, 218
Saccharum ravennae, 9
Saccharum sinensi, 212
Saccharum spontaneum (Kans grass), 5t, 6t,
    10, 13, 165, 212, 213, 220, 306
  in carbon sequestration, 219
  case study, 344
  chloroform extracted from flowers of,
    216f, 217
  ecology, 213
  as fodder ingredient, 216f, 217
  medical applications, 216f, 217
  morphology, 214f, 215
  propagation, 215
  stress tolerant genes, 218
  in wastelands ecorestoration, 218, 219
Saccharum spp., 79
  in bioethanol production, 216f, 218
  biomass and bioenergy production, 220
  in carbon sequestration, 219

case study, 344
  ecology, 212
  in fly ash dumps ecorestoration, 220
  geographic distribution, 215
  morphology, 213, 214f
  multiple uses, 216f
    fodder, 217
    medical applications, 217
    miscellaneous application, 218
    paper and pulp manufacturing, 217
    timber, 217
  overview, 211
  propagation, 215
  seeds of, 215
  in wastelands ecorestoration, 218, 219f
Sand-reed *(Calamovilfa longifolia)*, 305
Scanning electron microscopy (SEM)
  of leaf surface structure of *F. rubra* L.,
    145–147
Schizachyrium scoparium, 181
Secale, 119
Seed dispersal, *Phragmites australis*, 102
Seed production, *F. rubra* L., 129
Seed propagation. *See also* Propagation
  Miscanthus, 82
  reed canary grass, 168
Selenium (Se), 319
  in *F. rubra* L. grown on fly ash deposits,
    134t, 139
  role of, 139
Self-sustaining ecosystem, 227
SEM. *See* Scanning electron microscopy (SEM)
Semi-simultaneous saccharification
    fermentation (sSSF), 295
Sequestration, 313. *See also* Carbon
    sequestration
Serbia
  fly ash deposits and coalmine land in, 121f
    biorecultivation of *F. rubra* L. on
      (TENT, Serbia), 129
    grass-legume mixture, fertilization, and
      agrotechniques, 130t
  mining waste in, 120f
Setaria, 119
Setaria viridis, 120
Sewan grass *(Lasiurus sindicus* Henrard).
    See Lasiurus sindicus Henrard
    (Sewan grass)
Shannon index, 73

Silybum marianum, 98
Simultaneous saccharification and co-fermentation (SSCF), 295
Smilo grass *(Piptatherum miliaceum)*, 322t
Societal aspects, perennial grasses, 14
    cultural programs, 15
    as fodder, 14
    huts and animal shades, 14
    as raw material for crafts, 14
    rope manufacturing, 15
Socio-economic development, *C. flexuosus*, 204
Soil amendments/ameliorative technique, 128
Soil and water conservation
    Bermuda grass, 237
    vetiver grass, 45f, 48
Soil bacteria, 87. *See also* Microbes; role in grass-based phytoremediation
Soil erosion, 48
    perennial grasses, 13
    red fescue for, 127, 128
Soil(s)
    formation, *Phragmites* species in, 107
    metal uptake by grass roots from, 1, 2f
    oil contaminated, remediation vetiver grass in, 45
    as source of food, 115
Sonchus oleraceus, 98
Sorghastrum nutans (Indian grass), 181, 305
Sorghum, 119
Sorghum bicolor, 5t, 120
Sorghum drummondii, 120
Sorghum halepense (Johnson grass), 5t
Sorghum vulgare, 120
Soybean *(Glycine max)*, 322t
Spartina, 119
Spartina alterniflora, 305
Spartina patens, 305
Spartina pectinata (prairie cord grass), 9, 19
Sporobolus asper (dropseed), 305
SSCF. *See* Simultaneous saccharification and co-fermentation (SSCF)
St. Augustine grass *(Stenotaphrum secundatum)*, 5t
Steam distillation, 2
Stem(s), vetiver grass, 33
Stenotaphrum secundatum (St. Augustine grass), 5t
Stipa, 119

Stipa capillata (needle grass), 5t
Stipa spartea (needle grass), 305
Stipa viridula (green needlegrass), 305
Stolon(s), 307
    Phragmites australis, 99, 102f
Stress conditions, reed canary grass tolerance to, 170
Strontium (^{90}Sr), 45
Sulfur, lemon grass in ecorestoration of, dumping site rich in, 201
Sunflower *(Helianthus annuus)*, 308, 343
Sustainability. *See also* Nature sustainability
    biomass from grasses, 19
Swedish Agro-Fiber Project (1987-91), 172
Sweet grass *(Hierochloe ordorata)*, 18, 306
Switchgrass (*Panicum virgatum* L.). *See Panicum virgatum* L.(Switchgrass)
Switchgrass management, in Europe, 17
SWOT analysis, *Miscanthus* merits and demerits with, 89, 90f
Sympodial bamboo, 245

T

Tall fescue *(Festuca arundinacea). See Festuca arundinacea* (tall fescue)
Tetracycline, remediation, vetiver grass in, 45
Textile industry, 17
TF. *See* Translocation factor (TF)
Thar Desert (India), 63, 64, 73
Thelypteris dentata, 165
TI. *See* Tolerance Index (TI)
Timber, *Saccharum* spp. for, 216f, 217
Timothy grass, 16, 306
Tissue culture, 34
TNT (2,4,6-trinitrotoluene), 115, 318
    remediation, vetiver grass in, 37
Tobacco *(Nicotiana tabacum)*, 308
Tolerance Index (TI), 170
Top soil covering, 128
Total carotenoids, in *F. rubra* L. leaves grown on fly ash deposits, 139, 140, 141
Total petroleum hydrocarbons (TPHs), 9, 313
TPHs. *See* Total petroleum hydrocarbons (TPHs)
Transformation, 313
Translocation factor (TF), 117, 132
Trees, 4
"Trinpanchmool" *Saccharum* spp. roots in, 217

Triticum aestivum (wheat), 3–5, 120, 181
Tulsi *(Ocimum tenuiflorum)*, 238
Typha, 119
Typha domingensis, 98
Typha latifolia, 165

U

Underground leakingstorage tanks (USTs), 313
Urals (Russia), brown coal mine waste
    in, 122
Urban development, 115. *See also* Land
    contamination
US Department of Energy (DOE), 179, 182
Usher industries, 51

V

Vegetative propagation, 168. *See also*
    Propagation
  lemon grass, 198
  reed canary grass, 168, 170
Vertical subsurface flow constructed
    wetlands (VSF CW), 294
Vetiver grass. *See Vetiveria zizanioides*
    L. Nash (vetiver grass)
Vetiveria zizanioides L. Nash (vetiver grass),
    2, 4–6, 13, 21, 79, 119, 122, 303,
    307, 322t
  adaptive agricultural practices, 47
  in carbon sequestration, 46
  case study, 338
  contaminants removal, 31
  "ecological-climax" species, 35
  ecology, 35
  economic return (phytocommerce/
    phytoeconomics), 48, 50f
  as energy source, 51
  essential oils from, 17, 48
  extensive root system, 52
  features, 32, 52
  fly ash deposits, phytomanagement of, 45
  geographical distribution and expansion,
    35, 36f
  high biomass production, 52
  intercropping, 47
  limitations, 51
  long life span, 52
  as magic plant (properties), 50f
  medicinal uses, 50

  minimum competition for nutrient and
    moisture, 52
  morphology, 33f, 34t
    flowers, 33
    leaves, 33f
    roots, 33, 34t
    stems, 33
  nutrient loads removal, 46
  other uses, 51
  overview, 32
  in perfumery and cosmetics, 50
  physiology, 35
  in phytomanagement, 31
  phytoremediation and conservation
    properties, 38t
  pollutants, phytoremediation of, 36, 38t
    agrochemicals and antibiotics, 37
    heavy metals, 36, 307
    organic pollutants, 37, 307
    radioactive substances and oil
      contaminated soil, 45
  reproduction and propagation, 33, 34
  resistance to disease and pest, 52
  soil and water conservation, 47f, 48
  sterile and non-invasive, 52
  symbiotic association with microbes, 52
  xerophytic and hydrophytic properties, 52
  A-Vetivone, 48
  β-Vetivone, 48
Vicia faba (broad bean), 310, 312
VSF CW. *See* Vertical subsurface flow
    constructed wetlands (VSF CW)

W

Warm-season grasses. *See* C4 grasses
Wasteland(s)
  ecorestoration, *Saccharum* spp. in, 218, 219f
  restoration, Bermuda grass, 232
Water. *See also* Soil and water conservation
  polluted, detoxification of
    Phragmites species in, 106
Weeds, 304. *See also* Grass(es)
Weeping lovegrass *(Eragrostis curvula)*, 19
Western wheat grass *(Agropyron smithii)*,
    5t, 305
West Indian lemongrass, 197
Wheat *(Triticum aestivum)*. *See Triticum*
    aestivum (wheat)

Wild life shelter, perennial grasses and, 13
Wild rye *(Elymus virginicus)*, 305
World Bank, 32
Wruk Wona, 341

X

Xenobiotics, uptake efficiency, 303
Xerophytic properties, vetiver grass, 52

Y

Ylang-ylang *(Imperata cylindrica)*, 98

Z

Zea mays (maize), 4, 308, 320, 322t
ZIF1 protein, 138
Zinc (Zn), 115, 319
   in *F. rubra* L. grown on fly ash deposits,
     134t, 138
   role of, 137
ZIP1/3/4 transporters, 138
Ziziphus mauritiana, 165
Zn-MAs complex, 138
Zoysia japonica, 120

Printed in the United States
By Bookmasters